RACE, CLASS, and GENDER

An Anthology
Third Edition

Margaret L. Andersen
University of Delaware

Patricia Hill Collins
University of Cincinnati

Wadsworth Publishing Company   1998
IⓉP ™   An International Thomson Publishing Company

Belmont, CA•Albany, NY•Bonn•Boston•Cincinnati•Detroit•Johannesberg•London•Madrid
Melbourne • Mexico City • New York • Paris • Singapore • Tokyo • Toronto • Washington

Publisher: Eve Howard
Assistant Editors: Barbara Yien, Jennifer Burke
Marketing Manager: Chaun Hightower
Project Editor: Jerilyn Emori
Print Buyer: Karen Hunt
Permissions Editor: Robert Kauser
Production: Julie Kranhold/Ex Libris
Designer: Cynthia Schultz
Copy Editor: Steven Gray
Cover Designer: Laurie Anderson
Cover Art: *New World Flower*, 1990, by Frank La Pena.
Courtesy of American Indian Contemporary Arts—San Francisco
Compositor: G&S Typesetters, Inc.
Printer: The Maple-Vail Book Manufacturing Group

COPYRIGHT © 1998 by Wadsworth Publishing Company
A Division of International Thomson Publishing Inc.
I(T)P The ITP logo is a registered trademark under license.

Printed in the United States of America
1 2 3 4 5 6 7 8 9 10

For more information, contact: Wadsworth Publishing Company, 10 Davis Drive, Belmont, CA 94002, or
electronically at http://www.thomson.com/wadsworth.html

International Thomson Publishing Europe
Berkshire House 168–173
High Holborn
London, WC1V 7AA, England

Thomas Nelson Australia
102 Dodds Street
South Melbourne 3205
Victoria, Australia

Nelson Canada
1120 Birchmount Road
Scarborough, Ontario
Canada M1K 5G4

International Thomson Publishing GmbH
Königwinterer Strasse 418
53227 Bonn, Germany

International Thomson Editores
Campos Eliseos 385, Piso 7
Col. Polanco
11560 México D.F. México

International Thomson Publishing Asia
221 Henderson Road
#05-10 Henderson Building
Singapore 0315

International Thomson Publishing Japan
Hirakawacho Kyowa Building, 3F
2-2-1 Hirakawacho
Chiyoda-ku, Tokyo 102, Japan

International Thomson Publishing Southern Africa
Building 18, Constantia Park
240 Old Pretoria Road
Halfway House, 1685 South Africa

Library of Congress Cataloging-in-Publication Data
Race, class, and gender : an anthology / [compiled by] Margaret L.
Andersen, Patricia Hill Collins.—3rd ed.
    p.   cm.
  Includes bibliographical references.
  ISBN 0-534-52879-1
  1. United States—Social conditions—1980-   2. United States—Race
relations.   3. Social classes—United States.   4. Sex role—United
States.   5. Homosexuality—United States.   6. Discrimination—United
States.   I. Andersen, Margaret L.   II. Collins, Patricia Hill, 1948-   .
HN59.2.R32   1998
305—dc21   97-29045
                                                              CIP

Contents

IV Analyzing Social Issues   389

Preface

Race, Class, and Gender is an anthology that introduces students to how race, class, and gender shape the experiences of diverse groups in the United States. We want the book to help students see how the lives of different groups develop in the context of their race, class, and gender position in society. Central to the book is the idea that race, class, and gender are interrelated—and as a result, that experience depends on the particular configuration of these factors in a given group's experience. We are aware that people often interpret this to mean that only groups are victimized by race, class, and gender. Students and faculty often understandably resist that characterization, wanting instead to affirm group identity, especially for those who have been silenced, made invisible, and defined as "other." We are sympathetic to the resistance to a victimization perspective, and we do not want to make people appear to have no agency or choice in the formation of their lives; however, we also know that the structure of race, class, and gender in society has significant consequences for different groups, and we do not think this should be ignored. Thus, although our book is not just about victimization and while we want to affirm the value of diversity, our primary goal is for students to understand the structural arrangements in society and how they result in different systems of privilege and disadvantage. At the same time, we want students to learn how they can make a difference in their own lives and how they can help society at large.

Our introductory essay develops these ideas further and situates this book in the new race, class, gender studies of which we are a part. The introductory essay distinguishes models of race, class, and gender studies that are strictly additive from those that see race, class, and gender as interlocking systems of societal relationships. We want readers to conceptualize race, class, and gender as interactive systems, not merely as variables in sociological equations or separate cultural experiences. We want readers to see race, class, and gender in an analytical, not merely a descriptive way, although we recognize that description of group experience (both historical and current) is an important part of this process.

About the Third Edition

We have been pleased by the success of *Race, Class, and Gender* because it represents the commitment of many people to become more inclusive in their teaching and thinking. We appreciate the reactions people have shared with us, particularly those who have reviewed the book for Wadsworth.

One of the challenges of creating an anthology in a field so rich with new research and new writing is to include articles that will provide a full portrait of the complexity of race, class, and gender. We cannot cover every topic that might be included in such a book. We know that one book cannot do everything people want, but we hope that the third edition builds on the strengths of the first two editions and furthers people's thinking in race, class, gender studies. We want the book to influence students and help teachers with the hard work of building a more inclusive curriculum. We have expanded our original editorial introduction to develop the model of inclusive thinking that we hope we have established.

In the third edition, we have added articles reflecting some of the changes that have taken place in society, even in the short period of time since the second edition. For example, affirmative action, immigration, and welfare reform have emerged as major issues. Although the public discussion of these topics is typically uninformed by an understanding of race, class, and gender, we think race, class, and gender are central to addressing these issues meaningfully; thus, we have included selections in the third edition that we hope will provide a context for understanding these subjects more completely. We have included new articles to reflect current social changes and have asked authors, in some cases, to update their earlier contributions. Although globalization is not the focus of our book, we know that the structure of race, class, and gender in the United States must be understood in the broader international context. Our introductory essays reflect this context more, and we have elaborated points that needed fuller discussion.

We have made changes throughout the book, but most significantly in the sections on families, sexuality, and political activism. The sexuality section now includes pieces that strengthen the examination of how sexuality and studies of race, class, and gender are linked. As in the earlier editions, we do not want readers to conclude that heterosexual privilege is only significant when thinking about sexuality per se; as a result, we have included articles about gay and lesbian experience throughout the book.

We have also significantly revised our final section on political activism. Contemporary students often conclude that, even if they begin to understand the complexities of race, class, and gender relations, there is little they can do to change current realities. Our students often seem overwhelmed by the problems they see in society, and they feel powerless to do much about it. Books like ours, however well-intended, often present examples of activism that actually contribute to students' impressions that they can only be a significant agent of change by dedicating their entire life to change. We want students to think about social change differently, recognizing that people work for change in multiple ways—including in the context of their everyday lives. We also want students to understand that the responsibility for change does not fall exclusively on exceptionally committed individuals. Rather, all people share collective responsibility for change, including institutional change. One of the problems we see currently is the government's abandonment of social support for those in need, even while the state increases its support for wealthy individuals and corporate entities. We would like students to learn that creating social change is not just a matter of putting themselves on the line, but of insisting on a new model of social responsibility for society as a whole.

In revising the third edition, we hope we have maintained the strengths of the first and second editions—a strong sociological foundation, historical perspective provided by many authors, a conceptual grounding supplied in our introductions, inclusive attention to the interlocking dynamics of race, class, and gender, and a balance of different group experiences. Although not every group can be covered on every subject, we hope that our inclusion of different selections convey a sense of the multiple but related ways that race, class, and gender shape U.S. society and culture. We have maintained the second edition's theme of linking personal narratives with the analytical study of race, class, and gender. We are pleased when we hear that students enjoy the readings and find that they illuminate perspectives they had not thought about before in quite the same way.

This edition includes an *Instructor's Manual* with suggestions for class exercises, discussion and examination questions, and course assignments. We thank Pam Porter and Martha Thompson for their inspired work in developing it. We also thank the different faculty members who contributed materials for the *Instructor's Manual.*

This edition also premieres with online support via the Wadsworth Sociology Resource Center, representing how electronic technology is transforming the teaching profession. We encourage people to browse the Race, Class, & Gender page at the Wadsworth Web Site for related resources. The address is http://sociology.wadsworth.com.

A Note on Language

The reconstruction of existing ways of thinking to make them more inclusive requires many transformations. One transformation needed involves the language we use to describe and define different groups. This language reflects many assumptions about race, class, and gender; and for that reason, language changes and evolves as knowledge changes. The term *minority*, for example, marginalizes groups, making them seem somehow outside the mainstream or majority culture. Even worse, the phrase *non-White*, routinely used by social scientists, defines groups in terms of what they are not and assumes that Whites have the universal experiences against which the experiences of all other groups are measured. We have consciously avoided using both of these terms throughout this book.

We have capitalized *Black* in our writing because of the specific historical experience, varied as it is, of African Americans in the United States. We also capitalize *White* when referring to a particular group experience; however, we recognize that *White American* is no more a uniform experience than is *African American.* We realize that these are arguable points, but we want to make our decision apparent and explicit. For the benefit of purists who like to follow the rules, we note that the fourteenth edition of the *Chicago Manual of Style* recognizes that these grammatical questions are also political ones; this edition suggests that writers might want to capitalize *Black* and *White* to reflect the fact that they are referring to proper names of groups. We use "Hispanic" and "Latino" interchangeably, though we recognize that is not how groups necessarily define themselves. When citing data from other sources (typically government documents), we use "Hispanic," since that is how such data are reported.

Language becomes especially problematic when we want to talk about features of experience that different groups share. Using shortcut terms like *Hispanic, Latino, Native American,* and *women of color* homogenizes distinct historical experiences. Even the term *White* falsely unifies experiences across such factors as ethnicity, region, class, and gender, to name a few. At times, though, we want to talk of common experiences across different groups, so we have used labels like *Latino* and *women of color* to do so. Unfortunately, describing groups in this way reinforces basic categories of oppression. We do not know how to resolve this problem, but want readers to be aware of the limitations and significance of language as they try to think more inclusively about diverse group experiences.

Acknowledgments

An anthology rests on the efforts of more people than the editors alone. This book has been inspired by our work with scholars and teachers from around the country who are working to make their teaching and writing more inclusive and sensitive to the experiences of all groups. Over the years of our own collaboration, we have each been enriched by the work of those trying to make higher education a more equitable and fair institution. This book has grown from several of those projects, most notably the Memphis State Center for Research on Women Curriculum Workshops and the American Sociological Association Minority Opportunity Summer Training Program, now the Minority Opportunity Through School Transformation Program. These two programs provided the context for many of our early discussions about this book. Over the years, the networks that have emerged from these associations continue to sustain us. We thank Maxine Baca Zinn, Chuck Bonjean, Marion Coleman, Patrice Dickerson, Bonnie Thornton Dill, Valerie Hans, Elizabeth Higginbotham, Clarence Lo, Lionel Maldonado, Carole Marks, Cora Marrett, Howard Taylor, and Lynn Weber for inspiring our work and, in many cases, suggesting new ideas and articles to consider here. We also thank them for the companionship, encouragement, and vision that inspires our work. Several reviewers at Wadsworth suggested revisions that we found very helpful for the third edition. We would like to thank the following reviewers for their suggestions: Dennis Kalob, New England College; Donna Lee King, University of North Carolina/Wilmington; Patricia Turner, University of California at Davis; and Shirley Yee, University of Washington. Angela Block, Sacramento City College, Nicky Gonzalez-Yuen, De Anza College, Wendy Ng, San Jose State University, and C. Matthew Snipp, Stanford University, also participated in a focus group that sharpened our thinking about the third edition; we value the contributions they have made.

Many other people contributed to the development of this book. We especially thank Elana Messner and Wei Chen at the University of Delaware and Tina Beyene at the University of Cincinnati for their expert research assistance. In the earlier editions, Tina Dunhour and Rachel Levy Schiller, now graduates of the University of Delaware, advised us from students' perspectives about the articles we selected. We appreciate the support provided by our two institutions, with special thanks to Mel Schiavelli, Provost, University of Delaware; Joseph Caruso, Dean of the College of Arts and Sciences; Tony Perzigian, former Head of the Department of African American Studies; John Brackett, Head of the Department of African American Studies, and Robin Sheers, Head of the Center for Women's Studies, University of Cincinnati, for their support. We also thank Susan Phipps, Diana Kubasek, Anna Marie Brown, and Sadie Wright Oliver for their invaluable secretarial support. We hope they know how much we value them.

Eve Howard joined this project with great enthusiasm and commitment to the book, and we appreciate her guidance and support in helping us complete the third edition. She, Jenny Burke, and Barbara Yien all provided ideas and work that helped us complete the new edition. We sincerely appreciate their hard work and vision for the book and look forward to continued collaboration with them.

Developing this book has been an experience based on friendship, hard work, travel, and fun. We especially thank Valerie, Patrice, Roger, and Richard for giving us the support, love, and time we needed to do this work well. This book has deepened our friendship, as we have grown more committed to transformed ways of thinking and being. We are lucky to be working on a project that draws us closer together.

Margaret L. Andersen
Patricia Hill Collins

About the Editors

Margaret L. Andersen is Professor of Sociology and Women's Studies at the University of Delaware, where she is Interim Dean of the College of Arts & Science. She is the author of *Thinking About Women*, *Social Problems* (with Frank R. Scarpitti and Laura L. O'Toole), and a forthcoming book, *Sociology: Understanding a Diverse Society* (with Howard F. Taylor). She is President of the Eastern Sociological Society (1998–1999) and a former member of the Council of the American Sociological Association.

Patricia Hill Collins is Professor of African American Studies and Sociology at the University of Cincinnati. She is the author of *Fighting Words* (to be published in 1998) and *Black Feminist Thought: Knowledge, Consciousness, and the Politics of Empowerment*, which won the C. Wright Mills Award of the Society for the Study of Social Problems. She is a former member of the Council of the American Sociological Association.

About the Contributors

Randy Albelda is an Associate Professor of Economics in the College of Arts and Sciences at the University of Massachusetts in Boston. Her research and teaching focus is on women's economic status, income inequality, family structure, labor markets, and state and local finance. Albelda has been involved with Boston and Massachusetts groups concerned with women's economic well-being, economics education efforts, and tax reform. She has written widely for academic and popular audiences. She is an editorial board member of *Dollars and Sense* magazine and the coauthor of *Glass Ceilings and Bottomless Pits: Women, Income and Poverty in Massachusetts*, and *The War on the Poor: A Defense Manual*.

Paula Gunn Allen is Professor of English at UCLA. She was awarded the Native American Prize for Literature in 1990. That same year her anthology of short stories, *Spider Woman's Granddaughters*, was awarded the American Book Award, sponsored by the Before Columbus Foundation, and the Susan Koppleman Award. A major Native American poet, writer, and scholar, she has published eight volumes of poetry, a novel, a collection of essays, and three anthologies. Her prose and poetry appear widely in anthologies, journals, and scholarly publications. Her newest collection of essays will be published by Beacon Press in late 1998.

Teresa Amott is Associate Professor of Economics at Bucknell University. She is the author of *Race, Gender, and Work: A Multicultural Economic History* *of Women in the United States* (with Julie Matthaei) and *Caught in the Crisis: Women in the U.S. Economy Today*, along with numerous articles. She is an Editorial Associate with *Dollars and Sense* magazine, and is committed to sharing economic analysis with unions, welfare rights organizations, women's organizations, and other progressive groups.

Elijah Anderson is the Charles and William L. Day Professor of the Social Sciences, Professor of Sociology, and Director of the Philadelphia Ethnography Project at the University of Pennsylvania. An expert on the sociology of Black America, he is the author of the highly regarded sociological work *A Place on the Corner: A Study of Black Street Corner Men*, numerous articles on the Black experience, and the forthcoming *The Code of the Streets*. For his recently published ethnographic study, *Streetwise: Race, Class and Change in an Urban Community*, he was honored with the Robert E. Park Award by the American Sociological Association. An Associate Editor of *Qualitative Sociology*, he is a member of the Board of Directors of the American Academy of Political and Social Science.

John Anner is the founding editor and publisher of *Third Force*, a magazine that reports on and evaluates grass-roots labor and community organizing in communities of color. He is the editor of *Beyond Identity Politics: Emerging Social Justice Movements in Communities of Color*, and writes frequently about successful organizing.

Charon Asetoyer is the founder and Executve Director of the Native American Women's Health Education Resource Center, based on the Yankton Sioux Reservation in Lake Andes, South Dakota. She has been involved in community organizing and activism on women's health issues for the past twenty years.

Maxine Baca Zinn is Professor of Sociology at Michigan State University and Senior Research Associate in the Julian Samora Research Institute. She is coeditor of *Women of Color in U.S. Society* (with Bonnie Thornton Dill), co-author of *Diversity in Families, In Conflict and Order*, and *Social Problems* (with D. Stanley Eitzen), and the coeditor of *Through the Prism of Difference: A Sex and Gender Reader* (with Pierrette Hondagneu-Sotelo and Michael A. Messner).

Evelyn Torton Beck is Professor and Director of the Women's Studies Program at the University of Maryland at College Park, and a member of the Jewish Studies Program. She is the author of *Kafka and the Yiddish Theater*, *The Prism of Sex: Essays in the Sociology of Knowledge* (with Julia Sherman), and

Nice Jewish Girls: A Lesbian Anthology. She has lectured and written widely on Jewish women's studies, lesbian studies, anti-Semitism in the women's movement, feminist transformations of knowledge, and feminist pedagogy.

Alan Berkman is a political activist and physician who was imprisoned for anti-imperialist activities in the 1980s. While in prison, he wrote a number of articles reporting on his life as a prisoner. He was to be released in July 1992, but we have no further information about him.

Peter Blood is a Marriage and Family Therapist for the Institute for Christian Counseling and Therapy. He writes about and leads retreats that focus on men's issues. He is the Editor of *Rise Up Singing* and is Editorial Director for the *Sing Out* publication.

Tim Blunk became a student activist in the late 1970s. In 1986, he was sent to a maximum security federal penitentiary, accused by the FBI of escape conspiracy with other political activists. As a revolutionary, artist, and poet, he contributed to *Hauling Up the Morning* and *Cages of Steel.* He received Honorable Mention for his poetry in the 1990 Larry Neal Writers' Competition.

Olivia Castellano is Professor of English at California State University, Sacramento, where she teaches composition, Chicano literature, and creative writing. She was a Stanford Fellow in Modern Thought and Literature in 1976. She has published three collections of poetry: *Blue Mandolin–Yellow Field, Blue Horse of Madness,* and *Spaces That Time Missed.* Her fourth book of poetry, *West by Southwest,* is now ready for publication, and she is currently writing a novel about growing up in Southwest Texas, tentatively titled *Borderghost.*

Lisa Catanzarite received her Ph.D. in Sociology from Stanford University. She is Assistant Professor of Sociology and Urban Studies and Planning at the University of California, San Diego. Her work centers on social stratification and labor markets, especially on occupational segregation by gender and race/ethnicity. Her most recent research is on segregation of immigrant Latino workers and its implications for immigrants, native minorities, and native Whites.

Sucheng Chan is Professor of Asian American Studies at the University of California at Santa Barbara. She is the author of *This Bittersweet Soil: The Chinese in California Agriculture, 1860–1910, Asian Californians,* and *Asian Americans: An Interpretive History.* She has also edited eight books and written dozens of articles.

Grace Chang is completing her Ph.D. in Ethnic Studies at the University of California at Berkeley. She is coeditor of *Mothering: Ideology, Experience, and Agency* with Evelyn Nakano Glenn and Linda Rennie Forcey. Her essays on undocumented Latina workers have appeared in *Socialist Review* and *Radical America*. An essay on SAPs, welfare reform, and the global trade in Filipina migrant workers will appear in the forthcoming anthology *Dragon Ladies: Asian Feminists Breathe Fire*. Currently, she is working on *Gatekeeping and Housekeeping: The Politics of Regulating Migrant Women*, a comparative analysis of the regulation of migrant women's labor through U.S. welfare and immigration policies and global economic restructuring programs.

Sumi K. Cho is Assistant Professor at the DePaul University College of Law in Chicago where she pursues her research on Asian Pacific Americans and Critical Race Theory. She teaches Remedies, Employment Discrimination, Race, Racism & U.S. Law. As a campus-community organizer on issues of affirmative action, racial/gender discrimination, and sexual harassment, she strives to bring a multidisciplinary approach and community-based orientation to the study of race, gender, and law.

Ward Churchill (enrolled Keetoowah Band Cherokee) is a long-time native rights activist, acclaimed public speaker, and award-winning writer. A member of the Governing Council of the Colorado chapter of the American Indian Movement, he also serves as Professor of Ethnic Studies and Coordinator of American Indian Studies for the University of Colorado. He is a past national spokesperson for the Leonard Peltier Defense Committee and has served as a delegate to the United Nations Working Group on Indigenous Populations, as a Justice/Rapporteur for the 1993 International People's Tribunal on the Rights of Indigenous Hawaiians, and as an advocate/prosecutor of the First Nations International Tribunal for the Chiefs of Ontario. His many books include *Marxism and Native Americans, Agents of Repression, The COINTEL-PRO Papers, Critical Issues in Native North America, Fantasies of the Master Race, Struggle for the Land, Since Predator Came, From a Native Son,* and *A Little Matter of Genocide: Holocaust and Denial in the Americas*.

Johnnetta B. Cole is President Emerita of Spelman College. In 1987, she was named the first African American woman president of this historically Black women's college in Atlanta, Georgia. She held that position until 1997. In 1998 she will assume the position of Presidential Distinguished Professor of Anthropology, Women's Studies, and African American Studies.

Chuck Collins is the Coordinator of the Share the Wealth Project and works with the Tax Equity Alliance of Massachusetts. He contributes to the *Directory of Socially Responsible Investment* and writes regularly for *Dollars and Sense*.

Marc Cooper is a contributing editor for *The Nation*, and Host and Executive Producer of RadioNation. He is the author of *Roll Over, Che Guevara: Travels of a Radical Reporter* (Verso, 1994) and a PBS video, *Our Forgotten War*.

Bonnie Thornton Dill earned her Ph.D. at New York University after working for a number of years in anti-poverty and open-admissions programs in New York. She is currently Professor of Women's Studies at the University of Maryland, College Park, and was the founding director of the Center for Research on Women at Memphis State University. She has contributed articles to such journals as *Signs, Journal of Family History*, and *Feminist Studies* and is coeditor with Maxine Baca Zinn of the book *Women of Color in American Society* for Temple University Press. She is also conducting research on single mothers, race, and poverty in the rural South with a grant from the Aspen Institute and the Ford Foundation.

Michael Dyson, who has taught at Hartford Seminary, Chicago Theological Seminary, and Brown University, is presently a Professor of Communication Studies at the University of North Carolina, Chapel Hill. He is the author of the widely acclaimed *Reflecting Black: African-American Cultural Criticism*.

Barbara Ehrenreich writes feature articles, reviews, essays, and humor that have appeared in a wide range of publications, including *The Nation, The New York Times Magazine, The Atlantic Monthly, Harper's, The New Republic*, and *Time*. Her most recent book is *The Snarling Citizen*.

D. Stanley Eitzen is Professor Emeritus of Sociology at Colorado State University. With Maxine Baca Zinn, he has written *Diversity in Families, In Conflict and Order*, and *Social Problems*. He has also written *Elite Deviance* (with David R. Simon), *Sociology of North American Sport* (with George H. Sage), *Criminology* (with Doug A. Timmer), and *Paths to Homelessness* (with Doug A. Timmer and Kathryn Talley). His current book projects include *Demythologizing Sport* and *Where the Welfare State Works*.

Marilyn Frye teaches Philosophy and Women's Studies at Michigan State University. She received her Ph.D. at Cornell University in 1969 and is author of two books of essays, *The Politics of Reality* and *Willful Virgin*. Her recent philosophical work questions how we categorize women. With her partner Carolyn Shafer, she created and manages Bare Bones Studios for Women's Art—space and facilities for art-making in a wide range of media. She also edits a local community newsletter, *The Dyke Heights Dispatch*.

Theresa Funiciello is a political theorist, journalist, and activist. She is the author of the Pulitzer nominated *Tyranny of Kindness: Dismantling The Welfare*

System to End Poverty in America. In addition to her book, she has written for a variety of newspapers, journals, and magazines.

Rose Campbell Gibson is Faculty Associate at the Institute for Social Research and Institute of Gerontology, and Professor Emerita at the University of Michigan where she taught the Sociology of Aging and Research Methods. After receiving her Ph.D. from the University of Michigan, she was a Postdoctoral Fellow there in Statistics and Survey Research Design and Methodology on Minority Populations. She is the author of several books, including *Different Worlds: Inequality in The Aging Experience* (with Eleanor Palo Stoller); *Blacks in an Aging Society*; and *Health in Black America* (with James Jackson); and numerous articles and chapters on sociocultural issues in aging. She is the immediate past Editor-in-Chief of *The Gerontologist.* Her research on sociocultural factors in aging has been funded by the National Institute of Aging, the Administration on Aging, and private foundations. She has been a Distinguished Visiting Professor/Scholar at several universities, including Yonsei, Dong-A and Yeungnam Universities in South Korea, University of California at Berkeley, University of Georgia, and University of Iowa, where she was the Ida I. Beam Distinguished Visiting Professor.

Amy Gluckman is a social studies teacher in an alternative high school in Lowell, Massachusetts. She is part of an editorial collective that publishes the progressive economics magazine, *Dollars and Sense* and, with Betsy Reed, is the editor of a book, *Homo Economics*, a collection of articles about capitalism, economics, and gay and lesbian life.

Craig T. Greenlee, correspondent for *Black Issues in Higher Education*, is a community sports writer and copy editor/layout artist for the *News & Record* in Greensboro, North Carolina. The former *Upscale Magazine* sports editor received his master's and undergraduate degrees from Marshall University in West Virginia.

Elizabeth Higginbotham grew up in a working-class family in New York City. She is a Professor at the Center for Research on Women and the Department of Sociology and Social Work at The University of Memphis and will be joining the University of Delaware faculty in 1998. Higginbotham has published widely on issues of race, class, and gender. She is coeditor of *Women and Work: Exploring Race, Ethnicity, and Class* and has articles in *American Behavioral Scientist, Gender & Society, Women's Studies Quarterly*, and many edited collections. She is currently completing a book on social class differences in the experiences of college educated Black women titled *Too Much to Ask: The Cost of Black Female Success.*

Robert A. Hummer is Assistant Professor of Sociology and Research Associate at the Population Research Center at the University of Texas at Austin. His research examines the social and behavioral factors that affect health and mortality across the life span. Along with Richard Rogers, he is currently examining race and ethnic differences in U.S. adult mortality.

June Jordan is a Professor of African American studies at U.C. Berkeley, and the author of twenty-five books, to date. Her most recent titles include *I Was Looking at the Ceiling and Then I Saw the Sky* and *Kissing God Goodbye*. She is completing a new collection of political essays, affirmative acts, and a childhood memoir entitled *Portrait of the Poet as a Little Black Girl*.

Nazli Kibria is Associate Professor of Sociology at Boston University. She has written widely on the sociology of families, race and ethnicity, and gender. She has a special interest in Southeast Asia, and is the author of *Family Tightrope: The Changing Lives of Vietnamese Americans*.

Bruce Kokopeli is the author of *Leadership for Change: Toward a Feminist Model* (with George Lakey).

Celene Krauss is Associate Professor of Sociology and Women's Studies at Kean College in New Jersey. She has written extensively on women's issues and toxic waste protests, with a focus on race, ethnicity, and class. Her work has appeared in *Sociological Forum*, *Qualitative Sociology*, and other journals and books.

Tracy Lai teaches Multicultural History/Coordinated Studies at Seattle Central Community College. She is active in Seattle Community College's Federation of Teachers Local #1789 and the Asian Pacific American Labor Alliance. Her book reviews appear in *Pacific Reader*, an Asian Pacific North American review of books, published by the *International Examiner*. Her current projects include oral history interviews for Densho, the Japanese American Legacy Project, and writing an essay on women's activism in the Asian Pacific student movement.

George Lakey is a consultant and trainer, an activist and college teacher, and author of five books. He has conducted over 1000 workshops on five continents. He began leading men's conferences as early as the mid-1970s. He has initiated social change projects in civil rights, neighborhood development, economic conversion, and peace, and was a founder of the Movement for a New Society. An openly gay grandfather of four, he frequently lectures at major colleges and universities.

Donna Langston has worked at a variety of jobs—as secretary, waitress, factory worker, and on an oil refinery crew. She is coeditor of *Changing Our Power* and Associate Professor of Women's Studies at Mankato State University.

Audre Lorde, who passed away in 1992, grew up in the West Indian community of Harlem in the 1930s, the daughter of immigrants from Grenada. She attended Hunter College (later becoming Professor of English there), ventured to the American expatriate community in Mexico, and participated in the Greenwich Village scene of the early 1950s. She was a major figure in the lesbian and feminist movements. Among her works are *Sister Outsider*, *Zami: A New Spelling of My Name*, *Uses of the Erotic*, *Chosen Poems Old and New*, *The Black Unicorn*, and *From a Land Where Other People Live*.

Arturo Madrid is Murchison Distinguished Professor of the Humanities at Trinity University and the recipient of the Charles Frankel Prize in the Humanities, National Endowment for the Humanities, in 1996. From 1984 until 1993 he served as Founding President of The Tomás Rivera Center, a national institute for policy studies on Latino issues.

Patricia Yancey Martin, Daisy Parker Flory Alumni Professor of Sociology at Florida State University, teaches and studies gender and organizations. She coauthored (with Judith Lorber) a recent paper on "the social construction of the body" that explores weight, breast size, able-bodiedness, and athleticism, among other issues, from a feminist perspective. A forthcoming book on rape crisis centers asks how these organizations mobilize to help women while also cooperating with legal–justice officials. She is also writing a textbook with David Collinson on gender and organizations and a book on gender and race dynamics in large for-profit corporations.

Elizabeth Martinez is an editor for *Crossroads* magazine, a regular columnist for *Z* magazine, and author of *500 Years of Chicano History*. Martinez is a life-long activist in the civil rights and Chicano movements.

Peggy McIntosh is Associate Director of the Wellesley College Center for Research on Women. She is founder and codirector of the National SEED Project on Inclusive Curriculum and consults throughout the country and the world with college and school faculty who are creating gender-fair and multicultural curricula. She has written many articles on women's studies, curriculum change, and systems of unearned privilege. She has also taught at the Bearley School, Harvard University, Trinity College (Washington, D.C.), the University of Denver, the University of Durham (England), and Wellesley

College. She is contributing editor to *Women's Studies Quarterly* and a consulting editor to *SAGE: A Scholarly Journal on Black Women.*

Michael Messner is Associate Professor of Sociology and Gender Studies at the University of Southern California. He is past President of the North American Society for the Sociology of Sport. His books include *Power at Play: Sports and the Problem of Masculinity; Sex, Violence, and Power in Sports: Rethinking Masculinity* (with Donald F. Sabo); and *Politics of Masculinities: Men in Movements.*

Roslyn Arlin Mickelson is Professor of Sociology and Adjunct Professor of Women's Studies at the University of North Carolina at Charlotte. Her research examines the political economy of schooling, in particular how race, class, and gender shape educational processes and outcomes.

Barbara Miner is coauthor of *Selling Out Our Schools: Vouchers, Markets, and the Future of Public Education* and *False Choice: Why School Vouchers Threaten our Children's Future.* She is also Managing Editor of *Rethinking Schools,* a journal providing a grass-roots look at educational reform in urban schools.

Joan Moore is Distinguished Professor Emerita of Sociology at the University of Wisconsin in Milwaukee. She has written exclusively about urban poverty and about Latinos, with a particular concern for youth gangs.

Cherríe Moraga is a poet, playwright, and essayist. She is the coeditor of *This Bridge Called My Back: Writing by Radical Women of Color,* which won the Before Columbus American Book Award in 1986. She is the author of numerous plays including *Shadow of a Man,* winner of the 1990 Fund for New American Plays Award and *Heroes and Saints,* winner of the Will Glickman Prize and the Pen West Award in 1992. Her most recent book is a collection of poems and essays entitled *The Last Generation.* She is also a recipient of the National Endowment for the Arts' Theatre Playwrights' Fellowship.

Katherine Newman is Professor of Anthropology at Columbia University. She is the author of *Falling from Grace,* as well as other works. She received the Anthropology in Media Award from the American Anthropological Association in 1993.

Vilma Ortiz is Associate Professor of Sociology at University of California at Los Angeles. For more than 15 years, she has studied the socioeconomic experiences of Latinos in the United States. Currently she is directing (with Edward Telles) a 30-year longitudinal and two-generational study examining socioeconomic mobility and ethnic identity among Mexican Americans in Los Angeles and San Antonio.

Raquel Pinderhughes is Associate Professor of Urban Studies at San Francisco State University. She is the editor (with Joan Moore) of *In the Barrios: Latinos and the Underclass Debate*. Her research areas and interests include urban restructuring; urban poverty; the socioeconomic condition of Latino populations in the United States; the relationship between race, poverty, and environmental equity; community organizing around local environmental issues; and sustainable urban development. She recently completed studies on poverty and social inequality in San Francisco and on barriers to community organizing around small-source air polluters.

Bernice Johnson Reagon, Distinguished Professor of History at American University and Curator Emerita at the Smithsonian Institution, is also a composer and singer performing with Sweet Honey in the Rock, the ensemble she founded in 1973. Her recent works include *We Who Believe in Freedom: Sweet Honey in the Rock, We'll Understand It Better By and By*, and the re-mastered *Voices of the Civil Rights Movement, 1960–1965*, a two-CD collection with booklet anthology. Reagon was conceptual producer and narrator of the Peabody award-winning radio series "Wade in the Water: African American Sacred Music Traditions" produced by the Smithsonian Institution and National Public Radio and curator for a traveling exhibition of the same title. She received the 1995 Charles Frankel Prize, a Presidential medal, in recognition of her outstanding contribution to public understanding of the humanities.

Betsy Reed is a former Editor of *Dollars and Sense* and *Boston Review*. She is the Managing Editor of *The American Benefactor* and (with Amy Gluckman) coeditor of *Homo Economics*, a collection of articles about capitalism, economics, and gay and lesbian life.

Gabrielle Sándor is a frequent contributor to *American Demographics* magazine. She is the author of a number of articles on population, race, and gender.

Jason Schultz is the Director of Special Projects for Stir Fry Productions, an organization specializing in diversity training and racial conflict resolution. He was also the coordinator of the Duke Men's Project, a group for college men working on issues of gender and sexuality.

Stephen Samuel Smith is Associate Professor of Political Science at Winthrop University in Rock Hill, South Carolina. His interests include the politics of education, urban political economy, social movements, and program evaluation.

C. Matthew Snipp is a Professor of Sociology at Stanford University. He also has been a Professor of Rural Sociology and Sociology, and Director of the American Indian Studies Program at the University of Wisconsin in Madison. He has been a Research Fellow at the U.S. Bureau of the Census and a Fellow at the Center for Advanced Study in the Behavioral Sciences. Snipp has published numerous works on American Indian demography, economic development, poverty, and unemployment. He is the author of *American Indians: The First of This Land* and *Public Policy Impacts on American Indian Economic Development*. His current research and writing deals with poverty and unemployment on American Indian reservations, and American Indian ethnic identity. His tribal heritage is Oklahoma Cherokee and Choctaw.

Helena Stauber is a doctoral student in Counseling and Consulting Psychology at the Graduate School of Education, Harvard University. She is the coauthor of *Urban Girls* (with Ross Leadbeater and Niobe Way) and has research interests in high-risk behavior, personality, and cultural context.

Gloria Steinem is one of the most influential writers, editors, and activists of our time. She travels in this and other countries as a lecturer and feminist organizer, and is a frequent media interviewer and spokeswoman on issues of equality. Currently, she is a writer and consulting editor for *Ms. Magazine*, the international feminist bimonthly that she cofounded in 1972. Several of her books—*Moving Beyond Words, Revolution from Within: A Book of Self-Esteem*, and *Outrageous Acts and Everyday Rebellions*—have been best sellers. As an organizer, she has helped to found many associations and projects that empower women and girls. In 1993, her concern with child abuse led her to coproduce and narrate an Emmy award-winning TV documentary, "Multiple Personalities: The Search for Deadly Memories." As both a writer and an activist, she remains especially interested in the shared origins of caste systems based on sex and race; in gender roles and child abuse as the roots of violence; in nonviolent conflict resolution; in the cultures of indigenous people; and in organizing across national boundaries for peace and justice.

Eleanor Palo Stoller, Ph.D., is the Selah Chamberlain Professor of Sociology at Case Western Reserve University. Her research explores how older people and their families cope with health declines. She is the author of over forty articles and book chapters and coauthor (with Rose Campbell Gibson) of *Worlds of Difference, Inequality and the Aging Experience*. She is a member of the editorial boards of *The Gerontologist, The Journal of Gerontology: Social Sciences, Research on Aging, Journal of Applied Gerontology, Journal of Aging Studies,* and *Family Relations.* She also served on the Neuroscience, Behavior, and Sociology of Aging Review Committee of the National Institute on Aging. She is a GSA Fellow and a member of the Council of the Section on Aging of the American Sociological Association. She was named a Master Teacher by the Association for Gerontology in Higher Education.

Dana Y. Takagi is Associate Professor of Sociology at the University of California at Santa Cruz. Her book *The Retreat from Race: Asian American Admissions and Racial Politics* examines how discursive politics about Asian Americans can significantly shape contemporary debates about affirmative action. Her current work examines the politics of sovereignty in Hawaii.

Ronald Takaki is Professor of Ethnic Studies at the University of California at Berkeley. He served as Departmental Chairperson and also the Graduate Advisor of the Ethnic Studies Ph.D. program. He is the author of several books, including *Iron Cages: Race and Culture in 19th Century America, Strangers from a Different Shore: A History of Asian Americans,* and *A Different Mirror: A History of Multicultural America.* He helped to establish Berkeley's multicultural requirement for graduation.

Chris Tilly is Associate Professor of Regional Economic and Social Development at the University of Massachusetts at Lowell. He is a labor economist and author of *Half a Job: Bad and Good Part-Time Jobs in a Changing Labor Market, Glass Ceilings and Bottomless Pits: Women's Work, Women's Poverty* (with Randy Albelda), and *Work Under Capitalism* (with Charles Tilly). His research focuses on inequality, poverty, and the low end of the labor market. He is also a member of the editorial collective of *Dollars and Sense.*

Alan Tuttle is a contributor to *Off Their Backs* and *On Our Own Two Feet,* a publication on men and feminism.

Niobe Way is Assistant Professor of Applied Psychology at New York University. She received her doctorate in Human Development from Harvard University. She is the coeditor of *Urban Girls: Resisting Stereotypes, Creating Identities* (with Bonnie J. Leadbeater) and the author of a forthcoming book *Ordinary Courage: The Lives and Stories of Urban Youth.* Her work focuses on the development of urban, poor, and working class adolescents. She is currently at work on a study of friendships and family relationships among immigrant youth from low-income families.

Mary Waters is Professor of Sociology at Harvard University. She is the author of *Ethnic Options: Choosing Identities in America* and coauthor of *From Many Strands: Ethnic and Racial Groups in Contemporary America.* She has consulted with the Census Bureau on issues of measurement of race and ethnicity and was a member of the National Academy of Science's Study Panel on the Demographic and Economic Consequences of Immigration to the United States. She has been a Guggenheim Fellow, a Visiting Scholar at the Russell Sage Foundation, and is a member of the International Committee of the Social Science Research Council.

Lynn Weber is the Director of the Women's Studies Program at the University of South Carolina and a Professor of Sociology. Her research and teaching explores the intersections of race, class, gender, and sexuality particularly as they are manifest in the process of upward social mobility, in the work world, in women's mental health, and in the creation of an inclusive classroom environment. She is the coauthor of *The American Perception of Class* and is currently writing a book on race, class, gender, and sexuality in everyday life.

Cornel West is Professor of Afro-American Studies and the Philosophy of Religion at Harvard University. He is the author of numerous books and essays, including *Restoring Hope: Conversations on the Future of Black America* (with Shawn Sealey) and *Race Matters.*

Gaye Williams graduated in 1983 from Radcliffe-Harvard College. She has served as Acting Archivist at the Bethune Museum Archives National Historic Site. She is also an activist who tries to combine scholarship, politics, and art to affect change.

Linda Faye Williams is Associate Professor of Government and Politics at the University of Maryland at College Park. Prior to joining the faculty of the University of Maryland, she taught at a number of universities including Howard, Cornell, and Brandeis. She has also served as a Research Fellow at Harvard's Kennedy School of Government, as Acting Director of Maryland's Afro-American Studies Program, and as Director of the Congressional Black Caucus Foundation's Institute for Policy Research and Education. She has been a pollster for a number of surveys that oversample women and/or people of color. She is coeditor (along with Ralph C. Gomes) of *From Exclusion to Inclusion: The Long Struggle for Black Political Empowerment* and author of many articles on race and gender politics, public opinion, and social policy.

Naomi Wolf was born in San Francisco in 1962. She was educated at Yale University and New College, Oxford University, where she was a Rhodes Scholar. Her essays have appeared in various publications including *The New Republic, Wall Street Journal, Glamour, Ms., Esquire, The Washington Post*, and *The New York Times*. She lectures on college campuses and is the author of *The Beauty Myth* and, most recently, *Fire with Fire*. Currently she is writing *Promiscuities*, a history and memoir of female desire.

Deborah Woo is a sociologist and Associate Professor in Community Studies at the University of California at Santa Cruz. Her research has revolved around issues of cultural diversity, racial/ethnic stratification, and the ideological or theoretical framework that shapes social relationships within a particular institutional or organizational context. He most recent work focuses on glass ceiling issues affecting Asian Americans in the workplace.

Gloria Yamato is currently the Community Relations Associate for the Pacific Northwest Regional Office of the American Friends Service Committee. She is also the acting Executive Director for ACT/fra, an organization for community-based social/economic change. She has contributed essays to *Changing Our Power* and *Making Face, Making Soul/Haciendo Caras.*

Introduction

TOWARD INCLUSIVE THINKING THROUGH THE STUDY OF RACE, CLASS, AND GENDER

Anyone who takes even a cursory look at the contemporary United States has to acknowledge that this is one of the most diverse nations in the world. Different racial and ethnic groups populate American cities, suburbs, and small towns. In schools, students speak a variety of languages. At work, although different groups are still segregated into specific jobs, the presence of women, new immigrants, and racial minorities is growing. African Americans, Latinos, Asian Americans, and Native Americans together account for an increasing proportion of the total population—so much so that even the term *minority* is misleading, at least in the numerical sense. Furthermore, as shown in Figure 1, the population of the United States will become even more diverse in years to come, as African Americans, Latinos, Asian Americans, and American Indians become a larger proportion of the total population.

What does diversity mean? Because the public has become more conscious of the many differences among us, *diversity* has become a buzzword—a word popularly used, but loosely defined. People use *diversity* to mean cultural variety, numerical representation, changing social norms, and the inequalities that characterize the status of different groups. In thinking about diversity, people have recognized that gender, age, race, sexual orientation, and ethnicity matter; thus, groups who have previously been invisible, including people of color, gays and lesbians, old people, and immigrants, are now in some ways more visible. At the same time that diversity is more commonly recognized, however, these same groups continue to be defined as "other"; that is, they are perceived through dominant group values, treated in exclusionary ways, and subjected to social injustice and economic inequality. Moreover, dominant ways of thinking remain centered in the experiences of a few. As a result, groups who in total are the majority are still rendered invisible. Thus, while diversity is widely acknowledged, there is still a need to reconstruct how we think—and how we behave—if we are to create a more just society.

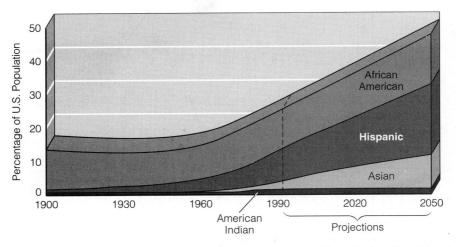

Source: Jeffrey S. Passel and Barry Edmonston, "Immigration and Race: Recent Trends in Immigration to the United States" (Washington, D.C.: The Urban Institute, 1992), and Bureau of the Census, *Current Population Reports* P-25, no. 1092 (Washington, D.C.: GPO, 1992).

FIGURE 1 Share of Minorities in the U.S. Population. 1900–2050.

Understanding the social dynamics of race, class, and gender is central to this new vision. Studying race, class, and gender brings new facts and new perspectives to light, and it illuminates familiar things in new ways. For example, as we explore in the introduction to Part II, college basketball can be seen as pure entertainment, but it can also be seen as a social system in which race, class, and gender play significant roles. Also, places that appear to be all-White or all-men can be understood as products of a social system that is premised on certain race, class, and gender arrangements. Understanding the significance of race, class, and gender puts the experience of all groups in the United States in a broader context. Furthermore, knowing how race, class, and gender operate within American national borders helps you see beyond those borders, developing an awareness of how the increasingly global basis of society influences the configuration of race, class, and gender relationships at home.

Thinking about race, class, and gender need not entail thinking about all groups at one time. Rather, it means understanding the experience of any group in context. So, even though writing exclusively about groups in the United States can be seen as writing from a singular place, understanding the United States in a global context means seeing the factors that shape any given experience more completely.

In this vein, studying diversity is not simply a matter of learning about other people's cultures, values, and ways of being; it involves discovering how race, class, and gender—along with factors like age, ethnicity, sexual orientation, and religion—frame people's lives. The point is not just that people are diverse, as if that were a nice and interesting fact of life, but that race, class, and gender are fundamental axes of society and, as such, are critical to understanding people's lives, institutional systems, contemporary social issues, and the possibilities for social change.

This book is part of the new effort to transform thinking by seeing how race, class, and gender operate together to create the experiences of diverse groups in society. In this introduction, we provide a partial framework for approaching the articles in this book. We want students and faculty—as well as our other readers—to understand fundamentally that race, class, and gender are linked experiences, no one of which is more important than the others; the three are interrelated and together configure the structure of this society.

NEW RACE, CLASS, AND GENDER STUDIES

Race, class, and gender shape the experiences of all people in the United States. This fact has been widely documented in research and, to some extent, is commonly understood. Thus, for years, social scientists have studied the consequences of race, class, and gender inequality for diverse groups in society. New race, class, and gender studies, however, have produced new understandings of how race, class, and gender operate together in people's lives. Fundamentally, these studies interpret race, class, and gender as interlocking categories of experience that affect all aspects of human life; thus, they simultaneously structure the experiences of all people in this society. At any moment, race, class, or gender may feel more salient or meaningful in a given person's life, but they are overlapping and cumulative in their effect on people's experience.

Because of their simultaneity in people's lives, in this book we conceptualize race, class, and gender as different but interrelated axes of social structure. We use the approach of a "matrix of domination" to analyze race, class, and gender. A matrix of domination posits multiple, interlocking levels of domination that stem from the societal configuration of race, class, and gender relations. This structural pattern affects individual consciousness, group interaction, and group access to institutional power and privileges (Collins 1990).

Approaching the study of race, class, and gender from this perspective differs from employing the usual "additive model," which focuses on the independent effects of race, class, and gender on human experience. The additive model is reflected in terms like *double* and *triple jeopardy*—terms used to describe the oppression of women of color by race, as well as by gender and class. We do not think of race and gender oppression in the simple additive terms implied by phrases like *double* and *triple jeopardy*. The effects of race, class, and gender do "add up," both over time and in intensity of impact, but seeing race, class, and gender only in additive terms misses the social structural connections between them and the particular ways in which different configurations of race, class, and gender affect group experience.

The additive model also tends to assess the impact of race, class, and gender exclusively on the experience of women and people of color, rather than seeing them as an integral part of social structures that shape the experience of all groups. The additive model also places people in "either/or" categories, as if one is either Black or White, oppressed or oppressor, powerful or powerless. We do not think that people can be dichotomized so easily. To begin with, race is not a fixed category, but a socially constructed category whose meaning shifts over time. This is not well understood, particularly in the United States, where race has been construed as a biological phenomenon. In other places, such as Great Britain, race is more typically understood as a political construct—as is ethnicity in national contexts where ethnicity is the basis for significant social conflict.

Although socially constructed, race nonetheless has actual consequences for different groups. Consider the U.S. census, a practice that involves counting and categorizing groups for purposes of representation in government, allocating of federal and state resources, and drawing political boundaries. How one is counted—indeed, whether one is counted—is of enormous importance. Underestimating the number of homeless, for example, affects the resources made available to assist them. Yet, many people do not see themselves as fitting neatly into the defined categories, as is the case for the growing number of multiracial people. Whether one even has the choice to define one's group membership is a question of state policy, which can change. The point is that race is a fluid, not a fixed, concept that develops in specific historic contexts, that defines power relations in society, and that develops within the particular circumstances faced by diverse groups.

New race, class, and gender studies also interpret the division between powerful and powerless, or between oppressor and oppressed, as not being clear cut. White women, for example, may be disadvantaged because of gender, but privileged by race and perhaps (but not necessarily) by class. Increasing class differentiation within racial-ethnic groups reminds us that race is not a monolithic category. Rather, race *and* class shape the experiences of, for example, Asian Americans, African Americans, and White Americans. For that matter, White Americans do not fit in "either/or" categories—either oppressor or oppressed. Isolating White Americans as if their experience stands alone ignores how White experience is intertwined with that of other groups.

Race, class, and gender are manifested differently, depending on their configuration with the others. Thus, although from the perspective of an additive model, one might say Black men are privileged as men, this makes no sense when their race *and* gender *and* class are taken into account. Otherwise, how can we possibly explain the particular disadvantages African American men experience *as men*—in the criminal justice system, in education, and in the labor market? For that matter, how can we explain the experience that Native American women experience *as Native American women*—disadvantaged by the unique experiences that they have based on race, class, and gender— none of which is isolated from the effects of the others?

DIVERSITY AND MULTICULTURALISM

Approaching the study of race, class, and gender from the perspective of the matrix of domination means that we see our work as differing somewhat from the concepts implicit in the language of "diversity" or "multiculturalism" suggests. Clearly, our work is part of these movements, but we find the framework of multiculturalism and diversity to be limited. Why do we find this problematic?

Diversity has become a catchword for trying to understand the complexities of race, class, and gender in the United States. The movement to promote diversity has made people more sensitive and aware of the intersections of race, class, and gender, and we are glad to be part of this. Thinking about diversity has also encouraged students and activists to see linkages to other forms of oppression, including sexuality, age, region, physical disability, national identity, and ethnicity; but "understanding diversity" is not the only point. The very term *diversity* implies that understanding race, class, and gender is simply a matter of recognizing the plurality of views and experiences in society—as if race, class, and gender oppression were just the sum total of different experiences. We want readers to be more analytical. Analyzing race, class, and gender involves more than "appreciating cultural diversity." It requires analysis and criticism of existing systems of power and privilege; otherwise, understanding diversity becomes just one more privilege for those with the greatest access to education—something that has always been a mark of the elite class. Analyzing race, class, and gender as they shape different group experiences also involves addressing issues of power, privilege, and equity. This means more than just knowing the cultures of an array of human groups. It means recognizing and analyzing the hierarchies and systems of domination that permeate society and that systematically exploit and control people.

In this sense, we distinguish "thinking comparatively" from "thinking relationally." People think comparatively when they learn about experiences other than their own and begin comparing and contrasting the experiences of different groups. This is a step beyond centering one's thinking on a single group (typically one's own), but it is nonetheless limited. For example, when students encounter studies of race, class, and gender for the first time, they often ask, "How is this group's experience like or not like my own?" This is an important question and a necessary first step, but it is not enough. For one thing, it tends to promote groups' ranking each other's oppression, as if the important thing were to determine who is most victimized. Furthermore, it frames one's understanding of different groups only within the context of other groups' experiences; thus, it can assume an artificial norm against which different groups are judged. Although thinking comparatively is an important step toward thinking more inclusively, we must go further.

Relational thinking involves seeing the interrelationships among diverse group experiences. When you think relationally, you see the social structures that simultaneously generate unique group histories and link them together in society. This does not mean that one group's experience is the same as another's, although finding commonalities is an important step toward more inclusive thinking. In thinking relationally, you untangle the workings of social systems that shape the experiences of different people and groups, and you move beyond just comparing (for example) gender oppression with race oppression, or the oppression of gays and lesbians with that of racial groups. When you recognize the systems of power that mark different groups' experiences, you possess the conceptual apparatus to think about changing the system, not just about documenting the effects of that system on different people.

This process also differs from simply adding up all the experiences of groups who have been defined as "other." Certainly, opening our awareness of distinct group experiences is important, but one need not know the experience of every single combination of oppressions to understand how the social system works. For example, some people in recent years have tended to think of race, class, and gender in terms of a "voices" metaphor—a metaphor that appropriately arose from challenging the silence surrounding many group experiences. In this framework, people think about diversity as "listening to the voices" of a multitude of previously silenced groups. This is an important part of coming to understand race, class, and gender, but it is not enough.

One problem is that people may begin hearing the voices as if they were disembodied from particular historical and social conditions—a framework that has been exacerbated by postmodernist trends in feminist theory. This perspective sees experience as a matter of competing discourses, personifying "voice" as if the voice or discourse itself constituted lived experience. Second, the "voices" framework suggests that any analysis is incomplete unless every voice is heard. In a sense, of course, this is true, since inclusion of silenced people is one of the goals of multicultural work. But an analysis grounded in race, class, and gender can be complete even if centered on the experience of a single group—as long as that group's experience is situated within a framework that recognizes the influence of race, class, and gender on group experience.

Finally, we see limits to the multicultural movement because we think it is important to challenge the idea that diversity is important only at the level of culture—an implication of the term *multiculturalism*. *Culture* is traditionally defined as the "total way of life" of a group of people. It encompasses both material and symbolic components and is an important dimension of understanding human life. Analysis of culture per se, however, tends to look at the group itself, rather than at the broader conditions within which the group lives. Of course, as anthropologists know, a sound analysis of culture situates group experience within these external conditions. Nonetheless, the focus on culture tends to ignore social conditions of power, privilege, and prestige. The result is that multicultural studies often seem tangled up with notions of cultural pluralism—as if knowing a culture other than one's own is the goal of a multicultural education.

We think it is important to see the diversity and plurality of different cultural forms; but in our view this perspective, taken in and of itself, misses the broader point of understanding how racism, sexism, homophobia, and class relations have shaped the experience of groups. Imagine, for example, trying to study the oppression of gays and lesbians in terms of gay culture only. Obviously, doing this turns attention onto the group itself, and away from the dominant society. Likewise, studying race in terms of Latino culture or Asian American culture or African American culture, or studying women's oppression only by looking at women's culture, encourages thinking that blames the victims for their own oppression. For all of these reasons, the focus of this book is on the institutional, or structural, bases for race, class, and gender relations.

ORGANIZATION OF THE BOOK

We have organized the book to follow this thinking. Part I, "Shifting the Center and Reconstructing Knowledge," contains personal reflections on how race, class, and gender shape individual and collective experiences. These articles provide a fresh beginning point for social thought by putting those who have traditionally been excluded at the center of our thought. These different accounts are not meant to represent everyone who has been excluded, but to ground readers in the practice of thinking about the lives of people who rarely occupy the center of dominant group thinking. When excluded groups are placed in the forefront of thought, they are typically stereotyped and distorted. When you look at society through the different perspectives of persons reflecting on their location within systems of race, class, and gender oppression, you begin to see just how much race, class, and gender really matter.

In Part II of the book, "Conceptualizing Race, Class, and Gender," we provide a basic conceptual grounding for the three major concepts being examined—race, class, and gender. Although we do not consider them separate systems, we treat them separately here for analytical purposes, so students will understand how each operates and then be better able to see their interlocking nature. We also want students to see the continuing effects of race, class, and gender on people's experiences: our introduction to this section and the articles within document those effects and point to the interrelationships between them.

In Part III, "Rethinking Institutions," we examine how race, class, and gender shape the organization of social institutions and how, as a result, these institutions affect group experience. Social scientists routinely document how different race, class, and gender groups are affected by institutional structures. We know this is true, but want to go beyond these analyses to scrutinize how institutions are constructed through race, class and gender relations. As contemporary feminist scholars argue, institutions are "gendered" (Acker 1993). We should add that they are also "racially formed" (Omi and Winant 1994) and organized as class systems.

Part IV, "Analyzing Social Issues," uses the inclusive perspective gained by understanding race, class, and gender to analyze contemporary social issues. We examine American identity, sexuality, and violence—social issues that are hotly contested in public discourse and that we think can only be properly understood if they are located within a race, class, and gender analysis. Indeed, we think that these social issues are hotly contested *because* they evolve from contemporary race, class, and gender politics. Although specific social issues may change from time to time, contemporary discussions of American identity and culture, sexual politics, and the causes and consequences of violence are framed by implicit and explicit assumptions that stem from competing ideologies of race, class, and gender. Understanding the social issues on the public agenda requires an analysis of race, class, and gender relations.

Finally, in Part V, "Making a Difference," we look at social change. When studying the effects of race, class, and gender, a person can all too easily think only in terms of how people are victimized and oppressed by these systems. Throughout the book, we have avoided a "social problems approach" because we want readers to move away from thinking exclusively in a problem-centered framework. Race, class, and gender do indeed form the basis of many social problems, but a problems-based approach has three serious limitations:

it tends to portray people primarily as victims, while ignoring their independent views of themselves and their society; it tends to see oppressed groups only from the perspective of the more privileged, relegating those who most suffer under race, class, and gender oppression to the status of "others" thereby reproducing the hierarchical viewpoints that have permeated traditional thinking; and it tends to leave students feeling that they have no power to make changes in the society.

People are not just victims; they are creative and visionary. As a result, people organize to resist oppression and to make liberating social changes. In fact, oppression generates resistance. In Part V we examine the meaning of activism and its connection to the conditions in which people live. We see these articles as providing a basis for understanding how people can make a difference—something that we find often puzzles students. These articles suggest that change takes place through new actions and new social structures, and sometimes through existing channels; but unless one has an inclusive framework from which to make change, one's action is limited. Resistance to oppression on behalf of one's own group is not enough. Truly multicultural change takes place only when people and groups build coalitions with others. Change also takes place over the long run. Short-term actions are needed, but the dedication needed to achieve long-term, institutional changes must be sustained by new visions. We hope the articles in this volume provide a visionary approach to the future.

In selecting articles for the anthology, we tried to find pieces that would show the intersections of race, class, and gender. Not every article does this, but we consider it equally important for students to learn to recognize connections between groups and social structural conditions even when these are not obvious. We searched for articles that are conceptually and theoretically informed, and at the same time accessible to undergraduate readers. Although it is important to think of race, class, and gender as analytical categories, we do not want to lose sight of how they affect human experiences and feelings; thus, we have included personal narratives that are reflective and analytical. We think that personal accounts generate empathy and also help us connect personal experiences to social structural conditions.

We have tried to be as inclusive of all groups as possible, representing the richness of difference and diversity within the United States. It is impossible to include all historically and presently marginalized groups in one book, so we encourage readers to explore the many other works available. We have selected materials that explain the relationships among race, class, and gender and that illuminate the experiences of many groups, not just the groups about whom an article is specifically written. As we begin to untangle the structure of race, class, and gender relations, we can better see both the commonalities and differences that various historical and contemporary experiences have generated.

We also want students to learn that thinking about race, class, and gender is not just a matter of studying victims. We remind students that race, class, and gender have affected the experiences of all groups. As a result, we do not think we should only talk about women when talking about gender, or people of color when talking about race. Because race, class, and gender affect the experience of all, it is important to study men when analyzing gender, to study whites when analyzing race, and to study the experience of the affluent when analyzing class. If we are thinking in an inclusive way, we will think about women, not just men, when studying race; Latinos and other people of color, when thinking about class; and women and men of color when studying gender.

In addition, we should not relegate the study of racial-ethnic groups, the working class, and women only to subjects marked explicitly as race, class, or gender studies. As categories of social experience, race, class, and gender shape all social institutions and systems of meaning; thus, it is important to think about people of color, different class experiences, and women in analyzing all social institutions and belief systems.

Developing truly inclusive thinking and teaching is a long-term process that involves personal, intellectual, and political change. We do not claim to be models of perfection in this regard. We have been pleased by the strong response to the first two editions of this book, and we are fascinated by how race, class, and gender studies have developed in the short period of time since the publication of the first edition. We know further work is needed. Our own teaching and thinking has been transformed by developing this book. We imagine many changes still to come.

References

Acker, Joan. 1992. "Gendered Institutions: From Sex Roles to Gendered Institutions." *Contemporary Sociology* 21 (September): 565–69.

Collins, Patricia Hill. 1990. *Black Feminist Thought: Knowledge, Consciousness, and Empowerment.* Cambridge: Unwin & Hyman.

Omi, Michael, and Howard Winant. 1994. *Racial Formation in the United States: From the 1960s to the 1980s,* 2d ed. New York: Rutledge.

I

Shifting the Center and Reconstructing Knowledge

We begin this book by asking, Who has been excluded from what is known, and how might we see the world differently if we were to acknowledge and value the experiences and thoughts of those who have been excluded? Many groups whose experiences have been vital in the formation of American society and culture have been silenced in the construction of knowledge about this society. The result is that what we know—about the experiences of both these silenced groups and the dominant culture—is distorted and incomplete.

As examples, think about the large number of social science studies that routinely make general conclusions about *the* population, when they have been based on research done only about and by men. Or, think about the content of history courses (and, for that matter, music, art, and literature courses) and how little the work of women, Asian Americans, Latinos, African Americans, Native Americans, gays, and lesbians is included. By excluding or marginalizing these groups, do we not communicate that they have no history or that their history is less important and less central to the development of American culture than is the history of White American men? What false conclusions does this exclusionary thinking generate? One example is the claim that democracy and egalitarianism were central cultural beliefs in the early history of the United States. If this is true, how do we explain the enslavement of millions of African Americans? the genocide of American Indians? the laws against intermarriage between Asian Americans and White Americans? Without more inclusive thinking, our knowledge of American society and history is limited or wrong, and we are left only with contemporary stereotypes and misleading information as the basis for our knowledge about each other.

"Shifting the center" means putting at the center of our thinking the experiences of groups who have formerly been excluded. Without doing so, many groups simply remain invisible. When they are seen, they are typically judged through the experiences of White people, rather than understood on their own terms; this establishes a false norm through which all groups are judged. This is well expressed by Scott Kurashige, who writes, "The public view of Asian Americans is a lot like that of Casper the Ghost: we're either white or we're invisible" (Kurashige 1994).

Shifting the center is a shift in stance that illuminates the experiences of not only the oppressed groups but also of those in the dominant culture. For example, the development of women's studies has changed what we know and how we think about women; at the same time, it has changed what we know and how we think about men. This does not mean that women's studies is about "male-bashing." It means that we take the experiences of women and men seriously and analyze the way that gender, race, and class have shaped the experiences of both—although in different but interrelated ways. Likewise, the study of racial-ethnic groups begins by learning the diverse histories and experiences of these groups, but in doing so, we transform our understanding of White experiences, too. The exclusionary thinking we have relied upon in the past simply does not reveal the intricate interconnections that exist between the different groups comprising American society.

Exclusionary thinking is increasingly being challenged by scholars and teachers who want to include the diversity of human experience in the construction and transmission of knowledge. Thinking more inclusively opens up the way the world is viewed, making the experience of previously excluded groups more visible and central in the construction of knowledge. Inclusive thinking shifts our perspective from the white, male-centered forms of thinking that have characterized much of Western thought, helping us better understand the intersections of race, class, and gender in the experiences of all groups, including those with privilege and power.

Why would we want to do this? Does reconstructing knowledge matter? To begin with, knowledge is not just some abstract thing—good to have, but not all that important. There are real consequences to having partial or distorted knowledge. First, learning about other groups helps you realize the partiality of your own perspective; furthermore, this is true for both dominant and subordinate groups. Only knowing the history of African Americans, for example, or seeing that history only in single-minded terms will not reveal the historical linkages between the oppression of African Americans and the exclusionary and exploitative treatment of other groups. This is well pointed out by Ronald Takaki in his essay on the multicultural history of American society included here.

Second, having misleading and incorrect knowledge gives us a poor social analysis and leads to the formation of bad social policy—policy that then reproduces, rather than solves, social problems. Finally, knowledge is not just about content and information; it provides an orientation to the world. What you know frames how you behave and how you think about yourself and others. If what you know is wrong because it is based on exclusionary thought, you are likely to act in exclusionary ways, thereby reproducing the racism, anti-Semitism, sexism, class oppression, and homophobia of society. This may not be because you are overtly racist, anti-Semitic, sexist, elitist, or homophobic (although it may encourage these beliefs), it may simply be because you do not know any better. Challenging oppressive race, class, and gender relations in society requires a reconstruction of knowledge so that we have some basis from which to change these damaging and dehumanizing systems of oppression.

As we said in the preface, developing inclusive thinking is more than just "understanding diversity" or valuing cultural pluralism. Inclusive thinking begins with the recognition that the United States is a multicultural and diverse society. Population data reveal that simple truth, but inclusive thinking is more than recognizing the plurality of experiences in this society. Understanding race, class, and gender means coming to see the systematic exclusion and exploitation of different groups. This is more than just adding in different group experiences to already established frameworks of thought. It means constructing new analyses that are focused on the centrality, of race, class, and gender in the experiences of us all.

You can begin to develop a more inclusive perspective by asking, How does the world look different if we put the experiences of those who have been excluded at the center of our thinking? At first, people might be tempted to simply assert the perspective and experience of their own group. Nationalist movements of different kinds have done this. They are valuable because they recognize the contributions and achievements of oppressed groups, but ultimately they are unidimensional frameworks because they are centered in a single group experience and they encourage, rather than discourage, exclusionary thought.

Seeing inclusively is more than just seeing the world through the perspective of any one group whose views have been distorted or ignored. Remember that group membership cuts across race, class, and gender categories: for example, one may be a working-class, Asian American woman or a Black, middle-class man or a gay, White, working-class woman. Inclusive thinking means seeing the interconnections between these experiences and not reducing a given person's or group's life to a single factor. In addition, developing inclusive thinking is more than just summing up the experiences of individual groups. Race, class, and gender are social structural categories. This means that they are embedded in the institutional structure of society. Understanding them requires a social structural analysis—by which we mean revealing the race, class, and gender patterns and processes that form the very framework of society.

We believe that shifting our perspective by thinking about the experiences of those who have been excluded from knowledge changes how we think about society, history, and culture. No longer do different groups seem just "different" or "deviant." Rather, specific patterns of the intersections of race, class, and gender are revealed, as are the connections that exist between groups. We then learn how our different experiences are linked, both historically and now.

Once we understand that race, class, and gender are simultaneous and intersecting systems of relationship and meaning, we come to see the different ways that other categories of experience intersect in society. Age, religion, sexual orientation, physical ability, region, and ethnicity also shape systems of privilege and inequality. We have tried to integrate these different experiences throughout the book, although we could not include as much as we would have liked.

Seeing these connections is what we think is most important about analyzing race, class, and gender. It is the reason for taking the perspective of people other than those like you. The purpose of the articles in "Shifting the Center" is not to appropriate others' experiences but to begin uncovering the structures of race, class, and gender that are embedded in the experiences of all people.

By providing personal accounts of what exclusion means and how it feels, the articles in Part I show the limits of existing knowledge. We have selected personal accounts that reflect the diverse experiences of race, class, gender, and/or sexual orientation. We intend for the personal nature of these accounts to build empathy between groups—an emotional stance that we think is critical to seeing linkages and connections. By describing how exclusion shapes individual and collective consciousness, these accounts also document the diversity of experience within the United States. As a result, we begin to see how diverse experiences can shape the concepts and theories we develop in the academy. Together, they suggest new possibilities for thinking inclusively and imagining a society that would be more enabling for everyone.

In "Missing People and Others," Arturo Madrid tells how exclusion and marginalization in the educational curriculum have affected us all. Madrid describes his schooling as a process of denial of the specific experience of Latinos, Asian Americans, Native Americans, and all other groups together considered to be "other." He asks us to consider what it feels like to be the missing person. As we read his account, we see how education has contributed to exclusion by marginalizing the history of Chicanos and groups considered to be "other."

In "La Güera," Cherríe Moraga writes about her developing awareness of her class, race, and sexual identity. Like the other authors here, Moraga tells us what exclusion has meant in her experience and how it feels. Her essay also reveals the intersections of race, class, and gender with the system of compulsory heterosexuality—that is, the institutionalized structures and beliefs that define and enforce heterosexual behavior as the only natural and permissible form of sexual expression. For Moraga, acknowledging her lesbian identity deepened her understanding of her mother's oppression as a poor Chicana. Moraga reminds us of the importance of breaking the silences that protect oppression.

Readers should keep the accounts of Madrid, and Moraga in mind as they read the remainder of this book. What new experiences, understandings, theories, histories, and analyses do these readings inspire? What does it take for a member of one group (say a White male) to be willing to learn from and value the experiences of another (for example, a Chicana lesbian)? June Jordan's essay, "Report from the Bahamas" is a model of this inclusive thinking. In it she begins from her own experience as a Black, middle-class, well-educated woman and reflects on her connections with other women in different locations in the race, class, gender system. In doing so, she finds connections in surprising places, while also understanding the differences that distinguish our experiences. She also shows her student, a White Catholic Irish woman, finding common ground with her South African classmate. There are no simple or singular answers to these questions, but recentering our frame of reference can reshape what we know, both formally and in every-day life. These essays show that, although we are caught in multiple systems, we can learn to see our connection to others. This is not just an intellec-tual exercise. As Paula Gunn Allen shows in her article on Native American women, "Angry Women Are Building," it is a matter of survival. Resisting systems of oppression means revising the ideas about ourselves and others that have been created as a part of a system of social control.

Reading these articles shows us the radical shift that thinking inclusively requires. Such a shift begins with valuing the experiences of those who have been excluded and questioning the assumptions made about all groups. When we begin to take a more inclusive perspective, divisions between privileged and underprivileged, oppressor and oppressed are challenged because we start to see race, class, and gender as intersecting systems of experience. For ex-ample, White women and women of color share some common experiences based on their gender, but their racial experiences are quite distinct; more-over, experiences within the race/gender system are further conditioned by one's class. (Johnnetta Cole points out in an article included in Part II that gender is manifested differently, depending on one's race and class position.)

Engaging oneself at the personal level is critical to this process of thinking inclusively. Changing one's mind is not just a matter of assessing facts and data, though that is important; it also requires examining one's feelings. That is why we begin with several personal narratives. Unlike more conventional forms of sociological data (such as surveys, interviews, and even direct observations), personal accounts are more likely to elicit emotional responses. Traditionally, social science has defined emotional engagement as an impediment to objectivity. Sociology, for example, has emphasized rational thought as the basis for social action and has often discouraged more personalized reflection, but the capacity to reflect on one's experience makes us distinctly human. Personal documents tap the private, reflective dimension of life, enabling us to see the inner lives of others and, in the process, revealing our own lives more completely. The idea that objectivity is best reached only through rational thought is a specifically Western and masculine way of thinking—one that we challenge throughout this book.

Including personal narratives is not meant to limit our level of understanding only to individuals. As sociologists, we study individuals in groups as a way of revealing the social structures shaping collective experiences. Through doing so, we discover our common experiences and see the impact of the social structures of race, class, and gender on our experiences. Marilyn Frye's article, "Oppression," introduces the structural perspective that lies at the heart of this book. She distinguishes oppression from suffering, pointing out that many individuals suffer in society, but that oppression is structured into the fabric of social institutions. Using the metaphor of a birdcage, Frye artfully explains the concept of social structure. Looking only at an individual wire in a cage does not reveal the network of wires that form a cage; likewise, social structure refers to the patterns of behavior, belief, resource distribution, and social control that constitute society.

Analysis of the multicultural basis of society is critical to understanding who we are as a society and a culture. The new focus on multiculturalism is, as Ronald Takaki says in "A Different Mirror," a debate over our national identity. Takaki makes a point of showing the common connections in the histories of African Americans, Chicanos, Irish Americans, Jews, and Indians.

He argues that only when we understand this multidimensional history will we see ourselves in the full complexity of our humanity.

As all of the authors in this section show, the strength and richness of our society lies in its diversity, but the potential of this diversity will only be realized through social structures that value and protect all human experiences. To reconstruct what we know, we begin with the experiences of those who have been excluded. In so doing, we begin to build a system of knowledge that ensures the survival of us all.

Reference

Kurashige, Scott. 1994. Cited in Karin Aguilar-San Juan, "Linking the Issues: From Identity to Activism." In Karin Aguilar-San Juan, ed., *The State of Asian America: Activism and Resistance in the 1990s* (Boston: South End Press), pp. 1–15.

Shifting the Center

MISSING PEOPLE AND OTHERS:

Joining Together to Expand the Circle

1

Arturo Madrid

I am a citizen of the United States, as are my parents and as were their parents, grandparents, and great-grandparents. My ancestors' presence in what is now the United States antedates Plymouth Rock, even without taking into account any American Indian heritage I might have.

I do not, however, fit those mental sets that define America and Americans. My physical appearance, my speech patterns, my name, my profession (a professor of Spanish) create a text that confuses the reader. My normal experience is to be asked, "And where are *you* from?"

My response depends on my mood. Passive-aggressive, I answer, "From here." Aggressive-passive, I ask, "Do you mean where am I originally from?" But ultimately my answer to those follow-up questions that ask about origins will be that we have always been from here.

Overcoming my resentment I will try to educate, knowing that nine times out of ten my words fall on inattentive ears. I have spent most of my adult life explaining who I am not. I am exotic, but—as Richard Rodriguez of *Hunger of Memory* fame so painfully found out—not exotic enough . . . not Peruvian, or Pakistani, or Persian, or whatever.

I am, however, very clearly *the other*, if only your everyday, garden-variety, domestic *other*. I've always known that I was *the other*, even before I knew the vocabulary or understood the significance of being *the other*.

From: *Change* 20 (May/June 1988): 55–59. Reprinted by permission.

I grew up in an isolated and historically marginal part of the United States, a small mountain village in the state of New Mexico, the eldest child of parents native to that region and whose ancestors had always lived there. In those vast and empty spaces, people who look like me, speak as I do, and have names like mine predominate. But the *americanos* lived among us: the descendants of those nineteenth-century immigrants who dispossessed us of our lands; missionaries who came to convert us and stayed to live among us; artists who became enchanted with our land and humanscape and went native; refugees from unhealthy climes, crowded spaces, unpleasant circumstances; and, of course, the inhabitants of Los Alamos, whose socio-cultural distance from us was moreover accentuated by the fact that they occupied a space removed from and proscribed to us. More importantly, however, they—*los americanos*—were omnipresent (and almost exclusively so) in newspapers, newsmagazines, books, on radio, in movies and, ultimately, on television.

Despite the operating myth of the day, school did not erase my otherness. It did try to deny it, and in doing so only accentuated it. To this day, schooling is more socialization than education, but when I was in elementary school—and given where I was—socialization was everything. School was where one became an American. Because there was a pervasive and systematic denial by the society that surrounded us that we were Americans. That denial was both explicit and implicit. My earliest memory of the former was that there were two kinds of churches: theirs and ours. The more usual was the implicit denial, our absence from the larger cultural, economic, political and social spaces— the one that reminded us constantly that we were *the other*. And school was where we felt it most acutely.

Quite beyond saluting the flag and pledging allegiance to it (a very intense and meaningful action, given that the U.S. was involved in a war and our brothers, cousins, uncles, and fathers were on the front lines) becoming American was learning English and its corollary—not speaking Spanish. Until very recently ours was a proscribed language—either *de jure* (by rule, by policy, by law) or *de facto* (by practice, implicitly if not explicitly; through social and political and economic pressure). I do not argue that learning English was not appropriate. On the contrary. Like it or not, and we had no basis to make any judgments on that matter, we were Americans by virtue of having been born Americans, and English was the common language of Americans. And there was a myth, a pervasive myth, that said that if we only learned to speak English well—and particularly without an accent—we would be welcomed into the American fellowship.

Senator Sam Hayakawa notwithstanding, the true text was not our speech, but rather our names and our appearance, for we would always have an accent, however perfect our pronunciation, however excellent our enunciation, however divine our diction. That accent would be heard in our pigmentation, our physiognomy, our names. We were, in short, *the other*.

Being *the other* means feeling different; is awareness of being distinct; is consciousness of being dissimilar. It means being outside the game, outside the circle, outside the set. It means being on the edges, on the margins, on the periphery. Otherness means feeling excluded, closed out, precluded, even disdained and scorned. It produces a sense of isolation, of apartness, of disconnectedness, of alienation.

Being *the other* involves a contradictory phenomenon. On the one hand being *the other* frequently means being invisible. Ralph Ellison wrote eloquently about that experience in his magisterial novel *The Invisible Man*. On the other hand, being *the other* sometimes involves sticking out like a sore thumb. What is she/he doing here?

If one is *the other*, one will inevitably be perceived unidimensionally; will be seen stereotypically; will be defined and delimited by mental sets that may not bear much relation to existing realities. There is a darker side to otherness as well. *The other* disturbs, disquiets, discomforts. It provokes distrust and suspicion. *The other* makes people feel anxious, nervous, apprehensive, even fearful. *The other* frightens, scares.

For some of us being *the other* is only annoying; for others it is debilitating; for still others it is damning. Many try to flee otherness by taking on protective colorations that provide invisibility, whether of dress or speech or manner or name. Only a fortunate few succeed. For the majority, otherness is permanently sealed by physical appearance. For the rest, otherness is betrayed by ways of being, speaking or of doing.

I spent the first half of my life downplaying the significance and consequences of otherness. The second half has seen me wrestling to understand its complex and deeply ingrained realities; striving to fathom why otherness denies us a voice or visibility or validity in American society and its institutions; struggling to make otherness familiar, reasonable, even normal to my fellow Americans.

I am also a missing person. Growing up in northern New Mexico I had only a slight sense of our being missing persons. *Hispanos*, as we called (and call) ourselves in New Mexico, were very much a part of the fabric of the society and there were Hispano professionals everywhere about me: doctors, lawyers, school teachers, and administrators. My people owned businesses, ran organizations and were both appointed and elected public officials.

To be sure, we did not own the larger businesses, nor at the time were we permitted to be part of the banking world. Other than that, however, people who looked like me, spoke like me, and had names like mine, predominated.

There was, to be sure, Los Alamos, but as I have said, it was removed from our realities.

My awareness of our absence from the larger institutional life of society became sharper when I went off to college, but even then it was attenuated by the circumstances of history and geography. The demography of Albuquerque still strongly reflected its historical and cultural origins, despite the influx of Midwesterners and Easterners. Moreover, many of my classmates at the University of New Mexico in Albuquerque were Hispanos, and even some of my professors were.

I thought that would obtain at UCLA, where I began graduate studies in 1960. Los Angeles already had a very large Mexican population, and that population was visible even in and around Westwood and on the campus. Many of the groundskeepers and food-service personnel at UCLA were Mexican. But Mexican-American students were few and mostly invisible, and I do not recall seeing or knowing a single Mexican-American (or, for that matter, black, Asian, or American Indian) professional on the staff or faculty of that institution during the five years I was there.

Needless to say, persons like me were not present in any capacity at Dartmouth College—the site of my first teaching appointment—and, of course, were not even part of the institutional or individual mind-set. I knew then that we—a "we" that had come to encompass American Indians, Asian Americans, black Americans, Puerto Ricans, and women—were truly missing persons in American institutional life.

Over the past three decades, the *de jure* and *de facto* segregations that have historically characterized American institutions have been under assault. As a consequence, minorities and women have become part of American institutional life, and although there are still many areas where we are not to be found, the missing persons phenomenon is not as pervasive as it once was.

However, the presence of *the other*, particularly minorities, in institutions and in institutional life, is, as we say in Spanish, *a flor de tierra*; spare plants whose roots do not go deep, a surface phenomenon, vulnerable to inclemencies of an economic, political, or social nature.

Our entrance into and our status in institutional life is not unlike a scenario set forth by my grandmother's pastor when she informed him that she and her family were leaving their mountain village to relocate in the Rio Grande Valley. When he asked her to promise that she would remain true to the faith and continue to involve herself in the life of the church, she assured him that she would and asked him why he thought she would do otherwise.

"Doña Trinidad," he told her, "in the Valley there is no Spanish church. There is only an American church." "But," she protested, "I read and speak English and would be able to worship there." Her pastor's response was: "It is possible that they will not admit you, and even if they do, they might not accept you. And that is why I want you to promise me that you are going to go to church. Because if they don't let you in through the front door, I want you to go in through the back door. And if you can't get in through the back door, go in the side door. And if you are unable to enter through the side door I want you to go in through the window. What is important is that you enter and that you stay."

Some of us entered institutional life through the front door; others through the back door; and still others through side doors. Many, if not most of us, came in through windows and continue to come in through windows. Of those who entered through the front door, some never made it past the lobby; others were ushered into corners and niches. Those who entered through back and side doors inevitably have remained in back and side rooms. And those who entered through windows found enclosures built around them. For despite the lip service given to the goal of the integration of minorities into institutional life, what has occurred instead is ghettoization, marginalization, isolation.

Not only have the entry points been limited: in addition, the dynamics have been singularly conflictive. Gaining entry and its corollary—gaining space—have frequently come as a consequence of demands made on institutions and institutional officers. Rather than entering institutions more or less passively, minorities have, of necessity, entered them actively, even aggressively. Rather than taking, they have demanded. Institutional relations have thus been adversarial, infused with specific and generalized tensions.

The nature of the entrance and the nature of the space occupied have greatly influenced the view and attitudes of the majority population within those institutions. All of us are put into the same box; that is, no matter what the individual reality, the assessment of the individual is inevitably conditioned by a perception that is held of the class. Whatever our history, whatever our record, whatever our validations, whatever our accomplishments, by and large we are perceived unidimensionally and are dealt with accordingly.

My most recent experience in this regard is atypical only in its explicitness. A few years ago I allowed myself to be persuaded to seek the presidency of a large and prestigious state university. I was invited for an interview and presented myself before the selection committee, which included members of the board of trustees. The opening question of the brief but memorable interview was directed at me by a member of that august body. "Dr. Madrid," he asked, "why does a one-dimensional person like you think he can be the president of a multi-dimensional institution like ours?"

If, as I happen to believe, the well-being of a society is directly related to the degree and extent to which all of its citizens participate in its institutions, we have a challenge before us. One of the strengths of our society—perhaps its main strength—has been a tradition of struggle against clubbishness, exclusivity, and restriction.

Today, more than ever, given the extraordinary changes that are taking place in our society, we need to take up that struggle again—irritating, grating, troublesome, unfashionable, unpleasant as it is. As educated and educator members of this society, we have a special responsibility for leading the struggle against marginalization, exclusion, and alienation.

Let us work together to assure that all American institutions, not just its precollegiate educational and penal institutions, reflect the diversity of our society. Not to do so is to risk greater alienation on the part of a growing segment of our society. It is to risk increased social tension in an already conflictive world. And ultimately it is to risk the survival of a range of institutions that, for all their defects and deficiencies, permit us the space, the opportunity, and the freedom to improve our individual and collective lot; to guide the course of our government, and to redress whatever grievances we have. Let us join together to expand, not to close the circle.

LA GÜERA   2

Cherríe Moraga

It requires something more than personal experience to gain a philosophy or point of view from any specific event. It is the quality of our response to the event and our capacity to enter into the lives of others that help us to make their lives and experiences our own.

Emma Goldman[1]

I am the very well-educated daughter of a woman who, by the standards in this country, would be considered largely illiterate. My mother was born

1. Alix Kates Shulman, "Was My Life Worth Living?" *Red Emma Speaks* (New York: Random House, 1972), p. 388.

From: Cherríe Moraga and Gloria Anzaldúa (eds.), *This Bridge Called My Back: Radical Writings by Women of Color* (New York: Kitchen Table Press, 1983), pp. 27–34. Copyright © 1983 by Cherríe Moraga. Reprinted by permission of the author and the publisher.

in Santa Paula, Southern California, at a time when much of the central valley there was still farm land. Nearly thirty-five years later, in 1948, she was the only daughter of six to marry an anglo, my father.

I remember all of my mother's stories, probably much better than she realizes. She is a fine story-teller, recalling every event of her life with the vividness of the present, noting each detail right down to the cut and color of her dress. I remember stories of her being pulled out of school at the ages of five, seven, nine, and eleven to work in the fields, along with her brothers and sisters; stories of her father drinking away whatever small profit she was able to make for the family; of her going the long way home to avoid meeting him on the street, staggering toward the same destination. I remember stories of my mother lying about her age in order to get a job as a hat-check girl at Agua Caliente Racetrack in Tijuana. At fourteen, she was the main support of the family. I can still see her walking home alone at 3 a.m., only to turn all of her salary and tips over to her mother, who was pregnant again.

The stories continue through the war years and on: walnut-cracking factories, the Voit Rubber factory, and then the computer boom. I remember my mother doing piecework for the electronics plant in our neighborhood. In the late evening, she would sit in front of the T.V. set, wrapping copper wires into the backs of circuit boards, talking about "keeping up with the younger girls." By that time, she was already in her mid-fifties.

Meanwhile, I was college-prep in school. After classes, I would go with my mother to fill out job applications for her, or write checks for her at the supermarket. We would have the scenario all worked out ahead of time. My mother would sign the check before we'd get to the store. Then, as we'd approach the checkstand, she would say—within earshot of the cashier—"oh honey, you go 'head and make out the check," as if she couldn't be bothered with such an insignificant detail. No one asked any questions.

I was educated, and wore it with a keen sense of pride and satisfaction, my head propped up with the knowledge, from my mother, that my life would be easier than hers. I was educated; but more than this, I was "la güera": fair-skinned. Born with the features of my Chicana mother, but the skin of my Anglo father, I had it made.

No one ever quite told me this (that light was right), but I knew that being light was something valued in my family (who were all Chicano, with the exception of my father). In fact, everything about my upbringing (at least what occurred on a conscious level) attempted to bleach me of what color I did have. Although my mother was fluent in it, I was never taught much Spanish at home. I picked up what I did learn from school and from over-heard snatches of conversation among my relatives and mother. She often called other lower-income Mexicans "braceros," or "wet-backs," referring to herself and her family as "a different class of people." And yet, the real story was that my family, too, had been poor (some still are) and farmworkers. My mother can remember this in her blood as if it were yesterday. But this is something she would like to forget (and rightfully), for to her, on a basic economic level, being Chicana meant being "less." It was through my mother's desire to protect her children from poverty and illiteracy that we became "anglocized"; the more effectively we could pass in the white world, the better guaranteed our future.

From all of this, I experience, daily, a huge disparity between what I was born into and what I was to grow up to become. Because, (as Goldman suggests) these stories my mother told me crept under my "güera" skin. I had no choice but to enter into the life of my mother. *I had no choice.* I took her life into my heart, but managed to keep a lid on it as long as I feigned being the happy, upwardly mobile heterosexual.

When I finally lifted the lid to my lesbianism, a profound connection with my mother reawakened in me. It wasn't until I acknowledged and confronted my own lesbianism in the flesh, that my heartfelt identification with and empathy for my mother's oppression—due to being poor, uneducated, and Chicana—was realized. My lesbianism is the avenue through which I have learned the most about silence and oppression, and it continues to be the most tactile reminder to me that we are not free human beings.

You see, one follows the other. I had known for years that I was a lesbian, had felt it in my bones, had ached with the knowledge, gone crazed with the knowledge, wallowed in the silence of it. Silence *is* like starvation. Don't be fooled. It's nothing short of that, and felt most sharply when one has had a full belly most of her life. When we are not physically starving, we have the luxury to realize psychic and emotional starvation. It is from this starvation that other starvations can be recognized—if one is willing to take the risk of making the connection—if one is willing to be responsible to the result of the connection. For me, the connection is an inevitable one.

What I am saying is that the joys of looking like a white girl ain't so great since I realized I could be beaten on the street for being a dyke. If my sister's being beaten because she's Black, it's pretty much the same principle. We're both getting beaten any way you look at it. The connection is blatant; and in the case of my own family, the difference in the privileges attached to looking white instead of brown are merely a generation apart.

In this country, lesbianism is a poverty—as is being brown, as is being a woman, as is being just plain poor. The danger lies in ranking the oppressions. *The danger lies in failing to acknowledge the specificity of the oppression.* The danger lies in attempting to deal with oppression purely from a theoretical base. Without an emotional, heartfelt grappling with the source of our own oppression, without naming the enemy within ourselves and outside of us, no authentic, nonhierarchical connection among oppressed groups can take place.

When the going gets rough, will we abandon our so-called comrades in a flurry of racist/heterosexist/what-have-you panic? To whose camp, then, should the lesbian of color retreat? Her very presence violates the ranking and abstraction of oppression. Do we merely live hand to mouth? Do we merely struggle with the "ism" that's sitting on top of our own heads?

The answer is: yes, I think first we do; and we must do so thoroughly and deeply. But to fail to move out from there will only isolate us in our own oppression—will only insulate, rather than radicalize us.

To illustrate: a gay male friend of mine once confided to me that he continued to feel that, on some level, I didn't trust him because he was male; that he felt, really, if it ever came down to a "battle of the sexes," I might kill him. I admitted that I might very well. He wanted to understand the source of my distrust. I responded, "You're not a woman. Be a woman for a day. Imagine being a woman." He confessed that the thought terrified him because, to him, being a woman meant being raped by men. He *had* felt raped by men; he wanted to forget what that meant. What grew from that discussion was the realization that in order for him to create an authentic alliance with me, he must deal with the primary source of his own sense of oppression. He must, first, emotionally come to terms with what it feels like to be a victim. If he— or anyone—were to truly do this, it would be impossible to discount the oppression of others, except by again forgetting how we have been hurt.

And yet, oppressed groups are forgetting all the time. There are instances of this in the rising Black middle class, and certainly an obvious trend of such "unconsciousness" among white gay men. Because to remember may mean giving up whatever privileges we have managed to squeeze out of this society by virtue of our gender, race, class, or sexuality.

Within the women's movement, the connections among women of different backgrounds and sexual orientations have been fragile, at best. I think this phenomenon is indicative of our failure to seriously address ourselves to some very frightening questions: How have I internalized my own oppression? How have I oppressed? Instead, we have let rhetoric do the job of poetry. Even the word "oppression" has lost its power. We need a new language, better words that can more closely describe women's fear of and resistance to one another; words that will not always come out sounding like dogma.

What prompted me in the first place to work on an anthology by radical women of color was a deep sense that I had a valuable insight to contribute, by virtue of my birthright and background. And yet, I don't really understand first-hand what it feels like being shitted on for being brown. I understand much more about the joys of it—being Chicana and having family are synonymous for me. What I know about loving, singing, crying, telling stories, speaking with my heart and hands, even having a sense of my own soul comes from the love of my mother, aunts, cousins . . . .

But at the age of twenty-seven, it is frightening to acknowledge that I have internalized a racism and classism, where the object of oppression is not only someone outside of my skin, but the someone inside my skin. In fact, to a large degree, the real battle with such oppression, for all of us, begins under the skin. I have had to confront the fact that much of what I value about being Chicana, about my family, has been subverted by anglo culture and my own cooperation with it. This realization did not occur to me overnight. For example, it wasn't until long after my graduation from the private college I'd attended in Los Angeles, that I realized the major reason for my total alienation from and fear of my classmates was rooted in class and culture. CLICK.

Three years after graduation, in an apple-orchard in Sonoma, a friend of mine (who comes from an Italian Irish working-class family) says to me, "Cherríe, no wonder you felt like such a nut in school. Most of the people there were white and rich." It was true. All along I had felt the difference, but not until I had put the words "class" and "color" to the experience, did my feelings make any sense. For years, I had berated myself for not being as "free" as my classmates. I completely bought that they simply had more guts than I did—to rebel against their parents and run around the country hitch-hiking, reading books and studying "art." They had enough privilege to be atheists, for chrissake. There was no one around filling in the disparity for me between their parents, who were Hollywood filmmakers, and my parents, who wouldn't know the name of a filmmaker if their lives depended on it (and precisely because their lives didn't depend on it, they couldn't be bothered). But I knew nothing about "privilege" then. White was right. Period. I could pass. If I got educated enough, there would never be any telling.

Three years after that, another CLICK. In a letter to Barbara Smith, I wrote:

> I went to a concert where Ntosake Shange was reading. There, everything exploded for me. She was speaking a language that I knew—in the deepest parts of me—existed, and that I had ignored in my own feminist studies and even in my own writing. What Ntosake caught in me is the realization that in my development as a poet, I have, in many ways, denied the voice of my brown mother—the brown in me. I have acclimated to the sound of a white language which, as my father represents it, does not speak to the emotions in my poems—emotions which stem from the love of my mother.
>
> The reading was agitating. Made me uncomfortable. Threw me into a week-long terror of how deeply I was affected. I felt that I had to start all over

again. That I turned only to the perceptions of white middle-class women to speak for me and all women. I am shocked by my own ignorance.

Sitting in that auditorium chair was the first time I had realized to the core of me that for years I had disowned the language I knew best—ignored the words and rhythms that were the closest to me. The sounds of my mother and aunts gossiping—half in English, half in Spanish—while drinking cerveza in the kitchen. And the hands—I had cut off the hands in my poems. But not in conversation; still the hands could not be kept down. Still they insisted on moving.

The reading had forced me to remember that I knew things from my roots. But to remember puts me up against what I don't know. Shange's reading agitated me because she spoke with power about a world that is both alien and common to me: "the capacity to enter into the lives of others." But you can't just take the goods and run. I knew that then, sitting in the Oakland auditorium (as I know in my poetry), that the only thing worth writing about is what seems to be unknown and, therefore, fearful.

The "unknown" is often depicted in racist literature as the "darkness" within a person. Similarly, sexist writers will refer to fear in the form of the vagina, calling it "the orifice of death." In contrast, it is a pleasure to read works such as Maxine Hong Kingston's *Woman Warrior*, where fear and alienation are described as "the white ghosts." And yet, the bulk of literature in this country reinforces the myth that what is dark and female is evil. Consequently, each of us—whether dark, female, or both—has in some way *internalized* this oppressive imagery. What the oppressor often succeeds in doing is simply *externalizing* his fears, projecting them into the bodies of women, Asians, gays, disabled folks, whoever seems most "other."

> call me
> roach and presumptuous
> nightmare on your white pillow
> your itch to destroy
> the indestructible
> part of yourself
>
> Audre Lorde[2]

But it is not really difference the oppressor fears so much as similarity. He fears he will discover in himself the same aches, the same longings as those of the people he has shitted on. He fears the immobilization threatened by his

---

2. From "The Brown Menace or Poem to the Survival of Roaches," *The New York Head Shop and Museum* (Detroit: Broadside, 1974), p. 48.

own incipient guilt. He fears he will have to change his life once he has seen himself in the bodies of the people he has called different. He fears the hatred, anger, and vengeance of those he has hurt.

This is the oppressor's nightmare, but it is not exclusive to him. We women have a similar nightmare, for each of us in some way has been both oppressed and the oppressor. We are afraid to look at how we have failed each other. We are afraid to see how we have taken the values of our oppressor into our hearts and turned them against ourselves and one another. We are afraid to admit how deeply "the man's" words have been ingrained in us.

To assess the damage is a dangerous act. I think of how, even as a feminist lesbian, I have so wanted to ignore my own homophobia, my own hatred of myself for being queer. I have not wanted to admit that my deepest personal sense of myself has not quite "caught up" with my "woman-identified" politics. I have been afraid to criticize lesbian writers who choose to "skip over" these issues in the name of feminism. In 1979, we talk of "old gay" and "butch and femme" roles as if they were ancient history. We toss them aside as merely patriarchal notions. And yet, the truth of the matter is that I have sometimes taken society's fear and hatred of lesbians to bed with me. I have sometimes hated my lover for loving me. I have sometimes felt "not woman enough" for her. I have sometimes felt "not man enough." For a lesbian trying to survive in a heterosexist society, there is no easy way around these emotions. Similarly, in a white-dominated world, there is little getting around racism and our own internalization of it. It's always there, embodied in some one we least expect to rub up against.

When we do rub up against this person, *there* then is the challenge. *There* then is the opportunity to look at the nightmare within us. But we usually shrink from such a challenge.

Time and time again, I have observed that the usual response among white women's groups when the "racism issue" comes up is to deny the difference. I have heard comments like, "Well, we're open to *all* women; why don't they (women of color) come? You can only do so much . . . ." But there is seldom any analysis of how the very nature and structure of the group itself may be founded on racist or classist assumptions. More importantly, so often the women seem to feel no loss, no lack, no absence when women of color are not involved; therefore, there is little desire to change the situation. This has hurt me deeply. I have come to believe that the only reason women of a privileged class will dare to look at *how* it is that *they* oppress, is when they've come to know the meaning of their own oppression. And understand that the oppression of others hurts them personally.

The other side of the story is that women of color and working-class women often shrink from challenging white middle-class women. It is much easier to rank oppressions and set up a hierarchy, rather than take responsibility for changing our own lives. We have failed to demand that white women, particularly those who claim to be speaking for all women, be accountable for their racism.

The dialogue has simply not gone deep enough.

I have many times questioned my right to even work on an anthology which is to be written "exclusively by Third World women." I have had to look critically at my claim to color, at a time when, among white feminist ranks, it is a "politically correct" (and sometimes peripherally advantageous) assertion to make. I must acknowledge the fact that, physically, I have had a *choice* about making that claim, in contrast to women who have not had such a choice, and have been abused for their color. I must reckon with the fact that for most of my life, by virtue of the very fact that I am white-looking, I identified with and aspired toward white values, and that I rode the wave of that Southern California privilege as far as conscience would let me.

Well, now I feel both bleached and beached. I feel angry about this—the years when I refused to recognize privilege, both when it worked against me, and when I worked it, ignorantly, at the expense of others. These are not settled issues. That is why this work feels so risky to me. It continues to be discovery. It has brought me into contact with women who invariably know a hell of a lot more than I do about racism, as experienced in the flesh, as revealed in the flesh of their writing.

I think: what is my responsibility to my roots—both white and brown, Spanish-speaking and English? I am a woman with a foot in both worlds; and I refuse the split. I feel the necessity for dialogue. Sometimes I feel it urgently.

But one voice is not enough, nor two, although this is where dialogue begins. It is essential that radical feminists confront their fear of and resistance to each other, because without this, there *will* be no bread on the table. Simply, we will not survive. If we could make this connection in our heart of hearts, that if we are serious about a revolution—better—if we seriously believe there should be joy in our lives (real joy, not just "good times"), then we need one another. We women need each other. Because my/your solitary, self-asserting "go-for-the-throat-of-fear" power is not enough. The real power, as you and I well know, is collective. I can't afford to be afraid of you, nor you of me. If it takes head-on collisions, let's do it: this polite timidity is killing us.

As Lorde suggests in the passage I cited earlier, it is in looking to the nightmare that the dream is found. There, the survivor emerges to insist on a future, a vision, yes, born out of what is dark and female. The feminist movement must be a movement of such survivors, a movement with a future.

REPORT FROM THE BAHAMAS   3

June Jordan

I am staying in a hotel that calls itself The Sheraton British Colonial. One of the photographs advertising the place displays a middle-aged Black man in a waiter's tuxedo, smiling. What intrigues me most about the picture is just this: while the Black man bears a tray full of "colorful" drinks above his left shoulder, both of his feet, shoes and trouserlegs, up to ten inches above his ankles, stand in the also "colorful" Caribbean salt water. He is so delighted to serve you he will wade into the water to bring you Banana Daquiris while you float! More precisely, he will wade into the water, fully clothed, oblivious to the ruin of his shoes, his trousers, his health, and he will do it with a smile.

I am in the Bahamas. On the phone in my room, a spinning complement of plastic pages offers handy index clues such as CAR RENTAL and CASINOS. A message from the Ministry of Tourism appears among these travelers' tips. Opening with a paragraph of "WELCOME," the message then proceeds to "A PAGE OF HISTORY," which reads as follows:

> New World History begins on the same day that modern Bahamian history begins—October 12, 1492. That's when Columbus stepped ashore—British influence came first with the Eleutherian Adventurers of 1647—After the Revolutions, American Loyalists fled from the newly independent states and settled in the Bahamas. Confederate blockade-runners used the island as a haven during the War between the States, and after the War, a number of Southerners moved to the Bahamas . . .

There it is again. Something proclaims itself a legitimate history and all it does is track white Mr. Columbus to the British Eleutherians through the Confederate Southerners as they barge into New World surf, land on New World turf, and nobody saying one word about the Bahamian people, the Black peoples, to whom the only thing new in their island world was this weird succession of crude intruders and its colonial consequences.

This is my consciousness of race as I unpack my bathing suit in the Sheraton British Colonial. Neither this hotel nor the British nor the long ago Italians nor the white Delta airline pilots belong here, of course. And every time

---

From: June Jordan, *On Call: Political Essays* (Boston: South End Press, 1985), pp. 39–49. Reprinted by permission of the author.

I look at the photograph of that fool standing in the water with his shoes on I'm about to have a West Indian fit, even though I know he's no fool; he's a middle-aged Black man who needs a job and this is his job—pretending himself a servile ancillary to the pleasures of the rich. (Compared to his options in life, I am a rich woman. Compared to most of the Black Americans arriving for this Easter weekend on a three nights four days' deal of bargain rates, the middle-aged waiter is a poor Black man.)

We will jostle along with the other (white) visitors and join them in the tee shirt shops or, laughing together, learn ruthless rules of negotiation as we, Black Americans as well as white, argue down the price of handwoven goods at the nearby straw market while the merchants, frequently toothless Black women seated on the concrete in their only presentable dress, humble themselves to our careless games:

"Yes? You like it? Eight dollar."

"Five."

"I give it to you. Seven."

And so it continues, this weird succession of crude intruders that, now, includes me and my brothers and my sisters from the North.

This is my consciousness of class as I try to decide how much money I can spend on Bahamian gifts for my family back in Brooklyn. No matter that these other Black women incessantly weave words and flowers into the straw hats and bags piled beside them on the burning dusty street. No matter that these other Black women must work their sense of beauty into these things that we will take away as cheaply as we dare, or they will do without food.

We are not white, after all. The budget is limited. And we are harmlessly killing time between the poolside rum punch and "The Native Show on the Patio" that will play tonight outside the hotel restaurant.

This is my consciousness of race and class and gender identity as I notice the fixed relations between these other Black women and myself. They sell and I buy or I don't. They risk not eating. I risk going broke on my first vacation afternoon.

We are not particularly women anymore; we are parties to a transaction designed to set us against each other.

"Olive" is the name of the Black woman who cleans my hotel room. On my way to the beach I am wondering what "Olive" would say if I told her why I chose The Sheraton British Colonial; if I told her I wanted to swim. I wanted to sleep. I did not want to be harassed by the middle-aged waiter, or his nephew. I did not want to be raped by anybody (white or Black) at all and I calculated that my safety as a Black woman alone would best be assured by a multinational hotel corporation. In my experience, the big guys take customer complaints more seriously than the little ones. I would suppose that's one reason why they're big; they don't like to lose money anymore than I like to be bothered when I'm trying to read a goddamned book underneath a palm tree I paid $264 to get next to. A Black woman seeking refuge in a multinational corporation may seem like a contradiction to some, but there you are. In this case it's a coincidence of entirely different self-interests: Sheraton/cash = June Jordan's short run safety.

Anyway, I'm pretty sure "Olive" would look at me as though I came from someplace as far away as Brooklyn. Then she'd probably allow herself one indignant query before righteously removing her vacuum cleaner from my room; "and why in the first place you come down you without your husband?"

I cannot imagine how I would begin to answer her.

My "rights" and my "freedom" and my "desire" and a slew of other New World values; what would they sound like to this Black woman described on the card atop my hotel bureau as "Olive the Maid"? "Olive" is older than I am and I may smoke a cigarette while she changes the sheets on my bed. Whose rights? Whose freedom? Whose desire?

And why should she give a shit about mine unless I do something, for real, about hers?

It happens that the book that I finished reading under a palm tree earlier today was the novel, *The Bread Givers*, by Anzia Yezierska. Definitely autobiographical, Yezierska lays out the difficulties of being both female and "a person" inside a traditional Jewish family at the start of the 20th century. That any Jewish woman became anything more than the abused servant of her father or her husband is really an improbable piece of news. Yet Yezierska managed such an unlikely outcome for her own life. In *The Bread Givers*, the heroine also manages an important, although partial, escape from traditional Jewish female destiny. And in the unpardonable, despotic father, the Talmudic scholar of that Jewish family, did I not see my own and hate him twice, again? When the heroine, the young Jewish child, wanders the streets with a filthy pail she borrows to sell herring in order to raise the ghetto rent and when she cries, "Nothing was before me but the hunger in our house, and no bread for the next meal if I didn't sell the herring. No longer like a fire engine, but like a houseful of hungry mouths my heart cried, 'herring—herring! Two cents apiece!'" who would doubt the ease, the sisterhood of conversation possible between that white girl and the Black women selling straw bags on the streets of paradise because they do not want to die? And is it not obvious that the wife of that Talmudic scholar and "Olive," who cleans my room here at the hotel, have more in common than I can claim with either one of them?

This is my consciousness of race and class and gender identity as I collect wet towels, sunglasses, wristwatch, and head towards a shower.

I am thinking about the boy who loaned this novel to me. He's white and he's Jewish and he's pursuing an independent study project with me, at the State University where I teach whether or not I feel like it, where I teach without stint because, like the waiter, I am no fool. It's my job and either I work or I do without everything you need money to buy. The boy loaned me the novel because he thought I'd be interested to know how a Jewish-American writer used English so that the syntax, and therefore the cultural habits of mind expressed by the Yiddish language, could survive translation. He did this because he wanted to create another connection between us on the basis of language, between his knowledge/his love of Yiddish and my knowledge/my love of Black English.

He has been right about the forceful survival of the Yiddish. And I had become excited by this further evidence of the written voice of spoken language protected from the monodrone of "standard" English, and so we had grown closer on this account. But then our talk shifted to student affairs more generally, and I had learned that this student does not care one way or the other about currently jeopardized Federal Student Loan Programs because, as he explained it to me, they do not affect him. He does not need financial help outside his family. My own son, however, is Black. And I am the only family help available to him and that means, if Reagan succeeds in eliminating Federal programs to aid minority students, he will have to forget about furthering his studies, or he or I or both of us will have to hit the numbers pretty big. For these reasons of difference, the student and I had moved away from each other, even while we continued to talk.

My consciousness turned to race, again, and class.

Sitting in the same chair as the boy, several weeks ago, a graduate student came to discuss her grade. I praised the excellence of her final paper; indeed it had seemed to me an extraordinary pulling together of recent left brain/right brain research with the themes of transcendental poetry.

She told me that, for her part, she'd completed her reading of my political essays. "You are so lucky!" she exclaimed.

"What do you mean by that?"

"You have a cause. You have a purpose to your life."

I looked carefully at this white woman; what was she really saying to me?

"What do you mean?" I repeated.

"Poverty. Police violence. Discrimination in general."

(Jesus Christ, I thought: Is that her idea of lucky?)

"And how about you?" I asked.

"Me?"

"Yeah, you. Don't you have a cause?"

"Me? I'm just a middle-aged woman: a housewife and a mother. I'm a nobody."

For a while, I made no response.

First of all, speaking of race and class and gender in one breath, what she said meant that those lucky preoccupations of mine, from police violence to nuclear wipe-out, were not shared. They were mine and not hers. But here she sat, friendly as an old stuffed animal, beaming good will or more "luck" in my direction.

In the second place, what this white woman said to me meant that she did not believe she was "a person" precisely because she had fulfilled the traditional female functions revered by the father of that Jewish immigrant, Anzia Yezierska. And the woman in front of me was not a Jew. That was not the connection. The link was strictly female. Nevertheless, how should that woman and I, another female connect, beyond this bizarre exchange?

If she believed me lucky to have regular hurdles of discrimination then why shouldn't I insist that she's lucky to be a middle-class white Wasp female who lives in such well-sanctioned and normative comfort that she even has the luxury to deny the power of the privileges that paralyze her life?

If she deserts me and "my cause" where we differ, if, for example, she abandons me to "my" problems of race, then why should I support her in "her" problems of housewifely oblivion?

Recollection of this peculiar moment brings me to the shower in the bathroom cleaned by "Olive." She reminds me of the usual Women's Studies curriculum because it has nothing to do with her or her job: you won't find "Olive" listed anywhere on the reading list. You will likewise seldom hear of Anzia Yezierska. But yes, you will find, from Florence Nightingale to Adrienne Rich, a white procession of independently well-to-do women writers. (Gertrude Stein/Virginia Woolf/Hilda Doolittle are standard names among the "essential" women writers.)

In other words, most of the women of the world—Black and First World and white who work because we must—most of the women of the world persist far from the heart of the usual Women's Studies syllabus.

Similarly, the typical Black history course will slide by the majority experience it pretends to represent. For example, Mary McLeod Bethune will scarcely receive as much attention as Nat Turner, even though Black women who bravely and efficiently provided for the education of Black people hugely outnumber those few Black men who led successful or doomed rebellions against slavery. In fact, Mary McLeod Bethune may not receive even honorable mention because Black history too often apes those ridiculous white history courses which produce such dangerous gibberish as The Sheraton British Colonial "history" of the Bahamas. Both Black and white history courses exclude from their central consideration those people who neither killed nor conquered anyone as the means to new identity, those people who took care of every one of the people who wanted to become "a person," those people who still take care of the life at issue: the ones who wash and who feed and who teach and who diligently decorate straw hats and bags with all of their historically unrequired gentle love— the women.

> *Oh the old rugged cross*
> *on a hill far away*
> *Well I cherish the old rugged cross*

It's Good Friday in the Bahamas. Seventy-eight degrees in the shade. Except for Sheraton territory, everything's closed.

It so happens that for truly secular reasons I've been fasting for three days. My hunger has now reached nearly violent proportions. In the hotel sandwich shop, the Black woman handling the counter complains about the tourists; why isn't the shop closed and why don't the tourists stop eating for once in their lives. I'm famished and I order chicken salad and cottage cheese and lettuce and tomato and a hard boiled egg and a hot cross bun and apple juice.

She eyes me with disgust.

To be sure, the timing of my stomach offends her serious religious practices. Neither one of us apologizes to the other. She seasons the chicken salad to the peppery max while I listen to the loud radio gospel she plays to console herself. It's a country Black version of "The Old Rugged Cross."

As I heave much chicken into my mouth, tears start. It's not the pepper. I am, after all, a West Indian daughter. It's the Good Friday music that dominates the humid atmosphere.

> *Well I cherish the old rugged cross*

And I am back, faster than a 747, in Brooklyn, in the home of my parents where we are wondering, as we do every year, if the sky will darken until Christ has been buried in the tomb. The sky should darken if God is in His heavens. And then, around 3 p.m., at the conclusion of our mournful church service at the neighborhood St. Phillips, and even while we dumbly stare at the black cloth covering the gold altar and the slender unlit candles, the sun should return through the high gothic windows and vindicate our waiting faith that the Lord will rise again, on Easter.

How I used to bow my head at the very name of Jesus: ecstatic to abase myself in deference to His majesty.

My mouth is full of salad. I can't seem to eat quickly enough. I can't think how I should lessen the offense of my appetite. The other Black woman on the premises, the one who disapprovingly prepared this very tasty break from my fast, makes no remark. She is no fool. This is a job that she needs. I suppose she notices that at least I included a hot cross bun among my edibles. That's something in my favor. I decide that's enough.

I am suddenly eager to walk off the food. Up a fairly steep hill I walk without hurrying. Through the pastel desolation of the little town, the road brings me to a confectionary pink and white plantation house. At the gates, an unnecessarily large statue of Christopher Columbus faces me down, or tries to. His hand is fisted to one hip. I look back at him, laugh without deference, and turn left.

It's time to pack it up. Catch my plane. I scan the hotel room for things not to forget. There's that white report card on the bureau.

"Dear Guests:" it says, under the name "Olive." "I am your maid for the day. Please rate me: Excellent. Good. Average. Poor. Thank you."

I tuck this momento from the Sheraton British Colonial into my notebook. How would "Olive" rate *me*? What would it mean for us to seem "good" to each other? What would that rating require?

But I am hastening to leave. Neither turtle soup nor kidney pie nor any conch shell delight shall delay my departure. I have rested, here, in the Bahamas, and I'm ready to return to my usual job, my usual work. But the skin on my body has changed and so has my mind. On the Delta flight home I realize I am burning up, indeed.

So far as I can see, the usual race and class concepts of connection, or gender assumptions of unity, do not apply very well. I doubt that they ever did. Otherwise why would Black folks forever bemoan our lack of solidarity when the deal turns real. And if unity on the basis of sexual oppression is something natural, then why do we women, the majority people on the planet, still have a problem?

The plane's ready for takeoff. I fasten my seatbelt and let the tumult inside my head run free. Yes: race and class and gender remain as real as the weather. But what they must mean about the contact between two individuals is less obvious and, like the weather, not predictable.

And when these factors of race and class and gender absolutely collapse is whenever you try to use them as automatic concepts of connection. They may serve well as indicators of commonly felt conflict, but as elements of connection they seem about as reliable as precipitation probability for the day after the night before the day.

It occurs to me that much organizational grief could be avoided if people understood that partnership in misery does not necessarily provide for partnership for change: *When we get the monsters off our backs all of us may want to run in very different directions.*

And not only that: even though both "Olive" and "I" live inside a conflict neither one of us created, and even though both of us therefore hurt inside that conflict, I may be one of the monsters she needs to eliminate from her universe and, in a sense, she may be one of the monsters in mine.

I am reaching for the words to describe the difference between a common identity that has been imposed and the individual identity any one of us will choose, once she gains that chance.

That difference is the one that keeps us stupid in the face of new, specific information about somebody else with whom we are supposed to have a connection because a third party, hostile to both of us, has worked it so that the two of us, like it or not, share a common enemy. *What happens beyond the idea of that enemy and beyond the consequences of that enemy?*

I am saying that the ultimate connection cannot be the enemy. The ultimate connection must be the need that we find between us. It is not only who you are, in other words, but what we can do for each other that will determine the connection.

I am flying back to my job. I have been teaching contemporary women's poetry this semester. One quandary I have set myself to explore with my students is the one of taking responsibility without power. We had been wrestling ideas to the floor for several sessions when a young Black woman, a South African, asked me for help, after class.

Sokutu told me she was "in a trance" and that she'd been unable to eat for two weeks.

"What's going on?" I asked her, even as my eyes startled at her trembling and emaciated appearance.

"My husband. He drinks all the time. He beats me up. I go to the hospital. I can't eat. I don't know what/anything."

In my office, she described her situation. I did not dare to let her sense my fear and horror. She was dragging about, hour by hour, in dread. Her husband, a young Black South African, was drinking himself into more and more deadly violence against her.

Sokutu told me how she could keep nothing down. She weighed 90 lbs. at the outside, as she spoke to me. She'd already been hospitalized as a result of her husband's battering rage.

I knew both of them because I had organized a campus group to aid the liberation struggles of Southern Africa.

Nausea rose in my throat. What about this presumable connection: this husband and this wife fled from that homeland of hatred against them, and now what? He was destroying himself. If not stopped, he would certainly murder his wife.

She needed a doctor, right away. It was a medical emergency. She needed protection. It was a security crisis. She needed refuge for battered wives and personal therapy and legal counsel. She needed a friend.

I got on the phone and called every number in the campus directory that I could imagine might prove helpful. Nothing worked. There were no institutional resources designed to meet her enormous, multifaceted, and ordinary woman's need.

I called various students. I asked the Chairperson of the English Department for advice. I asked everyone for help.

Finally, another one of my students, Cathy, a young Irish woman active in campus IRA activities, responded. She asked for further details. I gave them to her.

"Her husband," Cathy told me, "is an alcoholic. You have to understand about alcoholics. It's not the same as anything else. And it's a disease you can't treat any old way."

I listened, fearfully. Did this mean there was nothing we could do?

"That's not what I'm saying," she said. "But you have to keep the alcoholic part of the thing central in everybody's mind, otherwise her husband will kill her. Or he'll kill himself."

She spoke calmly. I felt there was nothing to do but to assume she knew what she was talking about.

"Will you come with me?" I asked her, after a silence. "Will you come with me and help us figure out what to do next?"

Cathy said she would but that she felt shy: Sokutu comes from South Africa. What would she think about Cathy?

"I don't know," I said. "But let's go."

We left to find a dormitory room for the young battered wife.

It was late, now, and dark outside.

On Cathy's VW that I followed behind with my own car, was the sticker that reads BOBBY SANDS FREE AT LAST. My eyes blurred as I read and reread the words. This was another connection: Bobby Sands and Martin Luther King Jr. and who would believe it? I would not have believed it; I grew up terrorized by Irish kids who introduced me to the word "nigga."

And here I was following an Irish woman to the room of a Black South African. We were going to that room to try to save a life together.

When we reached the little room, we found ourselves awkward and large. Sokutu attempted to treat us with utmost courtesy, as though we were honored guests. She seemed surprised by Cathy, but mostly Sokutu was flushed with relief and joy because we were there, with her.

I did not know how we should ever terminate her heartfelt courtesies and address, directly, the reason for our visit: her starvation and her extreme physical danger.

Finally, Cathy sat on the floor and reached out her hands to Sokutu.

"I'm here," she said quietly, "Because June has told me what has happened to you. And I know what it is. Your husband is an alcoholic. He has a disease.

I know what it is. My father was an alcoholic. He killed himself. He almost killed my mother. I want to be your friend."

"Oh," was the only small sound that escaped from Sokutu's mouth. And then she embraced the other student. And then everything changed and I watched all of this happen so I know that this happened: this connection.

And after we called the police and exchanged phone numbers and plans were made for the night and for the next morning, the young South African woman walked down the dormitory hallway, saying goodbye and saying thank you to us.

I walked behind them, the young Irish woman and the young South African, and I saw them walking as sisters walk, hugging each other, and whispering and sure of each other and I felt how it was not who they were but what they both know and what they were both preparing to do about what they know that was going to make them both free at last.

And I look out the windows of the plane and I see clouds that will not kill me and I know that someday soon other clouds may erupt to kill us all.

And I tell the stewardess No thanks to the cocktails she offers me. But I look about the cabin at the hundred strangers drinking as they fly and I think even here and even now I must make the connection real between me and these strangers everywhere before those other clouds unify this ragged bunch of us, too late.

ANGRY WOMEN ARE BUILDING: *Issues and Struggles Facing American Indian Women Today*

4

Paula Gunn Allen

The central issue that confronts American Indian women throughout the hemisphere is survival, *literal survival*, both on a cultural and biological level. According to the 1980 census, population of American Indians is just over one million. This figure, which is disputed by some American Indians, is

---

From: Paula Gunn Allen, *The Sacred Hoop: Recovering the Feminism in American Indian Traditions* (Boston: Beacon Press, 1986), pp. 189–193. Reprinted with permission.

probably a fair estimate, and it carries certain implications. [Editors' note: The 1995 population is over two million.]

Some researchers put our pre-contact population at more than 45 million, while others put it at around 20 million. The U.S. government long put it at 450,000—a comforting if imaginary figure, though at one point it was put at around 270,000. If our current population is around one million; if, as some researchers estimate, around 25 percent of Indian women and 10 percent of Indian men in the United States have been sterilized without informed consent; if our average life expectancy is, as the best-informed research presently says, 55 years; if our infant mortality rate continues at well above national standards; if our average unemployment for all segments of our population—male, female, young, adult, and middle-aged—is between 60 and 90 percent; if the U.S. government continues its policy of termination, relocation, removal, and assimilation along with the destruction of wilderness, reservation land, and its resources, and severe curtailment of hunting, fishing, timber harvesting and water-use rights—then existing tribes are facing the threat of extinction which for several hundred tribal groups has already become fact in the past five hundred years.

In this nation of more than 200 million, the Indian people constitute less than one-half of one percent of the population. [Editors' note: In 1995, this figure was one percent.] In a nation that offers refuge, sympathy, and billions of dollars in aid from federal and private sources in the form of food to the hungry, medicine to the sick, and comfort to the dying, the indigenous subject population goes hungry, homeless, impoverished, cut out of the American deal, new, old, and in between. Americans are daily made aware of the worldwide slaughter of native peoples such as the Cambodians, the Palestinians, the Armenians, the Jews—who constitute only a few groups faced with genocide in this century. . . . The American Indian people are in a situation comparable to the imminent genocide in many parts of the world today. The plight of our people north and south of us is no better; to the south it is considerably worse. Consciously or unconsciously, deliberately, as a matter of national policy, or accidentally as a matter of "fate," *every single government*, right, left, or centrist in the western hemisphere is consciously or subconsciously dedicated to the extinction of those tribal people who live within its borders.

Within this geopolitical charnel house, American Indian women struggle on every front for the survival of our children, our people, our self-respect, our value systems, and our way of life. The past five hundred years testify to our skill at waging this struggle: for all the varied weapons of extinction pointed at our heads, we endure.

We survive war and conquest; we survive colonization, acculturation, assimilation; we survive beating, rape, starvation, mutilation, sterilization, abandonment, neglect, death of our children, our loved ones, destruction of our land, our homes, our past, and our future. We survive, and we do more than just survive. We bond, we care, we fight, we teach, we nurse, we bear, we feed, we earn, we laugh, we love, we hang in there, no matter what.

Of course, some, many of us, just give up. Many are alcoholics, many are addicts. Many abandon the children, the old ones. Many commit suicide. Many become violent, go insane. Many go "white" and are never seen or heard from again. But enough hold on to their traditions and their ways so that even after almost five hundred brutal years, we endure. And we even write songs and poems, make paintings and drawings that say "We walk in beauty. Let us continue."

Currently our struggles are on two fronts: physical survival and cultural survival. For women this means fighting alcoholism and drug abuse (our own and that of our husbands, lovers, parents, children);[1] poverty; affluence—a destroyer of people who are not traditionally socialized to deal with large sums of money; rape, incest, battering by Indian men; assaults on fertility and other health matters by the Indian Health Service and the Public Health Service; high infant mortality due to substandard medical care, nutrition, and health information; poor educational opportunities or education that takes us away from our traditions, language, and communities; suicide, homicide, or similar expressions of self-hatred; lack of economic opportunities; substandard housing; sometimes violent and always virulent racist attitudes and behaviors directed against us by an entertainment and educational system that wants only one thing from Indians: our silence, our invisibility, and our collective death.

A headline in the *Navajo Times* . . . reported that rape was the number one crime on the Navajo reservation. In a professional mental health journal of the Indian Health Services, Phyllis Old Dog Cross reported that incest and rape are common among Indian women seeking services and that their incidence is increasing. "It is believed that at least 80 percent of the Native Women seen at the regional psychiatric service center (five state area) have experienced some sort of sexual assault."[2] Among the forms of abuse being suffered by Native American women, Old Dog Cross cites a recent phenomenon, something called "training." This form of gang rape is "a punitive act of a group of males who band together and get even or take revenge on a selected woman."[3]

These and other cases of violence against women are powerful evidence that the status of women within the tribes has suffered grievous decline since contact, and the decline has increased in intensity in recent years. The amount of violence against women, alcoholism, and violence, abuse, and neglect by women against their children and their aged relatives have all increased. These social ills were virtually unheard of among most tribes fifty years ago, popular American opinion to the contrary. As Old Dog Cross remarks:

> Rapid, unstable and irrational change was required of the Indian people if they were to survive. Incredible loss of all that had meaning was the norm. Inhuman treatment, murder, death, and punishment was a typical experience for all the tribal groups and some didn't survive.
>
> The dominant society devoted its efforts to the attempt to change the Indian into a white-Indian. No inhuman pressure to effect this change was overlooked. These pressures included starvation, incarceration and enforced education. Religious and healing customs were banished.
>
> In spite of the years of oppression, the Indian and the Indian spirit survived. Not, however, without adverse effect. One of the major effects was the loss of cultured values and the concomitant loss of personal identity . . . The Indian was taught to be ashamed of being Indian and to emulate the non-Indian. In short, "white was right." For the Indian male, the only route to be successful, to be good, to be right, and to have an identity was to be as much like the white man as he could.[4]

Often it is said that the increase of violence against women is a result of various sociological factors such as oppression, racism, poverty, hopelessness, emasculation of men, and loss of male self-esteem as their own place within traditional society has been systematically destroyed by increasing urbanization, industrialization, and institutionalization, but seldom do we notice that for the past forty to fifty years, American popular media have depicted American Indian men as bloodthirsty savages devoted to treating women cruelly. While traditional Indian men seldom did any such thing—and in fact among most tribes abuse of women was simply unthinkable, as was abuse of children or the aged—the lie about "usual" male Indian behavior seems to have taken root and now bears its brutal and bitter fruit.

Image casting and image control constitute the central process that American Indian women must come to terms with, for on that control rests our sense of self, our claim to a past and to a future that we define and that we build. Images of Indians in media and educational materials profoundly influence how we act, how we relate to the world and to each other, and how we value ourselves. They also determine to a large extent how our men act toward us, toward our children, and toward each other. The popular American media image of Indian people as savages with no conscience, no compassion, and no sense of the value of human life and human dignity was hardly true of the tribes—however true it was of the invaders. But as Adolf Hitler noted a little over fifty years ago, if you tell a lie big enough and often enough, it will be believed. Evidently, while Americans and people all over the world have been led into a deep and unquestioned belief that American Indians are cruel savages, a number of American Indian men have been equally deluded into internalizing that image and acting on it. Media images, literary images, and artistic images, particularly those embedded in popular culture, must be changed before Indian women will see much relief from the violence that destroys so many lives.

To survive culturally, American Indian women must often fight the United States government, the tribal governments, women and men of their tribe or their urban community who are virulently misogynist or who are threatened by attempts to change the images foisted on us over the centuries by whites. The colonizers' revisions of our lives, values, and histories have devastated us at the most critical level of all—that of our own minds, our own sense of who we are.

Many women express strong opposition to those who would alter our life supports, steal our tribal lands, colonize our cultures and cultural expressions, and revise our very identities. We must strive to maintain tribal status; we must make certain that the tribes continue to be legally recognized entities, sovereign nations within the larger United States, and we must wage this struggle in many ways—political, educational, literary, artistic, individual, and communal. We are doing all we can: as mothers and grandmothers; as family members and tribal members; as professionals, workers, artists, shamans, leaders, chiefs, speakers, writers, and organizers, we daily demonstrate that we have no intention of disappearing, of being silent, or of quietly acquiescing in our extinction.

NOTES

1. It is likely, say some researchers, that fetal alcohol syndrome, which is serious among many Indian groups, will be so serious among the White Mountain Apache and the Pine Ridge Sioux that if present trends continue, by the year 2000 some people estimate that almost one-half of all children born on those reservations will in some way be affected by FAS. (Michael Dorris, Native American Studies, Dartmouth College, private conversation. Dorris has done extensive research into the syndrome as it affects native populations in the United States as well as in New Zealand.)

2. Phyllis Old Dog Cross, "Sexual Abuse, a New Threat to the Native American Woman: An Overview," *Listening Post: A Periodical of the Mental Health Programs of Indian Health Services*, vol. 6, no. 2 (April 1982), p. 18.

3. Old Dog Cross, p. 18.

4. Old Dog Cross, p. 20.

OPPRESSION   5

Marilyn Frye

It is a fundamental claim of feminism that women are oppressed. The word "oppression" is a strong word. It repels and attracts. It is dangerous and dangerously fashionable and endangered. It is much misused, and sometimes not innocently.

The statement that women are oppressed is frequently met with the claim that men are oppressed too. We hear that oppressing is oppressive to those who oppress as well as to those they oppress. Some men cite as evidence of their oppression their much-advertised inability to cry. It is tough, we are told, to be masculine. When the stresses and frustrations of being a man are cited as evidence that oppressors are oppressed by their oppressing, the word "oppression" is being stretched to meaninglessness; it is treated as though its scope includes any and all human experience of limitation or suffering, no matter the cause, degree, or consequence. Once such usage has been put over on us, then if ever we deny that any person or group is oppressed, we seem to imply that we think they never suffer and have no feelings. We are accused of insensitivity; even of bigotry. For women, such accusation is particularly intimidating, since sensitivity is one of the few virtues that has been assigned to us. If we are found insensitive, we may fear we have no redeeming traits at all and perhaps are not real women. Thus are we silenced before we begin: the name of our situation drained of meaning and our guilt mechanisms tripped.

But this is nonsense. Human beings can be miserable without being oppressed, and it is perfectly consistent to deny that a person or group is oppressed without denying that they have feelings or that they suffer. . . .

The root of the word "oppression" is the element "press." *The press of the crowd; pressed into military service; to press a pair of pants; printing press; press the button.* Presses are used to mold things or flatten them or reduce them in bulk, sometimes to reduce them by squeezing out the gases or liquids in them. Something pressed is something caught between or among forces and barriers which are so related to each other that jointly they restrain, restrict or prevent the thing's motion or mobility. Mold. Immobilize. Reduce.

From: Marilyn Frye, *The Politics of Reality* (Trumansburg, NY: Crossing Press, 1983), pp. 1–16. Reprinted by permission.

The mundane experience of the oppressed provides another clue. One of the most characteristic and ubiquitous features of the world as experienced by oppressed people is the double bind—situations in which options are reduced to a very few and all of them expose one to penalty, censure, or deprivation. For example, it is often a requirement upon oppressed people that we smile and be cheerful. If we comply, we signal our docility and our acquiescence in our situation. We need not, then, be taken note of. We acquiesce in being made invisible, in our occupying no space. We participate in our own erasure. On the other hand, anything but the sunniest countenance exposes us to being perceived as mean, bitter, angry, or dangerous. This means, at the least, that we may be found "difficult" or unpleasant to work with, which is enough to cost one one's livelihood; at worst, being seen as mean, bitter, angry, or dangerous has been known to result in rape, arrest, beating, and murder. One can only choose to risk one's preferred form and rate of annihilation.

Another example: It is common in the United States that women, especially younger women, are in a bind where neither sexual activity nor sexual inactivity is all right. If she is heterosexually active, a woman is open to censure and punishment for being loose, unprincipled, or a whore. The "punishment" comes in the form of criticism, snide and embarrassing remarks, being treated as an easy lay by men, scorn from her more restrained female friends. She may have to lie and hide her behavior from her parents. She must juggle the risks of unwanted pregnancy and dangerous contraceptives. On the other hand, if she refrains from heterosexual activity, she is fairly constantly harassed by men who try to persuade her into it and pressure her to "relax" and "let her hair down"; she is threatened with labels like "frigid," "uptight," "man-hater," "bitch" and "cocktease." The same parents who would be disapproving of her sexual activity may be worried by her inactivity because it suggests she is not or will not be popular, or is not sexually normal. She may be charged with lesbianism. If a woman is raped, then if she has been heterosexually active she is subject to the presumption that she liked it (since her activity is presumed to show that she likes sex), and if she has not been heterosexually active, she is subject to the presumption that she liked it (since she is supposedly "repressed and frustrated"). Both heterosexual activity and heterosexual nonactivity are likely to be taken as proof that you wanted to be raped, and hence, of course, weren't *really* raped at all. You can't win. You are caught in a bind, caught between systematically related pressures.

Women are caught like this, too, by networks of forces and barriers that expose one to penalty, loss, or contempt whether one works outside the home or not, is on welfare or not, bears children or not, raises children or not, marries or not, stays married or not, is heterosexual, lesbian, both, or neither.

Economic necessity; confinement to racial and/or sexual job ghettos; sexual harassment; sex discrimination; pressures of competing expectations and judgments about *women, wives,* and *mothers* (in the society at large, in racial and ethnic subcultures, and in one's own mind); dependence (full or partial) on husbands, parents, or the state; commitment to political ideas; loyalties to racial or ethnic or other "minority" groups; the demands of self-respect and responsibilities to others. Each of these factors exists in complex tension with every other, penalizing or prohibiting all of the apparently available options. And nipping at one's heels, always, is the endless pack of little things. If one dresses one way, one is subject to the assumption that one is advertising one's sexual availability; if one dresses another way, one appears to "not care about oneself" or to be "unfeminine." If one uses "strong language," one invites categorization as a whore or slut; if one does not, one invites categorization as a "lady"—one too delicately constituted to cope with robust speech or the realities to which it presumably refers.

The experience of oppressed people is that the living of one's life is confined and shaped by forces and barriers which are not accidental or occasional and hence avoidable, but are systematically related to each other in such a way as to catch one between and among them and restrict or penalize motion in any direction. It is the experience of being caged in: all avenues, in every direction, are blocked or booby trapped.

Cages. Consider a birdcage. If you look very closely at just one wire in the cage, you cannot see the other wires. If your conception of what is before you is determined by this myopic focus, you could look at that one wire, up and down the length of it, and be unable to see why a bird would not just fly around the wire any time it wanted to go somewhere. Furthermore, even if, one day at a time, you myopically inspected each wire, you still could not see why a bird would have trouble going past the wires to get anywhere. There is no physical property of any one wire, *nothing* that the closest scrutiny could discover, that will reveal how a bird could be inhibited or harmed by it except in the most accidental way. It is only when you step back, stop looking at the wires one by one, microscopically, and take a macroscopic view of the whole cage, that you can see why the bird does not go anywhere; and then you will see it in a moment. It will require no great subtlety of mental powers. It is perfectly *obvious* that the bird is surrounded by a network of systematically related barriers, no one of which would be the least hindrance to its flight, but which, by their relations to each other, are as confining as the solid walls of a dungeon.

It is now possible to grasp one of the reasons why oppression can be hard to see and recognize: one can study the elements of an oppressive structure with great care and some good will without seeing the structure as a whole, and hence without seeing or being able to understand that one is looking at a cage and that there are people there who are caged, whose motion and mobility are restricted, whose lives are shaped and reduced.

The arresting of vision at a microscopic level yields such common confusion as that about the male door-opening ritual. This ritual, which is remarkably widespread across classes and races, puzzles many people, some of whom do and some of whom do not find it offensive. Look at the scene of the two people approaching a door. The male steps slightly ahead and opens the door. The male holds the door open while the female glides through. Then the male goes through. The door closes after them. "Now how," one innocently asks, "can those crazy womenslibbers say that is oppressive? The guy *removed* a barrier to the lady's smooth and unruffled progress." But each repetition of this ritual has a place in a pattern, in fact in several patterns. One has to shift the level of one's perception in order to see the whole picture.

The door-opening pretends to be a helpful service, but the helpfulness is false. This can be seen by noting that it will be done whether or not it makes any practical sense. Infirm men and men burdened with packages will open doors for able-bodied women who are free of physical burdens. Men will impose themselves awkwardly and jostle everyone in order to get to the door first. The act is not determined by convenience or grace. Furthermore, these very numerous acts of unneeded or even noisome "help" occur in counterpoint to a pattern of men not being helpful in many practical ways in which women might welcome help. What *women* experience is a world in which gallant princes charming commonly make a fuss about being helpful and providing small services when help and services are of little or no use, but in which there are rarely ingenious and adroit princes at hand when substantial assistance is really wanted either in mundane affairs or in situations of threat, assault, or terror. There is no help with the (his) laundry; no help typing a report at 4:00 a.m.; no help in mediating disputes among relatives or children. There is nothing but advice that women should stay indoors after dark, be chaperoned by a man, or when it comes down to it, "lie back and enjoy it."

The gallant gestures have no practical meaning. Their meaning is symbolic. The door-opening and similar services provided are services which really are needed by people who are for one reason or another incapacitated— unwell, burdened with parcels, etc. So the message is that women are incapable. The detachment of the acts from the concrete realities of what women need and do not need is a vehicle for the message that women's actual needs and interests are unimportant or irrelevant. Finally, these gestures imitate the behavior of servants toward masters and thus mock women, who are in most respects the servants and caretakers of men. The message of the false helpfulness of male gallantry is female dependence, the invisibility or insignificance of women, and contempt for women.

One cannot see the meanings of these rituals if one's focus is riveted upon the individual event in all its particularity, including the particularity of the individual man's present conscious intentions and motives and the individual woman's conscious perception of the event in the moment. It seems sometimes that people take a deliberately myopic view and fill their eyes with things seen microscopically in order not to see macroscopically. At any rate, whether it is deliberate or not, people can and do fail to see the oppression of women because they fail to see macroscopically and hence fail to see the various elements of the situation as systematically related in larger schemes.

As the cageness of the birdcage is a macroscopic phenomenon, the oppressiveness of the situations in which women live their various and different lives is a macroscopic phenomenon. Neither can be *seen* from a microscopic perspective. But when you look macroscopically you can see it—a network of forces and barriers which are systematically related and which conspire to the immobilization, reduction, and molding of women and the lives we live. . . .

A DIFFERENT MIRROR   6

Ronald T. Takaki

I had flown from San Francisco to Norfolk and was riding in a taxi to my hotel to attend a conference on multiculturalism. Hundreds of educators from across the country were meeting to discuss the need for greater cultural diversity in the curriculum. My driver and I chatted about the weather and the tourists. The sky was cloudy, and Virginia Beach was twenty minutes away. The rearview mirror reflected a white man in his forties. "How long have you been in this country?" he asked. "All my life," I replied, wincing. "I was born in the United States." With a strong southern drawl, he remarked: "I was

From: Ronald T. Takaki, *A Different Mirror: A History of Multicultural America* (Boston: Little, Brown, 1993), pp. 1–17. Copyright © 1993 by Ronald Takaki. Reprinted by permission of the publisher.

wondering because your English is excellent!" Then, as I had many times before, I explained: "My grandfather came here from Japan in the 1880s. My family has been here, in America, for over a hundred years." He glanced at me in the mirror. Somehow I did not look "American" to him; my eyes and complexion looked foreign.

Suddenly, we both became uncomfortably conscious of a racial divide separating us. An awkward silence turned my gaze from the mirror to the passing landscape, the shore where the English and the Powhatan Indians first encountered each other. Our highway was on land that Sir Walter Raleigh had renamed "Virginia" in honor of Elizabeth I, the Virgin Queen. In the English cultural appropriation of America, the indigenous peoples themselves would become outsiders in their native land. Here, at the eastern edge of the continent, I mused, was the site of the beginning of multicultural America. Jamestown, the English settlement founded in 1607, was nearby: the first twenty Africans were brought here a year before the Pilgrims arrived at Plymouth Rock. Several hundred miles offshore was Bermuda, the "Bermoothes" where William Shakespeare's Prospero had landed and met the native Caliban in *The Tempest*. Earlier, another voyager had made an Atlantic crossing and unexpectedly bumped into some islands to the south. Thinking he had reached Asia, Christopher Columbus mistakenly identified one of the islands as "Cipango" (Japan). In the wake of the admiral, many peoples would come to America from different shores, not only from Europe but also Africa and Asia. One of them would be my grandfather. My mental wandering across terrain and time ended abruptly as we arrived at my destination. I said good-bye to my driver and went into the hotel, carrying a vivid reminder of why I was attending this conference.

Questions like the one my taxi driver asked me are always jarring, but I can understand why he could not see me as American. He had a narrow but widely shared sense of the past—a history that has viewed American as European in ancestry. "Race," Toni Morrison explained, has functioned as a "metaphor" necessary to the "construction of Americanness": in the creation of our national identity, "American" has been defined as "white." [1]

But America has been racially diverse since our very beginning on the Virginia shore, and this reality is increasingly becoming visible and ubiquitous. Currently, one-third of the American people do not trace their origins to Europe; in California, minorities are fast becoming a majority. They already predominate in major cities across the country—New York, Chicago, Atlanta, Detroit, Philadelphia, San Francisco, and Los Angeles.

This emerging demographic diversity has raised fundamental questions about America's identity and culture. In 1990, *Time* published a cover story on "America's Changing Colors." "Someday soon," the magazine announced, "white Americans will become a minority group." How soon? By 2056, most Americans will trace their descent to "Africa, Asia, the Hispanic world, the Pacific Islands, Arabia—almost anywhere but white Europe." This dramatic change in our nation's ethnic composition is altering the way we think about ourselves. "The deeper significance of America's becoming a majority non-white society is what it means to the national psyche, to individuals' sense of themselves and their nation—their idea of what it is to be American." . . .[2]

What is fueling the debate over our national identity and the content of our curriculum is America's intensifying racial crisis. The alarming signs and symptoms seem to be everywhere—the killing of Vincent Chin in Detroit, the black boycott of a Korean grocery store in Flatbush, the hysteria in Boston over the Carol Stuart murder, the battle between white sportsmen and Indians over tribal fishing rights in Wisconsin, the Jewish-black clashes in Brooklyn's Crown Heights, the black-Hispanic competition for jobs and educational resources in Dallas, which *Newsweek* described as "a conflict of the have-nots," and the Willie Horton campaign commercials, which widened the divide between the suburbs and the inner cities.[3]

This reality of racial tension rudely woke America like a fire bell in the night on April 29, 1992. Immediately after four Los Angeles police officers were found not guilty of brutality against Rodney King, rage exploded in Los Angeles. Race relations reached a new nadir. During the nightmarish rampage, scores of people were killed, over two thousand injured, twelve thousand arrested, and almost a billion dollars' worth of property destroyed. The live televised images mesmerized America. The rioting and the murderous melee on the streets resembled the fighting in Beirut and the West Bank. The thousands of fires burning out of control and the dark smoke filling the skies brought back images of the burning oil fields of Kuwait during Desert Storm. Entire sections of Los Angeles looked like a bombed city. "Is this America?" many shocked viewers asked. "Please, can we get along here," pleaded Rodney King, calling for calm. "We all can get along. I mean, we're all stuck here for a while. Let's try to work it out."[4]

But how should "we" be defined? Who are the people "stuck here" in America? One of the lessons of the Los Angeles explosion is the recognition of the fact that we are a multiracial society and that race can no longer be defined in the binary terms of white and black. "We" will have to include Hispanics and Asians. While blacks currently constitute 13 percent of the Los Angeles population, Hispanics represent 40 percent. The 1990 census revealed that South Central Los Angeles, which was predominantly black in 1965 when the Watts rebellion occurred, is now 45 percent Hispanic. A majority of the first 5,438 people arrested were Hispanic, while 37 percent were black. Of the fifty-eight people who died in the riot, more than a third were

Hispanic, and about 40 percent of the businesses destroyed were Hispanic-owned. Most of the other shops and stores were Korean-owned. The dreams of many Korean immigrants went up in smoke during the riot: two thousand Korean-owned businesses were damaged or demolished, totaling about $400 million in losses. There is evidence indicating they were targeted. "After all," explained a black gang member, "we didn't burn our community, just *their* stores."[5]

"I don't feel like I'm in America anymore," said Denisse Bustamente as she watched the police protecting the firefighters. "I feel like I am far away." Indeed, Americans have been witnessing ethnic strife erupting around the world—the rise of neo-Nazism and the murder of Turks in Germany, the ugly "ethnic cleansing" in Bosnia, the terrible and bloody clashes between Muslims and Hindus in India. Is the situation here different, we have been nervously wondering, or do ethnic conflicts elsewhere represent a prologue for America? What is the nature of malevolence? Is there a deep, perhaps primordial, need for group identity rooted in hatred for the other? Is ethnic pluralism possible for America? But answers have been limited. Television reports have been little more than thirty-second sound bites. Newspaper articles have been mostly superficial descriptions of racial antagonisms and the current urban malaise. What is lacking is historical context; consequently, we are left feeling bewildered.[6]

How did we get to this point, Americans everywhere are anxiously asking. What does our diversity mean, and where is it leading us? *How* do we work it out in the post–Rodney King era?

Certainly one crucial way is for our society's various ethnic groups to develop a greater understanding of each other. For example, how can African Americans and Korean Americans work it out unless they learn about each other's cultures, histories, and also economic situations? This need to share knowledge about our ethnic diversity has acquired new importance and has given new urgency to the pursuit for a more accurate history. . . .

While all of America's many groups cannot be covered [here], the English immigrants and their descendants require attention, for they possessed inordinate power to define American culture and make public policy. What men like John Winthrop, Thomas Jefferson, and Andrew Jackson thought as well as did mattered greatly to all of us and was consequential for everyone. A broad range of groups [is important]: African Americans, Asian Americans, Chicanos, Irish, Jews, and Indians. While together they help to explain general patterns in our society, each has contributed to the making of the United States.

African Americans have been the central minority throughout our country's history. They were initially brought here on a slave ship in 1619. Actually, these first twenty Africans might not have been slaves; rather, like most of the white laborers, they were probably indentured servants. The transformation of Africans into slaves is the story of the "hidden" origins of slavery. How and when was it decided to institute a system of bonded black labor? What happened, while freighted with racial significance, was actually conditioned by class conflicts within white society. Once established, the "peculiar institution" would have consequences for centuries to come. During the nineteenth century, the political storm over slavery almost destroyed the nation. Since the Civil War and emancipation, race has continued to be largely defined in relation to African Americans—segregation, civil rights, the underclass, and affirmative action. Constituting the largest minority group in our society, they have been at the cutting edge of the Civil Rights Movement. Indeed, their struggle has been a constant reminder of America's moral vision as a country committed to the principle of liberty. Martin Luther King clearly understood this truth when he wrote from a jail cell: "We will reach the goal of freedom in Birmingham and all over the nation, because the goal of America is freedom. Abused and scorned though we may be, our destiny is tied up with America's destiny." [7]

Asian Americans have been here for over one hundred and fifty years, before many European immigrant groups. But as "strangers" coming from a "different shore," they have been stereotyped as "heathen," exotic, and unassimilable. Seeking "Gold Mountain," the Chinese arrived first, and what happened to them influenced the reception of the Japanese, Koreans, Filipinos, and Asian Indians as well as the Southeast Asian refugees like the Vietnamese and the Hmong. The 1882 Chinese Exclusion Act was the first law that prohibited the entry of immigrants on the basis of nationality. The Chinese condemned this restriction as racist and tyrannical. "They call us 'Chink,'" complained a Chinese immigrant, cursing the "white demons." "They think we no good! America cuts us off. No more come now, too bad!" This precedent later provided a basis for the restriction of European immigrant groups such as Italians, Russians, Poles, and Greeks. The Japanese painfully discovered that their accomplishments in America did not lead to acceptance, for during World War II, unlike Italian Americans and German Americans, they were placed in internment camps. Two-thirds of them were citizens by birth. "How could I as a 6-month-old child born in this country," asked Congressman Robert Matsui years later, "be declared by my own Government to be an enemy alien?" Today, Asian Americans represent the fastest-growing ethnic group. They have also become the focus of much mass media attention as "the Model Minority" not only for blacks and Chicanos, but also for whites on welfare and even middle-class whites experiencing economic difficulties. [8]

Chicanos represent the largest group among the Hispanic population, which is projected to outnumber African Americans. They have been in the United States for a long time, initially incorporated by the war against Mexico. The treaty had moved the border between the two countries, and the people of "occupied" Mexico suddenly found themselves "foreigners" in their "native land." As historian Albert Camarillo pointed out, the Chicano past is an integral part of America's westward expansion, also known as "manifest destiny." But while the early Chicanos were a colonized people, most of them today have immigrant roots. Many began the trek to El Norte in the early twentieth century. "As I had heard a lot about the United States," Jesus Garza recalled, "it was my dream to come here." "We came to know families from Chihuahua, Sonora, Jalisco, and Durango," stated Ernesto Galarza. "Like ourselves, our Mexican neighbors had come this far moving step by step, working and waiting, as if they were feeling their way up a ladder." Nevertheless, the Chicano experience has been unique, for most of them have lived close to their homeland—a proximity that has helped reinforce their language, identity, and culture. This migration to El Norte has continued to the present. Los Angeles has more people of Mexican origin than any other city in the world, except Mexico City. A mostly mestizo people of Indian as well as African and Spanish ancestries, Chicanos currently represent the largest minority group in the Southwest, where they have been visibly transforming culture and society.[9]

The Irish came here in greater numbers than most immigrant groups. Their history has been tied to America's past from the very beginning. Ireland represented the earliest English frontier: the conquest of Ireland occurred before the colonization of America, and the Irish were the first group that the English called "savages." In this context, the Irish past foreshadowed the Indian future. During the nineteenth century, the Irish, like the Chinese, were victims of British colonialism. While the Chinese fled from the ravages of the Opium Wars, the Irish were pushed from their homeland by "English tyranny." Here they became construction workers and factory operatives as well as the "maids" of America. Representing a Catholic group seeking to settle in a fiercely Protestant society, the Irish immigrants were targets of American nativist hostility. They were also what historian Lawrence J. McCaffrey called "the pioneers of the American urban ghetto," "previewing" experiences that would later be shared by the Italians, Poles, and other groups from southern and eastern Europe. Furthermore, they offer contrast to the immigrants from Asia. The Irish came about the same time as the Chinese, but they had a distinct advantage: the Naturalization Law of 1790 had reserved citizenship for "whites" only. Their compatible complexion allowed them to assimilate by blending into American society. In making their journey successfully into the mainstream, however, these immigrants from Erin pursued an Irish "ethnic" strategy: they promoted "Irish" solidarity in order to gain political power and also to dominate the skilled blue-collar occupations, often at the expense of the Chinese and blacks.[10]

Fleeing pogroms and religious persecution in Russia, the Jews were driven from what John Cuddihy described as the "Middle Ages into the Anglo-American world of the *goyim* 'beyond the pale.'" To them, America represented the Promised Land. This vision led Jews to struggle not only for themselves but also for other oppressed groups, especially blacks. After the 1917 East St. Louis race riot, the Yiddish *Forward* of New York compared this anti-black violence to a 1903 pogrom in Russia: "Kishinev and St. Louis—the same soil, the same people." Jews cheered when Jackie Robinson broke into the Brooklyn Dodgers in 1947. "He was adopted as the surrogate hero by many of us growing up at the time," recalled Jack Greenberg of the NAACP Legal Defense Fund. "He was the way we saw ourselves triumphing against the forces of bigotry and ignorance." Jews stood shoulder to shoulder with blacks in the Civil Rights Movement: two-thirds of the white volunteers who went south during the 1964 Freedom Summer were Jewish. Today Jews are considered a highly successful "ethnic" group. How did they make such great socioeconomic strides? This question is often reframed by neoconservative intellectuals like Irving Kristol and Nathan Glazer to read: if Jewish immigrants were able to lift themselves from poverty into the mainstream through self-help and education without welfare and affirmative action, why can't blacks? But what this thinking overlooks is the unique history of Jewish immigrants, especially the initial advantages of many of them as literate and skilled. Moreover, it minimizes the virulence of racial prejudice rooted in American slavery.[11]

Indians represent a critical contrast, for theirs was not an immigrant experience. The Wampanoags were on the shore as the first English strangers arrived in what would be called "New England." The encounters between Indians and whites not only shaped the course of race relations, but also influenced the very culture and identity of the general society. The architect of Indian removal, President Andrew Jackson told Congress: "Our conduct toward these people is deeply interesting to the national character." Frederick Jackson Turner understood the meaning of this observation when he identified the frontier as our transforming crucible. At first, the European newcomers had to wear Indian moccasins and shout the war cry. "Little by little," as they subdued the wilderness, the pioneers became "a new product" that was "American." But Indians have had a different view of this entire process. "The white man," Luther Standing Bear of the Sioux explained, "does not understand the Indian for the reason that he does not understand America."

Continuing to be "troubled with primitive fears," he has "in his consciousness the perils of this frontier continent. . . . The man from Europe is still a foreigner and an alien. And he still hates the man who questioned his path across the continent." Indians questioned what Jackson and Turner trumpeted as "progress." For them, the frontier had a different "significance": their history was how the West was lost. But their story has also been one of resistance. As Vine Deloria declared, "Custer died for your sins." [12]

By looking at these groups from a multicultural perspective, we can comparatively analyze their experiences in order to develop an understanding of their differences and similarities. Race, we will see, has been a social construction that has historically set apart racial minorities from European immigrant groups. Contrary to the notions of scholars like Nathan Glazer and Thomas Sowell, race in America has not been the same as ethnicity. A broad comparative focus also allows us to see how the varied experiences of different racial and ethnic groups occurred within shared contexts.

During the nineteenth century, for example, the Market Revolution employed Irish immigrant laborers in New England factories as it expanded cotton fields worked by enslaved blacks across Indian lands toward Mexico. Like blacks, the Irish newcomers were stereotyped as "savages," ruled by passions rather than "civilized" virtues such as self-control and hard work. The Irish saw themselves as the "slaves" of British oppressors, and during a visit to Ireland in the 1840s, Frederick Douglass found that the "wailing notes" of the Irish ballads reminded him of the "wild notes" of slave songs. The United States annexation of California, while incorporating Mexicans, led to trade with Asia and the migration of "strangers" from Pacific shores. In 1870, Chinese immigrant laborers were transported to Massachusetts as scabs to break an Irish immigrant strike; in response, the Irish recognized the need for interethnic working-class solidarity and tried to organize a Chinese lodge of the Knights of St. Crispin. After the Civil War, Mississippi planters recruited Chinese immigrants to discipline the newly freed blacks. During the debate over an immigration exclusion bill in 1882, a senator asked: If Indians could be located on reservations, why not the Chinese? [13]

Other instances of our connectedness abound. In 1903, Mexican and Japanese farm laborers went on strike together in California: their union officers had names like Yamaguchi and Lizarras, and strike meetings were conducted in Japanese and Spanish. The Mexican strikers declared that they were standing in solidarity with their "Japanese brothers" because the two groups had toiled together in the fields and were now fighting together for a fair wage. Speaking in impassioned Yiddish during the 1909 "uprising of twenty thousand" strikers in New York, the charismatic Clara Lemlich compared the abuse of Jewish female garment workers to the experience of blacks: "[The bosses] yell at the girls and 'call them down' even worse than I imagine the Negro slaves were in the South." During the 1920s, elite universities like Harvard worried about the increasing numbers of Jewish students, and new admissions criteria were instituted to curb their enrollment. Jewish students were scorned for their studiousness and criticized for their "clannishness." Recently, Asian-American students have been the targets of similar complaints: they have been called "nerds" and told there are "too many" of them on campus.[14]

Indians were already here, while blacks were forcibly transported to America, and Mexicans were initially enclosed by America's expanding border. The other groups came here as immigrants: for them, America represented liminality—a new world where they could pursue extravagant urges and do things they had thought beyond their capabilities. Like the land itself, they found themselves "betwixt and between all fixed points of classification." No longer fastened as fiercely to their old countries, they felt a stirring to become new people in a society still being defined and formed.[15]

These immigrants made bold and dangerous crossings, pushed by political events and economic hardships in their homelands and pulled by America's demand for labor as well as by their own dreams for a better life. "By all means let me go to America," a young man in Japan begged his parents. He had calculated that in one year as a laborer here he could save almost a thousand yen—an amount equal to the income of a governor in Japan. "My dear Father," wrote an immigrant Irish girl living in New York, "Any man or woman without a family are fools that would not venture and come to this plentyful Country where no man or woman ever hungered." In the shtetls of Russia, the cry "To America!" roared like "wildfire." "America was in everybody's mouth," a Jewish immigrant recalled. "Businessmen talked [about] it over their accounts; the market women made up their quarrels that they might discuss it from stall to stall; people who had relatives in the famous land went around reading their letters." Similarly, for Mexican immigrants crossing the border in the early twentieth century, El Norte became the stuff of overblown hopes. "If only you could see how nice the United States is," they said, "that is why the Mexicans are crazy about it."[16]

The signs of America's ethnic diversity can be discerned across the continent—Ellis Island, Angel Island, Chinatown, Harlem, South Boston, the Lower East Side, places with Spanish names like Los Angeles and San Antonio or Indian names like Massachusetts and Iowa. Much of what is familiar in America's cultural landscape actually has ethnic origins. The Bing cherry was developed by an early Chinese immigrant named Ah Bing. American Indians were cultivating corn, tomatoes, and tobacco long before the arrival of Columbus. The term *okay* was derived from the Choctaw word *oke*, meaning "it is so." There is evidence indicating that the name *Yankee* came from Indian terms for the English—from *eankke* in Cherokee and *Yankwis* in Delaware. Jazz and blues as well as rock and roll have African-American origins. The "Forty-Niners" of the Gold Rush learned mining techniques from the Mexicans; American cowboys acquired herding skills from Mexican *vaqueros* and adopted their range terms—such as *lariat* from *la reata*, *lasso* from *lazo*, and *stampede* from *estampida*. Songs like "God Bless America," "Easter Parade," and "White Christmas" were written by a Russian-Jewish immigrant named Israel Baline, more popularly known as Irving Berlin.[17]

Furthermore, many diverse ethnic groups have contributed to the building of the American economy, forming what Walt Whitman saluted as "a vast, surging, hopeful army of workers." They worked in the South's cotton fields, New England's textile mills, Hawaii's canefields, New York's garment factories, California's orchards, Washington's salmon canneries, and Arizona's copper mines. They built the railroad, the great symbol of America's industrial triumph. . . .

Moreover, our diversity was tied to America's most serious crisis: the Civil War was fought over a racial issue—slavery. . . .

. . . The people in our study have been actors in history, not merely victims of discrimination and exploitation. They are entitled to be viewed as subjects—as men and women with minds, wills, and voices.

> *In the telling and retelling*
> *of their stories,*
> *They create communities*
> *of memory.*

They also re-vision history. "It is very natural that the history written by the victim," said a Mexican in 1874, "does not altogether chime with the story of the victor." Sometimes they are hesitant to speak, thinking they are only "little people." "I don't know why anybody wants to hear my history," an Irish maid said apologetically in 1900. "Nothing ever happened to me worth the tellin'."[18]

But their stories are worthy. Through their stories, the people who have lived America's history can help all of us, including my taxi driver, understand that Americans originated from many shores, and that all of us are entitled to dignity. "I hope this survey do a lot of good for Chinese people," an immigrant told an interviewer from Stanford University in the 1920s. "Make American people realize that Chinese people are humans. I think very few American people really know anything about Chinese." But the remembering is also for the sake of the children. "This story is dedicated to the descendants of Lazar and Goldie Glauberman," Jewish immigrant Minnie Miller wrote in her autobiography. "My history is bound up in their history and the generations that follow should know where they came from to know better who they are." Similarly, Tomo Shoji, an elderly Nisei woman, urged Asian Americans to learn more about their roots: "We got such good, fantastic stories to tell. All our stories are different." Seeking to know how they fit into America, many young people have become listeners; they are eager to learn about the hardships and humiliations experienced by their parents and grandparents. They want to hear their stories, unwilling to remain ignorant or ashamed of their identity and past.[19]

The telling of stories liberates. By writing about the people on Mango Street, Sandra Cisneros explained, "the ghost does not ache so much." The place no longer holds her with "both arms. She sets me free." Indeed, stories may not be as innocent or simple as they seem to be. Native-American novelist Leslie Marmon Silko cautioned:

> *I will tell you something about stories . . .*
> *They aren't just entertainment.*
> *Don't be fooled.*

Indeed, the accounts given by the people in this study vibrantly re-create moments, capturing the complexities of human emotions and thoughts. They also provide the authenticity of experience. After she escaped from slavery, Harriet Jacobs wrote in her autobiography: "[My purpose] is not to tell you what I have heard but what I have seen—and what I have suffered." In their sharing of memory, the people in this study offer us an opportunity to see ourselves reflected in a mirror called history.[20]

In his recent study of Spain and the New World, *The Buried Mirror*, Carlos Fuentes points out that mirrors have been found in the tombs of ancient Mexico, placed there to guide the dead through the underworld. He also tells us about the legend of Quetzalcoatl, the Plumed Serpent: when this god was given a mirror by the Toltec deity Tezcatlipoca, he saw a man's face in the mirror and realized his own humanity. For us, the "mirror" of history can guide the living and also help us recognize who we have been and hence are. In *A Distant Mirror*, Barbara W. Tuchman finds "phenomenal parallels" between the "calamitous fourteenth century" of European society and our own era. We can, she observes, have "greater fellow-feeling for a distraught age" as we painfully recognize the "similar disarray," "collapsing assumptions," and "unusual discomfort."[21]

But what is needed in our own perplexing times is not so much a "distant" mirror, as one that is "different." While the study of the past can provide collective self-knowledge, it often reflects the scholar's particular perspective or view of the world. What happens when historians leave out many of America's peoples? What happens, to borrow the words of Adrienne Rich, "when someone with the authority of a teacher" describes our society, and "you are not in it"? Such an experience can be disorienting—"a moment of psychic disequilibrium, as if you looked into a mirror and saw nothing."[22]

Through their narratives about their lives and circumstances, the people of America's diverse groups are able to see themselves and each other in our common past. They celebrate what Ishmael Reed has described as a society "unique" in the world because "the world is here"—a place "where the cultures of the world crisscross." Much of America's past, they point out, has been riddled with racism. At the same time, these people offer hope, affirming the struggle for equality as a central theme in our country's history. At its conception, our nation was dedicated to the proposition of equality. What has given concreteness to this powerful national principle has been our coming together in the creation of a new society. "Stuck here" together, workers of different backgrounds have attempted to get along with each other.

> *People harvesting*
> *Work together unaware*
> *Of racial problems,*

wrote a Japanese immigrant describing a lesson learned by Mexican and Asian farm laborers in California.[23]

Finally, how do we see our prospects for "working out" America's racial crisis? Do we see it as through a glass darkly? Do the televised images of racial hatred and violence that riveted us in 1992 during the days of rage in Los Angeles frame a future of divisive race relations—what Arthur Schlesinger Jr., has fearfully denounced as the "disuniting of America"? Or will Americans of diverse races and ethnicities be able to connect themselves to a larger narrative? Whatever happens, we can be certain that much of our society's future will be influenced by which "mirror" we choose to see ourselves. America does not belong to one race or one group . . . Americans have been constantly redefining their national identity from the moment of first contact on the Virginia shore. By sharing their stories, they invite us to see ourselves in a different mirror.[24]

NOTES

1. Toni Morrison, *Playing in the Dark: Whiteness in the Literary Imagination* (Cambridge, Mass., 1992), p. 47.

2. William A. Henry III, "Beyond the Melting Pot," in "America's Changing Colors," *Time*, vol. 135, no. 15 (April 9, 1990), pp. 28–31.

3. "A Conflict of the Have-Nots," *Newsweek*, December 12, 1988, pp. 28–29.

4. Rodney King's statement to the press, *New York Times*, May 2, 1992, p. 6.

5. Tim Rutten, "A New Kind of Riot," *New York Times Review of Books*, June 11, 1992, pp. 52–53; Maria Newman, "Riots Bring Attention to Growing Hispanic Presence in South-Central Area," *New York Times*, May 11, 1992, p. A10; Mike Davis, "In L.A. Burning All Illusions," *The Nation*, June 1, 1992, pp. 744–745; Jack Viets and Peter Fimrite, "S.F. Mayor Visits Riot-Torn Area to Buoy Businesses," *San Francisco Chronicle*, May 6, 1992, p. A6.

6. Rick DelVecchio, Suzanne Espinosa, and Carl Nolte, "Bradley Ready to Lift Curfew," *San Francisco Chronicle*, May 4, 1992, p. A1.

7. Abraham Lincoln, "The Gettysburg Address," in *The Annals of America*, vol. 9, *1863–1865: The Crisis of the Union* (Chicago, 1968), pp. 462–463; Martin Luther King, *Why We Can't Wait* (New York, 1964), pp. 92–93.

8. Interview with old laundryman, in "Interviews with Two Chinese," circa 1924, Box 326, folder 325, Survey of Race Relations, Stanford University, Hoover Institution Archives; Congressman Robert Matsui, speech in the House of Representatives on the 442 bill for redress and reparations, September 17, 1987, *Congressional Record* (Washington, D.C., 1987), p. 7584.

9. Camarillo, *Chicanos in a Changing Society*, p. 2; Juan Nepornuceno Seguín, in David J. Weber (ed.), *Foreigners in Their Native Land: Historical Roots of the Mexican Americans* (Albuquerque, N. M., 1973), p. vi; Jesus Garza, in Manuel Garnio, *The Mexican Immigrant: His Life Story* (Chicago, 1931), p. 15; Ernesto Galarza, *Barrio Boy: The Story of a Boy's Acculturation* (Notre Dame, Ind., 1986), p. 200.

10. Lawrence J. McCaffrey, *The Irish Diaspora in America* (Washington, D.C., 1984), pp. 6, 62.

11. John Murray Cuddihy, *The Ordeal of Civility: Freud, Marx, Levi Strauss, and the Jewish Struggle with Modernity* (Boston, 1987), p. 165; Jonathan Kaufman, *Broken Alliance: The Turbulent Times between Blacks and Jews in America* (New York, 1989), pp. 28, 82, 83–84, 91, 93, 106.

12. Andrew Jackson, First Annual Message to Congress, December 8, 1829, in James D. Richardson (ed.), *A Compilation of the Messages and Papers of the Presidents, 1789–1897* (Washington, D.C., 1897), vol. 2, p. 457; Frederick Jackson Turner, "The Significance of the Frontier in American History," in *The Early Writings of Frederick Jackson Turner* (Madison, Wis., 1938), pp. 185ff.; Luther Standing Bear, "What the Indian Means to America," in Wayne Moquin (ed.), *Great Documents in American Indian History* (New York, 1973), p. 307; Vine Deloria, Jr., *Custer Died for Your Sins: An Indian Manifesto* (New York, 1969).

13. Nathan Glazer, *Affirmative Discrimination: Ethnic Inequality and Public Policy* (New York, 1978); Thomas Sowell, *Ethnic America: A History* (New York, 1981); David R. Roediger, *The Wages of Whiteness: Race and the Making of the American Working Class* (London, 1991), pp. 134–136; Dan Caldwell, "The Negroization of the Chinese Stereotype in California," *Southern California Quarterly*, vol. 33 (June 1971), pp. 123–131.

14. Thomas Almaguer, "Racial Domination and Class Conflict in Capitalist Agriculture: The Oxnard Sugar Beet Workers' Strike of 1903," *Labor History*, vol. 25, no. 3 (summer 1984), p. 347; Howard M. Sachar, *A History of the Jews in America* (New York, 1992), p. 183.

15. For the concept of liminality, see Victor Turner, *Dramas, Fields, and Metaphors: Symbolic Action in Human Society* (Ithaca, N.Y., 1974), pp. 232, 237; and Arnold Van Gennep, *The Rites of Passage* (Chicago, 1960). What I try to do is to apply liminality to the land called America.

16. Kazuo Ito, *Issei: A History of Japanese Immigrants in North America* (Seattle, 1973), p. 33; Arnold Schrier, *Ireland and the American Emigration, 1850–1900* (New York, 1970), p. 24; Abraham Cahan, *The Rise of David Levinsky* (New York, 1960; originally published in 1917), pp. 59–61; Mary Antin, quoted in Howe, *World of Our Fathers* (New York, 1983), p. 27; Lawrence A. Cardoso, *Mexican Emigration to the United States, 1897–1931* (Tucson, Ariz., 1981), p. 80.

17. Ronald Takaki, *Strangers from a Different Shore: A History of Asian Americans* (Boston, 1989), pp. 88–89; Jack Weatherford, *Native Roots: How the Indians Enriched America* (New York, 1991), pp. 210, 212; Carey McWilliams, *North from Mexico: The Spanish-Speaking People of the United States* (New York, 1968), p. 154; Stephan Themstrom (ed.), *Harvard Encyclopedia of American Ethnic Groups* (Cambridge, Mass., 1980), p. 22; Sachar, *A History of the Jews in America*, p. 367.

18. Weber (ed.), *Foreigners in Their Native Land*, p. vi; Hamilton Holt (ed.), *The Life Stories of Undistinguished Americans as Told by Themselves* (New York, 1906), p. 143.

19. "Social Document of Pany Lowe, interviewed by C. H. Burnett, Seattle, July 5, 1924," p. 6, Survey of Race Relations, Stanford University, Hoover Institution Archives; Minnie Miller, "Autobiography," private manuscript, copy from Richard Balkin; Tomo Shoji, presentation, Obana Cultural Center, Oakland, California, March 4, 1988.

20. Sandra Cisneros, *The House on Mango Street* (New York, 1991), pp. 109–110; Leslie Marmon Silko, *Ceremony* (New York, 1978), p. 2; Harriet A. Jacobs, *Incidents in the Life of a Slave Girl, written by herself* (Cambridge, Mass., 1987; originally published in 1857), p. xiii.

21. Carlos Fuentes, *The Buried Mirror: Reflections on Spain and the New World* (Boston, 1992), pp. 10, 11, 109; Barbara W. Tuchman, *A Distant Mirror: The Calamitous 14th Century* (New York, 1978), pp. xiii, xiv.

22. Adrienne Rich, *Blood, Bread, and Poetry: Selected Prose, 1979–1985* (New York, 1986), pp. 199.

23. Ishmael Reed, "America: The Multinational Society," in Rick Simonson and Scott Walker (eds.), *Multi-cultural Literacy* (St. Paul, 1988), p. 160; Ito, *Issei*, p. 497.

24. Arthur M. Schlesinger Jr., *The Disuniting of America: Reflections on a Multicultural Society* (Knoxville, Tenn., 1991); Carlos Bulosan, *America Is in the Heart: A Personal History* (Seattle, 1981), pp. 188–189.

Conceptualizing Race, Class, and Gender

Picture a typical college basketball game. It all seems pretty familiar: the players on the court, the cheerleaders arrayed at the side, the band playing, fans cheering, boosters watching from the best seats, and—if the team is ranked— perhaps a television crew. Everybody seems to have a place in the game. Everybody seems to be following the rules. But what are the "rules" of this game? What explains the patterns that we see and don't see?

Race clearly matters. The predominance of young African American men on many college basketball teams is noticeable. Why do so many young Black men play basketball? Some people argue that African Americans are better in areas requiring physical skills such as sports and are less capable of doing intellectual work in fields such as physics, law, and medicine. Others look to Black culture for explanations, suggesting that African Americans would be perfectly happy just playing ball and partying. But these perspectives fail to take into account the continuing effects of institutional racism. Lack of access to decent jobs, adequate housing, high-quality education, and adequate health care has resulted in higher rates of poverty for African Americans. In 1995, 29 percent of African Americans and 30 percent of Hispanics were poor, as compared to 11 percent of Whites (U.S. Census Bureau 1996*a*). For young Black men growing up in communities with few opportunities, sports like basketball become attractive as a way out of poverty.

Despite the importance of institutionalized racism, does a racial analysis fully explain the "rules" of the game of college basketball? Not really, because Black men are not the only players. White men also play college basketball, also raising questions about the significance of social class in explaining a basketball game. Who benefits from college basketball? Some observers argue that, because the players get scholarships and are offered a chance to earn college degrees, the players reap the rewards; but this misses the point of who really benefits. College athletics is big business, and the players make far less from it than many people believe. As amateur athletes, they are forbidden to take any payment for their skills. They are offered the hope of an NBA contract when they turn pro, or at least a college degree if they graduate. But, as Michael Messner shows in his "Masculinities and Athletic Careers," few actually turn pro; moreover, 45 percent of male college athletes actually graduate, compared to 54 percent of college students overall. These trends are accentuated among Black basketball players, only 39 percent of whom graduate from college. Interestingly, however, graduation rates among Black college athletes are higher than among nonathletes, perhaps because of the scholarship support they receive. The same is true for women athletes, who have higher graduation rates than nonathletes and higher graduation rates than male athletes. This is true for both Black and White women student athletes (National Collegiate Athletic Association, 1997; http://www.ncaa.org/i_graduationrates.html).[1]

So who actually benefits from college basketball? The colleges that recruit the athletes certainly benefit. For the university, winning teams garner increased admissions applications, alumni giving, corporate support, and television revenues. In athletics as a business, corporate sponsors want their names and products identified with winning teams and athletes; and advertisers want their products promoted by members of winning teams. Even though the athletes themselves are forbidden to promote the products, corporations create and market products in conjunction with prevailing excitement about basketball, sustained by the players' achievements. Products such as athletic shoes, workout clothing, cars, and beer all target the consumer dollars of those who enjoy watching basketball. Also, consider how many full-time jobs are supported by the revenues generated from the enterprise of college basketball. Referees, sports reporters—both at the games and on local media outlets—athletic trainers, coaches, and health personnel all benefit. Unlike the players, these people all get paid for their contributions to college basketball.

Do race and class fully explain the "rules" of basketball? Sometimes what we *don't* see can be just as revealing as what we do see. One other feature of the game on the court may escape attention because it is so familiar and, therefore, so unquestioned. All of the players are men. All of the coaches and support personnel are men. Probably, most of the camera crew and announcers are men. Where are the women? Those closest to the action on the court are probably cheerleaders—tumbling, dancing, and being thrown into the air in support of the exploits of the athletes. Others may be in the band. Some women are in the stands, cheering on the team—many of them accompanied by their husbands, boyfriends, parents, and children. Many work in the concession stands, fulfilling women's traditional roles of serving others. Still others are even more invisible, left to clean the restrooms, locker rooms, and stands after the crowd goes home. Women remain on the sidelines in other ways as well. The treatment of women basketball players markedly differs from that of their male counterparts: women have many fewer opportunities for scholarships and professional careers in athletics. The centrality of men's activities in basketball mirrors the centrality afforded men's activities in the society as a whole; thus, women's seeming invisibility in basketball ironically highlights the salience of gender.

Men's behavior, too, can reveal a gendered dimension to basketball, as in other sports. Where else are men able to put their arms around each other, slap one another's buttocks, hug each other, or cry in public without having their "masculinity" questioned? Sportscasters, too, bring gender into the play of sports, such as when they talk about men's heroic athletic achievements, but talk about women athletes' looks or their connection to children. For that matter, look at the prominence given to men's teams in sports pages of the daily newspaper, compared to sports news about women, who are typically relegated to the back pages—if their athletic accomplishments are reported at all.

This brief discussion of college basketball demonstrates how race, class, and gender each provide an important, yet partial, perspective on the action on the court. Race, class, and gender are so inextricably intertwined that gaining a comprehensive understanding of one basketball game requires that we think inclusively. If something as familiar and widespread as college basketball is embedded in race, class, and gender relations, we might ask to what extent other social practices, institutions, relations, and social issues are similarly structured.

Although some people believe that race, gender, and class divisions are relics of the past, we see them as deeply embedded in the structure of social institutions such as work, family, education, and the state. Race, class, and gender in turn shape the social issues that emerge from within as well as the institutions themselves. As we discussed in the introduction, they are interlocking systems of experience. In this part, however, we focus on each one to provide a basic grounding in how each operates, keeping in mind that they are connected and overlapping. Before you understand the intricacy of these intersections, though, you need to have a basic understanding of each concept.

Our purpose in this section is to conceptualize how each—race, class, and gender—shapes society and people's experience within it and how they are connected. We want readers to understand how race and class structure gender relations, as well as how gender and race structure class relations and how class and race structure gender relations. We look at each distinctly only to provide a foundation for understanding their interlocking nature.

RACE

In this volume we examine race, class, and gender relations from an institutional or structural perspective. Locating racial oppression in the structure of social institutions provides a different frame of analysis from what would be obtained by analyzing individuals only. *Individual racism* is one person's belief in the superiority of one race over another. Individual racism is related to *prejudice,* a hostile attitude toward a person who is presumed to have the negative characteristics associated with a group to which he or she belongs.

Institutional racism is more systematic than this. Institutions such as work, family, education, the state, and sports are "fairly stable social arrangements and practices" that persist over time (Knowles and Prewitt 1970: 5). *Institutional racism* is a system of behavior and beliefs by which a group defined as a race is oppressed, controlled, and exploited because of presumed cultural or biological characteristics (Blauner 1972). Racism is not the same thing as prejudice. Prejudice refers to people's attitudes. Racism is a system of power and privilege; it can be manifested in people's attitudes but is rooted in society's structure and is reflected in the different advantages and disadvantages that groups experience, based on their location in this societal system. The concept of institutional racism reminds us that racism is structured into society, not just into people's minds.

In this definition of institutional racism, notice first that racism is systematic: it is part of society's structure, not just present in individual bigots. As Gloria Yamato discusses in "Something About the Subject Makes It Hard to Name," racism can be intentional or unintentional. In a racist system, well-meaning White people benefit from racism even if they have no intention of acting or thinking like a "racist." Thus, institutional racism creates a built-in system of privilege. As Yamato suggests, different groups internalize it in different forms of consciousness. Peggy McIntosh's essay, "White Privilege and Male Privilege," describes how the system of racial privilege becomes invisible to those who benefit from it, even though it structures the everyday life of both White people and people of color.

Second, institutional racism shapes everyday social relations. In other words, despite its systematic nature, institutional racism depends on the presence of individual racists acting daily in order to continue. If you are a person of color, even being middle class does not protect you from the everyday realities of racism (Feagin and Sikes 1994). According to sociologist Philomena Essed (1991), it is important to identify practices, rather than individuals, as racist. Essed identifies numerous practices engaged in by White people as examples of everyday racism. For example, White people deny racism by failing to take a stand against racism, by refusing to admit racism, by acknowledging only extreme racism, and by expressing anger at Black people who point out racism, among other strategies. Although each act of everyday racism in and of itself may be relatively benign, institutional racism operates in our everyday lives as a sustained pattern of everyday actions such as these. The experience Michael Dyson describes in "The Plight of Black Men" offers another example of the practice of everyday racism. Dyson, an African American professor and philosopher, was using his credit card to get a cash advance when a bank manager in Princeton questioned him, nearly cut his card in half, and almost called the police. Dyson's experience is a familiar one to many African American men, who, given the dynamics of racism, are often perceived by Whites to be dangerous and threatening.

Practices of everyday racism are part of the edifice of institutional racism; yet we often misread their meaning. Many people believe that being nonracist means being color-blind—that is, refusing to recognize or treat as significant a person's racial background and identity. But to ignore the significance of race in a society where racial groups have distinct historical and contemporary experiences is to deny the reality of their group experience. Being color-blind in a society structured on racial privilege means assuming that everybody is "White," which is why people of color might be offended by friends who say, for example, "But I never think of you as Black." Such practices of everyday racism are powerful because, instead of seeing them as components of patterns of institutional racism, we experience these interactions as ordinary occurrences.

Another dimension of institutional racism is that its forms change over time. Racial discrimination is no longer legal, but racism is present nonetheless and continues to structure relations between groups and to differentiate the power that different groups have. The changing character of racism is also evident in the fact that specific racial group histories differ, but different racial groups share common experiences of racial oppression. Thus, Chinese Americans were never enslaved, but they experienced forced residential segregation and economic exploitation based on their presumed racial characteristics. Mexican Americans were never placed on federal reservations as Native Americans have been, but in some regards both groups share the experience of colonization by White settlers.

In a racist society, the very meaning of *race* reflects institutionalized racist practices and beliefs. Most people assume that race is biologically fixed, an assumption that has been fueled recently by popular arguments about the presumed biological basis for different forms of inequality. But the concept of race is more social than biological: it is given meaning and significance in specific social, historical, and political contexts. The significance of race comes from this context, not from whatever physical differences may exist between groups. This is what it means to say that race is a social-historical-political concept.

To understand this, think about how racial categories are created, by whom, and for what purposes. Racial classification systems reflect prevailing views of race, thereby establishing groups that are presumed to be "natural." These constructed racial categories then serve as the basis for allocating resources; furthermore, once defined, the categories frame political issues and conflicts (Omi and Winant 1994). Omi and Winant define *racial formation* as "the sociohistorical process by which racial categories are created, inhabited, transformed, and destroyed" (1994: 55). In Nazi Germany, Jews were considered to be a race—a social construction that became the basis for the Holocaust. In this understanding of racial formation, we see that societies construct race and that racial meanings constantly change as different groups contest prevailing racial definitions that empower some groups at the expense of others.

Consider, for example, the changing definitions of race in the U.S. census. Given the large number of multiracial groups and the increasing diversity brought about by immigration, we can no longer think of race in terms of mutually exclusive categories. Persons of mixed racial heritage, a small but growing group, present a challenge to the current system. In 1860, only three "races" were presumed to exist—Whites, Blacks, and mulattoes. By 1890, however, these original three "races" had been joined by five others—quadroon, octoroon, Chinese, Japanese, and Indian. Ten short years later, this list shrank to five "races"—White, Black, Chinese, Japanese, and Indian—a situation reflecting the growth of strict segregation in the South (O'Hare 1992). In the 1990 census, the U.S. government distinguished between race and ethnicity; thus, a person could be classified in one of five "racial" groups—White; Black African American; Asian or Pacific Islander; American Indian, Eskimo, or Aleut; and other—and one of two ethnicities, Hispanic or non-Hispanic. In this scheme, Hispanics may be of any race, even though many Hispanics do not identify with any of the racial categories established by the census. In fact, in the 1990 census, 43 percent of Hispanics identified their race as "other"; most people (96 percent) in the "other" race category were Hispanic (O'Hare 1992: 6).

In the coming census (to be conducted in the year 2000), the racial categories to be included are again the subject of dispute. The growing number of multiracial people in the United States poses a significant challenge for categorizing people. As Gabrielle Sándor points out in "The Other Americans," this is not just a matter of accurate statistics; it has significant consequences in the apportionment of societal resources. Data on racial groups are used to enforce voting rights, to regulate equal employment opportunities, and to determine various governmental supports, among other things. Although it may be easy for some people to say, "Race doesn't matter; we should be a color-blind society," Cornel West argues that race clearly matters and it matters a lot.

Shifting definitions of race are grounded in shifting relations of power. Recent decades have seen additional revisions to definitions of race. In particular, the experiences of Latino groups in the United States challenge long-standing racial categories of "Black" and "White." Elizabeth Martinez (in "Seeing More Than Black & White") notes that White–Black relations have defined racism in the United States for centuries, but that a rapidly changing population that includes diverse Latino groups is forcing Americans to reconsider the nature of racism. Color, she argues, has been the marker of race, but she challenges the dualistic thinking that has promoted this racist thinking.

The overarching structure of racial power relations means that placement in this structure leads to differences in outlook regarding the very presence of racism and what can be done about it. The reappearance of racial hostilities on college campuses is certainly evidence of the continuation of racist practices and beliefs; yet despite this and other evidence, Whites continue to be optimistic in their assessment of racial progress. They say that they are tired of hearing about racism and that they have done all they can to eliminate racial discrimination. Blacks are less sanguine about racial progress and are more aware of the nuances of racism. Although overt discrimination has been lessened, marked differences by race are still evident in employment, political representation, schooling, and other basic measures of group well-being.

Racism does not exist in a vacuum. As we have said, race, gender, and class are intersecting systems—experienced simultaneously, not separately. It is a mistake to think of any one category in the absence of the others. People's experiences with race and racism are framed by their location in this overarching system of race, class, and gender privileges and penalties. Race possesses not only objective dimensions that result from institutional racism; it also has subjective dimensions that relate to how people experience it. For example, some people of color grow up with class privilege; yet this does not eliminate racism, as Dyson's experience in the bank shows us. Although class differentiation has increased within racial groups and such class difference is significant, this does not mean that some people are immune from the effects of racism. Class differences within racial groups show how race and class together configure group experiences differently. As a consequence, the experiences of poor Black men in Michael Dyson's discussion differ from those of the Black women in Elizabeth Higginbotham's and Lynn Weber's analysis of differing mobility experiences ("Moving Up with Kin and Community"). All people of color encounter institutional racism, but their actual experiences with racism vary, depending on social class, gender, age, sexuality, and other markers of social position.

CLASS

Like institutional racism, the social class system is grounded in social institutions and practices. Rather than thinking of social class as an identity owned by an individual, think of social class as a series of relations that pervade the entire society and shape our social institutions and relationships with each other. In the United States, the social class system differentially structures group access to material resources, including economic, political, and social resources. This system results from patterns of capitalist development, especially as those patterns intersect with institutional racism and gender oppression.

In the United States, the social class system is marked by striking differences in wealth and income. *Wealth*, or net worth, is determined by adding all of one's financial assets and subtracting all debt. In 1993, White households had an average net worth of $45,740, as compared to $4,418 for Black households (U.S. Census Bureau 1997a). Household income data show similar patterns. In 1995, median income was $37,178 for White non-Hispanic households; $22,393 for Blacks; $22,860 for Hispanics; and $40,614 for Asian Americans/Pacific Islanders (U.S. Census Bureau 1996c: 1; see Figure 1).

Wealth is especially significant because it provides a *cumulative* advantage to those who have it. Wealth helps pay for college costs for children and down payments on houses; it can cushion the impact of emergencies, such as unexpected unemployment or sudden health problems. The median net worth of White households is more than ten times that of African American and Latino households. Buying a home, investing, being free of debt, sending one's children to college, and transferring economic assets to the next generation are all instances of class advantage that add up over time and produce advantage even beyond one's current income level. Sociologists Melvin Oliver

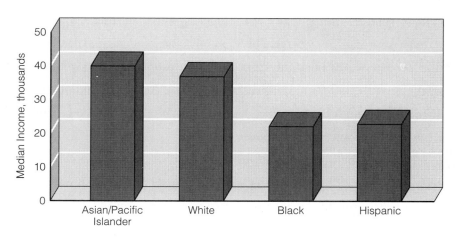

Source: U.S. Bureau of the Census, 1996. *Historical Income Tables—Families.*
World Wide Web site: http://www.census.gov:80/hhes/income/income95/in95sum

FIGURE 1 Median Household Income, 1995.

and Thomas Shapiro (1995) have found, for example, that Black and White Americans with the same educational and occupational assets still have a substantial difference in their financial assets—an average difference of $43,143 per year! This means that, even when earning the same income, the two groups are in quite different class situations, as reflected in tax breaks, the possibility of home ownership, the benefits of inheritance, and the accumulation of investments. Given these facts, Oliver and Shapiro conclude that "the long-term life prospects of black households are substantially poorer than those of whites in similar income brackets" (1995: 101). Furthermore, wealth produces more wealth, as perhaps best reflected in the ability—through inheritance—to transmit economic status from one generation to the next.

Aggregate data showing wide differences in economic status between Whites and people of color are revealing, but they can also obscure the workings of the social class system. We must be careful not to generalize from such data and see, for example, all Whites and Asians as wealthy and all African Americans and Latinos as poor. Consider the range of social class experiences just among Whites. Although on the average White households possess higher accumulated wealth and have higher incomes than Black and Hispanic households, large numbers of White households do not. Even though 8.5 percent of Whites were poor in 1995 (a figure lower than the 29.3 percent of African Americans, 14.6 percent of Asian Americans and Pacific Islanders, and 30.3 percent of Hispanics who are poor), White people account for 45 percent of the nation's poor (U.S. Census Bureau 1996a). Imagine what the range in White wealth must be to produce such a high White median income while so many White people remain in poverty.

Likewise, the aggregate data on Asian Americans show them as a group to be relatively well off. Stereotypes of the "model minority," however, obscure serious problems of poverty among many Asian Americans: 15 percent of Asian American/Pacific Islanders were poor in 1995, almost twice the percentage of White non-Hispanics (U.S. Census Bureau 1996a). The proportion of Asian Americans living in poverty has also increased substantially since the 1980s, particularly among the most recent immigrant groups, including Laotians, Cambodians, Vietnamese, Chinese, and Korean immigrants. Filipino and Asian Indian families had lower rates of poverty (Lee 1995).

In the U.S. class system, poverty is a problem for many, as the rates of poverty shown in Figure 2 show. Women and their children are especially hard hit by poverty; although they are not the only people who are poor, poverty in this group has reached alarming rates. Poverty rates among children are especially disturbing: 42 percent of African American children, 40 percent of Hispanic children, 20 percent of Asian/Pacific Islander children, and 16 percent of White children (those under 18 years of age) are poor— astonishing figures for one of the most affluent nations in the world (U.S. Census Bureau 1997b). Simplistic solutions suggested by current debates over welfare reform imply that women would not be poor if they would just get married or get a job. But as Lisa Catanzarite and Vilma Ortiz show in "Family Matters, Work Matters" the underlying causes of poverty for women are more

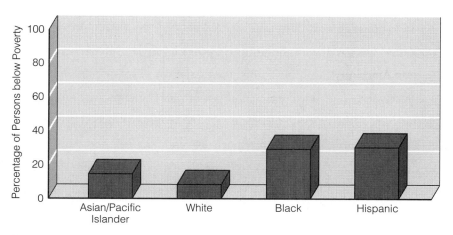

Source: U.S. Bureau of the Census, 1996. *Poverty 1995.*
World Wide Web site: http://www.census.gov:80/hhes/poverty/pov95/povest1.html

FIGURE 2 Living in Poverty, 1995.

complex than this and they are not the same for White women as for women of color.

In the United States, the social class system is also marked by differences in *power*. Social class is not just a matter of material difference; it is a pattern of domination in which some groups have more power than others. Power is the ability to influence and in some cases dominate others. This means not just interpersonal power but also the structural power that some groups hold because of their position of domination over others. Power is, of course, influenced by material well-being. Groups with vast amounts of wealth, for example, have the ability to influence systems like the media and the political process in ways that less powerful groups do not have. The power held by different groups varies enormously, depending on each group's position in the class system. Privilege in social class thus encompasses both a position of material advantage and the ability to control and influence others.

Like the system of institutional racism, the system of social class undergoes change. Changes in class are intimately linked to patterns of economic transformation in the U.S. political economy. Increasingly, these changes occur in a global context. Jobs are exported overseas as vast multinational corporations seek to enhance their profits by promoting new markets and cutting the cost of labor. Within the United States, specific economic changes include a shift from a manufacturing economy to a service economy, with corresponding changes in the types of jobs available and the wages attached to these jobs. As a result of mechanization, job export, deskilling of work, and other changes in the workplace, fewer skilled, high-paying manufacturing jobs exist today than in the past. This situation in turn has benefited some but left many people in jobs that do not pay enough, in part-time work, or without any work at all. One result is that a substantial number of people who are in poverty are employed. Close to half of the adult poor are employed (48 percent) and half of these are employed full-time (O'Hare 1996); moreover, the number of *working poor*—people whose wages for full-time work are below the federally defined poverty line—is increasing. The working poor now comprise 25 percent of all workers, with young workers especially likely to be among them. Nearly half of all women and over one-third of men aged 18 to 24 are in this group. One-quarter of Black workers and 31 percent of Hispanic workers (including all ages) earn poverty-level incomes, even though they are working full-time. This is in sharp contrast to popular stereotypes of the poor as lazy welfare queens who do not want to work.

As a result of these changes affecting the economic system, social class divisions in the United States are becoming more marked. In the nation's cities and towns, homelessness has become increasingly apparent even to casual observers. Middle-class people feel that their way of life is slipping away—a feeling supported by data showing increasing economic polarization. Yet as Barbara Ehrenreich points out in "The Silenced Majority," the remaining professional middle class is insulated from the realities of working-class experience. Dominant ideology renders the working class invisible, even though this group is the largest segment of the population.

If social class is so important in shaping life chances, why don't more people realize its significance? The answer lies in how dominant groups use ideology to explain the social class system and other systems of inequality. *Ideology* created by dominant groups refers to a system of beliefs that simultaneously distort reality and justify the status quo. As we learn in "Tired of Playing Monopoly?" by Donna Langston, the class system in the United States has been supported through the myth that we live in a classless society. This myth serves the dominant class, making class privilege seem like something that one earns, not something that is deeply embedded in the institutions of society. Langston also suggests that the system of privilege and inequality (by race, class, and gender) is least visible to those who are most privileged and who, in turn, control the resources to define the dominant cultural belief systems. Perhaps this is why the privileged, not the poor, are more likely to believe that one gets ahead through hard work. It may also explain why men more than women deny that patriarchy exists, and why Whites more than Blacks believe racism is disappearing.

In addition to giving some groups more privileges than others, class shapes social relationships. As Elizabeth Higginbotham and Lynn Weber point out in "Moving Up with Kin and Community," one of the best ways of seeing how social class shapes relationships among members of the same class and among people of different social classes is to examine the experiences of people who have undergone social mobility. As they also observe, mobility experiences significantly differ for men and for women—and for people of color and for Whites.

The social class system intersects with the system of institutional racism in shaping the experiences of men and women. The social class system also intersects with the system of gender oppression by placing men and women differently within this same overarching structure. For example, in 1995, White men's median income of $33,515 outstripped the income of all other groups of year-round, full-time workers (see Figure 3). Arrayed below are Asian American men, who earned $32,046; Asian American women, $25,505;

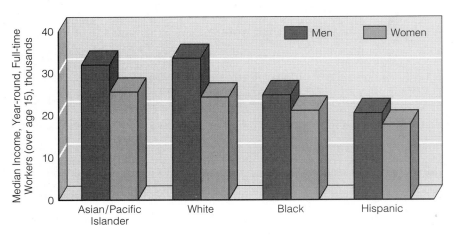

Source: U.S. Bureau of the Census, 1997. *Historical Income Tables—Persons.*
World Wide Web site: http://www.census.gov:80/hhes/income/histinc/p12.html

FIGURE 3 The Income Gap.

Black men, $24,798; White women, $24,264; Latino men, $20,553; Black women, $21,079; and Latinas, $17,855—most groups' earnings falling substantially below White men's earnings (U.S. Census Bureau 1997c). Overall, the effects of race, class, and gender manifest themselves in variable patterns of advantage or disadvantage. For example, in some ways—such as in rates of poverty and the racial gap in earnings in the labor market—women of color share a class position with men of color. In other ways, women of different races share a common class position: they all make less on average than men of their racial group. Together, these trends point to the need for inclusive thinking in developing a social class analysis.

GENDER

Gender, like race, is a socially constructed experience, not a biological imperative. Sociologists distinguish between the terms *sex* and *gender* to emphasize this point. *Sex* refers to one's biological identity as male or female; *gender* refers to the social identities attributed to women and men. Gender is rooted in social institutions and results in patterns within society that structure the relationships between women and men and that give them differing positions of advantage and disadvantage within institutions. As an identity, gender is learned; that is, through gender socialization, people construct definitions of themselves and others that are marked by gender. Like race, however, gender cannot be understood at the individual level alone. Gender is structured in social institutions, including work, families, mass media, and education. Changing gender relations is not just a matter of changing individual attitudes. As with race relations, change requires transformation of institutional structures.

The ideology of sexism supports the system of gender relations, just as the ideology of racism supports the system of race relations. Like racism, *sexism* is a system of beliefs and behaviors by which a group is oppressed, controlled, and exploited because of presumed gender differences. Sexism is manifested in individually held beliefs and in social behaviors, and it is embedded in cultural symbols. It supports the gender inequalities that are structured into social institutions.

Gender, however, is not a monolithic category. Audre Lorde points out in "Age, Race, Class, and Sex: Women Redefining Difference" that differences among women stem from racial group membership, social class location, and sexual orientation. Each of these produces lived realities for women: despite their similar status as women, women are not oppressed in the same ways by gender. As Johnnetta Cole notes in "Commonalities and Differences," women share many features of gender oppression, but the specific form of their experience depends on race and class as well. Tracy Lai in "Asian American Women" writes on the specific forms of gender oppression that Asian American women experience. Peggy McIntosh in "White Privilege and Male Privilege" reminds us that, although White women are oppressed by gender, they enjoy privileges of race and, in some cases, class and heterosexual identification.

Ideologies claiming that women are now equal to men have led many people to believe that sexism is disappearing. Seeing more women in the labor force, especially some in jobs traditionally reserved for men, can give the impression that women are doing well. But how real have these gains been? The gap between women's and men's income has closed slightly; however, most analysts agree that the narrowing of the gap reflects a drop in men's more than an increase in women's wages. The most dramatic gains in economic status have been made by younger women in professional jobs—those who become defined as the stereotypical "working woman." Despite this new image, though, most women remain concentrated in gender-segregated occupations with low wages, little opportunity for mobility, and stressful conditions. This is particularly true for women of color, who are more likely to be in occupations that are race and gender segregated. For women heading their

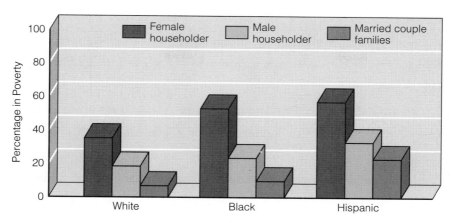

Source: U.S. Bureau of the Census, 1996. *Historical Poverty Tables—Persons.*
World Wide Web site: http://www.census.gov:80/hhes/poverty/histpov/histpov4.html

FIGURE 4 Families in Poverty with Children under 18, 1995.

own households, poverty persists at alarmingly high rates. Close to half of all
Black and Latino families (45.1 percent and 49.4 percent, respectively) and
26.6 percent of White families headed by women are officially categorized as
poor; that is, they earn incomes below the federal poverty line—$15,569 for
a family of four in 1995 (U.S. Census Bureau, 1996c). Female-headed households with children under 18 are even more in jeopardy, as shown in Figure 4.
Income and occupational data, however, do not tell the full story for women.
High rates of violence against women—whether in the home, on campus, in
the workplace, or on the streets—indicate the continuing devaluation of and
danger for women in this society.

Gender oppression also works with another major system of oppression,
that of sexuality. *Homophobia*—the fear and hatred of homosexuality—is part
of the system of social control that legitimates and enforces gender oppression. It supports the institutionalized power and privilege accorded to heterosexual behavior and identification. If only heterosexual forms of gender
identity are labeled "normal," then gays, lesbians, and bisexuals become ostracized, oppressed, and defined as "socially deviant." Homophobia affects heterosexuals as well because it is part of gender ideology used to define "normal"

men and women from those deemed deviant. Thus, oppression of lesbians and gay men is linked to the structure of gender in everyone's lives, even though the specific forms of oppression that stem from heterosexism are also shaped by race and class.

Institutionalized gender relations shape men's, as well as women's, experiences. As Peter Blood, Alan Tuttle, and George Lakey show in "Understanding and Fighting Sexism," not all men benefit equally from patriarchy. Depending on their race and class, men experience gender differently. Michael Messner in "Masculinities and Athletic Careers" shows how men's experiences and choices in sports are conditioned by race, class and gender.

Throughout Part II, you should keep the concept of *social structure* in mind. Remembering Marilyn Frye's analogy of the birdcage, be aware that race, class, and gender form a structure of societal relations. Although race, class, and gender are often discussed in terms of cultural difference, they are part of the institutional framework of society. A structural analysis studies the intersections of race, class, and gender within institutions and within individual's experiences in those institutions. Although we have divided the articles into sections on race, class, and gender, many theoretically fit in multiple categories. Thinking inclusively means recognizing that practices rarely belong to only one category. A more accurate way of viewing the world and the articles in Part II is to see them as interconnected.

NOTE

1. Data are not reported for other groups.

References

Blauner, Robert. 1972. *Racial Oppression in America*. New York: Harper & Row.

Essed, Philomena. 1991. *Understanding Everyday Racism: An Interdisciplinary Theory*. Newbury Park, CA: Sage.

Feagin, Joe R. and Melvin P. Sikes. 1994. *Living with Racism: The Black Middle-Class Experience*. Boston: Beacon Press.

Knowles, Louis L., and Kenneth Prewitt. 1970. *Institutional Racism in America*. Englewood Cliffs, NJ: Prentice-Hall.

Lee, Sharon M. 1995. "Poverty and the U.S. Asian Population." *Social Science Quarterly* 75 (September): 541–59.

O'Hare, William P. 1996. *A New Look at Poverty in America*. Washington, D.C.: Population Reference Bureau.

Oliver, Melvin L., and Thomas M. Shapiro. 1995. *Black Wealth/White Wealth: A New Perspective on Racial Inequality*. New York: Routledge.

Omi, Michael, and Howard Winant. 1994. *Racial Formation in the United States: From the 1960s to the 1980s*, 2d ed. New York: Routledge.

U.S. Census Bureau. 1996*a*. *Poverty 1995. Table A. Persons and Families in Poverty by Selected Characteristics: 1994 and 1995*. World Wide Web site: http://www.census.gov: 80/hhes/poverty/pov95/povest.html

U.S. Census Bureau. 1996*c*. *Historical Poverty Tables—Persons. Table 4. Poverty Status of Families, by Type of Family, Presence of Related Children, Race, and Hispanic Origin: 1959 to 1995*. World Wide Web site: http://www.census.gov:80/hhes/poverty/histpov/ hstpov.html

U.S. Census Bureau. 1997*a*. *Table F. Asset Ownership of Households: 1993*. World Wide Web site: http://www.census.gov:80/hhes/www/wealth/wlth93f.html

U.S. Census Bureau. 1997*b*. *Historical Income Tables—Persons. Table 3. Poverty Status of Persons, by Age, Race, and Hispanic Origin: 1959 to 1995*. World Wide Web site: http:// www.census.gov:80/hhes/poverty/histpov3f.html

U.S. Census Bureau. 1997*c*. *Historical Income Tables—Persons. Table P12. Year-Round, Full-Time Workers—Persons 15 Years Old and Over, by Median Income, Race, Hispanic Origin, and Sex: 1970 to 1995*.World Wide Web site: http://www.census.gov:80/hhes/ income/histinc/p12/html

Race and Racism

SOMETHING ABOUT THE SUBJECT MAKES IT HARD TO NAME

7

Gloria Yamato

Racism—simple enough in structure, yet difficult to eliminate. Racism—pervasive in the U.S. culture to the point that it deeply affects all the local town folk and spills over, negatively influencing the fortunes of folk around the world. Racism is pervasive to the point that we take many of its manifestations for granted, believing "that's life." Many believe that racism can be dealt with effectively in one hellifying workshop, or one hour-long heated discussion. Many actually believe this monster, racism, that has had at least a few hundred years to take root, grow, invade our space and develop subtle variations . . . this mind-funk that distorts thought and action, can be merely wished away. I've run into folks who really think that we can beat this devil, kick this habit, be healed of this disease in a snap. In a sincere blink of a well-intentioned eye, presto—poof—racism disappears. "I've dealt with my racism . . . (envision a laying on of hands) . . . Hallelujah! Now I can go to the beach." Well, fine. Go to the beach. In fact, why don't we all go to the beach and continue to work on the sucker over there? Cuz you can't even shave a little piece off this thing called racism in a day, or a weekend, or a workshop.

When I speak of *oppression*, I'm talking about the systematic, institutionalized mistreatment of one group of people by another for whatever reason.

From: Jo Whitehorse Cochran, Donna Langston, and Carolyn Woodward (eds.), *Changing Our Power: An Introduction to Women's Studies* (Dubuque, Iowa: Kendall-Hunt, 1988), pp. 3–6. Reprinted by permission.

The oppressors are purported to have an innate ability to access economic resources, information, respect, etc., while the oppressed are believed to have a corresponding negative innate ability. The flip side of oppression is *internalized oppression.* Members of the target group are emotionally, physically, and spiritually battered to the point that they begin to actually believe that their oppression is deserved, is their lot in life, is natural and right, and that it doesn't even exist. The oppression begins to feel comfortable, familiar enough that when mean ol' Massa lay down de whip, we got's to pick up and whack ourselves and each other. Like a virus, it's hard to beat racism, because by the time you come up with a cure, it's mutated to a "new cure-resistant" form. One shot just won't get it. Racism must be attacked from many angles.

The forms of racism that I pick up on these days are 1) aware/blatant racism, 2) aware/covert racism, 3) unaware/unintentional racism, and 4) unaware/self-righteous racism. I can't say that I prefer any one form of racism over the others, because they all look like an itch needing a scratch. I've heard it said (and understandably so) that the aware/blatant form of racism is preferable if one must suffer it. Outright racists will, without apology or confusion, tell us that because of our color we don't appeal to them. If we so choose, we can attempt to get the hell out of their way before we get the sweat knocked out of us. Growing up, aware/covert racism is what I heard many of my elders bemoaning "up north," after having escaped the overt racism "down south." Apartments were suddenly no longer vacant or rents were outrageously high, when black, brown, red, or yellow persons went to inquire about them. Job vacancies were suddenly filled, or we were fired for very vague reasons. It still happens, though the perpetrators really take care to cover their tracks these days. They don't want to get gummed to death or slobbered on by the toothless laws that supposedly protect us from such inequities.

Unaware/unintentional racism drives usually tranquil white liberals wild when they get called on it, and confirms the suspicions of many people of color who feel that white folks are just plain crazy. It has led white people to believe that it's just fine to ask if they can touch my hair (while reaching). They then exclaim over how soft it is, how it does not scratch their hand. It has led whites to assume that bending over backwards and speaking to me in high-pitched (terrified), condescending tones would make up for all the racist wrongs that distort our lives. This type of racism has led whites right to my doorstep, talking 'bout, "We're sorry/we love you and want to make things right," which is fine, and further, "We're gonna give you the opportunity to fix it while we sleep. Just tell us what you need. 'Bye!!"—which *ain't* fine. With the best of intentions, the best of educations, and the greatest generosity of heart, whites, operating on the misinformation fed to them from day one, will behave in ways that are racist, will perpetuate racism by being "nice" the way we're taught to be nice. You can just "nice" somebody to death with naïveté and lack of awareness of privilege. Then there's guilt and the desire to end racism and how the two get all tangled up to the point that people, morbidly fascinated with their guilt, are immobilized. Rather than deal with ending racism, they sit and ponder their guilt and hope nobody notices how awful they are. Meanwhile, racism picks up momentum and keeps on keepin' on.

Now, the newest form of racism that I'm hip to is unaware/self-righteous racism. The "good white" racist attempts to shame Blacks into being blacker, scorns Japanese-Americans who don't speak Japanese, and knows more about the Chicano/a community than the folks who make up the community. They assign themselves as the "good whites," as opposed to the "bad whites," and are often so busy telling people of color what the issues in the Black, Asian, Indian, Latino/a communities should be that they don't have time to deal with their errant sisters and brothers in the white community. Which means that people of color are still left to deal with what the "good whites" don't want to . . . racism.

Internalized racism is what really gets in my way as a Black woman. It influences the way I see or don't see myself, limits what I expect of myself or others like me. It results in my acceptance of mistreatment, leads me to believe that being treated with less than absolute respect, at least this once, is to be expected because I am Black, because I am not white. "Because I am (*you fill in the color*), you think, "Life is going to be hard." The fact is life may be hard, but the color of your skin is not the cause of the hardship. The color of your skin may be used as an excuse to mistreat you, but there is no reason or logic involved in the mistreatment. If it seems that your color is the reason, if it seems that your ethnic heritage is the cause of the woe, it's because you've been deliberately beaten down by agents of a greedy system until you swallowed the garbage. That is the internalization of racism.

Racism is the systematic, institutionalized mistreatment of one group of people by another based on racial heritage. Like every other oppression, racism can be internalized. People of color come to believe misinformation about their particular ethnic group and thus believe that their mistreatment is justified. With that basic vocabulary, let's take a look at how the whole thing works together. Meet "the Ism Family," racism, classism, ageism, adultism, elitism, sexism, heterosexism, physicalism, etc. All these ism's are systematic, that is, not only are these parasites feeding off our lives, they are also dependent on one another for foundation. Racism is supported and reinforced by classism, which is given a foothold and a boost by adultism, which also feeds sexism, which is validated by heterosexism, and so it goes on. You cannot have the "ism" functioning without first effectively installing its flip-side, the internal-ized version of the ism. Like twins, as one particular form of the ism grows in potency, there is a corresponding increase in its internalized form within the population. Before oppression becomes a specific ism like racism, usually all hell breaks loose. War. People fight attempts to enslave them, or to subvert their will, or to take what they consider theirs, whether that is territory or dignity. It's true that the various elements of racism, while repugnant, would not be able to do very much damage, but for one generally overlooked key piece: power/privilege.

While in one sense we all have power we have to look at the fact that, in our society, people are stratified into various classes and some of these classes have more privilege than others. The owning class has enough power and privilege to not have to give a good whinney what the rest of the folks have on their minds. The power and privilege of the owning class provides the ability to pay off enough of the working class and offer that paid-off group, the middle class, just enough privilege to make it agreeable to do various and sun-dry oppressive things to other working-class and outright disenfranchised folk, keeping the lid on explosive inequities, at least for a minute. If you're at the bottom of this heap, and you believe the line that says you're there because that's all you're worth, it is at least some small solace to believe that there are others more worthless than you, because of their gender, race, sexual prefer-ence . . . whatever. The specific form of power that runs the show here is the power to intimidate. The power to take away the most lives the quickest, and back it up with legal and "divine" sanction, is the very bottom line. It makes the difference between who's holding the racism end of the stick and who's getting beat with it (or beating others as vulnerable as they are) on the inter-nalized racism end of the stick. What I am saying is, while people of color are welcome to tear up their own neighborhoods and each other, everybody knows that you cannot do that to white folks without hell to pay. People of color can be prejudiced against one another and whites, but do not have an ice-cube's chance in hell of passing laws that will get whites sent to relocation camps "for their own protection and the security of the nation." People who have not thought about or refuse to acknowledge this imbalance of power/privilege often want to talk about the racism of people of color. But then that is one of the ways racism is able to continue to function. You look for someone to blame and you blame the victim, who will nine times out of ten accept the blame out of habit.

So, what can we do? Acknowledge racism for a start, even though and especially when we've struggled to be kind and fair, or struggled to rise above it all. It is hard to acknowledge the fact that racism circumscribes and pervades our lives. Racism must be dealt with on two levels, personal and societal, emotional and institutional. It is possible—and most effective—to do both at the same time. We must reclaim whatever delight we have lost in our own ethnic heritage or heritages. This so-called melting pot has only succeeded in turning us into fast-food gobbling "generics" (as in generic "white folks" who were once Irish, Polish, Russian, English, etc. and "black folks," who were once Ashanti, Bambara, Baule, Yoruba, etc.). Find or create safe places to actually *feel* what we've been forced to repress each time we were a victim of, witness to or perpetrator of racism, so that we do not continue, like puppets, to act out the past in the present and future. Challenge oppression. Take a stand against it. When you are aware of something oppressive going down, stop the show. At least call it. We become so numbed to racism that we don't even think twice about it, unless it is immediately life-threatening.

Whites who want to be allies to people of color: You can educate yourselves via research and observation rather than rigidly, arrogantly relying solely on interrogating people of color. Do not expect that people of color should teach you how to behave non-oppressively. Do not give into the pull to be lazy. Think, hard. Do not blame people of color for your frustration about racism, but do appreciate the fact that people of color will often help you get in touch with that frustration. Assume that your effort to be a good friend is appreciated, but don't expect or accept gratitude from people of color. Work on racism for your sake, not "their" sake. Assume that you are needed and capable of being a good ally. Know that you'll make mistakes and commit yourself to correcting them and continuing on as an ally, no matter what. Don't give up.

People of color, working through internalized racism: Remember always that you and others like you are completely worthy of respect, completely capable of achieving whatever you take a notion to do. Remember that the term "people of color" refers to a variety of ethnic and cultural backgrounds. These various groups have been oppressed in a variety of ways. Educate yourself about the ways different peoples have been oppressed and how they've resisted that oppression. Expect and insist that whites are capable of being good allies against racism. Don't give up. Resist the pull to give out the "people of color seal of approval" to aspiring white allies. A moment of appreciation is fine, but more than that tends to be less than helpful. Celebrate yourself. Celebrate yourself. Celebrate the inevitable end of racism.

WHITE PRIVILEGE

8

AND MALE PRIVILEGE: *A Personal Account of Coming to See Correspondences Through Work in Women's Studies (1988)*

Peggy McIntosh

Through work to bring materials and perspectives from Women's Studies into the rest of the curriculum, I have often noticed men's unwillingness to grant that they are overprivileged in the curriculum, even though they may grant that women are disadvantaged. Denials that amount to taboos surround the subject of advantages that men gain from women's disadvantages. These denials protect male privilege from being fully recognized, acknowledged, lessened, or ended.

Thinking through unacknowledged male privilege as a phenomenon with a life of its own, I realized that since hierarchies in our society are interlocking, there was most likely a phenomenon of white privilege that was similarly denied and protected, but alive and real in its effects. As a white person, I realized I had been taught about racism as something that puts others at a disadvantage, but had been taught not to see one of its corollary aspects, white privilege, which puts me at an advantage.

I think whites are carefully taught not to recognize white privilege, as males are taught not to recognize male privilege. So I have begun in an untutored way to ask what it is like to have white privilege. This paper is a partial record of my personal observations and not a scholarly analysis. It is based on my daily experiences within my particular circumstances.

I have come to see white privilege as an invisible package of unearned assets that I can count on cashing in each day, but about which I was "meant"

I have appreciated commentary on this paper from the Working Papers Committee of the Wellesley College Center for Research on Women, from members of the Dodge seminar, and from many individuals, including Margaret Andersen, Sorel Berman, Joanne Braxton, Johnnella Butler, Sandra Dickerson, Marnie Evans, Beverly Guy-Sheftall, Sandra Harding, Eleanor Hinton Hoytt, Pauline Houston, Paul Lauter, Joyce Miller, Mary Norris, Gloria Oden, Beverly Smith, and John Walter.

to remain oblivious. White privilege is like an invisible weightless knapsack of special provisions, assurances, tools, maps, guides, codebooks, passports, visas, clothes, compass, emergency gear, and blank checks.

Since I have had trouble facing white privilege, and describing its results in my life, I saw parallels here with men's reluctance to acknowledge male privilege. Only rarely will a man go beyond acknowledging that women are disadvantaged to acknowledging that men have unearned advantage, or that unearned privilege has not been good for men's development as human beings, or for society's development, or that privilege systems might ever be challenged and *changed*.

I will review here several types or layers of denial that I see at work protecting, and preventing awareness about, entrenched male privilege. Then I will draw parallels, from my own experience, with the denials that veil the facts of white privilege. Finally, I will list forty-six ordinary and daily ways in which I experience having white privilege, by contrast with my African American colleagues in the same building. This list is not intended to be generalizable. Others can make their own lists from within their own life circumstances.

Writing this paper has been difficult, despite warm receptions for the talks on which it is based.[1] For describing white privilege makes one newly accountable. As we in Women's Studies work reveal male privilege and ask men to give up some of their power, so one who writes about having white privilege must ask, "Having described it, what will I do to lessen or end it?"

The denial of men's overprivileged state takes many forms in discussions of curriculum change work. Some claim that men must be central in the curriculum because they have done most of what is important or distinctive in life or in civilization. Some recognize sexism in the curriculum but deny that it makes male students seem unduly important in life. Others agree that certain *individual* thinkers are male oriented but deny that there is any *systemic* tendency in disciplinary frameworks or epistemology to overempower men as a group. Those men who do grant that male privilege takes institutionalized and embedded forms are still likely to deny that male hegemony has opened doors for them personally. Virtually all men deny that male overreward alone can explain men's centrality in all the inner sanctums of our most powerful institutions. Moreover, those few who will acknowledge that male privilege systems have overempowered them usually end up doubting that we could dismantle these privilege systems. They may say they will work to improve women's status, in the society or in the university, but they can't or won't support the idea of lessening men's. In curricular terms, this is the point at which they say that they regret they cannot use any of the interesting new scholarship on women because the syllabus is full. When the talk turns to giving men less cultural room, even the most thoughtful and fair-minded of the men I know will tend to reflect, or fall back on, conservative assumptions about the inevitability of present gender relations and distributions of power, calling on precedent or sociobiology and psychobiology to demonstrate that male domination is natural and follows inevitably from evolutionary pressures. Others resort to arguments from "experience" or religion or social responsibility or wishing and dreaming.

After I realized, through faculty development work in Women's Studies, the extent to which men work from a base of unacknowledged privilege, I understood that much of their oppressiveness was unconscious. Then I remembered the frequent charges from women of color that white women whom they encounter are oppressive. I began to understand why we are justly seen as oppressive, even when we don't see ourselves that way. At the very least, obliviousness of one's privileged state can make a person or group irritating to be with. I began to count the ways in which I enjoy unearned skin privilege and have been conditioned into oblivion about its existence, unable to see that it put me "ahead" in any way, or put my people ahead, overrewarding us and yet also paradoxically damaging us, or that it could or should be changed.

My schooling gave me no training in seeing myself as an oppressor, as an unfairly advantaged person, or as a participant in a damaged culture. I was taught to see myself as an individual whose moral state depended on her individual moral will. At school, we were not taught about slavery in any depth; we were not taught to see slaveholders as damaged people. Slaves were seen as the only group at risk of being dehumanized. My schooling followed the pattern which Elizabeth Minnich has pointed out: whites are taught to think of their lives as morally neutral, normative, and average, and also ideal, so that when we work to benefit others, this is seen as work that will allow "them" to be more like "us." I think many of us know how obnoxious this attitude can be in men.

After frustration with men who would not recognize male privilege, I decided to try to work on myself at least by identifying some of the daily effects of white privilege in my life. It is crude work, at this stage, but I will give here a list of special circumstances and conditions I experience that I did not earn but that I have been made to feel are mine by birth, by citizenship, and by virtue of being a conscientious law-abiding "normal" person of goodwill. I have chosen those conditions that I think in my case *attach somewhat more to skin-color privilege* than to class, religion, ethnic status, or geographical location, though these other privileging factors are intricately intertwined. As far as I can see, my Afro-American co-workers, friends, and acquaintances with whom I come into daily or frequent contact in this particular time, place, and line of work cannot count on most of these conditions.

1. I can, if I wish, arrange to be in the company of people of my race most of the time.
2. I can avoid spending time with people whom I was trained to mistrust and who have learned to mistrust my kind or me.
3. If I should need to move, I can be pretty sure of renting or purchasing housing in an area which I can afford and in which I would want to live.
4. I can be reasonably sure that my neighbors in such a location will be neutral or pleasant to me.
5. I can go shopping alone most of the time, fairly well assured that I will not be followed or harassed by store detectives.
6. I can turn on the television or open to the front page of the paper and see people of my race widely and positively represented.
7. When I am told about our national heritage or about "civilization," I am shown that people of my color made it what it is.
8. I can be sure that my children will be given curricular materials that testify to the existence of their race.
9. If I want to, I can be pretty sure of finding a publisher for this piece on white privilege.
10. I can be fairly sure of having my voice heard in a group in which I am the only member of my race.
11. I can be casual about whether or not to listen to another woman's voice in a group in which she is the only member of her race.
12. I can go into a book shop and count on finding the writing of my race represented, into a supermarket and find the staple foods that fit with my cultural traditions, into a hairdresser's shop and find someone who can deal with my hair.
13. Whether I use checks, credit cards, or cash, I can count on my skin color not to work against the appearance that I am financially reliable.
14. I could arrange to protect our young children most of the time from people who might not like them.
15. I did not have to educate our children to be aware of systemic racism for their own daily physical protection.
16. I can be pretty sure that my children's teachers and employers will tolerate them if they fit school and workplace norms; my chief worries about them do not concern others' attitudes toward their race.
17. I can talk with my mouth full and not have people put this down to my color.

18. I can swear, or dress in secondhand clothes, or not answer letters, without having people attribute these choices to the bad morals, the poverty, or the illiteracy of my race.
19. I can speak in public to a powerful male group without putting my race on trial.
20. I can do well in a challenging situation without being called a credit to my race.
21. I am never asked to speak for all the people of my racial group.
22. I can remain oblivious to the language and customs of persons of color who constitute the world's majority without feeling in my culture any penalty for such oblivion.
23. I can criticize our government and talk about how much I fear its policies and behavior without being seen as a cultural outsider.
24. I can be reasonably sure that if I ask to talk to "the person in charge," I will be facing a person of my race.
25. If a traffic cop pulls me over or if the IRS audits my tax return, I can be sure I haven't been singled out because of my race.
26. I can easily buy posters, postcards, picture books, greeting cards, dolls, toys, and children's magazines featuring people of my race.
27. I can go home from most meetings of organizations I belong to feeling somewhat tied in, rather than isolated, out of place, outnumbered, unheard, held at a distance, or feared.
28. I can be pretty sure that an argument with a colleague of another race is more likely to jeopardize her chances for advancement than to jeopardize mine.
29. I can be fairly sure that if I argue for the promotion of a person of another race, or a program centering on race, this is not likely to cost me heavily within my present setting, even if my colleagues disagree with me.
30. If I declare there is a racial issue at hand, or there isn't a racial issue at hand, my race will lend me more credibility for either position than a person of color will have.
31. I can choose to ignore developments in minority writing and minority activist programs, or disparage them, or learn from them, but in any case, I can find ways to be more or less protected from negative consequences of any of these choices.
32. My culture gives me little fear about ignoring the perspectives and powers of people of other races.
33. I am not made acutely aware that my shape, bearing, or body odor will be taken as a reflection on my race.
34. I can worry about racism without being seen as self-interested or self-seeking.

35. I can take a job with an affirmative action employer without having my co-workers on the job suspect that I got it because of my race.
36. If my day, week, or year is going badly, I need not ask of each negative episode or situation whether it has racial overtones.
37. I can be pretty sure of finding people who would be willing to talk with me and advise me about my next steps, professionally.
38. I can think over many options, social, political, imaginative, or professional, without asking whether a person of my race would be accepted or allowed to do what I want to do.
39. I can be late to a meeting without having the lateness reflect on my race.
40. I can choose public accommodation without fearing that people of my race cannot get in or will be mistreated in the places I have chosen.
41. I can be sure that if I need legal or medical help, my race will not work against me.
42. I can arrange my activities so that I will never have to experience feelings of rejection owing to my race.
43. If I have low credibility as a leader, I can be sure that my race is not the problem.
44. I can easily find academic courses and institutions that give attention only to people of my race.
45. I can expect figurative language and imagery in all of the arts to testify to experiences of my race.
46. I can choose blemish cover or bandages in "flesh" color and have them more or less match my skin.

I repeatedly forgot each of the realizations on this list until I wrote it down. For me, white privilege has turned out to be an elusive and fugitive subject. The pressure to avoid it is great, for in facing it I must give up the myth of meritocracy. If these things are true, this is not such a free country; one's life is not what one makes it; many doors open for certain people through no virtues of their own. These perceptions mean also that my moral condition is not what I had been led to believe. The appearance of being a good citizen rather than a troublemaker comes in large part from having all sorts of doors open automatically because of my color.

A further paralysis of nerve comes from literary silence protecting privilege. My clearest memories of finding such analysis are in Lillian Smith's unparalleled *Killers of the Dream* and Margaret Andersen's review of Karen and Mamie Fields' *Lemon Swamp*. Smith, for example, wrote about walking toward black children on the street and knowing they would step into the gutter; Andersen contrasted the pleasure that she, as a white child, took on summer driving trips to the south with Karen Fields' memories of driving in a closed car stocked with all necessities lest, in stopping, her black family should suffer "insult, or worse." Adrienne Rich also recognizes and writes about daily experiences of privilege, but in my observation, white women's writing in this area is far more often on systemic racism than on our daily lives as light-skinned women.[2]

In unpacking this invisible knapsack of white privilege, I have listed conditions of daily experience that I once took for granted, as neutral, normal, and universally available to everybody, just as I once thought of a male-focused curriculum as the neutral or accurate account that can speak for all. Nor did I think of any of these perquisites as bad for the holder. I now think that we need a more finely differentiated taxonomy of privilege, for some of these varieties are only what one would want for everyone in a just society, and others give license to be ignorant, oblivious, arrogant, and destructive. Before proposing some more finely tuned categorization, I will make some observations about the general effects of these conditions on my life and expectations.

In this potpourri of examples, some privileges make me feel at home in the world. Others allow me to escape penalties or dangers that others suffer. Through some, I escape fear, anxiety, insult, injury, or a sense of not being welcome, not being real. Some keep me from having to hide, to be in disguise, to feel sick or crazy, to negotiate each transaction from the position of being an outsider or, within my group, a person who is suspected of having too close links with a dominant culture. Most keep me from having to be angry.

I see a pattern running through the matrix of white privilege, a pattern of assumptions that were passed on to me as a white person. There was one main piece of cultural turf; it was my own turf, and I was among those who could control the turf. I could measure up to the cultural standards and take advantage of the many options I saw around me to make what the culture would call a success of my life. *My skin color was an asset for any move I was educated to want to make.* I could think of myself as "belonging" in major ways and of making social systems work for me. I could freely disparage, fear, neglect, or be oblivious to anything outside of the dominant cultural forms. Being of the main culture, I could also criticize it fairly freely. My life was reflected back to me frequently enough so that I felt, with regard to my race, if not to my sex, like one of the real people.

Whether through the curriculum or in the newspaper, the television, the economic system, or the general look of people in the streets, I received daily signals and indications that my people counted and that others *either didn't exist or must be trying, not very successfully, to be like people of my race.* I was given cultural permission not to hear voices of people of other races or a tepid cultural tolerance for hearing or acting on such voices. I was also raised not to suffer seriously from anything that darker-skinned people might say about my group, "protected," though perhaps I should more accurately say *prohibited*, through the habits of my economic class and social group, from living in racially mixed groups or being reflective about interactions between people of differing races.

In proportion as my racial group was being made confident, comfortable, and oblivious, other groups were likely being made unconfident, uncomfortable, and alienated. Whiteness protected me from many kinds of hostility, distress, and violence, which I was being subtly trained to visit in turn upon people of color.

For this reason, the word "privilege" now seems to me misleading. Its connotations are too positive to fit the conditions and behaviors which "privilege systems" produce. We usually think of privilege as being a favored state, whether earned, or conferred by birth or luck. School graduates are reminded they are privileged and urged to use their (enviable) assets well. The word "privilege" carries the connotation of being something everyone must want. Yet some of the conditions I have described here work to systemically overempower certain groups. Such privilege simply *confers dominance*, gives permission to control, because of one's race or sex. The kind of privilege that gives license to some people to be, at best, thoughtless and, at worst, murderous should not continue to be referred to as a desirable attribute. Such "privilege" may be widely desired without being in any way beneficial to the whole society.

Moreover, though "privilege" may confer power, it does not confer moral strength. Those who do not depend on conferred dominance have traits and qualities that may never develop in those who do. Just as Women's Studies courses indicate that women survive their political circumstances to lead lives that hold the human race together, so "underprivileged" people of color who are the world's majority have survived their oppression and lived survivors' lives from which the white global minority can and must learn. In some groups, those dominated have actually become strong through *not* having all of these unearned advantages, and this gives them a great deal to teach the others. Members of so-called privileged groups can seem foolish, ridiculous, infantile, or dangerous by contrast.

I want, then, to distinguish between earned strength and unearned power conferred systemically. Power from unearned privilege can look like strength when it is, in fact, permission to escape or to dominate. But not all of the privileges on my list are inevitably damaging. Some, like the expectation that neighbors will be decent to you, or that your race will not count against you in court, should be the norm in a just society and should be considered as the entitlement of everyone. Others, like the privilege not to listen to less power-ful people, distort the humanity of the holders as well as the ignored groups. Still others, like finding one's staple foods everywhere, may be a function of being a member of a numerical majority in the population. Others have to do with not having to labor under pervasive negative stereotyping and mythology.

We might at least start by distinguishing between positive advantages that we can work to spread, to the point where they are not advantages at all but simply part of the normal civic and social fabric, and negative types of advan-tage that unless rejected will always reinforce our present hierarchies. For example, the positive "privilege" of belonging, the feeling that one belongs within the human circle, as Native Americans say, fosters development and should not be seen as privilege for a few. It is, let us say, an entitlement that none of us should have to earn; ideally it is an *unearned entitlement*. At present, since only a few have it, it is an *unearned advantage* for them. The negative "privilege" that gave me cultural permission not to take darker-skinned Others seriously can be seen as arbitrarily conferred dominance and should not be desirable for anyone. This paper results from a process of coming to see that some of the power that I originally saw as attendant on being a human being in the United States consisted in *unearned advantage* and *conferred domi-nance*, as well as other kinds of special circumstance not universally taken for granted.

In writing this paper I have also realized that white identity and status (as well as class identity and status) give me considerable power to choose whether to broach this subject and its trouble. I can pretty well decide whether to dis-appear and avoid and not listen and escape the dislike I may engender in other people through this essay, or interrupt, answer, interpret, preach, correct, criticize, and control to some extent what goes on in reaction to it. Being white, I am given considerable power to escape many kinds of danger or pen-alty as well as to choose which risks I want to take.

There is an analogy here, once again, with Women's Studies. Our male colleagues do not have a great deal to lose in supporting Women's Studies, but they do not have a great deal to lose if they oppose it either. They simply have the power to decide whether to commit themselves to more equitable distri-butions of power. They will probably feel few penalties whatever choice they make; they do not seem, in any obvious short-term sense, the ones at risk, though they and we are all at risk because of the behaviors that have been rewarded in them.

Through Women's Studies work I have met very few men who are truly distressed about systemic, unearned male advantage and conferred domi-nance. And so one question for me and others like me is whether we will be like them, or whether we will get truly distressed, even outraged, about unearned race advantage and conferred dominance and if so, what we will do to lessen them. In any case, we need to do more work in identifying how they actually affect our daily lives. We need more down-to-earth writing by people about these taboo subjects. We need more understanding of the ways in which white "privilege" damages white people, for these are not the same ways in which it damages the victimized. Skewed white psyches are an inseparable part of the picture, though I do not want to confuse the kinds of damage done to the holders of special assets and to those who suffer the deficits. Many, perhaps most, of our white students in the United States think that racism doesn't affect them because they are not people of color; they do not see "whiteness" as a racial identity. Many men likewise think that Women's Studies does not bear on their own existences because they are not female; they do not see themselves as having gendered identities. Insisting on the universal "effects" of "privilege" systems, then, becomes one of our chief tasks, and being more explicit about the *particular* effects in particular contexts is another. Men need to join us in this work.

In addition, since race and sex are not the only advantaging systems at work, we need to similarly examine the daily experience of having age advantage, or ethnic advantage, or physical ability, or advantage related to nationality, religion, or sexual orientation. Professor Marnie Evans suggested to me that in many ways the list I made also applies directly to heterosexual privilege. This is a still more taboo subject than race privilege: the daily ways in which heterosexual privilege makes some persons comfortable or powerful, providing supports, assets, approvals, and rewards to those who live or expect to live in heterosexual pairs. Unpacking that content is still more difficult, owing to the deeper imbeddedness of heterosexual advantage and dominance and stricter taboos surrounding these.

But to start such an analysis I would put this observation from my own experience: the fact that I live under the same roof with a man triggers all kinds of societal assumptions about my worth, politics, life, and values and triggers a host of unearned advantages and powers. After recasting many elements from the original list I would add further observations like these:

1. My children do not have to answer questions about why I live with my partner (my husband).
2. I have no difficulty finding neighborhoods where people approve of our household.
3. Our children are given texts and classes that implicitly support our kind of family unit and do not turn them against my choice of domestic partnership.

4. I can travel alone or with my husband without expecting embarrassment or hostility in those who deal with us.
5. Most people I meet will see my marital arrangements as an asset to my life or as a favorable comment on my likability, my competence, or my mental health.
6. I can talk about the social events of a weekend without fearing most listeners' reactions.
7. I will feel welcomed and "normal" in the usual walks of public life, institutional and social.
8. In many contexts, I am seen as "all right" in daily work on women because I do not live chiefly with women.

Difficulties and dangers surrounding the task of finding parallels are many. Since racism, sexism, and heterosexism are not the same, the advantages associated with them should not be seen as the same. In addition, it is hard to isolate aspects of unearned advantage that derive chiefly from social class, economic class, race, religion, region, sex, or ethnic identity. The oppressions are both distinct and interlocking, as the Combahee River Collective statement of 1977 continues to remind us eloquently.[3]

One factor seems clear about all of the interlocking oppressions. They take both active forms that we can see and embedded forms that members of the dominant group are taught not to see. In my class and place, I did not see myself as racist because I was taught to recognize racism only in individual acts of meanness by members of my group, never in invisible systems conferring racial dominance on my group from birth. Likewise, we are taught to think that sexism or heterosexism is carried on only through intentional, individual acts of discrimination, meanness, or cruelty, rather than in invisible systems conferring unsought dominance on certain groups. Disapproving of the systems won't be enough to change them. I was taught to think that racism could end if white individuals changed their attitudes; many men think sexism can be ended by individual changes in daily behavior toward women. But a man's sex provides advantage for him whether or not he approves of the way in which dominance has been conferred on his group. A "white" skin in the United States opens many doors for whites whether or not we approve of the way dominance has been conferred on us. Individual acts can palliate, but cannot end, these problems. To redesign social systems, we need first to acknowledge their colossal unseen dimensions. The silences and denials surrounding privilege are the key political tool here. They keep the thinking about equality or equity incomplete, protecting unearned advantage and conferred dominance by making these taboo subjects. Most talk by whites about equal opportunity seems to me now to be about equal opportunity to try to get into a position of dominance while denying that *systems* of dominance exist.

Obliviousness about white advantage, like obliviousness about male advantage, is kept strongly inculturated in the United States so as to maintain the myth of meritocracy, the myth that democratic choice is equally available to all. Keeping most people unaware that freedom of confident action is there for just a small number of people props up those in power and serves to keep power in the hands of the same groups that have most of it already. Though systemic change takes many decades, there are pressing questions for me and I imagine for some others like me if we raise our daily consciousness on the perquisites of being light-skinned. What will we do with such knowledge? As we know from watching men, it is an open question whether we will choose to use unearned advantage to weaken invisible privilege systems and whether we will use any of our arbitrarily awarded power to try to reconstruct power systems on a broader base.

NOTES

1. This paper was presented at the Virginia Women's Studies Association conference in Richmond in April, 1986, and the American Educational Research Association conference in Boston in October, 1986, and discussed with two groups of participants in the Dodge seminars for Secondary School Teachers in New York and Boston in the spring of 1987.

2. Andersen, Margaret, "Race and the Social Science Curriculum: A Teaching and Learning Discussion." *Radical Teacher*, November, 1984, pp. 17–20. Smith, Lillian, *Killers of the Dream*, New York: W. W. Norton, 1949.

3. "A Black Feminist Statement," The Combahee River Collective, pp. 13–22 in G. Hull, P. Scott, B. Smith, Eds., *All the Women Are White, All the Blacks Are Men, But Some of Us Are Brave: Black Women's Studies*, Old Westbury, NY: The Feminist Press, 1982.

THE "OTHER" AMERICANS   9

Gabrielle Sándor

Ada Nurie Pagán is a blonde, green-eyed, pale-skinned "person of color." When asked by affirmative-action officers, she says that her race is Hispanic, Latina, or Puerto Rican. The Census Bureau would say that Hispanic origin is an ethnicity, not a race, so it classifies Ada as white. But Ada doesn't want to give that answer.

"I can't check a circle that labels me black, white, Asian, or American Indian, because I'm not any of those things," she says. "My sister and I have the same parents, but she's much darker than me. If it came down to choosing black or white, I'd have to choose white and she couldn't. That wouldn't make any sense."

Ada's confusion stems from the ambiguity in current definitions of race and ethnicity. The Census Bureau sees race and ethnicity as two different demographic characteristics and has separate questions on census forms and surveys. But the people who fill out the forms sometimes find the two concepts impossible to separate.

The result is much more than a problem for census-takers. Ada's misgivings are part of a significant generation gap in attitudes toward race. The rules that govern government statistics on race and ethnicity are based on a 16-year-old definition, and young people like Ada live in a world where that definition seems arbitrary or even offensive.

People in their 20s today never saw sanctioned segregation. They read about the Civil Rights movement in history books, and they were taught in public schools that discrimination is wrong. Small wonder, then, that young adults are much more likely than older adults to object to racial categories or not to fit within them. In 1990, members of the "other race" category had a median age of about 24 years, compared with a national median of 33.

Some "other race" Americans are mixed-race children. Though relatively rare, they are often born to middle-class families that vote and organize. A larger share are recent immigrants to the U.S. They are likely to be young adults, and birth rates for most minority groups are higher than they are for

From: *American Demographics* 16 (June 1994): 36–40. Reprinted by permission.

non-Hispanic whites. As a result, younger Americans are more likely to be minorities.

Mixed-race children are also increasingly common, because the penalties once attached to interracial dating and marriage are slowly fading away. In 1991, 74 percent of Americans said that interracial marriage was acceptable for themselves or others, according to the Roper Organization, up from 70 percent who found it acceptable in 1986.

Nonwhite and Hispanic Americans will claim 28 percent of the total U.S. population in 2000, according to the Census Bureau. But at the same time, the share of minorities will be 36 percent among children under age 18 and 33 percent among young adults aged 18 to 34. If current trends continue, minorities will be approaching half of the total U.S. population as early as 2050. Today's children will be the first to see that society.

FOUR CATEGORIES

The conflict over racial statistics worries researchers. "The demand [for detailed data] is going in one direction, and the quality and consistency in counting is going the other way," says Greg Robinson, chief of population and analysis for the Census Bureau. "The quality of the race and ethnic data is deteriorating."

Young adults' attitudes toward race also raise important issues for businesses, because they clash with older rules. "Tolerance and diversity is absolutely the number-one shared value" of young adults, says Gwen Lipsky, MTV's senior vice president of research and planning. "They are much more [tolerant] than any previous generation."

But while young adults embrace tolerance and racial ambiguity, a rule requires federal agencies like the Census Bureau to fit all of their racial data into separate categories. The rule, known as Statistical Directive 15, was adopted in 1978 by the Office of Management and Budget (OMB). It creates four official racial categories—American Indian or Alaskan Native, Asian or Pacific Islander, black, and white. It also defines Hispanic origin as an ethnic category separate from race. The rule helps federal and state agencies share racial data, which are vital to affirmative action and other government programs.

The 1990 census questionnaire asked respondents to put themselves into a racial category and to indicate their ethnicity and ancestry elsewhere. Respondents who felt that they didn't fit into one of the four categories checked a box marked "other race." The number of people who checked that box increased 45 percent between the 1980 and 1990 censuses, to 9.8 million. That's about 1 in 25 Americans.

Most of the people who check "other race" probably have a clear racial identity, because 98 percent claim Hispanic origin on the ethnicity question. In other words, over 40 percent of the nation's 22 million Hispanics aren't willing to identify themselves as black or white.

"Most of them are probably indicating some version of multiracial," says Carlos Fernández, president of the Association of Multi-Ethnic Americans (AMEA) in San Francisco. "Mexicans, especially, regard themselves as mestizo—part Spanish, part indigenous. The large majority of Mexico is multiracial. It's almost the official culture. Mexico hasn't asked a race question on its census since 1921. So on the U.S. census, people who put Mexican for race know full well what it means. It's mestizo, and that is a racial designation, not a nationality."

Many Latin Americans trace the greatest part of their ancestry to native Indians, followed by Europeans and perhaps Africans. But the category on the U.S. census aplies only to North American Indians, so a Hispanic person's choices usually come down to black or white. Ada Pagán, whose great-great-grandmother was black, passes as white. But she knows exactly what her dark-skinned sister is. She's "trigueña," which Ada translates as "brown."

STATISTICAL HEADACHE

In 1990, about 400,000 people who marked "other race" were not Hispanic. Though their numbers are small, they are the ones raising their voices the loudest against the census's race question. Many of the protestors may be parents in racially mixed marriages who confront an obvious problem when answering the race question for their children.

The statistical headaches over race get worse every year, as the number of interracial marriages continues to rise. Interracial marriages were illegal in some states until 1967, when the Supreme Court struck down anti-miscegenation laws in the aptly titled case *Loving vs. Virginia.* Since 1980, the number of black and white interracial married couples has increased from 651,000 to 1.2 million, according to the Census Bureau. Children born to parents of different races were more than 3 percent of births in 1990, according to the National Center for Health Statistics, up from 1 percent in 1968.

Unfortunately for survey designers, the race problem can't be solved simply by adding new categories for mestizos and multiracials. The reasons are a pair of statistical bugaboos called "primacy" and "recency" effects. Studies show that when respondents see a long list of choices on a written survey, they are likely to pick the first choice that might apply to them instead of reading through the whole list. The reverse is true of telephone surveys, when people tend to pick the last choice. The more categories on the list, the more these effects complicate the results.

"We couldn't possibly get every person's idea of their race into the question," says Robert Tortora, associate director for statistical design, methodology, and standards at the Census Bureau. "There would be all kinds of methodological effects, and the logistics of it are daunting." Tortora agrees that there are problems with the ways race and ethnicity are currently measured, and he does see changes coming. "Basically, what you're asking is a person's judgment of their race or ethnicity. And since we don't have precise definitions of every possible category, people's perceptions of themselves and their answers can change from census to census."

Race statistics can become muddled in many ways. In 1970, for example, about 1 million residents of midwestern and southern states mistakenly identified themselves as Central or South Americans because they were confused by the census form. Such goofs are relatively easy to correct. When someone writes "German" for their race, for example, they are automatically reassigned to the "white" category by a census computer.

When "other race" is what the person meant to write, the result is a quiet tug-of-war between the respondent's preferences and Statistical Directive 15. If a person won't fit themselves into one of the four racial categories designated by the OMB, a census computer does it for them. Usually, computers scan other answers on the same form and assign the person to the category that seems best. The stubborn few who check "other race," write in "biracial," and give two races are assigned the first race they report.

If someone checks "other race" and writes "Latino," the census's computers will check to see if any other household members designated a race. If no one in the household did, the bureau will check the forms of neighboring households who wrote their race as "Latino" and assign a race on that basis. This process, called "hot-decking," is considered fairly accurate on a nationwide basis. But it may be responsible for serious inaccuracies in neighborhood-level census data.

The Census Bureau says that the number of blacks of Hispanic origin in Los Angeles increased from 17,000 in 1980 to nearly 60,000 in 1990. But two local experts can find no evidence of the change, and they suspect computer error. Latinos in the city's South Central neighborhood may have ignored the race question, says James P. Allen, a geographer at Cal State Northridge. The nonrespondents were probably assigned a race based on the skin color of their neighbors, who are likely to be black, says sociologist David Heer of the University of Southern California. The overcount of blacks may have affected government funds and redistricting plans in L.A., and it doesn't stop there.

"This is a terribly important error," Allen told *Buzz* magazine. "I strongly believe it has happened in Miami, Houston, Chicago, and all cities where Latinos are migrating into black neighborhoods."

YOUNG AND PROUD

With all of the confusion and the possibility of taking offense, some observers have wondered if collecting any racial statistics is a good idea. "As long as we differentiate, we will discriminate," writes Michael E. Tomlin, a professor of education at the University of Idaho, in a letter to this magazine. Tomlin describes himself as a native-born American of Scotch-Irish descent, but he refuses to answer any federal questions concerning ethnicity or race. "For our nation to truly earn its place as the land of the free," he continues, "we must free ourselves from placing others—and ourselves—in boxes."

Tomlin's objections might seem noble. But if he got his way, the federal government would have no way to fight racial discrimination. Racial data are used to enforce the Voting Rights Act, equal employment regulations, and affirmative action plans, to name just a few. These programs depend on a steady flow of data that conform to Statistical Directive 15, so they can be easily compared with data from other sources. State or federal agencies can ask for more detailed information on race, but the data must ultimately break down into the same four categories.

The OMB and Census Bureau are now reviewing Directive 15. Seven months of hearings in 1993 revealed three truths about racial statistics, according to Congressman Tom Sawyer, Chairman of the House Subcommittee on Census. First, many people agree on the need to continue collecting ethnic and racial data. Second, the data must be uniform across the government. Third, racial categories must be relevant to the public, or the public won't cooperate.

Groups such as the Association of Multi-Ethnic Americans say that the way to restore relevance to racial data is to add a new multiracial category to Directive 15. "People ask, 'Why don't we eliminate all categories?'" says the AMEA's Carlos Fernandez. "I say, that's nice, but you can't erase differences between people by ignoring them. When we get people from different communities intermarrying at a rate we would expect if there were no differences between them, then the categories will become irrelevant. Eventually it will happen."

The new generation's challenges to old racial rules extend from federal statistical policy to one's most intimate companions. Sunny Wang, an 18-year-old Harvard freshman whose parents are from Taiwan, doesn't only date Chinese men—but she hasn't dared to tell her parents. "My parents are completely against it," she says. "They say that when you marry a Caucasian, you will end up in a divorce. Your children will have no identity. My parents have lectured me on this for years. They say that I shouldn't marry anyone who isn't Chinese or at least Asian."

Wang understands why her parents feel that way, but she isn't letting their objections keep her from dating who she wants. "I was afraid to get in this relationship, but I don't think it's wrong. If it's a good relationship, it doesn't matter what you look like."

Wang says that her parents' objections to interracial dating aren't racial, but cultural. "I have a lot of respect for [my parents]. They aren't prejudiced. They don't want me to not marry a white guy because he's white. They don't want me to marry a white guy because they don't want me to be divorced, and that's what they think would happen."

MULTICULTURAL STYLE

The paradox is that young people are increasingly likely to marry outside their race at the same time that they are increasingly likely to celebrate ethnic traditions. "Before the civil rights movement, the whole effort was to pass [as white]," says Fernández. "Now, we're seeing the opposite—people wanting to identify the other way [as minorities]. Mixed-race people who once would have let themselves be considered white are insisting that they're black."

Young adults' attitudes about race also affect their consumer choices. The book *Modern Bride Wedding Celebrations* contains a section for interracial couples that offers advice, sources, case studies, and a list of wedding traditions for different ethnic groups.

The magazine *Interrace* serves 25,000 subscribers from its offices in Atlanta. A similar magazine, *New People*, is based in Oak Park, Michigan. *Interrace* is full of advertising for interracial-themed books, toys, dating and wedding services, and therapists who specialize in interracial couples. "Products for children are particularly hot," says Candy Mills, *Interrace*'s editor and publisher. "It's a market that has yet to be mined . . . I can't tell you how many people have written to us wishing there were companies making interracial wedding-cake decorations, toys, or greeting cards."

Gwen Lipsky of MTV says that the listening preferences of her young viewers are much more determined by personal choice than by ethnic background. "Rap and hip-hop albums are purchased primarily by white kids," she says. "More often, behavior is driven by taste."

Benetton, the clothing company based in Italy, helped create fashionable multiracialism with its famous "United Colors of Benetton" ad campaign. Years after the campaign began, the company still sells clothes by contrasting multicultural models with the dream of global unity. Of course, not all young adults will respond to messages like these in the same way. A Benetton-type ad might be far more popular with teenagers in San Francisco than with teens in a blue-collar Kentucky town.

Concern for racial tolerance drops off sharply among less-educated adults. When asked whether immigrants to this country should be prepared to assimilate, 88 percent of high school dropouts will say yes, compared with 84 percent of high school graduates and 76 percent of the college-educated, according to Roper. About one in six high school dropouts would prefer that no Hispanics live in their neighborhoods, compared with one in ten college graduates. And only 25 percent of dropouts consider interracial marriage acceptable for anyone including themselves, compared with 49 percent of college graduates. The bottom line is that more-educated, more-affluent adults are more likely to see racial tolerance as an important social goal.

If the government eventually allows individuals to declare themselves multiracial, the result will be more confusion for federal agencies, affirmative-action officers, and market researchers. But how people identify themselves depends on how they feel, and keeping in touch with those feelings is the key to staying in business. When a young Anglo-American decides to wear a fez and Navajo jewelry, she may be making a statement about racial harmony. But she will probably buy those accessories at the local mall.

SEEING MORE THAN BLACK & WHITE:
Latinos, racism, and the cultural divides

10

Elizabeth Martinez

A certain relish seems irresistible to this Latina as the mass media has been compelled to sit up, look south of the border, and take notice. Probably the Chiapas uprising and Mexico's recent political turmoil have won us no more

From: *Z Magazine* 7 (May 1994): 56–60. Reprinted with permission.

than a brief day in the sun. Or even less: liberal Ted Koppel still hadn't noticed the historic assassination of presidential candidate Colosio three days afterward. But it's been sweet, anyway.

When Kissinger said years ago "nothing important ever happens in the south," he articulated a contemptuous indifference toward Latin America, its people, and their culture which has long dominated U.S. institutions and attitudes. Mexico may be great for a vacation, and some people like burritos but the usual image of Latin America combines incompetence with absurdity in loud colors. My parents, both Spanish teachers, endured decades of being told kids were better off learning French.

U.S. political culture is not only Anglo-dominated but also embraces an exceptionally stubborn national self-centeredness, with no global vision other than relations of domination. The U.S. refuses to see itself as one nation sitting on a continent with 20 others all speaking languages other than English and having the right not to be dominated.

Such arrogant indifference extends to Latinos within the U.S. The mass media complain, "people can't relate to Hispanics"—or Asians, they say. Such arrogant indifference has played an important role in invisibilizing La Raza (except where we become a serious nuisance or a handy scapegoat). It is one reason the U.S. harbors an exclusively white-on-Black concept of racism. It is one barrier to new thinking about racism which is crucial today. There are others.

GOOD-BYE WHITE MAJORITY

In a society as thoroughly and violently racialized as the United States, white-Black relations have defined racism for centuries. Today the composition and culture of the U.S. are changing rapidly. We need to consider seriously whether we can afford to maintain an exclusively white/Black model of racism when the population will be 32 percent Latin/Asian/Pacific American and Native American—in short, neither Black nor white—by the year 2050. We are challenged to recognize that multi-colored racism is mushrooming, and then strategize how to resist it. We are challenged to move beyond a dualism comprised of two white supremacist inventions: Blackness and Whiteness.

At stake in those challenges is building a united anti-racist force strong enough to resist contemporary racist strategies of divde-and-conquer. Strong enough in the long run, to help defeat racism itself. Doesn't an exclusively Black/white model of racism discourage the perception of common interests among people of color and thus impede a solidarity that can challenge white supremacy? Doesn't it encourage the isolation of African Americans from potential allies? Doesn't it advise all people of color to spend too much energy understanding our lives in relation to Whiteness, and thus freeze us in a defensive, often self-destructive mode?

NO "OPPRESSION OLYMPICS"

For a Latina to talk about recognizing the multi-colored varieties of racism is not, and should not be, yet another round in the Oppression Olympics. We don't need more competition among different social groupings for that "Most Oppressed" gold. We don't need more comparisons of suffering between women and Blacks, the disabled and the gay, Latino teenagers and white seniors, or whatever. We don't need more surveys like the recent much publicized Harris Poll showing that different peoples of color are prejudiced toward each other—a poll patently designed to demonstrate that us coloreds are no better than white folk. (The survey never asked people about positive attitudes.)

Rather, we need greater knowledge, understanding, and openness to learning about each other's histories and present needs as a basis for working together. Nothing could seem more urgent in an era when increasing impoverishment encourages a self-imposed separatism among people of color as a desperate attempt at community survival. Nothing could seem more important as we search for new social change strategies in a time of ideological confusion.

My call to rethink concepts of racism in the U.S. today is being sounded elsewhere. Among academics, liberal foundation administrators, and activist-intellectuals, you can hear talk of the need for a new "racial paradigm" or model. But new thinking seems to proceed in fits and starts, as if dogged by a fear of stepping on toes, of feeling threatened, or of losing one's base. With a few notable exceptions, even our progressive scholars of color do not make the leap from perfunctorily saluting a vague multi-culturalism to serious analysis. We seem to have made little progress, if any, since Bob Blauner's 1972 book *Racial Oppression in America*. Recognizing the limits of the white-Black axis, Blauner critiqued White America's ignorance of and indifference to the Chicano/a experience with racism.

Real opposition to new paradigms also exists. There are academics scrambling for one flavor of ethnic studies funds versus another. There are politicians who cultivate distrust of others to keep their own communities loyal. When we hear, for example, of Black/Latino friction, dismay should be quickly followed by investigation. In cities like Los Angeles and New York, it may turn out that political figures scrapping for patronage and payola have played a narrow nationalist game, whipping up economic anxiety and generating resentment that sets communities against each other.

So the goal here, in speaking about moving beyond a bipolar concept of racism is to build stronger unity against white supremacy. The goal is to see our similarities of experience and needs. If that goal sounds naive, think about the hundreds of organizations formed by grassroots women of different colors coming together in recent years. Their growth is one of today's most energetic motions and it spans all ages. Think about the multicultural environmental justice movement. Think about the coalitions to save schools. Small rainbows of our own making are there, to brighten a long road through hellish times.

It is in such practice, through daily struggle together, that we are most likely to find the road to greater solidarity against a common enemy. But we also need a will to find it and ideas about where, including some new theory.

THE WEST GOES EAST

Until very recently, Latino invisibility—like that of Native Americans and Asian/Pacific Americans—has been close to absolute in U.S. seats of power, major institutions, and the non-Latino public mind. Having lived on both the East and West Coasts for long periods, I feel qualified to pronounce: an especially myopic view of Latinos prevails in the East. This, despite such data as a 24.4 percent Latino population of New York City alone in 1991, or the fact that in 1990 more Puerto Ricans were killed by New York police under suspicious circumstances than any other ethnic group. Latino populations are growing rapidly in many eastern cities and the rural South, yet remain invisible or stigmatized—usually both.

Eastern blinders persist. I've even heard that the need for a new racial paradigm is dismissed in New York as a California hangup. A black Puerto Rican friend in New York, when we talked about experiences of racism common to Black and brown, said "People here don't see Border Patrol brutality against Mexicans as a form of police repression," despite the fact that the Border Patrol is the largest and most uncontrolled police force in the U.S. It would seem that an old ignorance has combined with new immigrant bashing to sustain divisions today.

While the East (and most of the Midwest) usually remains myopic, the West Coast has barely begun to move away from its own denial. Less than two years ago in San Francisco, a city almost half Latino or Asian/Pacific American, a leading daily newspaper could publish a major series on contemporary racial issues and follow the exclusively Black-white paradigm. Although millions of TV viewers saw massive Latino participation in the April 1992 Los Angeles uprising, which included 18 out of 50 deaths and the majority of arrests, the mass media and most people labeled that event "a Black riot."

If the West Coast has more recognition of those who are neither Black nor white, it is mostly out of fear about the proximate demise of its white majority. A second, closely related reason is the relentless campaign by California Gov. Pete Wilson to scapegoat immigrants for economic problems and pass racist, unconstitutional laws attacking their health, education, and children's future. Wilson has almost single-handedly made the word "immigrant" mean Mexican or other Latino (and sometimes Asian). Who thinks of all the people coming from the former Soviet Union and other countries? The absolute racism of this has too often been successfully masked by reactionary anti-immigrant groups like FAIR blaming immigrants for the staggering African-American unemployment rate.

Wilson's immigrant bashing is likely to provide a model for other parts of the country. The five states with the highest immigration rates—California, Florida, New York, Illinois and Texas—all have a governor up for re-election in 1994. Wilson's tactics won't appear in every campaign but some of the five states will surely see intensified awareness and stigmatization of Latinos as well as Asian/Pacific Islanders.

As this suggests, what has been a regional issue mostly limited to western states is becoming a national issue. If you thought Latinos were just "Messicans" down at the border, wake up—they are all over North Carolina, Pennsylvania and 8th Avenue Manhattan now. A qualitative change is taking place. With the broader geographic spread of Latinos and Asian/Pacific Islanders has come a nationalization of racist practices and attitudes that were once regional. The west goes east, we could say.

Like the monster Hydra, racism is growing some ugly new heads. We will have to look at them closely.

THE ROOTS OF RACISM AND LATINOS

A bipolar model of racism—racism as white on Black—has never really been accurate. Looking for the roots of racism in the U.S. we can begin with the genocide against American Indians which made possible the U.S. land base, crucial to white settlement and early capitalist growth. Soon came the massive enslavement of African people which facilitated that growth. As slave labor became economically critical, "blackness" became ideologically critical; it provided the very source of "whiteness" and the heart of racism. Franz Fanon would write, "colour is the most outward manifestation of race."

If Native Americans had been a crucial labor force during those same centuries, living and working in the white man's sphere, our racist ideology might have evolved differently. "The tawny," as Ben Franklin dubbed them, might have defined the opposite of what he called "the lovely white." But with Indians decimated and survivors moved to distant concentration camps, they became unlikely candidates for this function. Similarly, Mexicans were concentrated in the distant West; elsewhere Anglo fear of them or need to control was rare. They also did not provide the foundation for a definition of whiteness.

Some anti-racist left activists have put forth the idea that only African Americans experience racism as such and that the suffering of other people of color results from national minority rather than racial oppression. From this viewpoint, the exclusively white/Black model for racism is correct. Latinos, then, experience exploitation and repression for reasons of culture and nationality—not for their "race." (It should go without saying in Z that while racism is an all-too-real social fact, race has no scientific basis.)

Does the distinction hold? This and other theoretical questions call for more analysis and more expertise than one article can offer. In the meantime, let's try on the idea that Latinos do suffer for their nationality and culture, especially language. They became part of the U.S. through the 1846–48 war on Mexico and thus a foreign population to be colonized. But as they were reduced to cheap or semi-slave labor, they quickly came to suffer for their "race"—meaning, as non-whites. In the Southwest of a super-racialized nation the broad parallelism of race and class embrace Mexicans ferociously.

The bridge here might be a definition of racism as "the reduction of the cultural to the biological," in the words of French scholar Christian Delacampagne now working in Egypt. Or: "racism exists wherever it is claimed that a given social status is explained by a given natural characteristic." We know that line: Mexicans are just naturally lazy and have too many children, so they're poor and exploited.

The discrimination, oppression and hatred experienced by Native Americans, Mexicans, Asian/Pacific Islanders, and Arab Americans are forms of racism. Speaking only of Latinos, we have seen in California and the Southwest, especially along the border, almost 150 years of relentless repression which today includes Central Americans among its targets. That history reveals hundreds of lynchings between 1847 and 1935, the use of counter-insurgency armed forces beginning with the Texas Rangers, random torture and murder by Anglo ranchers, forced labor, rape by border lawmen, and the prevailing Anglo belief that a Mexican life doesn't equal a dog's in value.

But wait. If color is so key to racial definition, as Fanon and others say, perhaps people of Mexican background experience racism less than national minority oppression because they are not dark enough as a group. For White

America, shades of skin color are crucial to defining worth. The influence of those shades has also been internalized by communities of color. Many Latinos can and often want to pass for whites; therefore, White America may see them as less threatening than darker sisters and brothers.

Here we confront more of the complexity around us today, with questions like: What about the usually poor, very dark Mexican or Central American of strong Indian or African heritage? (Yes, folks, 200,000–300,000 Africans were brought to Mexico as slaves, which is far, far more than the Spaniards who came.) And what about the effects of accented speech or foreign name, characteristics that may instantly subvert "passing"?

What about those cases where a Mexican-American is never accepted, no matter how light-skinned, well-dressed or well-spoken? A Chicano lawyer friend coming home from a professional conference in suit, tie and briefcase found himself on a bus near San Diego that was suddenly stopped by the Border Patrol. An agent came on board and made a beeline through the all-white rows of passengers direct to my friend. "Your papers." The agent didn't believe Jose was coming from a U.S. conference and took him off the bus to await proof. Jose was lucky; too many Chicanos and Mexicans end up killed.

In a land where the national identity is white, having the "wrong" nationality becomes grounds for racist abuse. Who would draw a sharp line between today's national minority oppression in the form of immigrant-bashing, and racism?

None of this aims to equate the African American and Latino experiences; that isn't necessary even if it were accurate. Many reasons exist for the persistence of the white/Black paradigm of racism; they include numbers, history, and the psychology of whiteness. In particular they include centuries of slave revolts, a Civil War, and an ongoing resistance to racism that cracked this society wide open while the world watched. Nor has the misery imposed on Black people lessened in recent years. New thinking about racism can and should keep this experience at the center.

A DEADLY DUALISM

The exclusively white/Black concept of race and racism in the U.S. rests on a western, Protestant form of dualism woven into both race and gender relations from earliest times. In the dualist universe there is only black and white. A disdain, indeed fear, of mixture haunts the Yankee soul; there is no room for any kind of multi-faceted identity, any hybridism.

As a people, La Raza combines three sets of roots—indigenous, European, and African—all in widely varying degrees. In short we represent a profoundly un-American concept: *mestizaje* (pronounced mess-tee-zah-hey), the mixing of peoples and emergence of new peoples. A highly racialized society like this one cannot deal with or allow room for *mestizaje*. It has never learned to do much more than hiss "miscegenation!" Or, like that Alabama high school principal who recently denied the right of a mixed-blood pupil to attend the prom, to say: "your parents made a mistake." Apparently we, all the millions of La Raza, are just that—a mistake.

Mexicans in the U.S. also defy the either-or, dualistic mind in that, on the one hand, we are a colonized people displaced from the ancestral homeland with roots in the present-day U.S. that go back centuries. Those ancestors didn't cross the border; the border crossed them. At the same time many of us have come to the U.S. more recently as "immigrants" seeking work. The complexity of Raza baffles and frustrates most Anglos; they want to put one neat label on us. It baffles many Latinos too, who often end up categorizing themselves racially as "Other" for lack of anything better. For that matter, the term "Latino" which I use here is a monumental simplification; it refers to 20-plus nationalities and a wide range of classes.

But we need to grapple with the complexity, for there is more to come. If anything, this nation will see more *mestizaje* in future, embracing innumerable ethnic combinations. What will be its effects? Only one thing seems certain: "white" shall cease to be the national identity.

A glimpse at the next century tells us how much we need to look beyond the white/Black model of race relations and racism. White/Black are real poles, central to the history of U.S. racism. We can neither ignore them nor stop there. But our effectiveness in fighting racism depends on seeing the changes taking place, trying to perceive the contours of the future. From the time of the Greeks to the present, racism around the world has had certain commonalties but no permanently fixed character. It is evolving again today, and we'd best labor to read the new faces of this Hydra-headed monster. Remember, for every head that Hydra lost it grew two more.

Sometimes the problem seems so clear. Last year I showed slides of Chicano history to a Oakland high school class with 47 African Americans and three Latino students. The images included lynchings and police beatings of Mexicans and other Latinos, and many years of resistance. At the end one Black student asked, "Seems like we have had a lot of experiences in common—so why can't Blacks and Mexicans get along better?" No answers, but there was the first step: asking the question.

A future article will look closer at the invisibilization of Latinos and the reasons for it.

RACE MATTERS

11

Cornel West

Since the beginning of the nation, white Americans have suffered from a deeper inner uncertainty as to who they really are. One of the ways that has been used to simplify the answer has been to seize upon the presence of black Americans and use them as a marker, a symbol of limits, a metaphor for the "outsider." Many whites could look at the social position of blacks and feel that color formed an easy and reliable gauge for determining to what extent one was or was not American. Perhaps that is why one of the first epithets that many European immigrants learned when they got off the boat was the term "nigger"—it made them feel instantly American. But this is tricky magic. Despite his racial difference and social status, something indisputably American about Negroes not only raised doubts about the white man's value system but aroused the troubling suspicion that whatever else the true American is, he is also somehow black.

Ralph Ellison, "What America Would Be Like Without Blacks" (1970)

What happened in Los Angeles in April of 1992 was neither a race riot nor a class rebellion. Rather, this monumental upheaval was a multiracial, trans-class, and largely male display of justified social rage. For all its ugly, xenophobic resentment, its air of adolescent carnival, and its downright barbaric behavior, it signified the sense of powerlessness in American society. Glib attempts to reduce its meaning to the pathologies of the black underclass, the criminal actions of hoodlums, or the political revolt of the oppressed urban masses miss the mark. Of those arrested, only 36 percent were black, more than a third had full-time jobs, and most claimed to shun political affiliation. What we witnessed in Los Angeles was the consequence of a lethal linkage of economic decline, cultural decay, and political lethargy in American life. Race was the visible catalyst, not the underlying cause.

The meaning of the earthshaking events in Los Angeles is difficult to grasp because most of us remain trapped in the narrow framework of the dominant liberal and conservative views of race in America, which with its worn-out vocabulary leaves us intellectually debilitated, morally disempowered, and personally depressed. The astonishing disappearance of the event from public dialogue is testimony to just how painful and distressing a serious engagement with race is. Our truncated public discussions of race suppress the best of who and what we are as a people because they fail to confront the complexity of the issue in a candid and critical manner. The predictable pitting of liberals against conservatives, Great Society Democrats against self-help Republicans, reinforces intellectual parochialism and political paralysis.

The liberal notion that more government programs can solve racial problems is simplistic—precisely because it focuses *solely* on the economic dimension. And the conservative idea that what is needed is a change in the moral behavior of poor black urban dwellers (especially poor black men, who, they say, should stay married, support their children, and stop committing so much crime) highlights immoral actions while ignoring public responsibility for the immoral circumstances that haunt our fellow citizens.

The common denominator of these views of race is that each still sees black people as a "problem people," in the words of Dorothy I. Height, president of the National Council of Negro Women, rather than as fellow American citizens with problems. Her words echo the poignant "unasked question" of W. E. B. Du Bois, who, in *The Souls of Black Folk* (1903), wrote:

> They approach me in a half-hesitant sort of way, eye me curiously or compassionately, and then instead of saying directly, How does it feel to be a problem? they say, I know an excellent colored man in my town. . . . Do not these Southern outrages make your blood boil? At these I smile, or am interested, or reduce the boiling to a simmer, as the occasion may require. To the real question, How does it feel to be a problem? I answer seldom a word.

Nearly a century later, we confine discussions about race in America to the "problems" black people pose for whites rather than consider what this way of viewing black people reveals about us as a nation.

This paralyzing framework encourages liberals to relieve their guilty consciences by supporting public funds directed at "the problems"; but at the same time, reluctant to exercise principled criticism of black people, liberals deny them the freedom to err. Similarly, conservatives blame the "problems" on black people themselves—and thereby render black social misery invisible or unworthy of public attention.

Hence, for liberals, black people are to be "included" and "integrated" into "our" society and culture, while for conservatives they are to be "well behaved" and "worthy of acceptance" by "our" way of life. Both fail to see that the presence and predicaments of black people are neither additions to nor defections from American life, but rather *constitutive elements of that life*.

To engage in a serious discussion of race in America, we must begin not with the problems of black people but with the flaws of American society—

flaws rooted in historic inequalities and longstanding cultural stereotypes. How we set up the terms for discussing racial issues shapes our perception and response to these issues. As long as black people are viewed as a "them," the burden falls on blacks to do all the "cultural" and "moral" work necessary for healthy race relations. The implication is that only certain Americans can define what it means to be American—and the rest must simply "fit in."

The emergence of strong black-nationalist sentiments among blacks, especially among young people, is a revolt against this sense of having to "fit in." The variety of black-nationalist ideologies, from the moderate views of Supreme Court Justice Clarence Thomas in his youth to those of Louis Farrakhan today, rest upon a fundamental truth: white America has been historically weak-willed in ensuring racial justice and has continued to resist fully accepting the humanity of blacks. As long as double standards and differential treatment abound—as long as the rap performer Ice-T is harshly condemned while former Los Angeles Police Chief Daryl F. Gates's antiblack comments are received in polite silence, as long as Dr. Leonard Jeffries's anti-Semitic statements are met with vitriolic outrage while presidential candidate Patrick J. Buchanan's anti-Semitism receives a genteel response—black nationalisms will thrive.

Afrocentrism, a contemporary species of black nationalism, is a gallant yet misguided attempt to define an African identity in a white society perceived to be hostile. It is gallant because it puts black doings and sufferings, not white anxieties and fears, at the center of discussion. It is misguided because—out of fear of cultural hybridization and through silence on the issue of class, retrograde views on black women, gay men, and lesbians, and a reluctance to link race to the common good—it reinforces the narrow discussions about race.

To establish a new framework, we need to begin with a frank acknowledgement of the basic humanness and Americanness of each of us. And we must acknowledge that as a people—*E Pluribus Unum*—we are on a slippery slope toward economic strife, social turmoil, and cultural chaos. If we go down, we go down together. The Los Angeles upheaval forced us to see not only that we are not connected in ways we would like to be but also, in a more profound sense, that this failure to connect binds us even more tightly together. The paradox of race in America is that our common destiny is more pronounced and imperiled precisely when our divisions are deeper. The Civil War and its legacy speak loudly here. And our divisions are growing deeper. Today, 86 percent of white suburban Americans live in neighborhoods that are less than 1 percent black, meaning that the prospects for the country depend largely on how its cities fare in the hands of a suburban electorate. There is no escape from our interracial interdependence, yet enforced racial hierarchy dooms us as a nation to collective paranoia and hysteria—the unmaking of any democratic order.

The verdict in the Rodney King case which sparked the incidents in Los Angeles was perceived to be wrong by the vast majority of Americans. But whites have often failed to acknowledge the widespread mistreatment of black people, especially black men, by law enforcement agencies, which helped ignite the spark. The verdict was merely the occasion for deep-seated rage to come to the surface. This rage is fed by the "silent" depression ravaging the country—in which real weekly wages of all American workers since 1973 have declined nearly 20 percent, while at the same time wealth has been upwardly distributed.

The exodus of stable industrial jobs from urban centers to cheaper labor markets here and abroad, housing policies that have created "chocolate cities and vanilla suburbs" (to use the popular musical artist George Clinton's memorable phrase), white fear of black crime, and the urban influx of poor Spanish-speaking and Asian immigrants—all have helped erode the tax base of American cities just as the federal government has cut its supports and programs. The result is unemployment, hunger, homelessness, and sickness for millions.

And a pervasive spiritual impoverishment grows. The collapse of meaning in life—the eclipse of hope and absence of love of self and others, the breakdown of family and neighborhood bonds—leads to the social deracination and cultural denudement of urban dwellers, especially children. We have created rootless, dangling people with little link to the supportive networks—family, friends, school—that sustain some sense of purpose in life. We have witnessed the collapse of the spiritual communities that in the past helped Americans face despair, disease, and death and that transmit through the generations dignity and decency, excellence and elegance.

The result is lives of what we might call "random nows," of fortuitous and fleeting moments preoccupied with "getting over"—with acquiring pleasure, property, and power by any means necessary. (This is not what Malcolm X meant by this famous phrase.) Post-modern culture is more and more a market culture dominated by gangster mentalities and self-destructive wantonness. This culture engulfs all of us—yet its impact on the disadvantaged is devastating, resulting in extreme violence in everyday life. Sexual violence against women and homicidal assaults by young black men on one another are only the most obvious signs of this empty quest for pleasure, property, and power.

Last, this rage is fueled by a political atmosphere in which images, not ideas, dominate, where politicians spend more time raising money than debating issues. The functions of parties have been displaced by public polls, and politicians behave less as thermostats that determine the climate of opinion than as thermometers registering the public mood. American politics has been rocked by an unleashing of greed among opportunistic public officials—who have followed the lead of their counterparts in the private sphere, where, as of 1989, 1 percent of the population owned 37 percent of the wealth and 10 percent of the population owned 86 percent of the wealth—leading to a profound cynicism and pessimism among the citizenry.

And given the way in which the Republican Party since 1968 has appealed to popular xenophobic images—playing the black, female, and homophobic cards to realign the electorate along race, sex, and sexual-orientation lines— it is no surprise that the notion that we are all part of one garment of destiny is discredited. Appeals to special interests rather than to public interests reinforce this polarization. The Los Angeles upheaval was an expression of utter fragmentation by a powerless citizenry that includes not just the poor but all of us.

What is to be done? How do we capture a new spirit and vision to meet the challenges of the post-industrial city, post-modern culture, and post-party politics?

First, we must admit that the most valuable sources for help, hope, and power consist of ourselves and our common history. As in the ages of Lincoln, Roosevelt, and King, we must look to new frameworks and languages to understand our multilayered crisis and overcome our deep malaise.

Second, we must focus our attention on the public square—the common good that undergirds our national and global destinies. The vitality of any public square ultimately depends on how much we *care* about the quality of our lives together. The neglect of our public infrastructure, for example—our water and sewage systems, bridges, tunnels, highways, subways, and streets— reflects not only our myopic economic policies, which impede productivity, but also the low priority we place on our common life.

The tragic plight of our children clearly reveals our deep disregard for public well-being. About one out of every five children in this country lives in poverty, including one out of every two black children and two out of every five Hispanic children. Most of our children—neglected by overburdened parents and bombarded by the market values of profit-hungry corporations— are ill-equipped to live lives of spiritual and cultural quality. Faced with these facts, how do we expect ever to constitute a vibrant society?

One essential step is some form of large-scale public intervention to ensure access to basic social goods—housing, food, health care, education, child care, and jobs. We must invigorate the common good with a mixture of government, business, and labor that does not follow any existing blueprint. After a period in which the private sphere has been sacralized and the public square gutted, the temptation is to make a fetish of the public square. We need to resist such dogmatic swings.

Last, the major challenge is to meet the need to generate new leadership. The paucity of courageous leaders—so apparent in the response to the events in Los Angeles—requires that we look beyond the same elites and voices that recycle the older frameworks. We need leaders—neither saints nor sparkling television personalities—who can situate themselves within a larger historical narrative of this country and our world, who can grasp the complex dynamics of our peoplehood and imagine a future grounded in the best of our past, yet who are attuned to the frightening obstacles that now perplex us. Our ideals of freedom, democracy, and equality must be invoked to invigorate all of us, especially the landless, propertyless, and luckless. Only a visionary leadership that can motivate "the better angels of our nature," as Lincoln said, and activate possibilities for a freer, more efficient, and stable America—only that leadership deserves cultivation and support.

This new leadership must be grounded in grass-roots organizing that highlights democratic accountability. Whoever *our* leaders will be as we approach the twenty-first century, their challenge will be to help Americans determine whether a genuine multiracial democracy can be created and sustained in an era of global economy and a moment of xenophobic frenzy.

Let us hope and pray that the vast intelligence, imagination, humor, and courage of Americans will not fail us. Either we learn a new language of empathy and compassion, or the fire this time will consume us all.

Class and Inequality

TIRED OF PLAYING MONOPOLY?   **12**

Donna Langston

I. Magnin, Nordstrom, The Bon, Sears, Penneys, K mart, Goodwill, Salvation Army. If the order of this list of stores makes any sense to you, then we've begun to deal with the first question which inevitably arises in any discussion of class here in the U.S.—huh? Unlike our European allies, we in the U.S. are reluctant to recognize class differences. This denial of class divisions functions to reinforce ruling class control and domination. America is, after all, the supposed land of equal opportunity where, if you just work hard enough, you can get ahead, pull yourself up by your bootstraps. What the old bootstraps theory overlooks is that some were born with silver shoe horns. Female-headed households, communities of color, the elderly, disabled and children find themselves, disproportionately, living in poverty. If hard work were the sole determinant of your ability to support yourself and your family, surely we'd have a different outcome for many in our society. We also, however, believe in luck and, on closer examination, it certainly is quite a coincidence that the "unlucky" come from certain race, gender and class backgrounds. In order to perpetuate racist, sexist and classist outcomes, we also have to believe that the current economic distribution is unchangeable, has always existed, and probably exists in this form throughout the known universe; i.e., it's "natural." Some people explain or try to account for poverty or class position by focusing on the personal and moral merits of an individual. If people are

From: Jo Whitehorse Cochran, Donna Langston, and Carolyn Woodward (eds.), *Changing Our Power: An Introduction to Women's Studies* (Dubuque, IA: Kendall-Hunt, 1988). Reprinted by permission.

poor, then it's something they did or didn't do; they were lazy, unlucky, didn't try hard enough, etc. This has the familiar ring of blaming the victims. Alternative explanations focus on the ways in which poverty and class position are due to structural, systematic, institutionalized economic and political power relations. These power relations are based firmly on dynamics such as race, gender, and class.

In the myth of the classless society, ambition and intelligence alone are responsible for success. The myth conceals the existence of a class society, which serves many functions. One of the main ways it keeps the working class and poor locked into a class-based system in a position of servitude is by cruelly creating false hope. It perpetuates the false hope among the working class and poor that they can have different opportunities in life. The hope that they can escape the fate that awaits them due to the class position they were born into. Another way the rags-to-riches myth is perpetuated is by creating enough visible tokens so that oppressed persons believe they, too, can get ahead. The creation of hope through tokenism keeps a hierarchical structure in place and lays the blame for not succeeding on those who don't. This keeps us from resisting and changing the class-based system. Instead, we accept it as inevitable, something we just have to live with. If oppressed people believe in equality of opportunity, then they won't develop class consciousness and will internalize the blame for their economic position. If the working class and poor do not recognize the way false hope is used to control them, they won't get a chance to control their lives by acknowledging their class position, by claiming that identity and taking action as a group.

The myth also keeps the middle class and upper class entrenched in the privileges awarded in a class-based system. It reinforces middle- and upper-class beliefs in their own superiority. If we believe that anyone in society really can get ahead, then middle- and upper-class status and privileges must be deserved, due to personal merits, and enjoyed—and defended at all costs. According to this viewpoint, poverty is regrettable but acceptable, just the outcome of a fair game: "There have always been poor people, and there always will be."

Class is more than just the amount of money you have; it's also the presence of economic security. For the working class and poor, working and eating are matters of survival, not taste. However, while one's class status can be defined in important ways in terms of monetary income, class is also a whole lot more—specifically, class is also culture. As a result of the class you are born into and raised in, class is your understanding of the world and where you fit in; it's composed of ideas, behavior, attitudes, values, and language; class is how you think, feel, act, look, dress, talk, move, walk; class is what stores you shop at, restaurants you eat in; class is the schools you attend, the education you attain; class is the very jobs you will work at throughout your adult life. Class even determines when we marry and become mothers. Working-class women become mothers long before middle-class women receive their bachelor's degrees. We experience class at every level of our lives; class is who our friends are, where we live and work even what kind of car we drive, if we own one, and what kind of health care we receive, if any. Have I left anything out? In other words, class is socially constructed and all-encompassing. When we experience classism, it will be because of our lack of money (i.e., choices and power in this society) and because of the way we talk, think, act, move—because of our culture.

Class affects what we perceive as and what we have available to us as choices. Upon graduation from high school, I was awarded a scholarship to attend any college, private or public, in the state of California. Yet it never occurred to me or my family that it made any difference which college you went to. I ended up just going to a small college in my town. It never would have occurred to me to move away from my family for school, because no one ever had and no one would. I was the first person in my family to go to college. I had to figure out from reading college catalogs how to apply—no one in my family could have sat down and said, "Well, you take this test and then you really should think about . . ." Although tests and high school performance had shown I had the ability to pick up white middle-class lingo, I still had quite an adjustment to make—it was lonely and isolating in college. I lost my friends from high school—they were at the community college, vo-tech school, working, or married. I lasted a year and a half in this foreign environment before I quit college, married a factory worker, had a baby and resumed living in a community I knew. One middle-class friend in college had asked if I'd like to travel to Europe with her. Her father was a college professor and people in her family had actually travelled there. My family had seldom been able to take a vacation at all. A couple of times my parents were able—by saving all year—to take the family over to the coast on their annual two-week vacation. I'd seen the time and energy my parents invested in trying to take a family vacation to some place a few hours away; the idea of how anybody ever got to Europe was beyond me.

If class is more than simple economic status but one's cultural background as well, what happens if you're born and raised middle class, but spend some of your adult life with earnings below a middle-class income bracket—are you then working-class? Probably not. If your economic position changes, you still have the language, behavior, educational background, etc., of the middle class, which you can bank on. You will always have choices. Men who consciously try to refuse male privilege are still male; whites who want to challenge white privilege are still white. I think those who come from middle-class backgrounds need to recognize that their class privilege does not float out with the rinse water. Middle-class people can exert incredible power just by being nice and polite. The middle-class way of doing things is the standard—they're always right, just by being themselves. Beware of middle-class people who deny their privilege. Many people have times when they struggle to get shoes for the kids, when budgets are tight, etc. This isn't the same as long-term economic conditions without choices. Being working class is also generational. Examine your family's history of education, work, and standard of living. It may not be a coincidence that you share the same class status as your parents and grandparents. If your grandparents were professionals, or your parents were professionals, it's much more likely you'll be able to grow up to become a yuppie, if your heart so desires, or even if you don't think about it.

How about if you're born and raised poor or working class, yet through struggle, usually through education, you manage to achieve a different economic level: do you become middle class? Can you pass? I think some working-class people may successfully assimilate into the middle class by learning to dress, talk, and act middle class—to accept and adopt the middle-class way of doing things. It all depends on how far they're able to go. To succeed in the middle-class world means facing great pressures to abandon working-class friends and ways.

Contrary to our stereotype of the working class—white guys in overalls—the working class is not homogeneous in terms of race or gender. If you are a person of color, if you live in a female-headed household, you are much more likely to be working class or poor. The experience of Black, Latino, American Indian or Asian American working classes will differ significantly from the white working classes, which have traditionally been able to rely on white privilege to provide a more elite position within the working class. Working-class people are often grouped together and stereotyped, but distinctions can be made among the working class, working poor, and poor. Many working-class families are supported by unionized workers who possess marketable skills. Most working-poor families are supported by non-unionized, unskilled men and women. Many poor families are dependent on welfare for their income.

Attacks on the welfare system and those who live on welfare are a good example of classism in action. We have a "dual welfare" system in this country whereby welfare for the rich in the form of tax-free capital gain, guaranteed loans, oil depletion allowances, etc., is not recognized as welfare. Almost everyone in America is on some type of welfare; but, if you're rich, it's in the form of tax deductions for "business" meals and entertainment, and if you're poor, it's in the form of food stamps. The difference is the stigma and humiliation connected to welfare for the poor, as compared to welfare for the rich, which is called "incentives." Ninety-three percent of AFDC (Aid to Families with Dependent Children, our traditional concept of welfare) recipients are women and children. Eighty percent of food stamp recipients are single mothers, children, the elderly and disabled. Average AFDC payments are $93 per person, per month.* Payments are so low nationwide that in only three states do AFDC benefits plus food stamps bring a household *up to* the poverty level. Food stamp benefits average $10 per person, per week (Sar Levitan, *Programs in Aid of the Poor for the 1980s*). A common focal point for complaints about "welfare" is the belief that most welfare recipients are cheaters—goodness knows there are no middle-class income tax cheaters out there. Imagine focusing the same anger and energy on the way corporations and big business cheat on their tax revenues. Now, there would be some dollars worth quibbling about. The "dual welfare" system also assigns a different degree of stigma to programs that benefit women and children, such as AFDC, and programs whose recipients are primarily male, such as veterans' benefits. The implicit assumption is that mothers who raise children do not work and therefore are not deserving of their daily bread crumbs.

Anti-union attitudes are another prime example of classism in action. At best, unions have been a very progressive force for workers, women and people of color. At worst, unions have reflected the same regressive attitudes which are out there in other social structures: classism, racism, and sexism. Classism exists within the working class. The aristocracy of the working class—unionized, skilled workers—have mainly been white and male and have viewed themselves as being better than unskilled workers, the unemployed, and the poor, who are mostly women and people of color. The white working class must commit itself to a cultural and ideological transformation of racist attitudes. The history of working people, and the ways we've resisted many types of oppressions, are not something we're taught in school. Missing from our education is information about workers and their resistance.

Working-class women's critiques have focused on the following issues:

Education: White middle-class professionals have used academic jargon to rationalize and justify classism. The whole structure of education is a classist system. Schools in every town reflect class divisions: like the store list at the beginning of this article, you can list schools in your town by what classes of

---

* *Editors' Note:* In 1994, prior to welfare reform, the average payment per family was $378, still far below the poverty line (U.S. Bureau of the Census, *Statistical Abstracts of the United States 1996.* Washington, D.C.: U.S. Government Printing Office, p. 381).

kids attend, and in most cities you can also list by race. The classist system is perpetuated in schools with the tracking system, whereby the "dumbs" are tracked into homemaking, shop courses and vocational school futures, while the "smarts" end up in advanced math, science, literature, and college-prep courses. If we examine these groups carefully, the coincidence of poor and working-class backgrounds with "dumbs" is rather alarming. The standard measurement of supposed intelligence is white middle-class English. If you're other than white middle class, you have to become bilingual to succeed in the educational system. If you're white middle class, you only need the language and writing skills you were raised with, since they're the standard. To do well in society presupposes middle-class background, experiences and learning for everyone. The tracking system separates those from the working class who can potentially assimilate to the middle class from all our friends, and labels us "college bound."

After high school, you go on to vocational school, community college, or college—public or private—according to your class position. Apart from the few who break into middle-class schools, the classist stereotyping of the working class as being dumb and inarticulate tracks most into vocational and low-skilled jobs. A few of us are allowed to slip through to reinforce the idea that equal opportunity exists. But for most, class position is destiny—determining our educational attainment and employment. Since we must overall abide by middle-class rules to succeed, the assumption is that we go to college in order to "better ourselves"—i.e., become more like them. I suppose it's assumed we have "yuppie envy" and desire nothing more than to be upwardly mobile individuals. It's assumed that we want to fit into their world. But many of us remain connected to our communities and families. Becoming college educated doesn't mean we have to, or want to, erase our first and natural language and value system. It's important for many of us to remain in and return to our communities to work, live, and stay sane.

Jobs: Middle-class people have the privilege of choosing careers. They can decide which jobs they want to work, according to their moral or political commitments, needs for challenge or creativity. This is a privilege denied the working class and poor, whose work is a means of survival, not choice (see Hartsock). Working-class women have seldom had the luxury of choosing between work in the home or market. We've generally done both, with little ability to purchase services to help with this double burden. Middle- and upper-class women can often hire other women to clean their houses, take care of their children, and cook their meals. Guess what class and race those "other" women are? Working a double or triple day is common for working-class women. Only middle-class women have an array of choices such as: parents put you through school, then you choose a career, then you choose when and if to have babies, then you choose a support system of working-class women to take care of your kids and house if you choose to resume your career. After the birth of my second child, I was working two part-time jobs—one loading trucks at night—and going to school during the days. While I was quite privileged because I could take my colicky infant with me to classes and the day-time job, I was in a state of continuous semi-consciousness. I had to work to support my family; the only choice I had was between school or sleep: Sleep became a privilege. A white middle-class feminist instructor at the university suggested to me, all sympathetically, that I ought to hire someone to clean my house and watch the baby. Her suggestion was totally out of my reality, both economically and socially. I'd worked for years cleaning other peoples' houses. Hiring a working-class woman to do the shit work is a middle-class woman's solution to any dilemma which her privileges, such as a career, may present her.

Mothering: The feminist critique of families and the oppressive role of mothering has focused on white middle-class nuclear families. This may not be an appropriate model for communities of class and color. Mothering and families may hold a different importance for working-class women. Within this context, the issue of coming out can be a very painful process for working-class lesbians. Due to the homophobia of working-class communities, to be a lesbian is most often to be excommunicated from your family, neighborhood, friends and the people you work with. If you're working class, you don't have such clearly demarcated concepts of yourself as an individual, but instead see yourself as part of a family and community that forms your survival structure. It is not easy to be faced with the risk of giving up ties which are so central to your identity and survival.

Individualism: Preoccupation with one's self—one's body, looks, relationships—is a luxury working-class women can't afford. Making an occupation out of taking care of yourself through therapy, aerobics, jogging, dressing for success, gourmet meals and proper nutrition, etc., may be responses that are directly rooted in privilege. The middle class has the leisure time to be preoccupied with their own problems, such as their waistlines, planning their vacations, coordinating their wardrobes, or dealing with what their mother said to them when they were five—my!

The white middle-class women's movement has been patronizing to working-class women. Its supporters think we don't understand sexism. What we don't understand is white middle-class feminism. They act as though they invented the truth, the light, and the way, which they merely need to pass along to us lower-class drudges. What they invented is a distorted form of what working-class women already know—if you're female, life sucks. Only at least we were smart enough to know that it's not just being female, but also being a person of color or class, which makes life a quicksand trap. The class system weakens all women. It censors and eliminates images of female strength. The idea of women as passive, weak creatures totally discounts the strength, self-dependence and inter-dependence necessary to survive as working-class and poor women. My mother and her friends always had a less-than-passive, less-than-enamoured attitude toward their spouses, male bosses, and men in general. I know from listening to their conversations, jokes and what they passed on to us, their daughters, as folklore. When I was five years old, my mother told me about how Aunt Betty had hit Uncle Ernie over the head with a skillet and knocked him out because he was raising his hand to hit her, and how he's never even thought about doing it since. This story was told to me with a good amount of glee and laughter. All the men in the neighborhood were told of the event as an example of what was a very acceptable response in the women's community for that type of male behavior. We kids in the neighborhood grew up with these stories of women giving husbands, bosses, the welfare system, schools, unions and men in general hell, whenever they deserved it. For me there were many role models of women taking action, control and resisting what was supposed to be their lot. Yet many white middle-class feminists continue to view feminism like math homework, where there's only supposed to be one answer. Never occurs to them that they might be talking algebra while working-class women might be talking metaphysics.

Women with backgrounds other than white middle-class experience compounded, simultaneous oppressions. We can't so easily separate our experiences by categories of gender, or race, or class, i.e., "I remember it well: on Saturday, June 3, I was experiencing class oppression, but by Tuesday, June 6, I was caught up in race oppression, then all day Friday, June 9, I was in the middle of gender oppression. What a week!" Sometimes, for example, gender and class reinforce each other. When I returned to college as a single parent after a few years of having kids and working crummy jobs—I went in for vocational testing. Even before I was tested, the white middle-class male vocational counselor looked at me, a welfare mother in my best selection from the Salvation Army racks, and suggested I quit college, go to vo-tech school and become a grocery clerk. This was probably the highest paying female working-class occupation he could think of. The vocational test results suggested I become an attorney. I did end up quitting college once again, not because of his suggestion, but because I was tired of supporting my children in ungenteel poverty. I entered vo-tech school for training as an electrician and, as one of the first women in a non-traditional field, was able to earn a living wage at a job which had traditionally been reserved for white working-class males. But this is a story for another day. Let's return to our little vocational counselor example. Was he suggesting the occupational choice of grocery clerk to me because of my gender or my class? Probably both. Let's imagine for a moment what this same vocational counselor might have advised, on sight only, to the following people:

1. A white middle-class male: doctor, lawyer, engineer, business executive.
2. A white middle-class female: close to the same suggestion as #1 if the counselor was not sexist, or, if sexist, then: librarian, teacher, nurse, social worker.
3. A middle-class man of color: close to the same suggestions as #1 if the counselor was not racist, or, if racist, then: school principal, sales, management, technician.
4. A middle-class woman of color: close to the same suggestions as #3 if counselor was not sexist; #2 if not racist; if not racist or sexist, then potentially #1.
5. A white working-class male: carpenter, electrician, plumber, welder.
6. A white working-class female—well, we already know what he told me, although he could have also suggested secretary, waitress and dental hygienist (except I'd already told him I hated these jobs).
7. A working-class man of color: garbage collector, janitor, fieldhand.
8. A working-class woman of color: maid, laundress, garment worker.

Notice anything about this list? As you move down it, a narrowing of choices, status, pay, working conditions, benefits and chances for promotions occurs. To be connected to any one factor, such as gender or class or race, can make life difficult. To be connected to multiple factors can guarantee limited economic status and poverty.

WAYS TO AVOID FACING CLASSISM

Deny Deny Deny: Deny your class position and the privileges connected to it. Deny the existence or experience of the working class and poor. You can even set yourself up (in your own mind) as judge and jury in deciding who qualifies as working class by your white middle-class standards. So if someone went to college, or seems intelligent to you, not at all like your stereotypes, they must be middle class.

Guilt Guilt Guilt: "I feel so bad, I just didn't realize!" is not helpful, but is a way to avoid changing attitudes and behaviors. Passivity—"Well, what can I do about it, anyway?"—and anger—"Well, what do they want!"—aren't too helpful either. Again, with these responses, the focus is on you and absolving the white middle class from responsibility. A more helpful remedy is to take action. Donate your time and money to local foodbanks. Don't cross picket lines. Better yet, go join a picket line.

HOW TO CHALLENGE CLASSISM

If you're middle class, you can begin to challenge classism with the following:

1. Confront classist behavior in yourself, others and society. Use and share the privileges, like time or money, which you do have.
2. Make demands on working-class and poor communities' issues—anti-racism, poverty, unions, public housing, public transportation, literacy and day care.
3. Learn from the skills and strength of working people—study working and poor people's history; take some Labor Studies, Ethnic Studies, Women Studies classes. Challenge elitism. There are many different types of intelligence: white middle-class, academic, professional intellectualism being one of them (reportedly). Finally, educate yourself, take responsibility and take action.

If you're working class, just some general suggestions (it's cheaper than therapy—free, less time-consuming and I won't ask you about what your mother said to you when you were five):

1. Face your racism! Educate yourself and others, your family, community, any organizations you belong to; take responsibility and take action. Face your classism, sexism, heterosexism, ageism, able-bodiness, adultism. . . .
2. Claim your identity. Learn all you can about your history and the history and experience of all working and poor peoples. Raise your children to be anti-racist, anti-sexist and anti-classist. Teach them the language and culture of working peoples. Learn to survive with a fair amount of anger and lots of humor, which can be tough when this stuff isn't even funny.
3. Work on issues which will benefit your community. Consider remaining in or returning to your communities. If you live and work in white middle-class environments, look for working-class allies to help you survive with your humor and wits intact. How do working-class people spot each other? We have antenna.

We need not deny or erase the differences of working-class cultures but can embrace their richness, their variety, their moral and intellectual heritage. We're not at the point yet where we can celebrate differences—not having money for a prescription for your child is nothing to celebrate. It's not time yet to party with the white middle class, because we'd be the entertainment ("Aren't they quaint? Just love their workboots and uniforms and the way they cuss!"). We need to overcome divisions among working people, not by ignoring the multiple oppressions many of us encounter, or by oppressing each other, but by becoming committed allies on all issues which affect working people: racism, sexism, classism, etc. An injury to one is an injury to all. Don't play by ruling-class rules, hoping that maybe you can live on Connecticut Avenue instead of Baltic, or that you as an individual can make it to Park Place and Boardwalk. Tired of Monopoly? Always ending up on Mediterranean Avenue? How about changing the game?

THE PLIGHT OF BLACK MEN   13

Michael Dyson

On a recent trip to Knoxville, I visited Harold's barber shop, where I had gotten my hair cut during college, and after whenever I had the chance. I had developed a friendship with Ike, a local barber who took great pride in this work. I popped my head inside the front door, and after exchanging friendly greetings with Harold, the owner, and noticing Ike missing, I inquired about his return. It had been nearly two years since Ike had cut my hair, and I was hoping to receive the careful expertise that comes from familiarity and repetition.

"Man, I'm sorry to tell you, but Ike got killed almost two years ago," Harold informed me. "He and his brother, who was drunk, got into a fight, and he stabbed Ike to death."

I was shocked, depressed, and grieved, these emotions competing in rapid-fire fashion for the meager psychic resources I was able to muster. In a daze of retreat from the fierce onslaught of unavoidable absurdity, I half-consciously slumped into Harold's chair, seeking solace through his story of Ike's untimely and brutal leave-taking. Feeling my pain, Harold filled in the

From: Michael Dyson, *Reflecting Black: African-American Cultural Criticism* (Minneapolis: University of Minnesota Press, 1993), pp. 182–94. Reprinted by permission of the author.

details of Ike's last hours, realizing that for me Ike's death had happened only yesterday. Harold proceeded to cut my hair with a methodical precision that was itself a temporary and all-too-thin refuge from the chaos of arbitrary death, a protest against the nonlinear progression of miseries that claim the lives of too many black men. After he finished, I thanked Harold, both of us recognizing that we would not soon forget Ike's life, or his terrible death.

This drama of tragic demise, compressed agony, nearly impotent commiseration, and social absurdity is repeated countless times, too many times, in American culture for black men. Ike's death forced to the surface a painful awareness that provides the chilling sound track to most black men's lives: it is still hazardous to be a human being of African descent in America.

Not surprisingly, much of the ideological legitimation for the contemporary misery of African-Americans in general, and black men in particular, derives from the historical legacy of slavery, which continues to assert its brutal presence in the untold suffering of millions of everyday black folk. For instance, the pernicious commodification of the black body during slavery was underwritten by the desire of white slave owners to completely master black life. The desire for mastery also fueled the severe regulation of black sexual activity, furthering the telos of southern agrarian capital by reducing black men to studs and black women to machines of production. Black men and women became sexual and economic property. Because of the arrangement of social relations, slavery was also the breeding ground for much of the mythos of black male sexuality that survives to this day: that black men are imagined as peripatetic phalluses with unrequited desire for their denied object—white women.

Also crucial during slavery was the legitimation of violence toward blacks, especially black men. Rebellion in any form was severely punished, and the social construction of black male image and identity took place under the disciplining eye of white male dominance. Thus healthy black self-regard and self-confidence were outlawed as punitive consequences were attached to their assertion in black life. Although alternative forms of resistance were generated, particularly those rooted in religious praxis, problems of self-hatred and self-abnegation persisted. The success of the American political, economic, and social infrastructure was predicated in large part upon a squelching of black life by white modes of cultural domination. The psychic, political, economic, and social costs of slavery, then, continue to be paid, but mostly by the descendants of the oppressed. The way in which young black men continue to pay is particularly unsettling.

Black men are presently caught in a web of social relations, economic conditions, and political predicaments that portray their future in rather bleak terms.[1] For instance, the structural unemployment of black men has reached virtually epidemic proportions, with black youth unemployment double that of white youth.* Almost one-half of young black men have had no work experience at all. Given the permanent shift in the U.S. economy from manufacturing and industrial jobs to high-tech and service employment and the flight of these jobs from the cities to the suburbs, the prospects for eroding the stubborn unemployment of black men appear slim.[2]

The educational front is not much better. Young black males are dropping out of school at alarming rates, due to a combination of severe economic difficulties, disciplinary entanglements, and academic frustrations. Thus the low level of educational achievement by young black men exacerbates their already precarious employment situation. Needless to say, the pool of high-school graduates eligible for college has severely shrunk, and even those who go on to college have disproportionate rates of attrition.

Suicide, too, is on the rise, ranking as the third leading cause of death among young black men. Since 1960, the number of black men who have died from suicide has tripled. The homicide rate of black men is atrocious. For black male teenagers and young adults, homicide ranks as the leading cause of death. In 1987, more young black men were killed within the United States in a single year than had been killed abroad in the entire nine years of the Vietnam War. A young black man has a one-in-twenty-one lifetime chance of being killed, most likely at the hands of another black man, belying the self-destructive character of black homicide.

Even with all this, a contemporary focus on the predicament of black males is rendered problematical and ironic for two reasons. First, what may be termed the "Calvin Klein" character of debate about social problems—which amounts to a "designer" social consciousness—makes it very difficult for the concerns of black men to be taken seriously. Social concern, like other commodities, is subject to cycles of production, distribution, and consumption. With the dwindling of crucial governmental resources to address a range of social problems, social concern is increasingly relegated to the domain of private philanthropic and nonprofit organizations. Furthermore, the selection of which problems merit scarce resources is determined, in part, by such philanthropic organizations, which highlight special issues, secure the services of prominent spokespersons, procure capital for research, and distribute the benefits of their information.

Unfortunately, Americans have rarely been able to sustain debate about pressing social problems over long periods of time. Even less have we been able to conceive underlying structural features that bind complex social issues

---

* *Editors' Note: Structural unemployment* is a technical term used by economists; here, Dyson is using the term only to refer to unemployment that is the result of societal conditions.

together. Such conceptualization of the intricate relationship of social problems would facilitate the development of broadly formed coalitions that address a range of social concerns. As things stand, problems like poverty, racism, and sexism go in and out of style. Black men, with the exception of star athletes and famous entertainers, are out of style.

Second, the the irony of the black male predicament is that it has reached its nadir precisely at the point when much deserved attention has been devoted to the achievements of black women like Alice Walker, Toni Morrison, and Terry McMillan.

The identification and development of the womanist tradition in African-American culture has permitted the articulation of powerful visions of black female identity and liberation. Michele Wallace, bell hooks, Alice Walker, Audre Lorde, and Toni Morrison have written in empowering ways about the disenabling forms of racism and patriarchy that persist in white and black communities. They have expressed the rich resources for identity that come from maintaining allegiances to multiple kinship groups defined by race, gender, and sexual orientation, while also addressing the challenges that arise in such membership.

Thus discussions about black men should not take place in an ahistorical vacuum, but should be informed by sensitivity to the plight of black women. To isolate and examine the pernicious problems of young black men does not privilege their perspectives or predicament. Rather, it is to acknowledge the decisively deleterious consequences of racism and classism that plague black folk, particularly young black males.

The aim of my analysis is to present enabling forms of consciousness that may contribute to the reconstitution of the social, economic, and political relations that continually consign the lives of black men to psychic malaise, social destruction, and physical death. It does not encourage or dismiss the sexism of black men, nor does it condone the patriarchal behavior that sometimes manifests itself in minority communities in the form of misdirected machismo. Above all, African-Americans must avoid a potentially hazardous situation that plays musical chairs with scarce resources allocated to black folk and threatens to inadvertently exacerbate already deteriorated relations between black men and women. The crisis of black inner-city communities is so intense that it demands our collective resources to stem the tide of violence and catastrophe that has besieged them.

I grew up as a young black male in Detroit in the 1960s and 1970s. I witnessed firsthand the social horror that is entrenched in inner-city communities, the social havoc wreaked from economic hardship. In my youth, Detroit had been tagged the "murder capital of the world," and many of those murders were of black men, many times by other black men. Night after night, the news media in Detroit painted the ugly picture of a homicide-ridden city caught in the desperate clutches of death, depression, and decay. I remember having recurring nightmares of naked violence, in which Hitchcockian vertigo emerged in Daliesque perspective to produce gun-wielding perpetrators of doom seeking to do me in.

And apart from those disturbing dreams, I was exercised by the small vignettes of abortive violence that shattered my circle of friends and acquaintances. My next-door neighbor, a young black man, was stabbed in the jugular vein by an acquaintance and bled to death in the midst of a card game. (Of course, one of the ugly statistics involving black-on-black crime is that many black men are killed by those whom they know.) Another acquaintance murdered a businessman in a robbery; another executed several people in a gangland-style murder.

At fourteen, I was at our corner store at the sales counter, when suddenly a jolt in the back revealed a young black man wielding a sawed-off, double-barreled shotgun, requesting, along with armed accomplices stationed throughout the store, that we hit the floor. We were being robbed. At the age of eighteen I was stopped one Saturday night at 10:30 by a young black man who ominously materialized out of nowhere, much like the .357 Magnum revolver that he revealed to me in a robbery attempt. Terror engulfed my entire being in the fear of imminent death. In desperation I hurled a protest against the asphyxiating economic hardships that had apparently reduced him to desperation, too, and appealed to the conscience I hoped was buried beneath the necessity that drove him to rob me. I proclaimed, "Man, you don't look like the type of brother that would be doin' something like this."

"I wouldn't be doin' this, man," he shot back, "but I got a wife and three kids, and we ain't got nothin' to eat. And besides, last week somebody did the same thing to me that I'm doin' to you." After convincing him that I really only had one dollar and thirty-five cents, the young man permitted me to leave with my life intact.

The terrain for these and so many other encounters that have shaped the lives of black males was the ghetto. Much social research and criticism has been generated in regard to the worse-off inhabitants of the inner city, the so-called underclass. From the progressive perspective of William Julius Wilson to the archconservative musings of Charles Murray, those who dwell in ghettos, or enclaves of civic, psychic, and social terror, have been the object of recrudescent interest within hallowed academic circles and governmental policy rooms.[3] In most cases they have not fared well and have borne the brunt of multifarious "blame the victim" social logics and policies.

One of the more devastating developments in inner-city communities is the presence of drugs and the criminal activity associated with their production, marketing, and consumption. Through the escalation of the use of the rocklike form of cocaine known as "crack" and intensified related gang activity, young black men are involved in a vicious subculture of crime. This subculture is sustained by two potent attractions: the personal acceptance and affirmation gangs offer and the possibility of enormous economic reward.

U.S. gang life had its genesis in the Northeast of the 1840s, particularly in the depressed neighborhoods of Boston and New York, where young Irishmen developed gangs to sustain social solidarity and to forge a collective identity based on common ethnic roots.[4] Since then, youth of every ethnic and racial origin have formed gangs for similar reasons, and at times have even functioned to protect their own ethnic or racial group from attack by harmful outsiders. Overall, a persistent reason for joining gangs is the sense of absolute belonging and unsurpassed social love that results from gang membership. Especially for young black men whose lives hold a low premium in America, gangs have fulfilled a primal need to possess a sense of social cohesion through group identity. Particularly when traditional avenues for the realization of personal growth, esteem, and self-worth, usually gained through employment and career opportunities, have been closed, young black men find gangs a powerful alternative.

Gangs also offer immediate material gratification through a powerful and lucrative underground economy. This underground economy is supported by exchanging drugs and services for money, or by barter. The lifestyle developed and made possible by the sale of crack presents often irresistible economic alternatives to young black men frustrated by their own unemployment. The death that can result from involvement in such drug- and gang-related activity is ineffective in prohibiting young black men from participating.

To understand the attraction such activity holds for black men, one must remember the desperate economic conditions of urban black life. The problems of poverty and joblessness have loomed large for African-American men, particularly in the Rust Belt, including New York, Chicago, Philadelphia, Detroit, Cleveland, Indianapolis, and Baltimore. From the 1950s to the 1980s, there was severe decline in manufacturing and in retail and wholesale trade, attended by escalating unemployment and a decrease in labor-force participation by black males, particularly during the 1970s.

During this three-decade decline of employment, however, there was not an expansion of social services or significant increase in entry-level service jobs. As William Julius Wilson rightly argues, the urban ghettos then became more socially isolated than at any other time. Also, with the mass exodus of black working- and middle-class families from the ghetto, the inner city's severe unemployment and joblessness became even more magnified. With black track from the inner city mimicking earlier patterns of white flight, severe class changes have negatively affected black ghettos. Such class changes have depleted communities of service establishments, local businesses, and stores that could remain profitable enough to provide full-time employment so that persons could support families, or even to offer youths part-time employment in order to develop crucial habits of responsibility and work. Furthermore, ghetto residents are removed from job networks that operate in more affluent neighborhoods. Thus, they are deprived of the informal contact with employers that results in finding decent jobs. All of these factors create a medium for the development of criminal behavior by black men in order to survive, ranging from fencing stolen goods to petty thievery to drug dealing. For many black families, the illegal activity of young black men provides their only income.

Predictably, then, it is in these Rust Belt cities, and other large urban and metropolitan areas, where drug and gang activity has escalated in the past decade. Detroit, Philadelphia, and New York have had significant gang and drug activity, but Chicago and Los Angeles have dominated of late. Especially in regard to gang-related criminal activity such as homicides, Chicago and L.A. form a terrible one-two punch. Chicago had 47 gang-related deaths in 1987, 75 in 1986, and 60 in 1988. L.A.'s toll stands at 400 for 1988.

Of course, L.A.'s gang scene has generated mythic interpretation in the Dennis Hopper film *Colors*. In the past decade, gang membership in L.A. has risen from 15,000 to almost 60,000 (with some city officials claiming as much as 80,000), as gang warfare claims one life per day. The ethnic composition of the groups include Mexicans, Armenians, Samoans, and Fijians. But gang life is dominated by South Central L.A. black gangs, populated by young black men willing to give their lives in fearless fidelity to their group's survival. The two largest aggregates of gangs, composed of several hundred microgangs, are the Bloods and the Crips, distinguished by the colors of their shoelaces, T-shirts, and bandannas.

The black gangs have become particularly dangerous because of their association with crack. The gangs control more than 150 crack houses in L.A., each of which does over five thousand dollars of business per day, garnering over half a billion dollars per year. Crack houses, which transform powdered cocaine into crystalline rock form in order to be smoked, offer powerful material rewards to gang members. Even young teens can earn almost a thousand dollars a week, often outdistancing what their parents, if they work at all, can earn in two months.

So far, most analyses of drug gangs and the black youth who comprise their membership have repeated old saws about the pathology of black culture and weak family structure, without accounting for the pressing economic realities and the need for acceptance that help explain such activity. As long as the poverty of young black men is ignored, the disproportionate number of black unemployed males is overlooked, and the structural features of racism and classism are avoided, there is room for the proliferation of social explanations that blame the victim. Such social explanations reinforce the misguided efforts of public officials to stem the tide of illegal behavior by state repression aimed at young blacks, such as the sweeps of L.A. neighborhoods resulting in mass arrests of more than four thousand black men, more than at any time since the Watts rebellion of 1965.

Helpful remedies must promote the restoration of job training (such as Neighborhood Youth Corps [NYC] and the Comprehensive Employment and Training Act [CETA]); the development of policies that support the family, such as child-care and education programs; a full employment policy; and dropout prevention in public schools. These are only the first steps toward the deeper structural transformation necessary to improve the plight of African-American men, but they would be vast improvements over present efforts.

Not to be forgotten, either, are forms of cultural resistance that are developed and sustained within black life and are alternatives to the crack gangs. An example that springs immediately to mind is rap music. Rap music provides space for cultural resistance to the criminal-ridden ethos that pervades segments of many underclass communities. Rap was initially a form of musical play that directed the creative urges of its producers into placing often humorous lyrics over the music of well-known black hits.

As it evolved, however, rap became a more critical and conscientious forum for visiting social criticism upon various forms of social injustice, especially racial and class oppression. For instance, Grandmaster Flash and the Furious Five pioneered the social awakening of rap with two rap records, "The Message" and "New York, New York." These rap records combined poignant descriptions of social misery and trenchant criticism of social problems as they remarked upon the condition of black urban America. They compared the postmodern city of crime, deception, political corruption, economic hardship, and cultural malaise to a "jungle." These young savants portrayed a chilling vision of life that placed them beyond the parameters of traditional African-American cultural resources of support: religious faith, communal strength, and familial roots. Thus, they were creating their own aesthetic of survival, generated from the raw material of their immediate reality, the black ghetto. This began the vocation of the rap artist, in part, as urban grist dispensing social and cultural critique.

Although rap music has been saddled with a reputation for creating violent outbursts by young blacks, especially at rap concerts, most of rap's participants have repeatedly spurned violence and all forms of criminal behavior as useless alternatives for black youth. Indeed rap has provided an alternative to patterns of identity formation provided by gang activity and has created musical vehicles for personal and cultural agency. A strong sense of self-confidence permeates the entire rap genre, providing healthy outlets for young blacks to assert, boast, and luxuriate in a rich self-conception based upon the achievement that their talents afford them. For those reasons alone, it deserves support. Even more, rap music, although its increasing expansion means being influenced by the music industry's corporate tastes and decisions, presents an economic alternative to the underground economy of crack gangs and the illegal activity associated with them.

However, part of the enormous difficulty in discouraging illegal activity among young black men has to do, ironically, with their often correct perception of the racism and classism still rampant in employment and educational opportunities open to upwardly mobile blacks. The subtle but lethal limits continually imposed upon young black professionals, for instance, as a result of the persistence of racist ideologies operating in multifarious institutional patterns and personal configurations, send powerful signals to young black occupants of the underclass that education and skill do not ward off racist, classist forms of oppression.

This point was reiterated to me upon my son's recent visit with me at Princeton near Christmas. Excited about the prospect of spending time together catching up on new movies, playing video games, reading, and the like, we dropped by my bank to get a cash advance on my MasterCard. I presented my card to the young service representative, expecting no trouble since I had just paid my bill a couple of weeks before. When he returned, he informed me that not only could I not get any money, but that he would have to keep my card. When I asked for an explanation, all he could say was that he was following the instructions of my card's bank, since my MasterCard was issued by a different bank.

After we went back and forth a few times about the matter, I asked to see the manager. "He'll just tell you the same thing that I've been telling you," he insisted. But my persistent demand prevailed, as he huffed away to the manager's office, resentfully carrying my request to his boss. My son, sitting next to me the whole time, asked what the problem was, and I told him that there must be a mistake, and that it would all be cleared up shortly. He gave me that confident look that says, "My dad can handle it." After waiting for about seven or eight minutes, I caught the manager's figure peripherally, and just as I turned, I saw him heading with the representative to an empty desk, opening the drawer and pulling out a pair of scissors. I could feel the blood begin to boil in my veins as I beseeched the manager, "Sir, if you're about to do what I fear you will, can we please talk first?" Of course, my request was to no avail, as he sliced my card in two before what had now become a considerable crowd. I immediately jumped up and followed him into his office, my son trailing close behind, crying now, tearfully pumping me with "Daddy, what's going on?"

I rushed into the manager's office and asked for the privacy of a closed door, to which he responded, "Don't let him close the door," as he beckoned three other employees into his office. I angrily grabbed the remnants of my card from his hands and proceeded to tell him that I was a reputable member of the community and a good customer of the bank and that if I had been wearing a three-piece suit (instead of the black running suit I was garbed in) and if I had been a white male (and not a black man) I would have been at least accorded the respect of a conversation, prior to a private negotiation of an embarrassing situation, which furthermore was the apparent result of a mistake on the bank's part.

His face flustered, the manager then prominently positioned his index finger beneath his desk drawer, and pushed a button, while declaring, "I'm calling the police on you." My anger now piqued, I was tempted to vent my rage on his defiant countenance, arrested only by the vision terrible that flashed before my eyes as a chilling premonition of destruction: I would assault the manager's neck; his co-workers would join the fracas, as my son stood by horrified by his helplessness to aid me; the police would come, and abuse me even further, possibly harming my son in the process. I retreated under the power of this proleptic vision, grabbing my son's hand as I marched out of the bank. Just as we walked through the doors, the policemen were pulling up.

Although after extensive protests, phone calls, and the like, I eventually received an apology from the bank's board and a MasterCard from their branch, this incident seared an indelible impression onto my mind, reminding me that regardless of how much education, moral authority, or personal integrity a black man possesses, he is still a "nigger," still powerless in many ways to affect his destiny.

The tragedy in all of this, of course, is that even when articulate, intelligent black men manage to rise above the temptations and traps of "the ghetto," they are often subject to continuing forms of social fear, sexual jealousy, and obnoxious racism. More pointedly, in the 1960s, during a crucial stage in the development of black pride and self-esteem, highly educated, deeply conscientious black men were gunned down in cold blood. This phenomenon finds paradigmatic expression in the deaths of Medgar Evers, Malcolm X, and Martin Luther King, Jr. These events of public death are structured deep in the psyches of surviving black men, and the ways in which these horrible spectacles of racial catastrophe represent and implicitly sanction lesser forms of social evil against black men remains hurtful to black America.

I will never forget the effect of King's death on me as a nine-year-old boy in Detroit. For weeks I could not be alone at night before an open door or window without fearing that someone would kill me, too. I thought that if they killed this man who taught justice, peace, forgiveness, and love, then they would kill all black men. For me, Martin's death meant that no black man in America was safe, that no black man could afford the gift of vision, that no black man could possess an intelligent fire that would sear the fierce edges of ignorance and wither to ashes the propositions of hate without being extinguished. Ultimately Martin's death meant that all black men, in some way, are perennially exposed to the threat of annihilation.

As we move toward the last decade of this century, the shadow of Du Bois's prophetic declaration that the twentieth century's problem would be the color line continues to extend itself in foreboding manner. The plight of black men, indeed, is a microcosmic reflection of the problems that are at the throat of all black people, an idiomatic expression of hurt drawn from the larger discourse of racial pain. Unless, however, there is vast reconstitution of our social, economic, and political policies and practices, most of which target black men with vicious specificity, Du Bois's words will serve as the frontispiece to the racial agony of the twenty-first century, as well.

NOTES

1. For a look at the contemporary plight of black men, especially black juvenile males, see *Young, Black, and Male in America: An Endangered Species*, ed. Jewelle Taylor Gibbs (Dover, Mass.: Auburn House, 1988).

2. See William Julius Wilson, *The Truly Disadvantaged: The Inner City, the Underclass, and Public Policy* (Chicago: University of Chicago Press, 1987).

3. See Wilson, *The Truly Disadvantaged*. For Charles Murray's views on poverty, welfare, and the ghetto underclass, see his influential book, *Losing Ground: American Social Policy, 1950–1980* (New York: Basic Books, 1984).

4. This section on gangs is informed by the work of Mike Davis in his *City of Quartz* (New York: Verso Press, 1991).

THE SILENCED MAJORITY:

14

Why the Average Working Person Has Disappeared from American Media and Culture

Barbara Ehrenreich

It is possible for a middle-class person today to read the papers, watch television, even go to college, without suspecting that America has any inhabitants other than white-collar people—and, of course, the annoyingly persistent "black underclass." The average American has disappeared—from the media, from intellectual concern, and from the mind of the American middle class. The producers of public affairs talk shows do not blush to serve up four upper-income professionals (all, incidentally, white, male, and conservative) to ponder the minimum wage or the possible need for national health insurance. Never, needless to say, an uninsured breadwinner or an actual recipient of the minimum wage. Working-class people are likely to cross the screen only as witnesses to crimes or sports events, never as commentators or—even when their own lives are under discussion—as "experts."

A quick definition: By "working class" I mean not only industrial workers in hard hats, but all those people who are not professionals, managers, or entrepreneurs; who work for wages rather than salaries; and who spend their working hours variously lifting, bending, driving, monitoring, typing, keyboarding, cleaning, providing physical care for others, loading, unloading, cooking, serving, etc. The working class so defined makes up 60 to 70 percent of the U.S. population.

By "middle class" I really mean the "professional middle class," or the "professional-managerial class." This group includes the journalists, professors, media executives, etc. who are responsible, in a day-to-day sense, for what we do or do not see or read about in the media. By this definition, the middle class amounts to no more than 20 percent of the U.S. population.

So when I say the working class is disappearing, I do not mean just a particular minority group favored, for theoretical reasons, by leftists. I mean the American majority. And I am laying the blame not only on the corporate sponsors of the media, but on many less wealthy and powerful people. Media

From: *Zeta Magazine* 2 (September 1989): 22. Reprinted by permission.

people for example. People who are, by virtue of their lifestyles and expectations, not too different from me, and possibly also you.

The disappearance of the working class reflects—and reinforces—the long-standing cultural insularity of the professional middle class. Compared to, say, a decade ago, the classes are less likely to mix in college (due to the decline of financial aid), in residential neighborhoods (due to the rise in real estate prices), or even in the malls (due to the now almost universal segmentation of the retail industry into upscale and downscale components).

In the absence of real contact or communication, stereotypes march on unchallenged; prejudices easily substitute for knowledge. The most intractable stereotype is of the working class (which is, in imagination, only white) as a collection of reactionaries and bigots—reflected, for example, in the use of the terms "hard hat" or "redneck" as class slurs. Even people who call themselves progressives are not immune to this prejudice.

The truth is that, statistically and collectively, the working class is far more reliably liberal than the professional middle class. It was more, not less, opposed to the war in Vietnam. It is more, not less, disposed to vote for a Democrat for president. And thanks to the careful, quantitative studies of Canadian historian Richard F. Hamilton, we know that the white working class (at least outside the South) is no more racist, and by some measures less so, than the white professional class.

Even deeper-rooted than the stereotype of the hard-hat bigot is the middle-class suspicion that the working class is dumb, inarticulate, and mindlessly loyal to old-fashioned values. In the entertainment media, for example, the working class is usually portrayed by macho exhibitionism (from *Saturday Night Fever* to *Working Girl*) or mental inferiority (*Married, With Children*). Mainstream sociologists have reinforced this prejudice with their emphasis on working-class "parochialism," as illustrated by this quote from a 1976 beginning sociology textbook: "The limited education, reading habits and associations isolate the lower class. . . . and this ignorance, together with their class position, makes them suspicious of [the] middle- and upper-class 'experts' and 'do-gooders'. . . ."

Finally, there is a level of prejudice that grows out of middle-class moralism about matters of taste and lifestyle. All privileged classes seek to differentiate themselves from the less-privileged through the ways they dress, eat, entertain themselves, and so on, and tend to see their own choices in these matters as inherently wiser, better, and more aesthetically inspired. In middle-class stereotype, the white working class, for example, is addicted to cigarettes, Budweiser, polyester, and network television. (In part this is true, and it is true in part because Bud is cheaper than Dos Equis and polyester is cheaper than linen.) Furthermore in the middle-class view, polyester and the like is "tacky"—a common code word for "lower class." Health concerns, plus a certain reverence for the "natural" in food and fiber, infuse these middle-class prejudices with a high-minded tone of moral indignation.

But I am alarmed by what seems to me to be the growing parochialism of the professional middle class—living in its own social and residential enclaves, condemned to hear only the opinions of its own members (or, of course, the truly rich), and cut off from the lives and struggles and insights of the American majority. This middle-class parochialism is insidiously self-reinforcing: The less "we" know about "them," the more likely "we" are to cling to our stereotypes—or forget "them" altogether.

FAMILY MATTERS, WORK MATTERS?

Poverty Among Women of Color and White Women

15

Lisa Catanzarite and Vilma Ortiz

Women's poverty is at the center of the current welfare reform debate. Women and their children do make up an increasing proportion of the population of poor people in the United States, and female-headed families are on the rise. The ample attention is also because of suspicion among some public policymakers that the welfare system itself may have exacerbated these trends (and this emphasis is consistent with the increasing popularity of "less government"). As a result, much of the discussion of policy innovations focuses on getting poor mothers off Aid to Families with Dependent Children (AFDC) and into the labor force. Additionally, a great deal of rhetoric has been devoted to "family values." Poor women (and perhaps women in general) are targeted with messages meant to reform behavior deemed inappropriate. The word is:

From: Diane Dujon and Ann Withorn, eds., *For Crying Out Loud: Women's Poverty in the United States* (Boston: South End Press, 1996), pp. 121–39.

This piece is an extension of a more technical article, "Racial/Ethnic Differences in the Impact of Work and Family on Women's Poverty," *Research in Politics and Society*, 1995 (5): 217–37. Reprinted by permission.

Get married or stay married; moreover, if you aren't married or won't stay married, then *do not* have children.

Female poverty, however, is complex. African-American and Latina women are more likely to be poor than are white women, and are likely to stay poor for longer stretches of time than are whites. Poverty policy will thus have a potentially greater effect on minority women and their children than on white women. In order to adequately design public policies that will help all poor women, we need a better understanding of the differences that are most important for poverty reduction and prevention. The key questions we address in this chapter are: Do work and family matter in the same way for different ethnic groups? Are working white women less likely to be poor than working Latinas with similar backgrounds? Does marriage reduce the likelihood of poverty more for white women than for comparable African Americans? How can public policy successfully address the critical issues that poor and working-poor women face?

FAMILY COMPOSITION, POVERTY, AND RACE/ETHNICITY

A number of factors are important in understanding poverty conditions among black, Latina, and white women. For example, minority women are more likely to have *grown up* poor than are whites. Lower socioeconomic backgrounds have a number of ramifications for upward mobility, such as depressed levels of educational attainment and (relatedly, in these days of rising college costs) fewer resources of the extended family network. Further, among some groups, particularly Latina immigrants, larger families are liable to contribute to higher poverty rates. Collective differences in poverty backgrounds of differing racial/ethnic groups result in part from underlying disparities in class background, family composition, English proficiency, and other characteristics. In order to get a clear understanding of racial and ethnic differences in poverty risk, we need to compare individuals of different ethnic groups who are otherwise similar. That is, we need to understand better whether racial and ethnic differences are due solely to factors such as education levels, work behavior, and the like, or whether such differences persist without these factors. Moreover, we must ask whether certain key poverty "prevention" strategies are more effective for whites than for women of color.

One reason commonly thought to explain higher poverty rates for minority women is their marital status and family composition, particularly their higher rates of family dissolution and female headship. Yet marital disruption is most significant in causing poverty for white women. Marital breakups are less likely to *cause* poverty for minority than for white women, since the women of color are more likely to be poor *within* marriages. Bane (1986) demonstrates that, among women who were impoverished after a transition from a male-headed to a female-headed household, only 24 percent of whites were indigent prior to the transition vs. a full 62 percent of African Americans. Leaving marriages is less of an economic disaster for black women than for white women because black women's economic status within marriages is already relatively poor. This is due in part to the lower average material resources of black than white men. While we expect that the same is true for Latinas, this has not been examined empirically. We expect that being married (vs. single) is less economically advantageous to African-American and Latina women than to whites. Conversely, being in a disrupted family (vs. being married) should be more harmful for whites than for minority women.

The discussion of black women's poverty has focused largely on the decline of nuclear families in the African-American community. A number of prominent authors have cited the rise in marital dissolution and female-headed families and connected these increases with the problem of minority male joblessness (Moynihan 1965; Garfinkel and McLanahan 1986) and the scarcity of "marriageable men" (Wilson and Neckerman 1986; Wilson 1987). Minority male joblessness and underemployment certainly contribute to indigence among minority women. Improving the employment opportunities of minority men will help to reduce women's poverty, but only indirectly, by providing women with better marriage prospects and male family heads. The underlying assumptions of these prescriptions are clearly patriarchal in nature. If women could only attach themselves to well-paid men, women's (and children's) poverty would be reduced. The reality, however, is that women, especially minority women, increasingly head families. And, they do so for a variety of reasons, only one of which is the employment status of minority men. Individual women may or may not prefer to be married. Rather than prescribing what should be the ideal family (as in recent Republican platforms) or lamenting changing patterns of family composition, discussion and policy concerning women's poverty should shift focus.

WORK, POVERTY, AND RACE/ETHNICITY

Given the reality of female-headed families, as well as the importance of women's earnings in poor two-parent and other families, we need greater attention to the conditions of women's employment—in particular, their jobs and wages. Certainly, much of the current debate about welfare—but not poverty—centers on forcing women to work more and receive lower AFDC benefits (see, Mead, 1986; more recently, California Governor Pete Wilson's 1996 State of the State address and proposed budget). But this policy, without a concomitant commitment to improving women's labor market locations and earnings, will not alleviate poverty, even if it does reduce welfare. In fact, it would surely worsen poverty for most recipients, as many women and children would be forced to forego medical insurance and scramble for childcare. Both welfare policy and labor market policy must recognize the increasing extent to which many women, especially lower-class women, have become critically important and often the primary economic providers for their families. Moreover, policymakers must begin to address the fact that women continue to be at a severe labor market disadvantage, their vital economic role notwithstanding.

Women's employment and earnings are critically important, both for married and unmarried women. The male "family wage" is no longer a reality (despite the fact that men's wages continue to be much higher than women's). With economic restructuring and the decline in real wages that began in the 1970s, most men now do not earn a wage sufficient to support a wife and children. For this and other reasons, the majority of married women, even those with infants under the age of one, are in the labor force.[1] Further, because of the rise in female-headed families, and the abysmal levels of child support payments, women are increasingly responsible not only for their own support, but also for that of their children. The problems of such families are heightened, as women face much worse prospects than men of securing a "family wage." Of course, this is even more important for African-American women than whites, since black women are much less likely to be married, and when married, contribute a higher proportion of family income than do white women.

Female and feminist scholars have argued that women's wages and labor market locations are critical to reducing poverty among female-headed families (see Smith 1984; Pearce 1987; Peterson 1987). A number of such authors have cited the problem of occupational segregation as a contributor to women's poverty (see Ehrenreich 1986; Pearce 1986; Amott and Matthaei 1986). And, while women in general are disadvantaged in the labor market, the disadvantage is more severe for minority women. However, the extent to which work effort has a greater impact on reducing indigence for white as opposed to minority women has not received sufficient attention. If a minority woman and a white woman (who "look" alike in every way except ethnicity) work the same number of hours, do they have the same chances of staying out of poverty? Further, the impact of occupational segregation on individual poverty has not been tested. If women work in low-level occupations with an over-representation of minority women, do they earn less than similar women in other occupations?

Certainly, work effort—working longer hours—improves earnings, and therefore lowers the likelihood of poverty. But, we expect that minority women get a lower payoff for their efforts than do white women. Wilson (1987) has argued that large scale economic changes have lessened opportunities for some segments of the minority population, particularly inner-city blacks. While traditional sociological theory predicts a direct relationship between level of work and economic status, we expect that this relationship is weaker among black and Latina women that among white women. More specifically, since minority women are likely to receive lower hourly wages than are white women with similar individual characteristics (for example, education and work experience), working longer hours will have a smaller payoff for minority than for white women as a poverty prevention strategy.

In addition, we think that minority women's heavier concentration than white women in marginal, female-dominated occupations contributes to their disadvantage at work.[2] Earnings tend to be lowest in occupations where minority women, particularly black, Latina, and Native American women, prevail (Dill, Cannon, and Vanneman 1987). Indeed, Dill (1987) demonstrate that the wage disadvantages, relative to workers' education and experience, are greater in occupations with a large contingent of minority women than in white-female-dominated occupations. Thus, concentration in poorly paid occupations contributes to the labor market disadvantage that minority women face. We expect that, among similar workers (those with similar hours, education, etc.), being in an occupation with a heavy concentration of minority women carries an increased risk of poverty, primarily because of low remuneration in such occupations.

OVERVIEW

In order to investigate these questions, we analyzed the incidence of poverty for black, Mexican-origin, "other" Latina, and white women in the Los Angeles metropolitan area using the 1980 census.[3] We focused only on women with 12 years of education or less because this is the population most at risk of impoverishment. The analysis had two main parts. First, we examined racial and ethnic differences in the impact of work and marital status on the probability of being needy. Then we looked only at employed women and asked whether or not working in a low-level, minority female occupation increases the likelihood of poverty. We looked for *net effects*, that is, we took into account other individual and occupational characteristics in estimating the effects of marital status, work effort, and occupational segregation. . . .

RESULTS

. . . [Poverty] is much more prevalent among African American women and recent immigrant Latinas (both Mexican and "other") than among whites and earlier-immigrant native Latinas. In our sample, black women and recent immigrant Latinas had poverty rates of 42–45 percent, while 15 percent of white women were impoverished. Earlier-immigrant native Latinas had poverty levels that fell between these two extremes (28 percent of Mexicans and 24 percent of other Latinas were poor). Hence, black women were almost 3 times as likely to be poor as whites; Mexicans had poverty rates between 2 and 3 times those for white women; and other Latinas had rates of impoverishment that were approximately 1.5 to 3 times higher than for whites. . . .[4]

These differences in poverty rates among women of different groups are due, in part, to divergences in other characteristics (for example, lower educational levels, more children among minority than white women). . . .

Recent immigrant Latinas had a large educational disadvantage. Not having completed high school was most common for these women (Mexicans, 87 percent; other Latinas, 70 percent). Native-born, earlier-immigrant Latinas did somewhat better (58 percent of Mexicans and 44 percent of other Latinas) than recent immigrants. Whites had far and away the highest rates of high school completion: only 26 percent had less than 12 years of education. And, black women were a distant second, with 40 percent (of those with 12 years of education or less) being high school dropouts.

Most women were in the labor force, and this was true for all groups. Not working was most common among Mexicans (49 percent of recent immigrants and 44 percent of other Mexicans were not in the labor force) and blacks (44 percent didn't work); staying out of the labor force was least common for other Latinas and white women (37–38 percent). The pattern of full-time work was the converse of this: rates of full-time work were lowest for Mexicans (18 percent among recent immigrants and 21 percent for natives/earlier immigrants), then blacks (23 percent), other Latinas (23–25 percent), and whites (26 percent). So, while working was the norm among these less-educated women, full-time employment was uncommon. Even among white women, only about one-fourth worked full-time.

The youngest group in the sample was Mexicans, followed by recent-immigrant other Latinas, African Americans, earlier-immigrant native other Latinas, and, finally, whites. The population of recent-immigrant Mexican women was overwhelmingly under 40 years old (81 percent), while just under half of white women (46 percent) were in this younger age group.[5]

African Americans were, by far, the least likely to be married—less than half of black women in our sample were wed (41 percent). Among all other groups, being married was the norm. For other Latinas, the prevalence of marriage was 59–62 percent (lower for recent immigrants); among whites, two-thirds were married; among Mexicans, 65 percent of earlier immigrants/ natives and 71 percent of recent immigrants were married. In conjunction with high marriage rates, Mexican women had the highest number of children (an average of 2 for recent immigrants, 1.4 among other Mexicans); next were other Latinas and blacks (just over 1 child), then whites (with an average of 0.7 children).

These descriptive statistics show large variations by ethnic/immigrant group. With respect to most characteristics thought to contribute to keeping women out of poverty, white women appear to enjoy an advantage. (The only exception is that marriage rates are higher among one group—recent-immigrant Mexicans—than among whites). Given these differences in background characteristics, it is perhaps no surprise that white women show lower rates of poverty than do minority women. But, are the racial and ethnic differences in poverty due solely to these differences, or are women of color still more likely to be impoverished? We now turn to the following questions:

1. Among women with similar characteristics (education, etc.), are minority women more likely to be poor than white women?
2. Do white women obtain greater economic benefits from marriage than do similarly situated minority women? Conversely, is marital disruption more economically damaging to whites than to women of color?
3. Do white women derive greater economic gains from work than do similarly situated minority women?
4. Among working women, does location in an occupation identified as minority, female contribute to poverty?

Black women were just over twice as likely to be poor as were similarly situated white women. Mexicans were 1.4 times as likely to live in poverty as their white counterparts, and other Latinas had indigence rates that were 1.6

times those for comparable whites. As expected, white women were considerably less likely to be impoverished than minority women with similar backgrounds, family composition, and work effort. . . .

. . . [S]ingle white women had poverty rates 8 times as high as their married counterparts, while single minority women were (only) 4 times as likely to be poor as married minorities. So getting married offered greater financial benefits for whites than for comparable African Americans and Latinas, as predicted.

Poverty risk was highest for separated women, and the divergence between whites and minorities was greatest for this group. White women who were separated had predicted poverty rates 11 times the rates for married whites; the contrast for minorities is a factor of 5. The same pattern holds, but is less pronounced, for divorced women. As we posited, the economic disadvantage that comes with a disrupted marriage is much greater for whites than for black or Latina women. This is presumably because the difference in family income resulting from a marital breakup is greater for whites than for women of color. The literature on the feminization of poverty has emphasized the devastating impact of separation and divorce on women's poverty. Our findings clearly suggest that this phenomenon is more pronounced for white than minority women. . . .

White women who were out of the labor force were 11 times more likely to live in poverty than whites who worked full-time, while relative risk for minorities was only 5 times that of their full-time counterparts. While the risk of poverty was strongly related to employment for both whites and minorities, the effect of working less than full-time (vs. full-time) was always higher for white women than for minorities. As predicted, white women's work effort had a greater impact on reducing the risk of poverty than was true for African-American and Latina women.

We also found that, among working women, the likelihood of being poor was higher for women in occupations with a prevalence of African-American women, without these factors. Additionally, women employed in occupations that are white-female-dominated were *less likely* to be needy than other, similarly situated women.

SUMMARY

Neither the marriage nor the labor market offers the same rewards to women of color as to white women. While family and work clearly matter a great deal for women's poverty, they matter *differently* for black and Latina women than for whites. In point of fact, they matter *less*. Minority women are more likely to be poor than are comparable whites of the same marital status at every level of work effort.

We found that marital status had less relevance in determining whether or not minority women were poor than was true for similar whites. Being married (vs. single) reduces indigence less for women of color than for white women. The smaller advantage of being married for blacks and Latinas is doubtlessly due largely to the fact that the earnings of minority husbands are generally lower than white men's pay. Similarly, because the change in family income resulting from a marital breakup tends to be less pronounced for minority than white women, we found that being separated or divorced (vs. being married) was less financially damaging for African-American and Latina women.

In addition to these differences in the relative advantages of marriage, we found that work effort was more effective in reducing the risk of privation for white women than for similar women of color, as predicted. The benefits of time spent working were clearly greater for white than minority women at every level of work effort.

Above and beyond individual-level factors, the occupational locations of different groups of women contributed to racial and ethnic differences in poverty risk. Working in an occupation with a large percentage of black women increased the likelihood of impoverishment, while being in a white-female-dominated occupation lowered this risk.

CONCLUSIONS

The two major routes out of poverty for women are work and marriage. Should it come as a surprise that the labor market is not a "level playing field" for women of color and white women? We think not. And, though we may not often conceptualize the *marriage* market in these terms, it appears that this market also is not a level field for women of different racial/ethnic groups.

Our findings underscore the importance of these differences. While marital status is related to privation for all women, the emphasis on marriage in discussions of women's poverty is less relevant to black women and Latinas than to white women. Accordingly, as a poverty prevention "strategy," getting married is less beneficial for African-American and Latina women than for whites. Further, popular claims that every woman is "just a divorce away" from penury are particularly misplaced for minority women, who experience relatively high rates of poverty even within marriages, and whose risk of indigence increases less dramatically with a family breakup than is true for whites.

Perhaps of greater importance, given the heavy emphasis on women's work in the current welfare reform debate, are our results regarding work effort: we find that minority women profit less from hours expended at work than do white women. Further, the occupational locations of women contribute to differences in poverty: women who work in heavily black female occupations have a higher risk of indigence than do similar women in comparable occupations. Occupational segregation contributes to minority women's disadvantage (in addition to lower returns to work effort for minorities). The fact that renumeration varies across occupations and tends to be depressed in low-level, minority female occupations highlights our contention that occupational segregation is a significant *poverty* issue for less-educated women.

Although seemingly paradoxical, working is of monumental importance to the economic standing of women of color—despite the fact that minorities derive less gain from time spent working than whites. This is precisely because *marriage* offers less monetary benefit to women of color than to white women. Particularly for less-educated African-American women—who are relatively unlikely to be married—work is a critical poverty prevention strategy.

Attention to the conditions of women's work is of utmost significance, particularly in the current environment, where pundits across the political spectrum emphasize work as the panacea for women's poverty. Our results strongly suggest that public policymakers concerned with indigence among working-age women and their children cannot simply focus on pushing women into the labor market and reducing dependency on public assistance. By doing so, they would simply replace an inadequate *welfare* check with an inadequate *pay* check. This would be further compounded by increased responsibility for medical insurance and childcare, whose costs are prohibitive for at-risk women. Clearly, any antipoverty policy that has women's work as its cornerstone must give attention to adequate childcare and affordable health care.

First and foremost, however, in order to be efficacious, an antipoverty strategy grounded on female employment should focus on *making women's work pay*. In the current climate, where the future of affirmative action is uncertain and employment discrimination legislation is being redefined and narrowed, we must not lose sight of the employment disadvantages encountered by women, particularly women of color. In order to effectively reduce poverty, it is essential that policymakers direct their attention to the labor market. Pivotal to the success of antipoverty efforts will be policies that break down occupational segregation; increase the minimum wage; reduce the financial penalty in marginal, female-dominated occupations; and increase women's—especially minority women's—access to jobs that offer better pay in return for women's work effort.

NOTES

1. By 1988, labor force participation (LFP) rates of married women with children under one were already 50.5 percent for whites and 71.5 percent for blacks (51.9 percent overall) (*Statistical Abstract of the United States*, 1990, p. 385). In light of recent data on the rising trend in LFP for this group (the latest historically to enter the labor force), rates are certainly higher now. Working has become the norm.

2. The heavy concentration of black women in female-dominated occupations is discussed in Malveaux and Wallace (1987) and Malveaux (1985). Catanzarite's (1990) analysis of national data for the 1970s suggests that Latinas may have occupied an intermediate position between black and white women in occupational segregation; that is, they appear to have been concentrated in better jobs than black women, but worse positions than whites. The relative positions of blacks and Latinas may have flipped with the recent increase in immigration.

3. The area includes Los Angeles, Orange, Riverside, San Bernardino, and Ventura counties. The Census file is the 5 percent Public Use Microdata Sample. The sample is comprised of 98,896 women, ages 20–64.

4. These figures are obtained by dividing minority women's poverty rates by those for white women.

5. Recall that the sample is limited to working-age women: 20- to 64 year-olds.

References

Amott, Teresa, and Julie Matthaei. 1986. "Comparable Worth, Incomparable Pay" in Rochelle Lefkowitz and Ann Withorn (eds.), *For Crying Out Loud: Women and Poverty in the United States*. New York: Pilgrim Press, pp. 314–23.

Bane, Mary Jo. 1986. "Household Composition and Poverty" in Sheldon Danziger and Daniel Weinberg (eds.), *Fighting Poverty: What Works and What Doesn't*. Cambridge: Harvard University Press, pp. 209–31.

Catanzarite, Lisa. 1990. *Job Characteristics and Occupational Segregation by Gender and Race/Ethnicity*. Ph.D. Dissertation, Department of Sociology, Stanford University, Stanford, California.

Catanzarite, Lisa, and Vilma Ortiz. 1995. "Racial/Ethnic Differences in the Impact of Work and Family on Women's Poverty." *Research in Politics and Society*. vol. 5, pp. 217–37.

Dill, Bonnie Thornton, Lynn Weber Cannon, and Reeve Vanneman. 1987. "Race and Gender in Occupational Segregation" in "National Committee on Pay Equity," *Pay Equity: An Issue of Race, Ethnicity and Sex*. Washington, D.C.: National Committee on Pay Equity, pp. 11–70.

Ehrenreich, Barbara. 1986. "What Makes Women Poor?" in Rochelle Lefkowitz and Ann Withorn (eds.), *For Crying Out Loud: Women and Poverty in the United States*. New York: Pilgrim Press, pp. 18–28.

Garfinkel, Irwin, and Sara S. McLanahan. 1986. *Single Mothers and Their Children: A New American Dilemma*. Washington, D.C.: The Urban Institute.

Malveaux, Julianne. 1985. "The Economic Interests of Black and White Women: Are They Similar?" *Review of Black Political Economy*. vol. 14 (Summer): pp. 5–28.

Malveaux, Julianne, and Phyllis Wallace. 1987. "Minority Women in the Workplace," in Karen S. Koziara, Michael H. Moskow and Lucretia D. Tanner (eds.), *Working Women: Past, Present and Future*. Washington, D.C.: Bureau of National Affairs, IRRA Series.

Moynihan, Daniel P. 1965. *The Negro Family: The Case for National Action*. Washington, D.C.: U.S. Department of Labor, Office of Policy Planning and Research.

Mead, Lawrence. 1986. *Beyond Entitlement: The Social Obligations of Citizenship*. New York: Free Press.

Pearce, Diana. 1986. "The Feminization of Poverty: Women, Work, and Welfare" in Rochelle Lefkowitz and Ann Withorn (eds.), *For Crying Out Loud: Women and Poverty in the United States*. New York: Pilgrim Press, pp. 29–46.

Pearce, Diana. 1987. "On the Edge: Marginal Women Workers and Employment Policy" in Christine Bose and Glenna Spitze (eds.), *Ingredients for Women's Employment Policy*. Albany: State University of New York Press, pp. 197–210.

Petersen, Trond. 1985. "A Comment on Presenting Results from Logit and Probit Models." *American Sociological Review*. vol. 509 (1): pp. 130–31.

Peterson, Janice. 1987. "The Feminization of Poverty," *Journal of Economic Issues*. vol. 21(1): pp. 329–37.

Security Pacific National Bank, 1979. *The Sixty Mile Circle*. Los Angeles: Security Pacific National Bank.

Smith, Joan. 1984. "The Paradox of Women's Poverty: Wage-Earning Women and Economic Transformation." *Signs* (special issue on Women and Poverty), vol. 10 (2): pp. 291–310.

Tienda, Marta and Jennifer Glass. 1985. "Household Structure and Labor Force Participation of Black, Hispanic and White Mothers." *Demography* 22(3): 381–94.

Tienda, Marta and Lief Jensen. 1988. "Poverty and Minorities: A Quarter-Century Profile of Color and Socioeconomic Disadvantage" in Gary Sandefur and Marta Tienda (eds.), *Divided Opportunities and Minorities, Poverty, and Social Policy*. New York: Plenum.

Wilson, William Julius, and Kathryn M. Neckerman. 1986. "Poverty and Family Structure: The Widening Gap Between Evidence and Public Policy Issues" in Sheldon Danziger and Daniel Weinberg (eds.), *Fighting Poverty: What Works and What Doesn't*. Cambridge: Harvard University Press, pp. 232–59.

Wilson, William Julius. 1987. *The Truly Disadvantaged: The Inner City, The Underclass, and Public Policy*. Chicago: University of Chicago Press.

MOVING UP WITH KIN AND COMMUNITY: *Upward Social Mobility for Black and White Women*

16

Elizabeth Higginbotham and Lynn Weber

... When women and people of color experience upward mobility in America, they scale steep structural as well as psychological barriers. The long process of moving from a working-class family of origin to the professional-managerial class is full of twists and turns: choices made with varying degrees of information and varying options; critical junctures faced with support and encouragement or disinterest, rejection, or active discouragement; and interpersonal relationships in which basic understandings are continuously negotiated and renegotiated. It is a fascinating process that profoundly shapes the lives of those who experience it, as well as the lives of those around them. Social mobility is also a process engulfed in myth. One need only pick up any newspaper or turn on the television to see that the myth of upward mobility remains firmly entrenched in American culture: With hard work, talent, determination, and some luck, just about anyone can "make it." ...

The image of the isolated and detached experience of mobility that we have inherited from past scholarship is problematic for anyone seeking to understand the process for women or people of color. Twenty years of scholarship in the study of both race and gender has taught us the importance of interpersonal attachments to the lives of women ... and a commitment to racial uplift among people of color. ...

... Lacking wealth, the greatest gift a Black family has been able to give to its children has been the motivation and skills to succeed in school. Aspirations for college attendance and professional positions are stressed as *family* goals, and the entire family may make sacrifices and provide support. ... Black women have long seen the activist potential of education and have sought it as a cornerstone of community development—a means of uplifting the race. When women of color or white women are put at the center of the analysis of

Authors' note: The research reported here was supported by National Institute for Mental Health Grant MH38769. All the names used in this article are pseudonyms.

From: *Gender & Society* 6(3) (September 1992): 416–40. © 1992 Sociologists for Women in Society. Reprinted by permission of Sage Publications, Inc., and the authors.

upward mobility, it is clear that different questions will be raised about social mobility and different descriptions of the process will ensue. . . .

RESEARCH DESIGN

These data are from a study of full-time employed middle-class women in the Memphis metropolitan area. This research is designed to explore the processes of upward social mobility for Black and white women by examining differences between women professionals, managers, and administrators who are from working- and middle-class backgrounds—that is, upwardly mobile and middle-class stable women. In this way, we isolate subjective processes shared among women who have been upwardly mobile from those common to women who have reproduced their family's professional-managerial class standing. Likewise, we identify many experiences in the attainment process that are shared by women of the same race, be they upwardly mobile or stable middle class. Finally, we specify some ways in which the attainment process is unique for each race-class group. . . .

. . . We rely on a model of social class basically derived from the work of Poulantzas (1974), Braverman (1974), Ehrenreich and Ehrenreich (1979), and elaborated in Vanneman and Cannon (1987). These works explicate a basic distinction between social class and social status. Classes represent bounded categories of the population, groups set in a relation of opposition to one another by their roles in the capitalist system. The middle class, or professional-managerial class, is set off from the working class by the power and control it exerts over workers in three realms: economic (power through ownership), political (power through direct supervisory authority), and ideological (power to plan and organize work; Poulantzas 1974; Vanneman and Cannon 1987).

In contrast, education, prestige, and income represent social statuses—hierarchically structured relative rankings along a ladder of economic success and social prestige. Positions along these dimensions are not established by social relations of dominance and subordination but, rather, as rankings on scales representing resources and desirability. In some respects, they represent both the justification for power differentials vested in classes and the rewards for the role that the middle class plays in controlling labor.

Our interest is in the process of upward social class mobility, moving from a working-class family of origin to a middle-class destination—from a position of working-class subordination to a position of control over the working class. Lacking inherited wealth or other resources, those working-class people who attain middle-class standing do so primarily by obtaining a college education and entering a professional, managerial, or administrative occupation.

Thus we examine carefully the process of educational attainment not as evidence of middle-class standing but as a necessary part of the mobility process for most working-class people.

Likewise, occupation alone does not define the middle class, but professional, managerial, and administrative occupations capture many of the supervisory and ideologically based positions whose function is to control workers' lives. Consequently, we defined subjects as *middle class* by virtue of their employment in either a professional, managerial, or administrative occupation. . . . Classification of subjects as either professional or managerial-administrative was made on the basis of the designation of occupations in the U.S. Bureau of the Census's (1983) "Detailed Population Characteristics: Tennessee." Managerial occupations were defined as those in the census categories of managers and administrators; professionals were defined as those occupations in the professional category, excluding technicians, whom Braverman (1974) contends are working class.

Upwardly mobile women were defined as those women raised in families where neither parent was employed as a professional, manager, or administrator. Typical occupations for working-class fathers were postal clerk, craftsman, semi-skilled manufacturing worker, janitor, and laborer. Some working-class mothers had clerical and sales positions, but many of the Black mothers also worked as private household workers. *Middle-class stable* women were defined as those women raised in families where *either* parent was employed as a professional, manager, or administrator. Typical occupations of middle-class parents were social worker, teacher, and school administrator as well as high-status professionals such as attorneys, physicians, and dentists. . . .

FAMILY EXPECTATIONS
FOR EDUCATIONAL ATTAINMENT

Four questions assess the expectations and support among family members for the educational attainment of the subjects. First, "Do you recall your father or mother stressing that you attain an education?" Yes was the response of 190 of the 200 women. Each of the women in this study had obtained a college degree, and many have graduate degrees. It is clear that for Black and white women, education was an important concern in their families. . . .

The comments of Laura Lee, a 39-year-old Black woman who was raised middle class, were typical:

> Going to school, that was never a discussable issue. Just like you were born to live and die, you were going to go to school. You were going to prepare yourself to do something.

It should be noted, however, that only 86 percent of the white working-class women answered yes, compared to 98 percent of all other groups. Although this difference is small, it foreshadows a pattern where white women raised in working-class families received the least support and encouragement for educational and career attainment.

"When you were growing up, how far did your father expect you to go in school?" While most fathers expected college attendance from their daughters, differences also exist by class of origin. Only 70 percent of the working-class fathers, both Black and white, expected their daughters to attend college. In contrast, 94 percent of the Black middle-class and 88 percent of the white middle-class women's fathers had college expectations for their daughters.

When asked the same question about mother's expectations, 88 percent to 92 percent of each group's mothers expected their daughters to get a college education, except the white working-class women, for whom only 66 percent of mothers held such expectations. In short, only among the white working-class women did a fairly substantial proportion (about one-third) of both mothers and fathers expect less than a college education from their daughters. About 30 percent of Black working-class fathers held lower expectations for their daughters, but not the mothers; virtually all middle-class parents expected a college education for their daughters.

Sara Marx is a white, 33-year-old director of counseling raised in a rural working-class family. She is among those whose parents did not expect a college education for her. She was vague about the roots of attending college:

> It seems like we had a guest speaker who talked to us. Maybe before our exams somebody talked to us. I really can't put my finger on anything. I don't know where the information came from exactly.

"Who provided emotional support for you to make the transition from high school to college?" While 86 percent of the Black middle-class women indicated that family provided that support, 70 percent of the white middle class, 64 percent of the Black working class, and only 56 percent of the white working class received emotional support from family.

"Who paid your college tuition and fees?" Beyond emotional support, financial support is critical to college attendance. There are clear class differences in financial support for college. Roughly 90 percent of the middle-class respondents and only 56 percent and 62 percent of the Black and white working-class women, respectively, were financially supported by their families. These data also suggest that working-class parents were less able to give emotional or financial support for college than they were to hold out the expectation that their daughters should attend.

FAMILY EXPECTATIONS
FOR OCCUPATION OR CAREER

When asked, "Do you recall your father or mother stressing that you should have an occupation to succeed in life?", racial differences appear. Ninety-four percent of all Black respondents said yes. In the words of Julie Bird, a Black woman raised-middle-class junior high school teacher:

> My father would always say, "You see how good I'm doing? Each generation should do more than the generation before." He expects me to accomplish more than he has.

Ann Right, a 36-year-old Black attorney whose father was a janitor, said:

> They wanted me to have a better life than they had. For all of us. And that's why they emphasized education and emphasized working relationships and how you get along with people and that kind of thing.

Ruby James, a Black teacher from a working-class family, said:

> They expected me to have a good-paying job and to have a family and be married. Go to work every day. Buy a home. That's about it. Be happy.

In contrast, only 70 percent of the white middle-class and 56 percent of the white working-class women indicated that their parents stressed that an occupation was needed for success. Nina Pentel, a 26-year-old white medical social worker, expressed a common response: "They said 'You're going to get married but get a degree, you never know what's going to happen to you.' They were pretty laid back about goals."

When the question focuses on a career rather than an occupation, the family encouragement is lower and differences were not significant, but similar patterns emerged. We asked respondents, "Who, if anyone, encouraged you to think about a career?" Among Black respondents, 60 percent of the middle-class and 56 percent of the working-class women answered that family encouraged them. Only 40 percent of the white working-class women indicated that their family encouraged them in their thinking about a career, while 52 percent of the white middle-class women did so. . . .

When working-class white women seek to be mobile through their own attainments, they face conflicts. Their parents encourage educational attainment, but when young women develop professional career goals, these same parents sometimes become ambivalent. This was the case with Elizabeth

Marlow, who is currently a public interest attorney—a position her parents never intended her to hold. She described her parents' traditional expectations and their reluctance to support her career goals fully.

> My parents assumed that I would go to college and meet some nice man and finish, but not necessarily work after. I would be a good mother for my children. I don't think that they ever thought I would go to law school. Their attitude about my interest in law school was, "You can do it if you want to, but we don't think it is a particularly practical thing for a woman to do."

Elizabeth is married and has three children, but she is not the traditional housewife of her parents' dreams. She received more support outside the family for her chosen life-style.

Although Black families are indeed more likely than white families to encourage their daughters to prepare for careers, like white families, they frequently steer them toward highly visible traditionally female occupations, such as teacher, nurse, and social worker. Thus many mobile Black women are directed toward the same gender-segregated occupations as white women. . . .

MARRIAGE

Although working-class families may encourage daughters to marry, they recognize the need for working-class women to contribute to family income or to support themselves economically. To achieve these aims, many working-class girls are encouraged to pursue an education as preparation for work in gender-segregated occupations. Work in these fields presumably allows women to keep marriage, family, and child rearing as life goals while contributing to the family income and to have "something to fall back on" if the marriage does not work out. This interplay among marriage, education, financial need, and class mobility is complex (Joslin 1979).

We asked, "Do you recall your mother or father emphasizing that marriage should be your primary life goal?" While the majority of all respondents did not get the message that marriage was the *primary life goal*, Black and white women's parents clearly saw this differently. Virtually no Black parents stressed marriage as the primary life goal (6 percent of the working class and 4 percent of the middle class), but significantly more white parents did (22 percent of the working class and 18 percent of the middle class).

Some white women said their families expressed active opposition to marriage, such as Clare Baron, a raised-working-class nursing supervisor, who said, "My mother always said, 'Don't get married and don't have children!'"

More common responses recognized the fragility of marriage and the need to support oneself. For example, Alice Page, a 31-year-old white raised-middle-class librarian, put it this way:

> I feel like I am really part of a generation that for the first time is thinking, "I don't want to have to depend on somebody to take care of me because what if they say they are going to take care of me and then they are not there? They die, or they leave me or whatever." I feel very much that I've got to be able to support myself and I don't know that single women in other eras have had to deal with that to the same degree.

While white working-class women are often raised to prepare for work roles so that they can contribute to family income and, if necessary, support themselves, Black women face a different reality. Unlike white women, Black women are typically socialized to view marriage separately from economic security, because it is not expected that marriage will ever remove them from the labor market. As a result, Black families socialize all their children—girls and boys—for self-sufficiency (Clark 1986; Higginbotham and Cannon 1988)....

... Fairly substantial numbers of each group had never married by the time of the interview, ranging from 20 percent of the white working-class to 34 percent of the Black working-class and white middle-class respondents. Some of the women were pleased with their singlehood, like Alice Page, who said:

> I am single by choice. That is how I see myself. I have purposely avoided getting into any kind of romantic situation with men. I have enjoyed going out but never wanted to get serious. If anyone wants to get serious, I quit going out with him.

Other women expressed disappointment and some shock that they were not yet married. When asked about her feeling about being single, Sally Ford, a 32-year-old white manager, said:

> That's what I always wanted to do: to be married and have children. To me, that is the ideal. I want a happy, good marriage with children. I do not like being single at all. It is very, very lonesome. I don't see any advantages to being single. None! . . .

SUBJECTIVE SENSE OF DEBT TO KIN AND FRIENDS

McAdoo (1978) reports that upwardly mobile Black Americans receive more requests to share resources from their working-class kin than do middle-class

Black Americans. Many mobile Black Americans feel a "social debt" because their families aided them in the mobility process and provided emotional support. When we asked the white women in the study the following question: "Generally, do you feel you owe a lot for the help given to you by your family and relatives?" many were perplexed and asked what the question meant. In contrast, both the working- and middle-class Black women tended to respond immediately that they felt a sense of obligation to family and friends in return for the support they had received. Black women, from both the working class and the middle class, expressed the strongest sense of debt to family, with 86 percent and 74 percent, respectively, so indicating. White working-class women were least likely to feel that they owed family (46 percent), while 68 percent of white middle-class women so indicated. In short, upwardly mobile Black women were almost twice as likely as upwardly mobile white women to express a sense of debt to family.

Linda Brown, an upwardly mobile Black woman, gave a typical response, "Yes, they are there when you need them." Similar were the words of Jean Marsh, "Yes, because they have been supportive. They're dependable. If I need them I can depend upon them."

One of the most significant ways in which Black working-class families aided their daughters and left them with a sense of debt related to care for their children. Dawn March expressed it thus:

> They have been there more so during my adult years than a lot of other families that I know about. My mother kept all of my children until they were old enough to go to day care. And she not only kept them, she'd give them a bath for me during the daytime and feed them before I got home from work. Very, very supportive people. So, I really would say I owe them for that.

Carole Washington, an upwardly mobile Black woman occupational therapist, also felt she owed her family. She reported:

> I know the struggle that my parents have had to get me where I am. I know the energy they no longer have to put into the rest of the family even though they want to put it there and they're willing. I feel it is my responsibility to give back some of that energy they have given to me. It's self-directed, not required.

White working-class women, in contrast, were unlikely to feel a sense of debt and expressed their feelings in similar ways. Irma Cox, part owner of a computer business, said, "I am appreciative of the values my parents instilled in me. But I for the most part feel like I have done it on my own." Carey Mink, a 35-year-old psychiatric social worker, said, "No, they pointed me in a direction and they were supportive, but I've done a lot of the work myself." Debra Beck, a judge, responded, "No, I feel that I've gotten most places on my own." . . .

COMMITMENT TO COMMUNITY

The mainstream "model of community stresses the rights of individuals to make decisions in their own self-interest, regardless of the impact on the larger society" (Collins 1990, 52). This model may explain relations to community of origin for mobile white males but cannot be generalized to other racial and gender groups. In the context of well-recognized structures of racial oppression, America's racial-ethnic communities develop collective survival strategies that contrast with the individualism of the dominant culture but ensure the community's survival. . . . Widespread community involvement enables mobile people of color to confront and challenge racist obstacles in credentialing institutions, and it distinguishes the mobility process in racial-ethnic communities from mobility in the dominant culture. For example, Lou Nelson, now a librarian, described the support she felt in her southern, segregated, inner-city school. She said:

> There was a closeness between people and that had a lot to do with neighborhood schools. I went to Tubman High School with people that lived in the Tubman area. I think that there was a bond, a bond between parents, the PTA . . . I think that it was just that everybody felt that everybody knew everybody. And that was special.

Family and community involvement and support in the mobility process means that many Black professionals and managers continue to feel linked to their communities of origin. Lillian King, a high-ranking city official who was raised working class, discussed her current commitment to the Black community. She said:

> Because I have more opportunities, I've got an obligation to give more back and to set a positive example for Black people and especially for Black women. I think we've got to do a tremendous job in building self-esteem and giving people the desire to achieve.

Judith Moore is a 34-year-old single parent employed as a health investigator. She has been able to maintain her connection with her community, and that is a source of pride.

> I'm proud that I still have a sense of who I am in terms of Black people. That's very important to me. No matter how much education or professional status I get, I do not want to lose touch with where I've come from. I think that you need to look back and that kind of pushes you forward. I think the degree and other things can make you lose sight of that, especially us Black folks, but I'm glad that I haven't and I try to teach that [commitment] to my son.

For some Black women, their mobility has enabled them to give to an even broader community. This is the case with Sammi Lewis, a raised-working-class woman who is a director of a social service agency. She said, "I owe a responsibility to the entire community, and not to any particular group." . . .

CROSSING THE COLOR LINE

Mobility for people of color is complex because in addition to crossing class lines, mobility often means crossing racial and cultural ones as well. Since the 1960s, people of color have increasingly attended either integrated or predominantly white schools. Only mobile white ethnics have a comparable experience of simultaneously crossing class and cultural barriers, yet even this experience is qualitatively different from that of Black and other people of color. White ethnicity can be practically invisible to white middle-class school peers and co-workers, but people of color are more visible and are subjected to harsher treatment. Our research indicates that no matter when people of color first encounter integrated or predominantly white settings, it is always a shock. The experience of racial exclusion cannot prepare people of color to deal with the racism in daily face-to-face encounters with white people.

For example, Lynn Johnson was in the first cohort of Black students at Regional College, a small private college in Memphis. The self-confidence and stamina Lynn developed in her supportive segregated high school helped her withstand the racism she faced as the first female and the first Black to graduate in economics at Regional College. Lynn described her treatment:

> I would come into class and Dr. Simpson (the Economics professor) would alphabetically call the roll. When he came to my name, he would just jump over it. He would not ask me any questions, he would not do anything. I stayed in that class. I struggled through. When it was my turn, I'd start talking. He would say, "Johnson, I wasn't talking to you" [because he never said *Miss* Johnson). I'd say, "That's all right, Dr. Simpson, it was my turn. I figured you just overlooked me. I'm just the littlest person in here. Wasn't that the right answer?" He would say, "Yes, that was the right answer." I drove him mad, I really did. He finally got used to me and started to help me.

In southern cities, where previous interaction between Black and white people followed a rigid code, adjustments were necessary on both sides. It was clear to Lynn Johnson and others that college faculty and students had to adapt to her small Black cohort at Regional College.

Wendy Jones attended a formerly predominantly white state university that had just merged with a formerly predominantly Black college. This new institution meant many adjustments for faculty and students. As a working-class person majoring in engineering, she had a rough transition. She recalled:

> I had never gone to school with white kids. I'd always gone to all Black schools all my life and the Black kids there [at the university] were snooty. Only one friend from high school went there and she flunked out. The courses were harder and all my teachers were men and white. Most of the kids were white. I was in classes where I'd be the only Black and woman. There were no similarities to grasp for. I had to adjust to being in that situation. In about a year I was comfortable where I could walk up to people in my class and have conversations.

For some Black people, their first significant interaction with white people did not come until graduate school. Janice Freeman described her experiences:

> I went to a Black high school, a Black college and then worked for a Black man who was a former teacher. Everything was comfortable until I had to go to State University for graduate school. I felt very insecure. I was thrown into an environment that was very different—during the 1960s and 1970s there was so much unrest anyway—so it was extremely difficult for me.

It was not in graduate school but on her first job as a social worker that Janice had to learn to work *with* white people. She said, "After I realized that I could hang in school, working at the social work agency allowed me to learn how to work *with* white people. I had never done that before and now I do it better than anybody."

Learning to live in a white world was an additional hurdle for all Black women in this age cohort. Previous generations of Black people were more likely to be educated in segregated colleges and to work within the confines of the established Black community. They taught in segregated schools, provided dental and medical care to the Black communities, and provided social services and other comforts to members of their own communities. They also lived in the Black community and worshiped on Sunday with many of the people they saw in different settings. As the comments of our respondents reveal, both Black and white people had to adjust to integrated settings, but it was more stressful for the newcomers.

SUMMARY AND CONCLUSIONS

Our major aim in this research was to reopen the study of the subjective experience of upward social mobility and to begin to incorporate race and gender into our vision of the process. In this exploratory work, we hope to raise issues and questions that will cast a new light on taken-for-granted assumptions about the process and the people who engage in it. The experiences of these women have certainly painted a different picture from the one we were left some twenty years ago. First and foremost, these women are not detached, isolated, or driven solely by career goals. Relationships with family of origin, partners, children, friends, and the wider community loom large in the way they envision and accomplish mobility and the way they sustain themselves as professional and managerial women.

Several of our findings suggest ways that race and gender shape the mobility process for baby boom Black and white women. Education was stressed as important in virtually all of the families of these women; however, they differed in how it was viewed and how much was desired. The upwardly mobile women, both Black and white, shared some obstacles to attainment. More mobile women had parents who never expected them to achieve a college education. They also received less emotional and financial support for college attendance from their families than the women in middle-class families received. Black women also faced the unique problem of crossing racial barriers simultaneously with class barriers.

There were fairly dramatic race differences in the messages that the Black and white women received from family about what their lives should be like as adults. Black women clearly received the message that they needed an occupation to succeed in life and that marriage was a secondary concern. Many Black women also expressed a sense that their mobility was connected to an entire racial uplift process, not merely an individual journey.

White upwardly mobile women received less clear messages. Only one-half of these women said that their parents stressed the need for an occupation to succeed, and 20 percent said that marriage was stressed as the primary life goal. The most common message seemed to suggest that an occupation was necessary, because marriage could not be counted on to provide economic survival. Having a career, on the other hand, could even be seen as detrimental to adult happiness.

Upward mobility is a process that requires sustained effort and emotional and cognitive, as well as financial, support. The legacy of the image of mobility that was built on the white male experience focuses on credentialing institutions, especially the schools, as the primary place where talent is recognized and support is given to ensure that the talented among the working class are mobile. Family and friends are virtually invisible in this portrayal of the mobility process.

Although there is a good deal of variation in the roles that family and friends play for these women, they are certainly not invisible in the process. Especially among many of the Black women, there is a sense that they owe a great debt to their families for the help they have received. Black upwardly mobile women were also much more likely to feel that they give more than they receive from kin. Once they have achieved professional managerial employment, the sense of debt combines with their greater access to resources to put them in the position of being asked to give and of giving more to both family and friends. Carrington (1980) identifies some potential mental health hazards of such a sense of debt in upwardly mobile Black women's lives.

White upwardly mobile women are less likely to feel indebted to kin and to feel that they have accomplished alone. Yet even among this group, connections to spouses and children played significant roles in defining how women were mobile, their goals, and their sense of satisfaction with their life in the middle class.

These data are suggestive of a mobility process that is motivated by a desire for personal, but also collective, gain and that is shaped by interpersonal commitments to family, partners and children, community, and the race. Social mobility involves competition, but also cooperation, community support, and personal obligations. Further research is needed to explore fully this new image of mobility and to examine the relevance of these issues for white male mobility as well.

REFERENCES

Braverman, Harry. 1974. *Labor and monopoly capital.* New York: Monthly Review Press.

Carrington, Christine. 1980. Depression in Black women: A theoretical appraisal. In *The Black woman*, edited by La Frances Rodgers Rose. Beverly Hills, CA: Sage.

Clark, Reginald. 1986. *Family life and school achievement.* Chicago: University of Chicago Press.

Collins, Patricia Hill. 1990. *Black feminist thought: Knowledge, consciousness, and the politics of empowerment.* Boston: Routledge.

Ehrenreich, Barbara, and John Ehrenreich. 1979. The professional-managerial class. In *Between labor and capital*, edited by Pat Walker. Boston: South End Press.

Higginbotham, Elizabeth, and Lynn Weber Cannon. 1988. *Rethinking mobility: Towards a race and gender inclusive theory.* Research Paper no. 8. Center for Research on Women, Memphis State University.

Joslin, Daphne. 1979. Working-class daughters, middle-class wives: Social identity and self-esteem among women upwardly mobile through marriage. Ph.D. diss., New York University, New York.

McAdoo, Harriette Pipes. 1978. Factors related to stability in upwardly mobile Black families. *Journal of Marriage and the Family* 40:761–76.

Poulantzas, Nicos. 1974. *Classes in contemporary capitalism.* London: New Left Books.

U.S. Bureau of the Census. 1983. Detailed population characteristics: Tennessee. Census of the Population, 1980. Washington, D.C.: GPO.

Vanneman, Reeve, and Lynn Weber Cannon. 1987. *The American perception of class.* Philadelphia: Temple University Press.

Gender and Sexism

COMMONALITIES AND DIFFERENCES   17

Johnnetta B. Cole

If you see one woman, have you seen them all? Does the heavy weight of patriarchy level all differences among US women? Is it the case, as one woman put it, that "there isn't much difference between having to say 'Yes suh Mr. Charlie' and 'Yes dear'?" Does "grandmother" convey the same meaning as "abuela," as "buba," as "gran'ma"? Is difference a part of what we share, or is it, in fact, *all* that we share? As early as 1970, Toni Cade Bambara asked: "How relevant are the truths, the experiences, the findings of white women to black women? Are women after all simply women?" (Bambara 1970: 9).

Are US women bound by our similarities or divided by our differences? The only viable response is *both*. To address our commonalities without dealing with our differences is to misunderstand and distort that which separates as well as that which binds us as women. Patriarchal oppression is not limited to women of one race or of one particular ethnic group, women in one class, women of one age group or sexual preference, women who live in one part of the country, women of any one religion, or women with certain physical abilities or disabilities. Yet, while oppression of women knows no such limitations, we cannot, therefore, conclude that the oppression of all women is identical.

Among the things which bind women together are the assumptions about the way that women think and behave, the myths—indeed the stereotypes—about what is common to all women. For example, women will be asked nicely in job interviews if they type, while men will not be asked such a question. In response to certain actions, the expression is used: "Ain't that just like a woman?" Or during a heated argument between a man and a woman, as the voice of each rises and emotions run high, the woman makes a particularly good point. In a voice at the pitch of the ongoing argument, the man screams at her: "You don't have to get hysterical!"

In an interesting form of "what goes around comes around," as Malcolm X put it, there is the possibility that US women are bound together by our assumptions, attitudes toward, even stereotypes of the other gender. Folklorist Rayna Green, referring to women of the southern setting in which she grew up, says this:

> Southern or not, women everywhere talk about sex. . . . In general men are more often the victims of women's jokes than not. Tit for tat, we say. Usually the subject for laughter is men's boasts, failures, or inadequacies ("comeuppance for lack of upcommance," as one of my aunts would say). Poking fun at a man's sexual ego, for example, might never be possible in real social situations with the men who have power over their lives, but it is possible in a joke. (Green 1984: 23–24)

That which US women have in common must always be viewed in relation to the particularities of a group, for even when we narrow our focus to one particular group of women it is possible for differences within that group to challenge the primacy of what is shared in common. For example, what have we said and what have we failed to say when we speak of "Asian American women"? As Shirley Hune notes (1982), Asian American women as a group share a number of characteristics. Their participation in the work force is higher than that of women in any other ethnic group. Many Asian American women live life supporting others, often allowing their lives to be subsumed by the needs of the extended family. And they are subjected to stereotypes by the dominant society: the sexy but "evil dragon lady," the "neuter gender," the "passive/demure" type, and the "exotic/erotic" type.

However, there are many circumstances when these shared experiences are not sufficient to accurately describe the condition of particular Asian American women. Among Asian American women there are those who were born in the United States, fourth and fifth generation Asian American women with firsthand experience of no other land, and there are those who recently arrived in the United States. Asian American women are diverse in their heritage or country of origin: China, Japan, the Philippines, Korea, India, Vietnam, Cambodia, Thailand, or another country in Asia. If we restrict ourselves to Asian American women of Chinese descent, are we referring to those women who are from the People's Republic of China or those from

Taiwan, those from Hong Kong or those from Vietnam, those from San Francisco's Chinatown or those from Mississippi? Are we subsuming under "Asian American" those Pacific Island women from Hawaii, Samoa, Guam, and other islands under U.S. control? Although the majority of Asian American women are working class—contrary to the stereotype of the "ever successful" Asians—there are poor, "middle-class," and even affluent Asian American women (Hune 1982: 1–2, 13–14).

It has become very common in the United States today to speak of "Hispanics," putting Puerto Ricans, Chicanos, Dominicans, Cubans, and those from every Spanish-speaking country in the Americas into one category of people, with the women referred to as Latinas or Hispanic women. Certainly there is a language, or the heritage of a language, a general historical experience, and certain cultural traditions and practices which are shared by these women. But a great deal of harm can be done by sweeping away differences in the interest of an imposed homogeneity.

Within one group of Latinas there is, in fact, considerable variation in terms of self-defined ethnic identity, such that some women refer to themselves as Mexican Americans, others as *Chicanas*, others as Hispanics, and still others as Americans. Among this group of women are those who express a commitment to the traditional roles of women and others who identify with feminist ideals. Some Chicanas are monolingual—in Spanish or English— and others are bilingual. And there are a host of variations among Chicanas in terms of educational achievements, economic differences, rural or urban living conditions, and whether they trace their ancestry from women who lived in this land well before the United States forcibly took the northern half of Mexico, or more recently arrived across the border that now divides the nations called Mexico and the United States.

Women of the Midwest clearly share a number of experiences which flow from living in the U.S. heartland, but they have come from different places, and they were and are today part of various cultures.

> Midwestern women are the Native American women whose ancestors were brought to the plains in the mid-nineteenth century to be settled on reservations, the black women whose forebears emigrated by the thousands from the South after Reconstruction. They are the descendants of the waves of Spanish, French, Norwegian, Danish, Swedish, Bohemian, Scottish, Welsh, British, Irish, German, and Russian immigrants who settled the plains, the few Dutch, Italians, Poles, and Yugoslavs who came with them. (Boucher 1982: 3)

There is another complexity: when we have identified a commonality among women, cutting across class, racial, ethnic, and other major lines of

difference, the particular ways that commonality is acted out and its consequences in the larger society may be quite diverse. Ostrander makes this point
in terms of class:

> When women stroke and soothe men, listen to them and accommodate
> their needs, men of every class return to the workplace with renewed energies.
> When women arrange men's social lives and relationships, men of every class
> are spared investing the time and energy required to meet their social needs.
> When women run the households and keep family concerns in check, men of
> every class are freer than women to pursue other activities, including work, out-
> side the home. But upper-class women perform these tasks for men at the very
> top of the class structure. . . . Supporting their husbands as individuals, they
> support and uphold the very top of the class structure. In this way they distin-
> guish themselves from women of other social classes. (Ostrander 1984: 146)

Suppose that we can accurately and exclusively identify the characteristics
shared by one particular group of women. For each of the women within that
group, into how many other groups does she want to, or is she forced to, fit?
Or can we speak of similarities only with respect to a group such as Puerto
Rican women who are forty-three years old, were born in San Juan, Puerto
Rico, migrated to New York City when they were five years old, work as
eighth-grade school teachers, attend a Catholic church, are heterosexual, married, with two male and two female children, and have no physical disabilities?

Then there is that unpredictable but often present quality of individuality,
the idiosyncrasies of a particular person. Shirley Abbott, describing experiences of growing up in the South, contrasts her mother's attitude and behavior
toward the black woman who was her maid with what was the usual stance of
"southern white ladies."

> I don't claim that my mother's way of managing her black maid was typical.
> Most white women did not help their laundresses hang the washing on the
> line. . . . Compulsive housewifery had some part in it. So did her upbringing. . . .
> There was another motive too. . . . Had she used Emma in just the right way,
> Mother could have become a lady. But Mother didn't want to be a lady. Some-
> thing in her was against it, and she couldn't explain what frightened her, which
> was why she cried when my father ridiculed her. (Abbott 1983: 78–79)

Once we have narrowed our focus to one specific group of women (Arme-
nian American women, or women over sixty-five, or Arab American women, or
black women from the Caribbean, or Ashkenazi Jewish women), the oppression that group of women experiences may take different forms at different
times. Today, there is no black woman in the United States who is the legal
slave of a white master: "chosen" for that slave status because of her race,

forced to give her labor power without compensation because of the class arrangements of the society, and subjected to the sexual whims of her male master because of her gender. But that does not mean that black women today are no longer oppressed on the basis of race, class, and gender.

There are also groups of women who experience intense gender discrimination today, but in the past had a radically different status in their society. Contrary to the popular image of female oppression as being both universal and as old as human societies, there is incontestable evidence of egalitarian societies in which men and women related in ways that did not involve male dominance and female subjugation. Eleanor Leacock is the best known of the anthropologists who have carried out the kind of detailed historical analysis which provides evidence on gender relations in precolonial North American societies. In discussing the debate on the origins and spread of women's oppression, Leacock points out that women's oppression is a reality today in virtually every society, and while socialist societies have reduced it, they have not eliminated gender inequality. However, it does not follow that women's oppression has always existed and will always exist. What such arguments about universal female subordination do is to project onto the totality of human history the conditions of today's world. Such an argument also "affords an important ideological buttress for those in power" (Leacock 1979: 10–11).

Studies of precolonial societies indicate considerable variety in terms of gender relations.

> Women retained great autonomy in much of the pre-colonial world, and related to each other and to men through public as well as private procedures as they carried out their economic and social responsibilities and protected their rights. Female and male modalities of various kinds operated reciprocally within larger kin and community contexts before the principle of male dominance within individual families was taught by missionaries, defined by legal status, and solidified by the economic relations of colonialism. (Leacock 1979: 10–11)

Even when there is evidence of female oppression among women of diverse backgrounds, it is important to listen to the individual assessment which each women makes of her own condition, rather than assume that a synonymous experience of female oppression exists among all women. As a case in point, Sharon Burmeister Lord, in describing what it was like to grow up "Appalachian style," speaks of the influence of female role models in shaping the conditions of her development. In Williamson, West Virginia, she grew up knowing women whose occupations were Methodist preacher, elementary school principal, county sheriff, and university professor. Within her own family, her mother works as a secretary, writes poetry and songs, and "swims faster than any boy"; her aunt started her own seed and hardware store; one grandmother is a farmer and the other runs her own boarding house. Summarizing the effect of growing up among such women, Lord says:

> When a little girl has had a chance to learn strength, survival tactics, a firm grasp of reality, and an understanding of class oppression from the women around her, it doesn't remove oppression from her life, but it does give her a fighting chance. And that's an advantage! (Lord 1979: 25)

Finally, if it is agreed that today, to some extent, all women are oppressed, to what extent can a woman, or a group of women, also act as oppressor? Small as the numbers may be, there are some affluent black women. . . . Is it not possible that among this very small group of black women there are those who, while they experience oppression because of their race, act in oppressive ways toward other women because of their class? Does the experience of this society's heterosexism make a Euro-American lesbian incapable of engaging in racist acts toward women of color? The point is very simply that privilege can and does coexist with oppression (Bulkin et al. 1984: 99) and being a victim of one form of discrimination does not make one immune to victimizing someone else on a different basis. . . .

REFERENCES

Abbott, S. 1983. *Womenfolks: Growing Up Down South*. New York: Ticknor and Fields.

Bambara, T. C., ed. 1970. *The Black Woman: An Anthology*. New York: Signet.

Boucher, S. 1982. *Heartwomen: An Urban Feminist Odyssey Home*. New York: Harper & Row.

Bulkin, E., M. E. Pratt, and B. Smith, eds. 1984. *Yours in Struggle*. Brooklyn, N.Y.: Long Haul Press.

Green, R. 1984. "Magnolias Grow in Dirt: The Bawdy Lore of Southern Women." In *Speaking for Ourselves*, M. Alexander, ed., pp. 20–28. New York: Pantheon.

Hune, S. 1982. Asian American Women: Past and Present, Myth and Reality. Unpublished manuscript prepared for conference on Black Women's Agenda for the Feminist Movement in the 80s, Williams College, Williamstown, Mass., November 12–14, 1982.

Leacock, E. 1979. "Women, Development and Anthropological Facts and Fictions." In *Women in Latin America: An Anthology from Latin American Perspectives*, pp. 7–16. Riverside, Calif.: Latin American Perspectives.

Lord, S. B. 1979. "Growin' Up—Appalachian, Female, and Feminist." In *Appalachian Women: A Learning/Teaching Guide*, S. B. Lord and C. Patton-Crowder, eds., pp. 22–25. Knoxville, Tenn.: University of Tennessee.

Ostrander, S. A. 1984. *Women of the Upper Class*. Philadelphia: Temple University Press.

UNDERSTANDING AND FIGHTING SEXISM:

18

A Call to Men

Peter Blood, Alan Tuttle, and George Lakey

PART 1: UNDERSTANDING THE ENEMY: HOW SEXISM WORKS IN THE U.S.A.

What Is Sexism?

Sexism is much more than a problem with the language we use, our personal attitudes, or individual hurtful acts toward women. Sexism in our country is a complex mesh of practices, institutions, and ideas which have the overall effect of giving more power to men than to women. By "power" we mean the ability to influence important decisions—political decisions of government on every level, economic decisions (jobs, access to money, choice of priorities), and a wide variety of other life areas down to the most personal concerns, such as whether two people are going to make love on a given night or not. The word "patriarchy" is sometimes used to refer to the actual power structure built around men's domination of women. Two key areas where women are denied power are the area of jobs and the area of violence directed toward women.

Women have much less earning power in our labor market than men do. Reasons for this include the fact that much of women's labor is unwaged (housecleaning, childrearing, little services to please bosses or lovers); the low status and pay of most of the traditionally women's jobs that are waged (secretary, sales clerk, childcare, nursing home attendant); the non-union status of most women workers; and the discriminatory practices such as the recent Supreme Court decision allowing companies to exclude pregnancy from their medical insurance and sick leave benefits.

Women face a constant threat of physical violence and sexual aggression in our society. As men we are rarely aware of how pervasive this is or the powerful effect it has on women's outlook on themselves and the world. Actual rape or sadistic violence is the tip of the iceberg. Physical abuse of wives and

From: *Off Their Backs . . . and on Our Own Two Feet* (Philadelphia: New Society Publishers, 1983), pp. 1–8. Reprinted by permission.

lovers is common and rarely publicized. A majority of women have probably experienced some form of sexual abuse as children. The memory of these experiences often gets suppressed because they feel so humiliated and scared, and because adults deny repeatedly that such a thing could happen. Society is filled with messages pressuring women to provide men with sexual pleasure.

All of the above combine with differences in physical strength, voice, acculturated ways of dealing with anger, and the very concrete power men hold in other areas of life to keep many women intimidated, passive, and unable to even acknowledge their own fear openly. The rapist is the shock trooper for an overall system of unequal power.

Patriarchy is not just a power structure "out there"; it is mainly enforced by our own acceptance of its character ideals for our lives. The character ideal which is held up for men to reach toward is "masculinity." A masculine man is supposed to be tough, good at abstract reasoning, hard-working, unfeeling except for anger and sexual desire, and habitually taking the initiative. Masculinity exists only in contrast to femininity, the model for women. Feminine characteristics include cooperativeness, emotionality, patience, passivity, nurturance, and sexual appeal.

We all know that human characteristics are *not* distributed neatly between the sexes that way. A nursery school will often include girls who do abstract reasoning and get into fights and boys who cry easily. We also know that the culture does not leave them alone—the tomboy usually learns to become a lady, and the gentle boy develops armor to protect him from the jibes of his mates. They also learn that masculinity is valued in our culture more than femininity, especially when it comes to gaining power. In fact, the characteristics which are assigned to men by the patriarchy are the power-linked characteristics. In other words, by accepting masculinity as an ideal for ourselves, men buy into a system which keeps women down.

SEXISM AND OUR ECONOMIC SYSTEM

Why does sexism exist? Why is it so hard to root it out? Who really benefits from it? There is a variety of theories as to where sexism originated. Some say it started with the advent of class society; others say it preceded all other forms of oppression.

Regardless of its historic roots, it is clear that sexism today is intimately connected with our economic system, which is called "corporate capitalism."

In our society, a small group of people own and control almost all of the factories, financial institutions, networks of transport and sales of food and clothing. What this means, first of all, is that though we as men generally have more power than women do, the great majority of us have relatively little real power. Most of us are given very little chance to make the major decisions which affect our lives (for example, whether society will emphasize public transportation or private cars, how decisions are made at the factory where we work, what quality air we breathe, what is taught to our children in school). In fact, this kind of power is concentrated in the hands of a relatively small number of white, middle-aged to older men.

However much power the system has given to each of us, this power is only useful in keeping things going according to the present rules, not in changing things in any basic way. The moment one of us tries to use power conveyed by the system to make fundamental changes, we find it taken away from us by that system. (All of us do possess the power to influence the course of history and bring about fundamental changes if we work together. This power comes from a very different source: from our power to work collectively to change the world around us, by "grass roots" power.)

Corporate capitalism is supported by sexism in many ways. Perhaps the most important way is that women function as a surplus labor force, where they can be pushed in and out of employment in keeping with current needs of the economy (depression, expansion, wartime, or a period of union-busting). In effect, the prevalent attitude in this society might be: "A woman's place is in the home . . . *until she's needed in the factory.*" Women's low wages keep wages lower for all: our boss holds the implicit threat that he can replace us with lower-paid women workers if we get too uppity or demand too high wages. Job role stereotyping also helps get women to do unpleasant forms of work which men would rather not do.

There are other ways sexism helps maintain capitalism. Women's unwaged labor of reproducing and taking care of the work force is called "women's work" and taken for granted. Women's servicing of men at home keeps our feelings of alienation and frustration concerning our jobs in check, which might otherwise lead to rebellion or burnout. Men's role as sole breadwinner in many families makes them reluctant to take militant stands on safety, wages, or other issues at work. It is threatening for a man not to be filling the breadwinner role—and he knows his wife will have a very difficult time finding work that pays enough to support a family. Women and men are played off against each other in competing for scarce jobs rather than sharing the fruits of a non-competitive, non-wasteful economy. Finally, excess consumption is encouraged by the lifeless, unsatisfying sex roles which we get shunted into on the job and at home.

The capitalist economic-political structure works to preserve sexism through indoctrination in our schools, a constant flow of brainwashing (subtle or otherwise) in advertising, and through its control of the mass media (what news gets reported and what does not, what programs get chosen for airing . . .). It also maintains it through control of legislation (influence through money), through the policies of our large corporations, and through failure to make use of leadership positions to educate people about the ways sexism affects us and ways it can be overcome.

Sexism and capitalism are so intertwined, in fact, that we believe there is no way that sexism can be thoroughly uprooted from our society without the total remaking of our economic system. . . .

Setting up a socialist society in no way guarantees an end to sexism (as can also be seen in the persistence of sexism to one degree or another in [many] countries). The patriarchy is a power arrangement which has a life of its own independent of capitalism's indoctrination. The struggle against sexism, therefore, must be a high priority for any movement which is offering leadership in creating a new society. The style pervading socialist movements in this country and the groups which are leading socialist countries is all too often determined by the masculine conditioning of domination and competition. This conditioning needs to be struggled against much more forcefully than it has been if we are ever to achieve a world free from sexism.

ARE WE MEN THE ENEMY?

Some people say that men are the enemy when it comes to fighting sexism. We do not agree; blame and guilt don't help in understanding why people function as they do or in getting them to change.

Does this mean we are not responsible for what is happening? Not at all! As men, we are all involved in the oppression women experience, and we benefit from it each day. Yet this is no reason to fix blame on ourselves as "the oppressor" (or, for that matter, to place blame on any woman for "failing to fight back"). Over many years society has forced men and women into these roles of domination and submission.

How did society do this? It is clear we were pushed into these roles when we were young children and especially vulnerable. At that time, all of us were hurt in many ways through the expectations put on our sex. These hurts came through ridicule and threats directed at us and also through having to watch others get hurt while feeling ourselves incapable (because of lack of information or strength) of stopping that hurt. This experience of powerlessness in turn has reduced our ability to see the world clearly and act accordingly. The ingrained fear from these experiences tends to lead to either of two responses: feeling powerless and playing the victim role, or turning the experience around and acting out an oppressor role.

These dynamics do not just happen between men and women, but also between races, age groups, gays and non-gays, healthy and physically challenged people, and a number of other groups. As a matter of fact, everyone in our society has probably ended up playing both of these roles at one time or another. Even the richest, most powerful man was once powerless as a child in relation to his parents. Even a poor black woman could experience some privilege in relation to a very young person or a gay person.

The point is that we were all taught very early not to go beyond the expected stereotyped behavior. The sooner we recognize the effects of this kind of conditioning on us, the sooner we can effectively change the way things happen. This may mean stopping our domination of others or ceasing to accept oppression ourselves. It is good to recognize that stepping out of the old roles usually feels uncomfortable and may well require years of painful struggle, but to give up before the process is over is to miss the rich rewards that are down the road. We can travel that road successfully if we join with others for strength and support.

WHY CHANGE?

What rewards are there for us in this process? Much of what society thinks of as being manly is really a way of hardening up so we can dominate or coerce other people. We pay a high price for this hardening, however; it means giving up much of our humanity. We men are conditioned to suppress our feelings and don't learn how to give and receive support, nurturance, and affection awarely. We are taught to take all the responsibility for a situation on our own shoulders. We are taught to have heavy expectations of accomplishment for ourselves and for others. The direct result of this is a high level of tension and anxiety; the indirect result is a high disease rate and early death.

So, as men we have a lot to gain by fighting sexism. From what we have seen and experienced, men (at least in the long run) feel relief and joy just from being freed from the roles that lead to the oppression of women.

But this in no way means that we change easily. Conditioning is far too strong, and the temptation to keep the privilege too great. Nor does it mean, as some people suggest, that "men and women are both oppressed equally by sexism." However much men are hurt or limited by sex roles in this country, the fact remains that they are *not* systematically denied power simply because of being born a certain sex, as women are. Tremendous amounts of struggle will be required—as well as lots of loving support, especially from other men—to undergo the process of change.

RACISM, AGEISM, AND THE OPPRESSION
OF LESBIANS AND GAY MEN

All forms of oppression in our society are closely connected both to each other and to our economic system. Racism, for example, has many close parallels with sexism. Racial minorities, like women, function as a "reserve labor force," shunted into unattractive jobs with little reward or decision-making power, pushed in and out of employment as benefits the system. Racism, like sexism, divides working people from one another, preventing them from looking at who holds most of the power in our country. Racial minorities, like women, are thought to be unintelligent and to have less motivation to achieve and work hard than white males. Both groups are kept in place by violent intimidation and are referred to with abusive words. Both frequently get deflected from militant struggle against the oppression of their group by cooptation of a few of their leaders into the lower levels of the power structure. In addition, sexual myths and paranoia about blacks and Latinos play an especially vicious role in undergirding white people's racial fears and stereotypes.

People are just beginning to have a glimpse of what oppression based on age involves. The fact is that our society is almost totally blind to the dignity and capacities of the very young and the very old. Children are like women in being considered helpless, dependent, and cute—creatures to be cherished and taken care of, but not full human beings to be deeply respected and trusted with significant power. They experience 10–15 years of unpaid labor and brainwashing in our current form of education. Older people are looked at as children—except that they often find themselves without anyone interested in cherishing or taking care of them. For men, growing up is associated with taking on more and more of the hard masculine traits we mentioned before. Crying, acting afraid, and showing too much tenderness are all considered shameful because they are "childish" or "woman-like." ("You gotta stop that crying—don't be a sissy.")

Most of us know gay people, but usually we do not know who they are! Lesbians and gay men are so frequently hurt in this society that they usually do not feel safe to come out, even to their friends. An invisible colony of 20,000,000 people in our midst, lesbians and gay men generally work to service the status quo—even the Hollywood illusions of heterosexual romance—while their own dignity and security are denied.

People are still beaten and even killed for being homosexual in America, and the memory of the mass deaths of gays in Nazi concentration camps remains vivid. Now the "new right" is cutting back some recent gains lesbians and gay men have made in rights to jobs and housing. The fact that the same forces are opposing the Equal Rights Amendment and gay rights is a tip-off to the intimate connection between sexism and heterosexism. For one thing, gay men are mistakenly seen as taking the role of women in heterosexual relationships, and therefore not being "real men." For another, lesbians are seen as "uppity" because they act with a freedom not usually found when dependent upon men for loving.

Gay oppression is one of the ways the potential unity of all workers is prevented. It is also one of the cornerstones of the American nuclear family, which in turn is used to promote consumption and to teach sexist division of labor and sex roles. Lesbians and gay men are also exploited by being forced into ghettos and by the commercialized culture which profits from them.

Everyone is hurt by gay oppression. The fear of being considered gay limits and distorts everyone's life choices and relationships. Men are often afraid to get close to their male friends because it might imply gayness—and might even reveal a half-suspected gay dimension of themselves. An essential prop for sexism, in keeping people within their accustomed sex roles, is this fear of homosexuality, or homophobia. Because of this, women's liberation and men's liberation depend partly on gay liberation. . . .

AGE, RACE, CLASS, AND SEX:
Women Redefining Difference

19

Audre Lorde

Much of Western European history conditions us to see human differences in simplistic opposition to each other: dominant/subordinate, good/bad, up/down, superior/inferior. In a society where the good is defined in terms of profit rather than in terms of human need, there must always be some group of people who, through systematized oppression, can be made to feel surplus, to occupy the place of the dehumanized inferior. Within this society, that

Paper delivered at the Copeland Colloquium, Amherst College, April 1980.

From: Audre Lorde, *Sister Outsider* (Freedom, Calif.: Crossing Press, 1984), pp. 114–23. Reprinted by permission.

group is made up of Black and Third World people, working-class people, older people, and women.

As a forty-nine-year-old Black lesbian feminist socialist mother of two, including one boy, and a member of an interracial couple, I usually find myself a part of some group defined as other, deviant, inferior, or just plain wrong. Traditionally, in american society, it is the members of oppressed, objectified groups who are expected to stretch out and bridge the gap between the actualities of our lives and the consciousness of our oppressor. For in order to survive, those of us for whom oppression is as american as apple pie have always had to be watchers, to become familiar with the language and manners of the oppressor, even sometimes adopting them for some illusion of protection. Whenever the need for some pretense of communication arises, those who profit from our oppression call upon us to share our knowledge with them. In other words, it is the responsibility of the oppressed to teach the oppressors their mistakes. I am responsible for educating teachers who dismiss my children's culture in school. Black and Third World people are expected to educate white people as to our humanity. Women are expected to educate men. Lesbians and gay men are expected to educate the heterosexual world. The oppressors maintain their position and evade responsibility for their own actions. There is a constant drain of energy which might be better used in redefining ourselves and devising realistic scenarios for altering the present and constructing the future.

Institutionalized rejection of difference is an absolute necessity in a profit economy which needs outsiders as surplus people. As members of such an economy, we have *all* been programmed to respond to the human differences between us with fear and loathing and to handle that difference in one of three ways: ignore it, and if that is not possible, copy it if we think it is dominant, or destroy it if we think it is subordinate. But we have no patterns for relating across our human differences as equals. As a result, those differences have been misnamed and misused in the service of separation and confusion.

Certainly there are very real differences between us of race, age, and sex. But it is not those differences between us that are separating us. It is rather our refusal to recognize those differences, and to examine the distortions which result from our misnaming them and their effects upon human behavior and expectation.

Racism, the belief in the inherent superiority of one race over all others and thereby the right to dominance. Sexism, the belief in the inherent superiority of one sex over the other and thereby the right to dominance. Ageism. Heterosexism. Elitism. Classism.

It is a lifetime pursuit for each one of us to extract these distortions from our living at the same time as we recognize, reclaim, and define those differences upon which they are imposed. For we have all been raised in a society where those distortions were endemic within our living. Too often, we pour the energy needed for recognizing and exploring difference into pretending those differences are insurmountable barriers, or that they do not exist at all. This results in a voluntary isolation, or false and treacherous connections. Either way, we do not develop tools for using human difference as a springboard for creative change within our lives. We speak not of human difference, but of human deviance.

Somewhere, on the edge of consciousness, there is what I call a *mythical norm*, which each one of us within our hearts knows "that is not me." In america, this norm is usually defined as white, thin, male, young, heterosexual, christian, and financially secure. It is with this mythical norm that the trappings of power reside within this society. Those of us who stand outside that power often identify one way in which we are different, and we assume that to be the primary cause of all oppression, forgetting other distortions around difference, some of which we ourselves may be practicing. By and large within the women's movement today, white women focus upon their oppression as women and ignore differences of race, sexual preference, class, and age. There is a pretense to a homogeneity of experience covered by the word *sisterhood* that does not in fact exist.

Unacknowledged class differences rob women of each others' energy and creative insight. Recently a women's magazine collective made the decision for one issue to print only prose, saying poetry was a less "rigorous" or "serious" art form. Yet even the form our creativity takes is often a class issue. Of all the art forms, poetry is the most economical. It is the one which is the most secret, which requires the least physical labor, the least material, and the one which can be done between shifts, in the hospital pantry, on the subway, and on scraps of surplus paper. Over the last few years, writing a novel on tight finances, I came to appreciate the enormous differences in the material demands between poetry and prose. As we reclaim our literature, poetry has been the major voice of poor, working class, and Colored women. A room of one's own may be a necessity for writing prose, but so are reams of paper, a typewriter, and plenty of time. The actual requirements to produce the visual arts also help determine, along class lines, whose art is whose. In this day of inflated prices for material, who are our sculptors, our painters, our photographers? When we speak of a broadly based women's culture, we need to be aware of the effect of class and economic differences on the supplies available for producing art.

As we move toward creating a society within which we can each flourish, ageism is another distortion of relationship which interferes without vision. By ignoring the past, we are encouraged to repeat its mistakes. The "generation gap" is an important social tool for any repressive society. If the younger members of a community view the older members as contemptible or suspect or excess, they will never be able to join hands and examine the living memories of the community, nor ask the all important question, "Why?" This gives rise to a historical amnesia that keeps us working to invent the wheel every time we have to go to the store for bread.

We find ourselves having to repeat and relearn the same old lessons over and over that our mothers did because we do not pass on what we have learned, or because we are unable to listen. For instance, how many times has this all been said before? For another, who would have believed that once again our daughters are allowing their bodies to be hampered and purgatoried by girdles and high heels and hobble skirts?

Ignoring the differences of race between women and the implications of those differences presents the most serious threat to the mobilization of women's joint power.

As white women ignore their built-in privilege of whiteness and define *woman* in terms of their own experience alone, then women of Color become "other," the outsider whose experience and tradition is too "alien" to comprehend. An example of this is the signal absence of the experience of women of Color as a resource for women's studies courses. The literature of women of Color is seldom included in women's literature courses and almost never in other literature courses, nor in women's studies as a whole. All too often, the excuse given is that the literatures of women of Color can only be taught by Colored women, or that they are too difficult to understand, or that classes cannot "get into" them because they come out of experiences that are "too different." I have heard this argument presented by white women of otherwise quite clear intelligence, women who seem to have no trouble at all teaching and reviewing work that comes out of the vastly different experiences of Shakespeare, Molière, Dostoyefsky, and Aristophanes. Surely there must be some other explanation.

This is a very complex question, but I believe one of the reasons white women have such difficulty reading Black women's work is because of their reluctance to see Black women as women and different from themselves. To examine Black women's literature effectively requires that we be seen as whole people in our actual complexities—as individuals, as women, as human—rather than as one of those problematic but familiar stereotypes provided in this society in place of genuine images of Black women. And I believe this holds true for the literatures of other women of Color who are not Black.

The literatures of all women of Color recreate the textures of our lives, and many white women are heavily invested in ignoring the real differences. For as long as any difference between us means one of us must be inferior, then the recognition of any difference must be fraught with guilt. To allow women of Color to step out of stereotypes is too guilt provoking, for it threatens the complacency of those women who view oppression only in terms of sex.

Refusing to recognize difference makes it impossible to see the different problems and pitfalls facing us as women.

Thus, in a patriarchal power system where whiteskin privilege is a major prop, the entrapments used to neutralize Black women and white women are not the same. For example, it is easy for Black women to be used by the power structure against Black men, not because they are men, but because they are Black. Therefore, for Black women, it is necessary at all times to separate the needs of the oppressor from our own legitimate conflicts within our communities. This same problem does not exist for white women. Black women and men have shared racist oppression and still share it, although in different ways. Out of that shared oppression we have developed joint defenses and joint vulnerabilities to each other that are not duplicated in the white community, with the exception of the relationship between Jewish women and Jewish men.

On the other hand, white women face the pitfall of being seduced into joining the oppressor under the pretense of sharing power. This possibility does not exist in the same way for women of Color. The tokenism that is sometimes extended to us is not an invitation to join power; our racial "otherness" is a visible reality that makes that quite clear. For white women there is a wider range of pretended choices and rewards for identifying with patriarchal power and its tools.

Today, with the defeat of ERA, the tightening economy, and increased conservatism, it is easier once again for white women to believe the dangerous fantasy that if you are good enough, pretty enough, sweet enough, quiet enough, teach the children to behave, hate the right people, and marry the right men, then you will be allowed to co-exist with patriarchy in relative peace, at least until a man needs your job or the neighborhood rapist happens along. And true, unless one lives and loves in the trenches it is difficult to remember that the war against dehumanization is ceaseless.

But Black women and our children know the fabric of our lives is stitched with violence and with hatred, that there is no rest. We do not deal with it only on the picket lines, or in dark midnight alleys, or in the places where we dare to verbalize our resistance. For us, increasingly, violence weaves through the daily tissues of our living—in the supermarket, in the classroom, in the elevator, in the clinic and the schoolyard, from the plumber, the baker, the saleswoman, the bus driver, the bank teller, the waitress who does not serve us.

Some problems we share as women, some we do not. You fear your children will grow up to join the patriarchy and testify against you, we fear our children will be dragged from a car and shot down in the street, and you will turn your backs upon the reasons they are dying.

The threat of difference has been no less blinding to people of Color. Those of us who are Black must see that the reality of our lives and our struggle does not make us immune to the errors of ignoring and misnaming difference. Within Black communities where racism is a living reality, differences among us often seem dangerous and suspect. The need for unity is often misnamed as a need for homogeneity, and a Black feminist vision mistaken for betrayal of our common interests as a people. Because of the continuous battle against racial erasure that Black women and Black men share, some Black women still refuse to recognize that we are also oppressed as women, and that sexual hostility against Black women is practiced not only by the white racist society, but implemented within our Black communities as well. It is a disease striking the heart of Black nationhood, and silence will not make it disappear. Exacerbated by racism and the pressures of powerlessness, violence against Black women and children often becomes a standard within our communities, one by which manliness can be measured. But these woman-hating acts are rarely discussed as crimes against Black women.

As a group, women of Color are the lowest paid wage earners in america. We are the primary targets of abortion and sterilization abuse, here and abroad. In certain parts of Africa, small girls are still being sewed shut between their legs to keep them docile and for men's pleasure. This is known as female circumcision, and it is not a cultural affair as the late Jomo Kenyatta insisted, it is a crime against Black women.

Black women's literature is full of the pain of frequent assault, not only by a racist patriarchy, but also by Black men. Yet the necessity for and history of shared battle have made us, Black women, particularly vulnerable to the false accusation that anti-sexist is anti-Black. Meanwhile, womanhating as a recourse of the powerless is sapping strength from Black communities, and our very lives. Rape is on the increase, reported and unreported, and rape is not aggressive sexuality, it is sexualized aggression. As Kalamu ya Salaam, a Black male writer points out, "As long as male domination exists, rape will exist. Only women revolting and men made conscious of their responsibility to fight sexism can collectively stop rape."*

Differences between ourselves as Black women are also being misnamed and used to separate us from one another. As a Black lesbian feminist comfortable with the many different ingredients of my identity, and a woman committed to racial and sexual freedom from oppression, I find I am con-

---

* From "Rape: A Radical Analysis, An African-American Perspective" by Kalamu ya Salaam in *Black Books Bulletin*, vol. 6, no. 4 (1980).

stantly being encouraged to pluck out some one aspect of myself and present this as the meaningful whole, eclipsing or denying the other parts of self. But this is a destructive and fragmenting way to live. My fullest concentration of energy is available to me only when I integrate all the parts of who I am, openly, allowing power from particular sources of my living to flow back and forth freely through all my different selves, without the restrictions of externally imposed definition. Only then can I bring myself and my energies as a whole to the service of those struggles which I embrace as part of my living.

A fear of lesbians, or of being accused of being a lesbian, has led many Black women into testifying against themselves. It has led some of us into destructive alliances, and others into despair and isolation. In the white women's communities, heterosexism is sometimes a result of identifying with the white patriarchy, a rejection of that interdependence between women-identified women which allows the self to be, rather than to be used in the service of men. Sometimes it reflects a die-hard belief in the protective coloration of heterosexual relationships, sometimes a self-hate which all women have to fight against, taught us from birth.

Although elements of these attitudes exist for all women, there are particular resonances of heterosexism and homophobia among Black women. Despite the fact that woman-bonding has a long and honorable history in the African and African-american communities, and despite the knowledge and accomplishments of many strong and creative women-identified Black women in the political, social and cultural fields, heterosexual Black women often tend to ignore or discount the existence and work of Black lesbians. Part of this attitude has come from an understandable terror of Black male attack within the close confines of Black society, where the punishment for any female self-assertion is still to be accused of being a lesbian and therefore unworthy of the attention or support of the scarce Black male. But part of this need to misname and ignore Black lesbians comes from a very real fear that openly women-identified Black women who are no longer dependent upon men for their self-definition may well reorder our whole concept of social relationships.

Black women who once insisted that lesbianism was a white woman's problem now insist that Black lesbians are a threat to Black nationhood, are consorting with the enemy, are basically un-Black. These accusations, coming from the very women to whom we look for deep and real understanding, have served to keep many Black lesbians in hiding, caught between the racism of white women and the homophobia of their sisters. Often, their work has been ignored, trivialized, or misnamed, as with the work of Angelina Grimke, Alice Dunbar-Nelson, Lorraine Hansberry. Yet women-bonded women have always been some part of the power of Black communities, from our unmarried aunts to the amazons of Dahomey.

And it is certainly not Black lesbians who are assaulting women and raping children and grandmothers on the streets of our communities.

Across this country, as in Boston during the spring of 1979 following the unsolved murders of twelve Black women, Black lesbians are spearheading movements against violence against Black women.

What are the particular details within each of our lives that can be scrutinized and altered to help bring about change? How do we redefine difference for all women? It is not our differences which separate women, but our reluctance to recognize those differences and to deal effectively with the distortions which have resulted from the ignoring and misnaming of those differences.

As a tool of social control, women have been encouraged to recognize only one area of human difference as legitimate, those differences which exist between women and men. And we have learned to deal across those differences with the urgency of all oppressed subordinates. All of us have had to learn to live or work or coexist with men, from our fathers on. We have recognized and negotiated these differences, even when this recognition only continued the old dominant/subordinate mode of human relationship, where the oppressed must recognize the masters' difference in order to survive.

But our future survival is predicated upon our ability to relate within equality. As women, we must root out internalized patterns of oppression within ourselves if we are to move beyond the most superficial aspects of social change. Now we must recognize differences among women who are our equals, neither inferior nor superior, and devise ways to use each others' difference to enrich our visions and our joint struggles.

The future of our earth may depend upon the ability of all women to identify and develop new definitions of power and new patterns of relating across difference. The old definitions have not served us, nor the earth that supports us. The old patterns, no matter how cleverly rearranged to imitate progress, still condemn us to cosmetically altered repetitions of the same old exchanges, the same old guilt, hatred, recrimination, lamentation, and suspicion.

For we have, built into all of us, old blueprints of expectation and response, old structures of oppression, and these must be altered at the same time as we alter the living conditions which are a result of those structures. For the master's tools will never dismantle the master's house.

As Paulo Freire shows so well in *The Pedagogy of the Oppressed*,* the true focus of revolutionary change is never merely the oppressive situations which we seek to escape, but that piece of the oppressor which is planted deep within each of us, and which knows only the oppressors' tactics, the oppressors' relationships.

---

* Seabury Press, New York, 1970.

Change means growth, and growth can be painful. But we sharpen self-definition by exposing the self in work and struggle together with those whom we define as different from ourselves, although sharing the same goals. For Black and white, old and young, lesbian and heterosexual women alike, this can mean new paths to our survival.

> We have chosen each other
> and the edge of each others battles
> the war is the same
> if we lose
> someday women's blood will congeal
> upon a dead planet
> if we win
> there is no telling
> we seek beyond history
> for a new and more possible meeting.*

MASCULINITIES AND ATHLETIC CAREERS

20

Michael Messner

. . . It is now widely accepted in sport sociology that social institutions such as the media, education, the economy, and (a more recent and controversial addition to the list) the black family itself all serve to systematically channel disproportionately large numbers of young black men into football, basketball, boxing, and baseball, where they are subsequently "stacked" into low-prestige and high-risk positions, exploited for their skills, and finally, when

---

* From "Outlines," unpublished poem.

Author's note: Parts of this article were presented as papers at the American Sociological Association Annual Meeting, Chicago, in August 1987, and at the North American Society for the Sociology of Sport Annual Meeting in Edmonton, Alberta, in November 1987. I thank Maxine Baca Zinn, Bob Blauner, Bob Dunn, Pierrette Hondagneu-Sotelo, Carol Jacklin, Michael Kimmel, Judith Lorber, Don Sabo, Barrie Thorne, and Carol Warren for constructive comments on earlier versions of this article.

From: *Gender & Society* 3 (March 1989): 71–88. Reprinted by permission of Sage Publications, Inc.

their bodies are used up, excreted from organized athletics at a young age with no transferable skills with which to compete in the labor market (Edwards 1984; Eitzen and Purdy 1986; Eitzen and Yetman 1977).

While there are racial differences in involvement in sports, class, age, and educational differences seem more significant. Rudman's (1986) initial analysis revealed profound differences between whites' and blacks' orientations to sports. Blacks were found to be more likely than whites to view sports favorably, to incorporate sports into their daily lives, and to be affected by the outcome of sporting events. However, when age, education, and social class were factored into the analysis, Rudman found that race did not explain whites' and blacks' different orientations. Black's affinity to sports is best explained by their tendency to be clustered disproportionately in lower-income groups.

The 1980s has ushered in what Wellman (1986, p. 43) calls a "new political linguistics of race," which emphasizes cultural rather than structural causes (and solutions) to the problems faced by black communities. The advocates of the cultural perspective believe that the high value placed on sports by black communities has led to the development of unrealistic hopes in millions of black youths. They appeal to family and community to bolster other choices based upon a more rational assessment of "reality." Visible black role models in many other professions now exist, they say, and there is ample evidence which proves that sports careers are, at best, a bad gamble.

Critics of the cultural perspective have condemned it as conservative and victim blaming. But it can also be seen as a response to the view of black athletes as little more than unreflexive dupes of an all-powerful system, which ignores the importance of agency. Gruneau (1983) has argued that sports must be examined within a theory that views human beings as active subjects who are operating within historically constituted structural constraints. Gruneau's reflexive theory rejects the simplistic views of sports as either a realm of absolute oppression or an arena of absolute freedom and spontaneity. Instead, he argues, it is necessary to construct an understanding of how and why participants themselves actively make choices and construct and define meaning and a sense of identity within the institutions that they find themselves.

None of these perspectives considers the ways that gender shapes men's definitions of meaning and choices. Within the sociology of sport, gender as a process that interacts with race and class is usually ignored or taken for granted—except when it is *women* athletes who are being studied. Sociologists who are attempting to come to grips with the experiences of black men in general, and in organized sports in particular, have almost exclusively focused their analytic attention on the variable "black," while uncritically taking "men" as a given. Hare and Hare (1984), for example, view masculinity as a biologically determined tendency to act as a provider and protector that is thwarted for black men by socioeconomic and racist obstacles. Staples (1982) does view masculinity largely as a socially produced script, but he accepts this script as a given, preferring to focus on black men's blocked access to male role fulfillment. These perspectives on masculinity fail to show how the male role itself, as it interacts with a constricted structure of opportunity, can contribute to locking black men into destructive relationships and life-styles (Franklin 1984; Majors 1986).

This article will examine the relationships among male identity, race, and social class by listening to the voices of former athletes. I will first briefly describe my research. Then I will discuss the similarities and differences in the choices and experiences of men from different racial and social class backgrounds. Together, these choices and experiences help to construct what Connell (1987) calls "the gender order." Organized sports, it will be suggested, is a practice through which men's separation from and power over women is embodied and naturalized at the same time that hegemonic (white, heterosexual, professional-class) masculinity is clearly differentiated from marginalized and subordinated masculinities.

DESCRIPTION OF RESEARCH

Between 1983 and 1985, I conducted 30 open-ended, in-depth interviews with male former athletes. My purpose was to add a critical understanding of male gender identity to Levinson's (1978) conception of the "individual life-course"—specifically, to discover how masculinity develops and changes as a man interacts with the socially constructed world of organized sports. Most of the men I interviewed had played the U.S. "major sports"—football, basketball, baseball, track. At the time of the interview, each had been retired from playing organized sports for at least 5 years. Their ages ranged from 21 to 48, with the median, 33. Fourteen were black, 14 were white, and 2 were Hispanic. Fifteen of the 16 black and Hispanic men had come from poor or working-class families, while the majority (9 of 14) of the white men had come from middle-class or professional families. Twelve had played organized sports through high school, 11 through college, and 7 had been professional athletes. All had at some time in their lives based their identities largely on their roles as athletes and could therefore be said to have had athletic careers.

MALE IDENTITY AND ORGANIZED SPORTS

. . . For the men in my study, the rule-bound structure of organized sports became a context in which they struggled to construct a masculine positional identity.

All of the men in this study described the emotional salience of their earliest experiences in sports in terms of relationships with other males. It was not winning and victories that seemed important at first; it was something "fun" to do with fathers, older brothers or uncles, and eventually with same-aged peers. As a man from a white, middle-class family said, "The most important thing was just being out there with the rest of the guys—being friends." A 32-year-old man from a poor Chicano family, whose mother had died when he was 9 years old, put it more succinctly:

> What I think sports did for me is it brought me into kind of an instant family. By being on a Little League team, or even just playing with kids in the neighborhood, it brought what I really wanted, which was some kind of closeness.

Though sports participation may have initially promised "some kind of closeness," by the ages of 9 or 10, the less skilled boys were already becoming alienated from—or weeded out of—the highly competitive and hierarchical system of organized sports. Those who did experience some early successes received recognition from adult males (especially fathers and older brothers) and held higher status among peers. As a result, they began to pour more and more of their energies into athletic participation. It was only after they learned that they would get recognition from other people for being a good athlete—indeed, that this attention was contingent upon *being a winner*—that performance and winning (the dominant values of organized sports) became extremely important. For some, this created pressures that served to lessen or eliminate the fun of athletic participation (Messner 1987a, 1987b).

While feminist psychoanalytic and developmental theories of masculinity are helpful in explaining boys' early attraction and motivations in organized sports, the imperatives of core gender identity do not fully determine the contours and directions of the life course. As Rubin (1985) and Levinson (1978) have pointed out, an understanding of the lives of men must take into account the processual nature of male identity as it unfolds through interaction between the internal (psychological ambivalences) and the external (social, historical, and institutional) contexts.

To examine the impact of the social contexts, I divided my sample into two comparison groups. In the first group were 10 men from higher-status backgrounds, primarily white, middle-class, and professional families. In the second group were 20 men from lower-status backgrounds, primarily minority, poor, and working-class families. While my data offered evidence for the similarity of experiences and motivations of men from poor backgrounds, independent of race, I also found anecdotal evidence of a racial dynamic that operates independently of social class. However, my sample was not large enough to separate race and class, and so I have combined them to make two status groups.

In discussing these two groups, I will focus mainly on the high school years. During this crucial period, the athletic role may become a master status for a young man, and he is beginning to make assessments and choices about his future. It is here that many young men make a major commitment to—or begin to back away from—athletic careers.

Men from Higher-Status Backgrounds

The boyhood dream of one day becoming a professional athlete—a dream shared by nearly all the men interviewed in this study—is rarely realized. The sports world is extremely hierarchical. The pyramid of sports careers narrows very rapidly as one climbs from high school, to college, to professional levels of competition (Edwards 1984; Harris and Eitzen 1978; Hill and Lowe 1978). In fact, the chances of attaining professional status in sports are approximately 4/100,000 for a white man, 2/100,000 for a black man, and 3/100,000 for a Hispanic man in the United States (Leonard and Reyman 1988). For many young athletes, their dream ends early when coaches inform them that they are not big enough, strong enough, fast enough, or skilled enough to compete at the higher levels. But six of the higher-status men I interviewed did not wait for coaches to weed them out. They made conscious decisions in high school or in college to shift their attentions elsewhere—usually toward educational and career goals. Their decision not to pursue an athletic career appeared to them in retrospect to be a rational decision based on the growing knowledge of how very slim their chances were to be successful in the sports world. For instance, a 28-year-old white graduate student said:

> By junior high I started to realize that I was a good player—maybe even one of the best in my community—but I realized that there were all these people all over the country and how few will get to play pro sports. By high school, I still dreamed of being a pro—I was a serious athlete, I played hard—but I knew it wasn't heading anywhere. I wasn't going to play pro ball.

A 32-year-old white athletic director at a small private college had been a successful college baseball player. Despite considerable attention from professional scouts, he had decided to forgo a shot at a baseball career and to enter graduate school to pursue a teaching credential. As he explained this decision:

> At the time I think I saw baseball as pissing in the wind, really. I was married, I was 22 years old with a kid. I didn't want to spend 4 or 5 years in the

> minors with a family. And I could see I wasn't a superstar; so it wasn't really worth it. So I went to grad school. I thought that would be better for me.

Perhaps most striking was the story of a high school student body president and top-notch student who was also "Mr. Everything" in sports. He was named captain of his basketball, baseball, and football teams and achieved All-League honors in each sport. This young white man from a middle-class family received attention from the press and praise from his community and peers for his athletic accomplishments, as well as several offers of athletic scholarships from universities. But by the time he completed high school, he had already decided to quit playing organized sports. As he said:

> I think in my own mind I kind of downgraded the stardom thing. I thought that was small potatoes. And sure, that's nice in high school and all that, but on a broad scale, I didn't think it amounted to all that much. So I decided that my goal's to be a dentist, as soon as I can.

In his sophomore year of college, the basketball coach nearly persuaded him to go out for the team, but eventually he decided against it:

> I thought, so what if I can spend two years playing basketball? I'm not going to be a basketball player forever and I might jeopardize my chances of getting into dental school if I play.

He finished college in three years, completed dental school, and now, in his mid-30s, is again the epitome of the successful American man: a professional with a family, a home, and a membership in the local country club.

How and why do so many successful male athletes from higher-status backgrounds come to view sports careers as "pissing in the wind," or as "small potatoes"? How and why do they make this early assessment and choice to shift from sports and toward educational and professional goals? The white, middle-class institutional context, with its emphasis on education and income, makes it clear to them that choices exist and that the pursuit of an athletic career is not a particularly good choice to make. Where the young male once found sports to be a convenient institution within which to construct masculine status, the postadolescent and young adult man from a higher-status background simply *transfers* these same strivings to other institutional contexts: education and careers.

For the higher-status men who had chosen to shift from athletic careers, sports remained important on two levels. First, having been a successful high school or college athlete enhances one's adult status among other men in the community —but only as a badge of masculinity that is *added* to his professional status. In fact, several men in professions chose to be interviewed in their offices, where they publicly displayed the trophies and plaques that attested to their earlier athletic accomplishments. Their high school and college athletic careers may have appeared to them as "small potatoes," but many successful men speak of their earlier status as athletes as having "opened doors" for them in their present professions and in community affairs. Similarly, Farr's (1988) research on "Good Old Boys Sociability Groups" shows how sports, as part of the glue of masculine culture, continues to facilitate "dominance bonding" among privileged men long after active sports careers end. The college-educated, career-successful men in Farr's study rarely express overtly sexist, racist, or classist attitudes; in fact, in their relationships with women, they "often engage in expressive intimacies" and "make fun of exaggerated 'machismo'" (p. 276). But though they outwardly conform more to what Pleck (1982) calls "the modern male role," their informal relationships within their sociability groups, in effect, affirm their own gender and class status by constructing and clarifying the boundaries between themselves and women and lower-status men. This dominance bonding is based largely upon ritual forms of sociability (camaraderie, competition), "the superiority of which was first affirmed in the exclusionary play activities of young boys in groups" (Farr 1988, p. 265).

In addition to contributing to dominance bonding among higher-status adult men, sports remains salient in terms of the ideology of gender relations. Most men continued to watch, talk about, and identify with sports long after their own disengagement from athletic careers. Sports as a mediated spectacle provides an important context in which traditional conceptions of masculine superiority—conceptions recently contested by women—are shored up. As a 32-year-old white professional-class man said of one of the most feared professional football players today:

> A woman can do the same job as I can do—maybe even be my boss. But I'll be *damned* if she can go out on the football field and take a hit from Ronnie Lott.

Violent sports as spectacle provide linkages among men in the project of the domination of women, while at the same time helping to construct and clarify differences among various masculinities. The statement above is a clear identification with Ronnie Lott *as a man*, and the basis of the identification is the violent male body. As Connell (1987, p. 85) argues, sports is an important organizing institution for the embodiment of masculinity. Here, men's power over women becomes naturalized and linked to the social distribution of violence. Sports, as a practice, suppresses natural (sex) similarities, constructs differences, and then, largely through the media, weaves a structure of symbol and interpretation around these differences that naturalizes them (Hargreaves 1986, p. 112). It is also significant that the man who made the above statement about Ronnie Lott was quite aware that he (and perhaps 99 percent of the rest of the U.S. male population) was probably as incapable as most women of taking a "hit" from someone like Lott and living to tell of it. For middle-class men, the "tough guys" of the culture industry—the Rambos, the Ronnie Lotts who are fearsome "hitters," who "play hurt"—are the heroes who "prove" that "we men" are superior to women. At the same time, they play the role of the "primitive other," against whom higher-status men define themselves as "modern" and "civilized."

Sports, then, is important from boyhood through adulthood for men from higher-status backgrounds. But it is significant that by adolescence and early adulthood, most of these young men have concluded that sports *careers* are not for them. Their middle-class cultural environment encourages them to decide to shift their masculine strivings in more "rational" directions: education and nonsports careers. Yet their previous sports participation continues to be very important to them in terms of constructing and validating their status within privileged male peer groups and within their chosen professional careers. And organized sports, as a public spectacle, is a crucial locus around which ideologies of male superiority over women, as well as higher-status men's superiority over lower-status men, are constructed and naturalized.

Men from Lower-Status Backgrounds

For the lower-status young men in this study, success in sports was not an added proof of masculinity; it was often their only hope of achieving public masculine status. A 34-year-old black bus driver who had been a star athlete in three sports in high school had neither the grades nor the money to attend college, so he accepted an offer from the U.S. Marine Corps to play on their baseball team. He ended up in Vietnam, where a grenade blew four fingers off his pitching hand. In retrospect, he believed that his youthful focus on sports stardom and his concomitant lack of effort in academics made sense:

> You can go anywhere with athletics—you don't have to have brains. I mean, I didn't feel like I was gonna go out there and be a computer expert, or something that was gonna make a lot of money. The only thing I could do and live comfortably would be to play sports—just to get a contract—doesn't matter if you play second or third team in the pros, you're gonna make big bucks. That's all I wanted, a confirmed livelihood at the end of my ventures, and the only way I could do it would be through sports. So I tried. It failed, but that's what I tried.

Similar, and even more tragic, is the story of a 34-year-old black man who is now serving a life term in prison. After a career-ending knee injury at the age of 20 abruptly ended what had appeared to be a certain road to professional football fame and fortune, he decided that he "could still be rich and famous" by robbing a bank. During his high school and college years, he said, he was nearly illiterate:

> I'd hardly ever go to classes and they'd give me *C*s. My coaches taught some of the classes. And I felt, "So what? They *owe* me that! I'm an *athlete!* I thought that was what I was born to do—to play sports—and everybody understood that.

Are lower-status boys and young men simply duped into putting all their eggs into one basket? My research suggested that there was more than "hope for the future" operating here. There were also immediate psychological reasons that they chose to pursue athletic careers. By the high school years, class and ethnic inequalities had become glaringly obvious, especially for those who attended socioeconomically heterogeneous schools. Cars, nice clothes, and other signs of status were often unavailable to these young men, and this contributed to a situation in which sports took on an expanded importance for them in terms of constructing masculine identities and status. A white, 36-year-old man from a poor, single-parent family who later played professional baseball had been acutely aware of his low-class status in his high school:

> I had one pair of jeans, and I wore them every day. I was always afraid of what people thought of me—that this guy doesn't have anything, that he's wearing the same Levi's all the time, he's having to work in the cafeteria for his lunch. What's going on? I think that's what made me so shy. . . . But boy, when I got into sports, I let it all hang out—[laughs]—and maybe that's why I became so good, because I was frustrated, and when I got into that element, they gave me my uniform in football, basketball, and baseball, and I didn't have to worry about how I looked, because then it was *me* who was coming out, and not my clothes or whatever. And I think that was the drive.

Similarly, a 41-year-old black man who had a 10-year professional football career described his insecurities as one of the few poor blacks in a mostly white, middle-class school and his belief that sports was the one arena in which he could be judged solely on his merit:

> I came from a poor family, and I was very sensitive about that in those days. When people would say things like "Look at him—he has dirty pants on," I'd think about it for a week. [But] I'd put my pants on and I'd go out on the football field with the intention that I'm gonna do a job. And if that calls on me to

> hurt you, I'm gonna do it. It's as simple as that. I demand respect just like
> everybody else.

"Respect" was what I heard over and over when talking with the men from lower-status backgrounds, especially black men. I interpret this type of respect to be a crystallization of the masculine quest for recognition through public achievement, unfolding within a system of structured constraints due to class and race inequities. The institutional context of education (sometimes with the collusion of teachers and coaches) and the constricted structure of opportunity in the economy made the pursuit of athletic careers appear to be the most rational choice to these young men.

The same is not true of young lower-status women. Dunkle (1985) points out that from junior high school through adulthood, young black men are far more likely to place high value on sports than are young black women, who are more likely to value academic achievement. There appears to be a gender dynamic operating in adolescent male peer groups that contributes toward their valuing sports more highly than education. Franklin (1986, p. 161) has argued that many of the normative values of the black male peer group (little respect for nonaggressive solutions to disputes, contempt for nonmaterial culture) contribute to the constriction of black men's views of desirable social positions, especially through education. In my study, a 42-year-old black man who did succeed in beating the odds by using his athletic scholarship to get a college degree and eventually becoming a successful professional said:

> By junior high, you either got identified as an athlete, a thug, or a book-
> worm. It's very important to be seen as somebody who's capable in some area.
> And you *don't* want to be identified as a bookworm. I was very good with books,
> but I was kind of covert about it. I was a closet bookworm. But with sports, I was
> *somebody;* so I worked very hard at it.

For most young men from lower-status backgrounds, the poor quality of their schools, the attitudes of teachers and coaches, as well as the anti-education environment within their own male peer groups, made it extremely unlikely that they would be able to succeed as students. Sports, therefore, became *the* arena in which they attempted to "show their stuff." For these lower-status men, as Baca Zinn (1982) and Majors (1986) argued in their respective studies of Chicano men and black men, when institutional resources that signify masculine status and control are absent, physical presence, personal style, and expressiveness take on increased importance. What Majors (1986, p. 6) calls "cool pose" is black men's expressive, often aggressive, assertion of masculinity. This self-assertion often takes place within a social context in which the young man is quite aware of existing social inequities. As the black bus driver, referred to above, said of his high school years:

> See, the rich people use their money to do what they want to do. I use my ability. If you wanted to be around me, if you wanted to learn something about sports, I'd teach you. But you're gonna take me to lunch. You're gonna let me use your car. See what I'm saying? In high school I'd go where I wanted to go. I didn't have to be educated. I was well-respected. I'd go somewhere, and they'd say, "Hey, that's Mitch Harris,[1] yeah, that's a bad son of a bitch!"

Majors (1986) argues that although "cool pose" represents a creative survival technique within a hostile environment, the most likely long-term effect of this masculine posturing is educational and occupational dead ends. As a result, we can conclude, lower-status men's personal and peer-group responses to a constricted structure of opportunity—responses that are rooted, in part, in the developmental insecurities and ambivalences of masculinity—serve to lock many of these young men into limiting activities such as sports.

SUMMARY AND CONCLUSIONS

This research has suggested that within a social context that is stratified by social class and by race, the choice to pursue—or not to pursue—an athletic career is explicable as an individual's rational assessment of the available means to achieve a respected masculine identity. For nearly all of the men from lower-status backgrounds, the status and respect that they received through sports was temporary—it did not translate into upward mobility. Nonetheless, a strategy of discouraging young black boys and men from involvement in sports is probably doomed to fail, since it ignores the continued existence of structural constraints. Despite the increased number of black role models in nonsports professions, employment opportunities for young black males have actually deteriorated. . . .

But it would be a mistake to conclude that we simply need to breed socioeconomic conditions that make it possible for poor and minority men to mimic the "rational choices" of white, middle-class men. If we are to build an appropriate understanding of the lives of all men, we must critically analyze white middle-class masculinity, rather than uncritically taking it as a normative standard. To fail to do this would be to ignore the ways in which organized sports serves to construct and legitimate gender differences and inequalities among men and women.

Feminist scholars have demonstrated that organized sports gives men from all backgrounds a means of status enhancement that is not available to young women. Sports thus serve the interests of all men in helping to construct and legitimize their control of public life and their domination of women (Bryson 1987; Hall 1987; Theberge 1987). Yet concrete studies are suggesting that men's experiences within sports are not all of a piece. Brian

Pronger's (1990) research suggests that gay men approach sports differently than straight men do, with a sense of "irony." And my research suggests that although sports are important for men from both higher- and lower-status backgrounds, there are crucial differences. In fact, it appears that the meaning that most men give to their athletic strivings has more to do with competing for status among men than it has to do with proving superiority over women. How can we explain this seeming contradiction between the feminist claim that sports links all men in the domination of women and the research findings that different groups of men relate to sports in very different ways?

. . . Connell's (1987) concept of the "gender order" is useful. The gender order is a dynamic process that is constantly in a state of play. Moving beyond static gender-role theory and reductionist concepts of patriarchy that view men as an undifferentiated group which oppresses women, Connell argues that at any given historical moment, there are competing masculinities—some hegemonic, some marginalized, some stigmatized. Hegemonic masculinity (that definition of masculinity which is culturally ascendant) is constructed in relation to various subordinated masculinities as well as in relation to femininities. The project of male domination of women may tie all men together, but men share very unequally in the fruits of this domination.

These are key insights in examining the contemporary meaning of sports. Utilizing the concept of the gender order, we can begin to conceptualize how hierarchies of race, class, age, and sexual preference among men help to construct and legitimize men's overall power and privilege over women. And how, for some black, working-class, or gay men, the false promise of sharing in the fruits of hegemonic masculinity often ties them into their marginalized and subordinate statuses within hierarchies of intermale dominance. For instance, black men's development of what Majors (1986) calls "cool pose" within sports can be interpreted as an example of creative resistance to one form of social domination (racism); yet it also demonstrates the limits of an agency that adopts other forms of social domination (masculinity) as its vehicle. As Majors (1990) points out:

> Cool Pose demonstrates black males' potential to transcend oppressive conditions in order to express themselves *as men*. [Yet] it ultimately does not put black males in a position to live and work in more egalitarian ways with women, nor does it directly challenge male hierarchies.

Indeed, as Connell's (1990) analysis of an Australian "Iron Man" shows, the commercially successful, publicly acclaimed athlete may embody all that is valued in present cultural conceptions of hegemonic masculinity—physical strength, commercial success, supposed heterosexual virility. Yet higher-status men, while they admire the public image of the successful athlete, may also look down on him as a narrow, even atavistic, example of masculinity. For these higher-status men, their earlier sports successes are often status enhancing and serve to link them with other men in ways that continue to exclude women. Their decisions not to pursue athletic careers are equally important signs of their status vis-à-vis other men. Future examinations of the contemporary meaning and importance of sports to men might take as a fruitful point of departure that athletic participation, and sports as public spectacle, serve to provide linkages among men in the project of the domination of women, while at the same time helping to construct and clarify differences and hierarchies among various masculinities.

NOTE

1. "Mitch Harris" is a pseudonym.

REFERENCES

Bryson, L. 1987. "Sport and the Maintenance of Masculine Hegemony." *Women's Studies International Forum* 10:349–60.

Connell, R. W. 1987. *Gender and Power.* Stanford, CA: Stanford University Press.

———. 1990. "An Iron Man: The Body and Some Contradictions of Hegemonic Masculinity." In *Sport, Men, and the Gender Order: Critical Feminist Perspectives,* edited by M. A. Messner and D. S. Sabo. Champaign, IL: Human Kinetics.

Dunkle, M. 1985. "Minority and Low-Income Girls and Young Women in Athletics." *Equal Play* 5 (Spring-Summer):12–13.

Edwards, H. 1971. "The Myth of the Racially Superior Athlete." *The Black Scholar* 3 (November).

———. 1973. *The Sociology of Sport.* Homewood, IL: Dorsey.

———. 1984. "The Collegiate Athletic Arms Race: Origins and Implications of the 'Rule 48' Controversy." *Journal of Sport and Social Issues* 8:4–22.

Eitzen, D. S. and D. A. Purdy, 1986. "The Academic Preparation and Achievement of Black and White College Athletes." *Journal of Sport and Social Issues* 10:15–29.

Eitzen, D. S. and N. B. Yetman. 1977. "Immune From Racism?" *Civil Rights Digest* 9:3–13.

Farr, K. A. 1988. "Dominance Bonding Through the Good Old Boys Sociability Group." *Sex Roles* 18:259–77.

Franklin, C. W. II. 1984. *The Changing Definition of Masculinity.* New York: Plenum.

———. 1986. "Surviving the Institutional Decimation of Black Males: Causes, Consequences, and Intervention." in *The Making of Masculinities: The New Men's Studies,* edited by H. Brod, pp. 155–70. Winchester, MA: Allen & Unwin.

Gruneau, R. 1983. *Class, Sports, and Social Development.* Amherst: University of Massachusetts Press.

Hall, M. A. (ed.). 1987. "The Gendering of Sport, Leisure, and Physical Education." *Women's Studies International Forum* 10:361–474.

Hare, N. and J. Hare. 1984. *The Endangered Black Family: Coping with the Unisexualization and Coming Extinction of the Black Race.* San Francisco, CA: Black Think Tank.

Hargreaves, J. A. 1986. "Where's the Virtue? Where's the Grace? A Discussion of the Social Production of Gender Through Sport." *Theory, Culture and Society* 3: 109–21.

Harris, D. S. and D. S. Eitzen. 1978. "The Consequences of Failure in Sport." *Urban Life* 7:177–88.

Hill, P. and B. Lowe. 1978. "The Inevitable Metathesis of the Retiring Athlete." *International Review of Sport Sociology* 9:5–29.

Leonard, W. M. II and J. M. Reyman. 1988. "The Odds of Attaining Professional Athlete Status: Refining the Computations." *Sociology of Sport Journal* 5:162–69.

Lever, J. 1976. "Sex Differences in the Games Children Play." *Social Problems* 23: 478–87.

Levinson, D. J. 1978. *The Seasons of a Man's Life.* New York: Ballantine.

Majors, R. 1986. "Cool Pose: The Proud Signature of Black Survival." *Changing Men: Issues in Gender, Sex, and Politics* 17:5–6.

———. 1990. "Cool Pose: Black Masculinity in Sports." In *Sport, Men, and the Gender Order: Critical Feminist Perspectives,* edited by M. A. Messner and D. S. Sabo. Champaign, IL: Human Kinetics.

Messner, M. 1985. "The Changing Meaning of Male Identity in the Lifecourse of the Athlete." *Arena Review* 9:31–60.

———. 1987a. "The Meaning of Success: The Athletic Experience and the Development of Male Identity." in *The Making of Masculinities: The New Men's Studies,* edited by H. Brod, pp. 193–209. Winchester, MA: Allen & Unwin.

———. 1987b. "The Life of a Man's Seasons: Male Identity in the Lifecourse of the Athlete." in *Changing Men: New Directions in Research on Men and Masculinity,* edited by M. S. Kimmel, pp. 53–67. Newbury Park, CA: Sage.

Pleck, J. H. 1982. *The Myth of Masculinity.* Cambridge: MIT Press.

Pronger, B. 1990. "Gay Jocks: A Phenomenology of Gay Men in Athletics." In *Sport, Men, and the Gender Order: Critical Feminist Perspectives,* edited by M. A. Messner and D. S. Sabo. Champaign, IL: Human Kinetics.

Rubin, L. B. 1985. *Just Friends: The Role of Friendship in Our Lives.* New York: Harper & Row.

Rudman, W. J. 1986. "The Sport Mystique in Black Culture." *Sociology of Sport Journal* 3:305–19.

Theberge, N. 1987. "Sport and Women's Empowerment." *Women's Studies International Forum* 10:387–93.

Zinn, M. Baca. 1982. "Chicano Men and Masculinity." *Journal of Ethnic Studies* 10: 29–44.

ASIAN AMERICAN WOMEN:
Not for Sale

21

Tracy Lai

Asian American women are not for sale. We will not be bought off, materially or otherwise. We say this to the white men who use the mail-order bride catalogues, hoping to buy an obedient Asian wife. But we also say this to anyone who participates in the long history of stereotyping Asian peoples and cultures as inferior and exotic. Stereotypes dehumanize people and turn them into objects to be manipulated.

It is a struggle to be an Asian in America. Historically, Asians have been denied political, economic and social equality in America. The very term "Asian American" carries a political assertion that Asians have been and continue to be a legitimate and integral part of American society. We are not forever foreign with an identity attached only to an Asian country. The Asian American movement has its origins in the social uprising of the 1960s and '70s, inspired by the black liberation movement. Chinese Americans, Japanese Americans, Pilipino Americans,[1] and Korean Americans united in recognition of a common history of oppression and common goals of community empowerment and pride. Asian American women played a leading role, drawing from the strength of their Vietnamese sisters in the National Liberation Front. The racist war abroad in Vietnam had its counterpart back home. Asian American communities fought the enemy at home, raising such issues as bilingual-bicultural education, ethnic studies, and low-income housing. Today, that quest for equality and political power continues. And Asian American women continue to be an important part of that struggle. The urgency of this struggle is reflected in the rising incidence of anti-Asian violence. *Pacific Citizen*, newspaper of the Japanese American Citizens League, reported the tragic death in February, 1984, of Ly Yung Cheung, a seamstress in New York's Chinatown. She was pushed into the path of an oncoming subway train and was decapitated. Her attacker, John Cardinale, reportedly shouted, "we're even" and later based his defense on a "psychotic phobia about Orientals."

What makes this struggle even more complicated is the die-hard myth that Asians have made it. We are told we have overcome our oppression, and that therefore we are the model minority. *Model* refers to the cherished dictum of capitalism that "pulling hard on your bootstraps" brings due rewards. The lesson drawn is that if you work hard enough, you will succeed—and if you don't succeed, you must not be working hard enough. High-profile Asian American women such as Connie Chung, a national television newscaster with a six-figure salary, are promoted as examples of Asian American success stories. But such examples, while certainly remarkable, do little to illuminate the actual conditions of the majority of Asian Americans. Such examples conceal the more typical Asian American experience of unemployment, underemployment and struggle to survive.

The model minority myth thus classically scapegoats Asian Americans. It labels us in a way that dismisses the real problems that many do face, while at the same time pitting Asians against other oppressed people of color. The fact that Asian Americans lack political representation, power and community control conveniently disappears. Model minority labelling has also given rise to an insidious cultural hierarchy. The concept of cultural deprivation implies that black, Latino, and American Indian cultures lack the cultural reinforcements that lead to successful achievement in education and career advancement, and hence to higher socioeconomic levels. For example, Asians are claimed to value education more than other minorities and to have special intellectual affinities for math and science. In fact, this is a racist rationale implying the intellectual inferiority of other minorities, while ignoring important historical and class differences in backgrounds. The cultural deprivation and model minority analyses also fail to examine capitalism as an economic system that thrives on exploitation by class, sex and race. The lack of success of other minority groups and women of all races is deliberate and necessary under capitalism. Profits for capitalism come from the low wages justified in sexist and racist terms.

A MULTIFACETED COMMUNITY

The forces shaping the experience of Asian American women come out of the historical policies of the U.S. toward Asians. These policies were largely aimed at using and discarding Asians as a temporary, cheap labor pool. Thus, young Asian men were desirable while Asian women were not, especially since women would bear children who could legally claim citizenship rights in the United States. Asian Americans are united by our oppression. The racist claim that all Asians look alike could more accurately be stated, "treat all Asians as if they were alike." We are denied respect for our cultural identities, and in the American eye, Asia and the Pacific merge into a single cultural entity. No matter how many generations we have lived in America, we are assumed to be foreigners. The 1980 Census counted more than 20 separate Asian and Pacific nationalities in the United States. The total number of Asian Americans is approximately 3.7 million. . . . [Editors' note: In 1995, there were 9.7 million Asian Americans.] Each [nationality] has a unique history and culture. However, common to all has been the history of U.S. intervention. Beginning in the nineteenth century, the U.S. followed Britain's example, forcing China and then Japan to open their doors to trade and immigration. Eventually, the U.S. sought to dominate the Pacific and Asia for trade and military purposes. As a consequence, the U.S. annexed, occupied or otherwise dominated Hawaii, the Micronesian Islands, the Philippines, South Korea, Vietnam, and Kampuchea.

Immigration of Asian and Pacific people has been heavily shaped by the nature of U.S. involvement in their countries. . . . The U.S. military bases [in many of these countries] have a devastating impact on the local Asian women, one that spills over to Asian American women as American military men bring back stereotypes and expectations of conquest. For instance, U.S. bases generate a huge prostitution business, reinforcing the stereotype that Asian women seek fulfillment through serving men.

Historically, the U.S. has also maintained discriminatory immigration policies explicitly aimed at controlling and eliminating the Asian population in the United States. The first wave of Chinese immigrants arrived in the 1850s, but by 1882, an organized movement of trade unions and opportunistic politicians secured the Chinese Exclusion Act, which prohibited immigration of Chinese laborers and their wives. Thus, families were curtailed, wives effectively abandoned in China, and an entire generation of Chinese "bachelors" were trapped in the United States. Violent killings and expulsions followed in the wake of this legislation, such as the 1885–1886 expulsions of the Chinese communities from Tacoma and Seattle, Washington. The Chinese were wanted for their labor but they were considered undesirable and unassimilable for settlement. The Chinese and, later, other Asians were accused of stealing white workers' jobs and lowering the standard of living. In fact, Asians performed work that white workers refused to do, and Asians had to be recruited to fill the labor shortage of the westward expansion. Their labor laid the foundation for the industrial and agricultural wealth of the West today.

Japanese, Korean and Pilipino workers followed in successive waves, filling one another's footsteps in low-paying, low-status jobs. In turn, each faced similarly hostile accusations and blame for downturns in the economy. Recruited for the sugar plantations in the 1800s, the Japanese immigrated to Hawaii, and subsequently to the mainland. The anti-Chinese movement reorganized itself as the Asiatic Exclusion League in 1905, dedicated to the preservation of the Caucasian race upon American soil. In 1924 the League successfully passed the National Origins Act, which barred immigration of all Asians. Only the Pilipinos remained problematic. Pilipinos were considered nationals, since the U.S. had seized the Philippines during the Spanish-American War of 1898. As nationals, they were exempt from exclusionary immigration laws. After 1924, Pilipinos were a primary source of labor and, as their numbers increased, so did white hostility. In 1929 and 1930, anti-Pilipino riots erupted all along the west coast. The Tydings-McDuffie Act of 1934 ostensibly granted independence to the Philippines, but its real purpose was to limit the immigration quota of Pilipinos to 50 per year.

In many ways, these exclusionary laws were aimed at Asian women to prevent the development of families and communities. By keeping Asian women from immigrating, and by passing anti-miscegenation laws, it was hoped that the largely male Asian labor force would eventually die out. A common theme in the anti-Asian propaganda threatened destruction of America through an invasion of the "yellow hordes" or the "yellow peril." Asian American women were described as breeding like rats. This stereotype continues into the present in the form of the "Oriental Beauty" who has extraordinary sexual powers. To raise families and to build communities under such conditions become acts of resistance.

A more recent example of U.S. foreign policy impacting Asian Americans is the resettlement in the United States of 760,854 Southeast Asians between 1975–1985. Over half are Vietnamese; the rest are nearly evenly divided between Laotians and Kampucheans. They are all refugees created by the U.S. imperialist war in Southeast Asia, which raged from the 1950s to 1975. These newer Asian Americans have borne the brunt of the anti-Asian violence, as well as a specifically anti-refugee/anti-Southeast Asian hostility. Refugee women are especially vulnerable to attacks such as robbery, rape and intimidation. The perpetrators seem to believe that these attacks are acceptable, even deserved, because the U.S. lost the war in Vietnam, and war can now be made on its victims/refugees. Refugee women are perceived as likely targets, easier to physically overpower and intimidate. Because refugees are unfamiliar with American behavior, language and laws, it is difficult for them to fight back. At the same time, the pressure to assimilate falls heavily on the women, even as the community's traditional values and roles are fundamentally undermined and their children become strangers to them.

Historically, the tendency has been to reject and exclude Asians from participating fully in American society. Besides the exclusionary immigration laws, in many states, Asians could not be naturalized, vote, own property, or marry persons of other races. Their lives and jobs were restricted in every way. Although most of the discriminatory laws have eventually changed, deep-rooted stereotypes and hostility towards Asians have not. Throughout, there has been a clear pattern of violence used to intimidate and eliminate Asians. Asians, like other minorities, have been expendable: their lives have not been valued as highly as white lives. Burned into Asian American consciousness is the violent uprooting of more than 110,000 Japanese who were interned in American concentration camps during World War II. Western Defense Commander General John L. DeWitt declared that "the Japanese race is an enemy race,"[2] thus providing the racist rationale that all Japanese on the west coast must be locked up as a military necessity (see notes). They were marched away at gunpoint and imprisoned behind barbed wire, arbitrarily and illegally stripped of all civil rights. They were stripped of their dignity and pride as a people and of their dreams. Most lost their businesses, homes and possessions, an estimated value of as much as $400 million (Weglyn, p. 276). Japanese American women fiercely resisted these attacks on family and community integrity, and the demoralization and shame. They turned the prisons into homes and transformed the anger into the will to survive. Today, the camps remain a warning that the old anti-Asian hysteria could strike at any time; you are not safe behind the yellow face.

ASIAN AMERICAN WOMEN: DANGERS WITHIN AND WITHOUT

We are triply oppressed: as Asian Americans, as Asian American women, and as Asian American women workers. Racism has been and continues to be a primary force in shaping our oppression as women and as workers. The model minority stereotype has ominous overtones when applied to Asian American women. Asian American women are described as being desirable because they are cute (as in doll-like), quiet rather than militant, and unassuming rather than assertive. In a word, non-threatening. This is the image being sold as a commodity on the front page of the *Wall Street Journal*, January 25, 1984, headlined: "American Men Find Asian Brides Fill the Unliberated Bill— Mail Order Firms Help Them Look for the Ideal Women They Didn't Find at Home."

There are about 50 mail order bride services in the U.S., carrying names such as "Cherry Blossom" and "Love Overseas." For a fee, men receive photo catalogs of Asian women, primarily from poor families in Malaysia and the Philippines. Descriptions of the women include statements such as "They love to do things to make their husbands happy," and "Most, if not all, are very feminine, loyal, loving—and virgins!" (as quoted in *Pacific Citizen*). The men who use these services are often middle-aged or older, disillusioned and divorced. Many have served in the U.S. military overseas in Asia. They blame previous marriage failures on the women's liberation movement. The mail-order services claim to sell unliberated women who will supposedly be satisfied homemakers and be subservient to their husbands. The reality is that this market is a by-product of the U.S. military and economic domination in Asia. The mail-order phenomenon is a threat not only to Asian and Asian American women, but to all women. It promotes a degrading view of women's roles and the acceptability of selling women.

The stereotypes also pay off in the form of inflated profits extracted by superexploiting Asian American women workers. Businesses want docile, subservient workers who will not complain, file grievances, or organize unions. Many businesses purposely seek immigrant workers with limited English skills as further insurance against backtalk. Asian American women are also stereotyped as having special dexterity and endurance for routine, thus making them fit for assembly work of various types. They are thought to be "loyal, diligent and attentive to detail," again good qualities for subordinates, but certainly not for supervisors. Asian American women continue to be hired mainly in low-profile, low-status, low-paying occupations, such as clerical and service work. Asian American families tend to have multiple wage earners, to support larger, often extended, families, and to live in urban areas with a relatively high cost of living. These factors skew the Asian American family median income upwards. This higher figure has been used to suggest wrongly that Asian American families are as or more successful than white families. Comparisons of median income levels within specific cities (instead of using national averages) reveal that Asian Americans, like other minorities, consistently earn less than whites.

The force of history and the conditions of our lives demand many battles, but Asian American women must also simultaneously wage an inner struggle with feudal and religious cultural baggage. In Asian cultures that have been heavily influenced by Confucianism, women are regarded as secondary to men, existing for their service. The Spanish imposed Catholicism on the Philippines with similar results. While the Asian experience in America has modified some of these ideas, every wave of immigration tends to revive the old cultures. Asian American women must be able to reject negative traditions without feeling like they are rejecting their whole Asian American identity. Assimilation appears to demand this same rejection, but it is from the standpoint of shame and self-hatred. No culture is static, and this struggle to consciously develop and redefine the best in Asian American culture is based on pride and love of our people.

Writers Merle Woo and Kitty Tsui have eloquently articulated this many-fronted struggle, becoming a voice for other Asian American lesbians. As lesbians, they have faced a painful rejection from the Asian American community. In "Letter to Ma," Merle directly addresses homophobia in the community: "If my reaction to being a Yellow Woman is different than yours was, please know that that is not a judgment on you, a criticism or denial of you, your worth" (Woo, p. 146). Merle explains that being a Yellow Feminist does not mean "'separatism,' either by cutting myself off from non-Asians or men . . . it means changing the economic class system and psychological forces (sexism, racism, and homophobia) that really hurt all of us" (p. 142). Kitty describes herself as a warrior who grapples with those same three many-headed demons. In "The Words of A Woman Who Breathes Fire," she affirms: "I am a woman who loves women, / I am a woman who loves myself" (p. 52). Self-affirmation as a source of boldness and vision becomes a strength for Asian American women, lesbian and straight.

Writers/activists Sasha Hohri, Miya Iwataki and Janice Mirikitani are a few more of the unsung Asian American women continuing the strong tradition of political activism and organizing. Their issues range from redress/reparations for the Japanese interned during World War II to racist violence against Asians and the Rainbow Coalition for political power. They continue in the spirit of earlier Korean and Pilipino women who organized and continue to organize in the U.S. for independence in their homelands. A Japanese American woman, Mitsuye Endo, had partial success in the Supreme Court in 1944, challenging the internment of concededly loyal American citizens (Weglyn, p. 227). More recently, some of the largest labor rallies and significant employer concessions have been won by striking Chinese American garment workers. In 1982, 10,000 Chinese American women garment workers rallied in New York, winning a union contract that addressed their substandard working conditions. In 1986, following the shutdown of P & L Sportswear, 300 Chinese American women garment workers forced the city of Boston and the state of Massachusetts to implement the required retraining.

But while the Chinese American community is celebrating these victories, the women's movement has yet to recognize the significance of this achievement. Asian American women are trying to organize one of the least organized sectors of labor. They are fighting for basic working conditions denied to them precisely because they are Asian American women. Asian American women are often most actively involved in their communities and workplaces because those conditions directly determine their future. If we do not fight our own battles, who will? In *East Wind: Focus on Asian Women*, Sasha Hohri and Sadie Lum analyze Asian American women's oppression as intrinsically linked to class and race issues. We cannot separate ourselves from any one part. Our liberation is linked to that of our communities and ultimately, to that of our whole society. Liberation requires revolution.

As yet, feminism has not provided sufficient analysis and direction to the basic struggle of survival facing Asian Americans in this country. Feminism appears to have a more limited agenda, one concerned primarily with women's oppression. However, women's oppression is irresolvable in a society which is inherently unequal. A capitalist society means that a few will profit while the majority will not. Feminism must deal with the structure of capitalism and its exploitation of people by race and class, as well as the way this exploitation parallels and compounds women's oppression. As feminists broaden their perspective on what issues are of priority to women of all colors, more unity can be forged with Asian American and other sisters of color who are moving ahead, organizing and surviving. We cannot choose to stop struggling. We can only choose how we work together.

NOTES

1. "Pilipino"; Filipino is the anglicized form and symbolizes the colonization and domination of the Philippines by foreign powers such as the United States. In Tagalog, the national language of the Philippines, the word is pronounced with a "p" sound, hence "Pilipino."

2. General DeWitt's statement is part of his February 1942 recommendation to Secretary Stimson on exclusion of the Japanese, as quoted in *Personal Justice Denied*, p. 6.

References

"American Men Find Asian Brides Fill the Unliberated Bill," Raymond A. Joseph, *Wall Street Journal*, January 25, 1984, pp. 1, 22.

"Chinese Garment Workers Shake Up New York," August 20, 1982 and "Chinese Garment Workers Win Retraining in Boston," September 12, 1986 in *Unity*, Unity Publications: Oakland, California.

East Wind: Politicians and Culture of Asians in the U.S. Focus: Asian Women, vol. 2 no. 1, Spring/Summer 1983, Oakland, California: Getting Together Publications.

"JACL Report on Asian Bride Catalogs," *Pacific Citizen*, February 22, 1985, pp. 10–11.

"Letter to Ma," Merle Woo, in *This Bridge Called My Back*, Cherríe Moraga and Gloria Anzaldúa, editors, Watertown, Massachusetts: Persephone Press, 1981.

Personal Justice Denied, Report of the Commission on Wartime Relocation and Internment of Civilians, U.S. Government Printing Office, Washington, D.C., 1982.

Recent Activities Against Citizens and Residents of Asian Descent, U.S. Commission on Civil Rights, Clearinghouse Publication No. 88.

With Silk Wings, Elaine H. Kim with Janice Otani, Asian Women United of California, 1983.

The Words of a Woman Who Breathes Fire, Kitty Tsui, Argyle, New York: Spinsters, Ink, 1983.

Years of Infamy, Michi Weglyn, New York: William Morrow and Co., 1976.

Rethinking Institutions

Social institutions exert a powerful influence on our everyday lives. They are also powerful channels for societal penalties and privileges. The type of work you do, the structure of your family, the kind of education you receive, and how you are treated by the state are all shaped by the institutional structure of society. Because institutions are patterned by race, class, and gender, their effect differs, depending on who you are. We rely on institutions to meet our needs, although they do so better for some groups than others. When a specific institution (such as the economy) fails us, we often appeal to another institution (such as the state) for redress. In this sense, institutions are both sources of support and sources of oppression.

The concept of an institution is abstract, since there is no thing or object that one can point to as an institution. *Social institutions* are the established societal patterns of behavior organized around particular purposes. The economy is an institution, as are the family, education, and the state—the four societal institutions examined here. Each is organized around a specific purpose; in the case of the economy, for example, that purpose is the production, distribution, and consumption of goods and services. Within a given institution, there may be various patterns, such as different family structures, but as a whole, institutions are general patterns of behavior that emerge because of the specific societal conditions in which groups live. Institutions do change over time, both as societal conditions evolve and as groups challenge specific institutional structures, but they are also enduring and persistent, even in the face of active efforts to change them. Institutions confront us from birth and live on after we die.

Social institutions are the fundamental conduits for race, class, and gender oppression in this society, even though they are often presented as entities far removed from these experiences. The American creed portrays institutions as neutral in their treatment of different groups; indeed, the liberal framework of the law allegedly makes access to public institutions (like education and work) gender- and race-blind. Still, institutions differentiate on the basis of race, class, and gender. As the articles included here show, institutions are actually structured on the basis of race, class, and gender relations.

As an example, think of the economy. Economic institutions in this society are founded on capitalism—an economic system based on the pursuit of profit and the principle of private ownership. Such a system creates class inequality since, in simple terms, the profits of some stem from the exploitation of the labor of others. The U.S. capitalist economy is further divided by race and class, resulting in a *dual labor market.* The dual labor market includes: (1) a primary labor market characterized by relatively high wages, opportunities for advancement, employee benefits, and rules of due process that protect workers' rights; and (2) a secondary labor market (where most women and minorities are located) characterized by low wages, little opportunity for advancement, few benefits, and little protection for workers.

One result of the dual labor market is the persistent wage gap between men and women, and between Whites and people of color—even when they have the same level of education (see Figure 1). Women tend to be clustered in jobs that employ mostly women workers; such jobs have been economically devalued as a result. Gender segregation and race segregation intersect in the

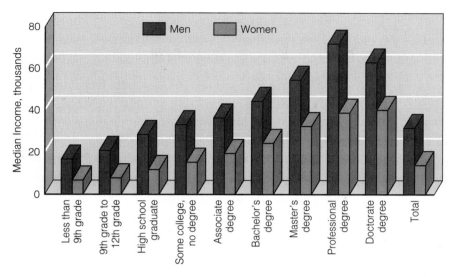

Source: U.S. Bureau of the Census, 1996. *Historical Income Tables—Persons.*
World Wide Web site: http://www.census.gov:80/hhes/income/histinc/p07.html

FIGURE 1 The Gender Gap in Income.

dual labor market, so women of color are most likely to be working in occupations where most of the other workers are also women of color. At the same time, men of color are segregated into particular segments of the market, too. Indeed, there is a direct connection between gender and race segregation and wages, since wages are lowest in occupations where women of color predominate (U.S. Department of Labor 1997).

The dual labor market is embedded in a larger economic institution that rests on race, class, and gender inequality. This is what it means to say that institutions are structured by race, class, and gender: institutions are built from and then reflect the historical and contemporary patterns of race, class, and gender relations in society.

As a second example, think of the state. The *state* refers to the organized system of power and authority in society. This includes the government, the police, the military, and the law (as reflected in both social policy and the civil and criminal justice system). The state is supposed to protect all citizens, regardless of their race, class, and gender (as well as other characteristics, such as disability and age); yet, like other "gendered" institutions, the state can be described as "male" (MacKinnon 1982). This means not only that the majority of powerful people in the state are men (such as elected officials, judges, police, and the military) but, just as important, that the state works to protect men's interests. Policies about reproductive rights provide a good example. They are largely enacted by men, but have a particularly profound effect on women. Similarly, welfare policies designed to encourage people to work are based on the model of men's experiences, since they presume that staying home to care for one's children is not working. In this sense, the state is a "gendered" institution.

A gendered institution embeds the specific characteristics associated with gender into its structure (Acker 1992). The family provides a good example. Historically, the family has been presumed to be the world of women: the ideology of the family (i.e., dominant belief systems about the family) purports that families are places for nurturing, love, and support—characteristics that have been associated with women. This ideal identifies women with the private world of the family and men with the public sphere of work. In this sense, the family ideal identifies the family as a gendered institution. Family ideology, of course, only projects an ideal, since we know that few families actually fit the presumed ideal. Nonetheless, the ideology of the family provides a standard against which all families are judged.

This example demonstrates an important point: institutions are not only gendered, they are also structured by race and class. In the preceding example, the ideology of the family is class- and race-specific; that is, it ignores and distorts the family experiences of African Americans, Latinos, and most Whites. Bonnie Thornton Dill's essay, "Our Mothers' Grief," shows that for African American, Chinese American, and Mexican American women, family structure is deeply impacted by the relationships of families to the structures of race, class, and gender.

Most people do not examine the institutional structure of society when thinking about race, class, and gender. The individualist framework of the dominant culture sees race, class, and gender as attributes of individuals, instead of seeing them as embedded in institutional structures. People do, of course, have race, class, and gender identities; and these identities have an enormous impact on individual experience. However, seeing race, class, and gender from an individualist viewpoint overlooks how profoundly embedded these identities are in the structure of American institutions.

In this section of the book we examine how institutions structure race, class, and gender experiences and, in turn, how race, class, and gender structure institutions. We look at four major institutions: work and the economic system, families, education, and the state and social policy. Each can be shown to have a unique impact on different groups (such as the discriminatory treatment of African American men by the criminal justice system). At the same time, each institution and its interrelationship with other institutions can be seen as specifically structured through the dynamics of race, class, and gender relations. Moving historically marginalized groups to the center of analysis clarifies the importance of social institutions as links between individual experience and larger structures of race, class, and gender.

WORK AND ECONOMIC INSTITUTIONS

Structural transformations in the economy have dramatically changed the conditions under which people work. As Maxine Baca Zinn and D. Stanley Eitzen show in "Economic Restructuring and Systems of Inequality," four basic transformations currently affect the character of work: the development of new technologies, global economic interdependence, capital flight, and the growth of the service sector. These changes have brought new opportunities for some groups—those positioned to benefit from the changes—while, for others, they have meant massive economic dislocation. For example, as Teresa Amott shows in "Shortchanged," women have been uniquely affected by these changes; furthermore, different groups of women have been differently affected. Her analysis shows the effect of each race, gender, and class in shaping economic problems for women. High unemployment rates among African American, Asian, and Latino male workers also demonstrate the impact of economic restructuring on different groups according to race, class, and gender.

One of the groups most affected by such changes, but also typically hidden from public view, are the 30 million people who comprise the working poor. We have already seem (in the introduction to Part II) that a large proportion of people living in poverty are employed, but at wages too low to bring their standard of living above the poverty level. Katherine Newman points out in "Working Poor, Working Hard" that the working poor, many of whom are mothers with children, seldom have either job benefits or the government supports provided to those with even less income. In the context of welfare reforms that require welfare recipients to work, the working poor are even more vulnerable. Under new federal welfare laws, most adults on welfare must spend thirty hours a week in a "work experience"; thus, at least 4 million new people will be competing for jobs in the lowest levels of the labor market. Companies can receive federal subsidies for hiring those on welfare, providing business owners financial incentives to displace the working poor and hire people in the welfare system at a lower wage (Cooper 1997).

The articles in this section also show that economic restructuring has unique effects on different groups. In "The Latino Population" Joan Moore and Raquel Pinderhughes document specific effects of economic restructuring on the urban Latino population. They point to the decline in traditional manufacturing and the growth of low-wage service and manufacturing industries (most notably, the electronics and garment industries) as most affecting the employment of Latino workers. These industries rely on using cheap labor, thus encouraging migration among Latinos. In addition, the growth of "global cities," where internationalization of the labor force occurs, creates a unique mix of high- and low-paid service jobs in major cities and encourages the development of urban labor markets with a distinct race, class, and gender structure.

In the midst of these changes, numerous myths abound regarding who succeeds and why. Deborah Woo, in "The Gap Between Striving and Achieving," refutes one such myth: the recurring claim that current economic changes allow Asian Americans, the so-called model minority, to escape racial discrimination. Woo shows that the myth of the model minority obscures knowledge about the large number of Asian Americans in working-class jobs and the structural barriers to Asian American success. Taken together, these essays demonstrate the interactive relationship of race, class, and gender in shaping economic institutions.

FAMILIES

Families are another primary social institution profoundly influenced by systems of race, class, and gender. Bonnie Thornton Dill's historical analysis of racial-ethnic women and their families in "Our Mothers' Grief" examines diverse patterns of family organization directly influenced by a group's placement in the larger political economy. Just as the political economy of the nineteenth century affected women's experience in families, the political economy of the late twentieth century shapes family relations for women and men of all races.

The articles in this section illustrate several points that have emerged from feminist studies of families, as identified by Barrie Thorne (1992). First, the family is not monolithic. The now widely acknowledged diversity among families refutes the idea that there is a normative family: White, middle-class, with children, organized around a heterosexual married couple (preferably with the wife not employed), and needing little support from relatives or neighbors. As the experiences of African Americans, Latinos, lesbians, Asian Americans, and others reveal, this so-called normal family actually represents a minority perspective. In "The Diversity of American Families," Eleanor Palo Stoller and Rose Campbell Gibson show how traditional myths about the American family mask the diversity of families in the United States (see also Figure 2). By focusing on older women, they also show us the different contexts of family experience of people in different generations.

Second, the best way to analyze families is to look at the underlying structures of "gender, generation, sexuality, race, and class" (Thorne 1992: 5). Families, for example, have been presumed to be formed around a hetero-

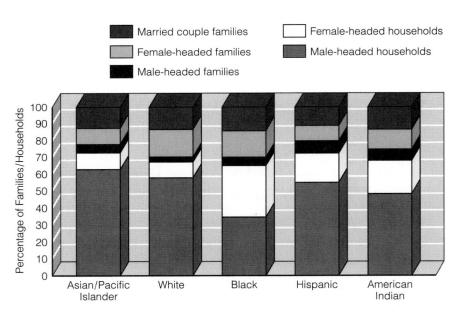

Note: In the U.S. census, families are defined as two or more people living together and related by birth or marriage; households are residential units where people may or may not be related.

Source: U.S. Bureau of the Census, 1996. *Projected Number of Households by Type, Race, and Hispanic Origin: 1995 to 2010, Series 3.* World Wide Web site: http://www.census.gov:80/population/projections/nation/hh-fam/table 4n.t

FIGURE 2 Diversity in U.S. Families and Households.

sexual norm, as Stoller and Campbell point out; but gay and lesbian partners often form permanent relationships, sometimes including children. Although such partnerships are not legally recognized, they are part of the diversity among families. Raising children in diverse family environments so that they appreciate not only their own families, but also diverse ways of forming families, is a challenge in a society that devalues gays and lesbians and people of color.

Third, Thorne points out that the dominant ideology of the family has glorified the family, hiding the underlying conflicts that are embedded in family systems. Although we do not examine family violence directly here, we know that violence stems from the gender, race, and class conflicts that are encouraged in a sexist, racist, and class-based society. In addition, we think of family violence not only as that which occurs between family members, but also as what happens to families as they confront the institutional structures of race, class, and gender oppression.

Fourth, Thorne argues that the boundaries separating family, work, welfare, and other institutions are more fluid than was previously believed. No institution is isolated from another. Changes in the economy, for example, affect family structures; moreover, this effect is reciprocal, since changes in the family generate changes in the economy. We think the interrelationships among institutions are especially apparent when studied in the context of race, class, and gender. Race, class, and gender oppression rests on a network of interconnected social institutions. Understanding the interconnections between institutions helps us see that we are all part of one historically created system that finds structural form in interconnected social institutions.

Finally, Thorne's fifth point—that the presumed dichotomy between the public and private spheres is false—is especially evident in light of the experience of people of color. Racial-ethnic families have rarely been provided the protection and privacy that the alleged split between public and private assumes. From welfare policy to reproductive rights, this dichotomy is based on the myth that families are insulated from the society around them.

Recentering one's thinking about the family by understanding the inter-connections of race, class, and gender reveals new understandings about families and the myths that have pervaded assumptions about family experience. Niobe Way and Helena Stauber demonstrate this well in their research on absent fathers. Their article "Are 'Absent Fathers' Really Absent?" takes a longstanding myth—that urban adolescents raised in female-headed households suffer from the lack of a fathering figure in the family—and shows that it is untrue. Their interviews with adolescent girls in an urban area find that, contrary to the myth of the absent father, most of these girls' fathers are very engaged in their daughters' lives. Many of the young women describe open and warm relationships with their fathers; from these interviews we learn about the complexity of these relationships. The young women's words likely strike a chord in the experiences of other adolescents.

Just as Way's and Stauber's research brings new perspectives to understanding father-daughter relationships in the context of race, class, and gender, Grace Chang and Nazlia Kibria generate new insights about how race, class, and gender structure family experiences. Chang's article "Undocumented Latinas" connects the experiences of more privileged families to the lives of private household workers, many of whom are recent Latina immigrants. Chang shows how the advance of White, middle-class women in the labor market is often linked to the exploitation of poor, immigrant women—many of whom leave their own children to care for middle-class children. Chang's analysis of the Immigration Reform and Control Act also shows the role of the state in promoting this system of labor and family work. Her study provides a good example of how the interconnections among race, class, and gender shape state policies.

Despite the structural problems that systems of work and family produce, people are not just passive victims of these forces. Nazlia Kibria's research ("Migration and Vietnamese American Women") shows how Vietnamese American women and men negotiate with each other to adjust to the specific experiences they face as immigrants. By examining gender roles in Vietnamese American families, Kibria shows how women rebuild families to provide economic and social support during the transition to a new community. Her analysis reminds us of how people shape viable family networks and support systems that enable them to survive even in the face of cultural, economic, and social assaults. Taken as a whole, the articles in this section reveal the complex relationship of all families to systems of race, class, and gender oppression.

EDUCATION

Next we turn to education. Education has recently and frequently been described as an institution in crisis. People worry that children are not learning in school. High school dropout rates, especially among students from poor and working-class families, reveal deep problems in the system of education. The sheer cost of post-secondary education makes it inaccessible to many. All of these factors contribute to the continuing disparity in the educational attainment of different groups (see Figure 3).

School violence indicates that the conflicts of race, class, and gender in the society at large are also played out within the schools. The persistence of inequality in educational opportunity because of race, class, and gender mean that many disadvantaged groups are less able to succeed. All groups whose experience differs from the White norm are ignored and distorted in the educational curriculum. However they are described, the problems in education are a logical outgrowth of the race, class, and gender inequities that exist in society at large.

In "Can Education Eliminate Race, Class, and Gender Inequality?" Roslyn Mickelson and Stephen Samuel Smith show how schools promote inequalities of race, class, and gender. Their analysis reveals the interplay between school policies that perpetuate existing hierarchies (such as credential

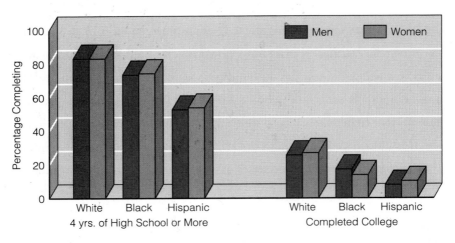

Source: U.S. Bureau of the Census, May 1997. Table 18.
World Wide Web site: http://www.census.gov:80/population/socdemo/education/table 18.1

FIGURE 3 Race. Gender. and Education.

inflation) and belief systems about race, class, and gender that are embedded in the so-called hidden curriculum. Students encounter these structures and belief systems while young, when their identities assume major importance. Schools thus become key sites where the inequalities of race, class, and gender are fostered and resisted.

Ironically, however, while schools reproduce inequality, education is also a source of social mobility. This is well illustrated in "Reminiscence of a Post-Integration Kid" by Gaye Williams and in "Canto, Locura y Poesia" by Olivia Castellano. These compelling narratives explain how African American and Latina women experience their schooling, yet manage to cope and often excel. Both women identify their struggle for a self-defined consciousness as fundamental to their survival.

Recently contested discussions about the content of the curriculum also show how race, class, and gender are embedded in the politics of education. Transforming the curriculum is not a matter of hollow-minded "political correctness"; it is fundamental to addressing the problems in education, since students otherwise "learn without focus." These articles show that education cannot be improved without addressing the arrangements of race, class, and gender in the society as a whole. Craig Greenlee's discussion of Title IX, the federal law forbidding gender discrimination in education, provides evidence that resolving race, class, and gender inequities is not without conflict. The benefits to women that Title IX has produced may come at the expense of African American men, if, as he suggests, producing equity in women's athletics means reducing the educational benefits that African American men gain through athletic scholarships. His essay is a powerful reminder that we cannot afford to improve the status of one oppressed group at the expense of another—a theme that runs through several of the articles in this section.

THE STATE

Finally, we include a section on the state and systems of social control. As we noted earlier, the state is the system of legitimated power and authority in society. Race, class, and gender are fundamental in the construction of the state and its policies, just as the state is fundamental in shaping and defining these categories. State policies determine people's rights; and as the history of Jim Crow legislation shows, the state has often been the basis for extreme exclusionary action. Matthew Snipp's opening essay, "The First Americans," details the historical role of the state in forcing Native Americans from their homes and subordinating them to a new system of state control. Genocide, forced relocation, and regulation by state law have shaped Native American experience; state processes, whether in the form of war, legislation, or social policy, have contributed to the current place of Native Americans in U.S. society.

The experience of Native Americans shows the oppression of the state relative to racial-ethnic groups in U.S. history. Likewise, Alan Berkman's and Tim Blunk's analysis of political prisoners in "Thoughts on Class, Race, and Prison" shows how the state controls those judged to be a threat to the status quo. Their discussion gives context to the fact that disproportionate numbers of Black, Latino, and Native American men are confined in prison. In fact, the criminal justice system reflects the differential treatment that individuals receive based on race, class, and gender. Prisons are thus systems of social control.

At the same time, as Snipp notes that groups are resilient as they face adversity and often turn to the state for assistance. The history of civil rights legislation shows that the state can be an invaluable resource for addressing the wrongs of race, class, and gender injustice. In this sense, the state is both a source of oppression *and* an avenue for seeking justice. For example, state policy has done much to eliminate earlier structures of inequality. At least in law (*de jure*), citizens' rights are race, class, and gender blind; in practice (*de facto*), however, these inequities continue.

In addition, confronting the state can be overwhelming, especially for the most disenfranchised groups in society. Theresa Funiciello shows this in "The Brutality of the Bureaucracy" where she painfully describes the brutality of the welfare state in her discussion of welfare mothers confronting the state bureaucracy. Her description will be familiar to anyone who has been affronted by an impersonal and obstinate state bureaucracy; but as she shows, the indignity is even greater when compounded by the insult of presumed race, class, and gender inferiority.

Social policies can be developed on behalf of disadvantaged groups, and historically the state has had a significant role in developing such policies. The state can negotiate groups' rights and organize people's access to state and societal resources. States can solve problems that individuals cannot solve on their own, although many people are now cynical about the state's ability to do so. Public confidence in government has eroded, with many observers believing that the government is incapable of solving our social problems. Indeed, in recent years, the state has been more involved in dismantling social supports than in providing assistance for those in need.

Welfare reform is a particularly vivid example of this. As Randy Albelda and Chris Tilly point out in "It's a Family Affair," welfare mothers have been reviled in the dominant culture and blamed for problems not of their making. Stereotypes of welfare recipients as preferring to be on the dole and not wanting to make an honest living are especially ironic when you see which groups really depend on state subsidies, as Chuck Collins shows in "Aid to Dependent Corporations." This title is especially ironic given the abandonment of state support for programs like Aid to Families with Dependent Children, the historic welfare system; instead, the state provides extensive entitlements to groups of a different race, class, and gender than popular stereotypes presume.

State action has been an important component of programs meant to address historic patterns of discrimination. As Linda Faye Williams shows in "Tracing the Politics of Affirmative Action," such policies are often contested—especially in an economic context, where different race, class, and gender groups compete for jobs. Williams notes how the globalization of the economy has exacerbated conflict over policies like affirmative action, now a highly divisive issue, even though its original intent was to benefit both White women and racial minorities. The debate over affirmative action shows that race, class, and gender have political dimensions (often enacted via the state) that pit groups against each other.

Institutions are powerful mechanisms for perpetuating the race, class, and gender inequities in society. Who controls them, who benefits from institutional resources, and who is best able to negotiate their way through institutional structures all reveal patterns of race, class, and gender inequity. Still, power does not operate only in a "top-down" fashion. Challenges to institutions come in many ways, but change begins with analysis.

References

Acker, Joan. 1992. "Gendered Institutions: From Sex Roles to Gendered Institutions." *Contemporary Sociology* 21 (September): 565–69.

Cooper, Marc. 1997. "When Push Comes to Shove: Who Is Welfare Reform Really Helping?" *The Nation* 264 (June 2): 11–15.

MacKinnon, Catharine. 1982. "Feminism, Marxism, Method, and the State: An Agenda for Theory." *Signs* 7 (Spring): 515–44.

Thorne, Barrie, with Marilyn Yalom. 1992. *Rethinking the Family: Some Feminist Questions*, rev. ed. Boston: Northeastern University Press.

U.S. Department of Labor. 1997 (January). *Employment and Earnings*. Washington, DC: U.S. Government Printing Office.

Work and Economic Transformation

ECONOMIC RESTRUCTURING AND SYSTEMS OF INEQUALITY

22

Maxine Baca Zinn and D. Stanley Eitzen

The present generation is in the midst of economic changes that are more far-reaching and are occurring faster than at any other time in human history. While all people in the United States are affected by the technological and economically based reorganization of society, the magnitude of this structural transformation is different throughout society. The old inequalities of class, race, and gender are thriving. New and subtle forms of discrimination are becoming prevalent throughout society as the economic base changes and settles.

Four factors are at work here: new technologies based primarily on the computer chip, global economic interdependence, capital flight, and the dominance of the information and service sectors over basic manufacturing industries. Together these factors have reinforced the unequal placement of individuals and families in the larger society. They have deepened patterns of social inequality, and they have formed new patterns of domination in which the affluent control the poor, whites dominate people of color, and men subjugate women.

The disproportionate effects of economic and industrial change are most visible in three trends: (1) structural unemployment; (2) the changing distribution and organization of jobs; and (3) the low income-generating capacity

Reprinted by permission of the authors.

of jobs. Each of these three trends has significant consequences for the hierarchies of class, race, and gender.

CLASS

Two major developments stand out when the emerging class structure is examined. The first is the growing gap between the rich and the poor since 1970. The second is the decline of the middle class.

The distribution of income is very unequal and widening. From 1970 to 1993 the income share of the highest quintile rose from 43.3 to 48.2 percent, while the bottom fifth fell from 4.1 percent to 3.6 percent of all income (U.S. Bureau of the Census 1995: 41). While salaries and stock options plus a rising stock market have increased the wealth of those at the top, the wages of the majority have stagnated and benefits (medical and pensions) have declined.

Many Americans have experienced a slowdown of income growth. Since the early 1970s wages have fallen for 80 percent of the workforce, while the richest 20 percent have become richer by far. The following facts demonstrate the contours of this growing disparity:

- "Between 1972 and 1994 the percentage of men aged 25 to 34 with incomes *below* the poverty level for a family of four increased from 14 percent to 32 percent" (Coontz 1997: 139).
- "While 20 million households stayed . . . ahead of inflation from 1991 to 1993 . . . 78 million households lost ground" (Coontz 1997: 126).
- "Between 1973 and 1993 weekly earnings of nonsupervisory workers (adjusted for inflation) fell from $315.38 per week to $254.87, a decline of more than 19 percent" (Kuttner 1995: 5).
- "A young man under 25 years of age employed full time in 1994 earned 31 percent less per week than what his same-aged counterpart earned in 1973" (Sum, Fogg, and Taggert 1996: 83–84).
- In 1970, the average chief executive officer (CEO) earned about as much as 41 factory workers. In 1992 that CEO made as much as 157 factory workers (Sklar 1995). The gap continues to widen. In 1995 the average compensation for CEOs (salary, bonus, and stock options) increased by 26.9 percent compared to the 2.8 percent increase in wages for the average worker (CNBC 1996). Thus, in 1995 CEOs were earning more than 200 times as much as the average worker (Uchitelle 1996). The gap continues to widen: in 1996 the CEOs for the top 365 corporations received compensation increases averaging 54 percent over the previous year. In a most telling statistic, the CEOs at the 30 corporations with the biggest layoffs averaged increases of 67.3 percent (United for a Fair Economy; reported in Gordon 1997).

In sum, many Americans have experienced a sharp slowdown in income with the advent of corporate downsizing, the increased use of temporary workers (who do not, typically, receive extra benefits such as medical insurance and pension plans), the loss of unions in membership and clout, the decline in the number of well-paid industry jobs, and the emergence of a bipolar wage structure in high tech and service work. These data indicate a shrinkage of the middle class, a trend heretofore unknown in U.S. history. The changing of the economy, in effect, has closed many of the old avenues to social mobility resulting, for the first time, in a rate of downward mobility that exceeds the rate of upward mobility.

RACE

Technology, foreign competition, and the changing distribution of jobs are having devastating effects on minority communities across the United States. The employment status of minorities is falling in all regions. It is worse, however, in areas of industrial decline and in segregated urban cores, where race and class intersect. Sociologist William Julius Wilson provides the data:

> In the ghetto census tracts of the nation's one hundred largest central cities, there were only 65.6 employed persons for every hundred adults who did not hold a job in a typical week in 1990. In contrast, the nonpoverty areas contained 182.3 employed persons for every hundred of those not working. In other words, the ratio of employed to jobless persons was three times greater in census tracts not marked by poverty (Wilson 1996: 19).

Race alone is significant because white job applicants continue to be disproportionately chosen over equally qualified minority ones. A 1991 study by the Urban Institute, for example, using carefully matched and trained pairs of white and black men applying for entry-level jobs, found that discrimination against blacks is "entrenched and widespread" (reported in Sklar 1993: 54). Similarly, studies in Chicago, Tampa, Atlanta, Denver, and other metropolitan areas found consistently that African Americans and Latinos face discrimination in housing, pay more than whites for similar housing, have greater difficulty than whites in receiving business and home loans, pay higher insurance rates, and pay more for consumer goods (Timmer, Eitzen, and Talley 1994: 94–95).

Latinos and African Americans have suffered disproportionately from industrial job loss, declining manufacturing employment, and low-wage production jobs. By every measure including employment rates, occupational standing, and wage rates, the labor market status of racial minorities has deteriorated relative to whites. In 1995, for example, the official employment rate for Latinos was 9.4 percent compared with 10 percent for African Americans and 4.8 percent for whites (Thurm 1995). Not only are minorities twice as likely as whites to be unemployed, they are more likely to work in dead-end jobs. About a quarter of all African Americans and a fifth of all Latinos work at the low end of the occupational ladder, in poorly paid service jobs (Folbre 1995: 4–9).

GENDER

Women and men are affected differently by the transformation of the economy from its manufacturing base to a base in service and high technology. Industrial jobs, traditionally filled by men, are being replaced with service jobs that are increasingly filled by women. Since 1980, women have taken 80 percent of the new jobs created in the economy, but the overall degree of gender segregation has not changed much since 1900 (Herz and Wootton 1996: 56). Unlike other forms of inequality, sexism in U.S. society has not become more intense as a result of the economic transformation. Instead, sexism has taken new forms as women are propelled into the labor market. Women have continued to move from the private sphere (family) to public arenas, but they have done so under conditions of labor market discrimination that have always plagued women. Today, the "typical" job is a non-union, service sector, low-paying job occupied by a woman. Despite the growing feminization of the work force, male domination has been more firmly entrenched in the social organization of work. The rise of the contingent work force (including part-time work, temporary agencies, and subcontracted work) offers many advantages to employers but at considerable cost to women workers.

Women workers do not approach earnings parity with men even when they work in similar occupations and have similar levels of education. Current estimates are that women earn about 72 percent of the wage rate of men (Dunn 1996). The full impact of economic restructuring on women must take into account the relatively low wage levels and limited opportunities for advancement that characterize their work in the new economy.

NEW INTERSECTIONS OF CLASS, RACE, AND GENDER

The hierarchies of class, race, and gender are simultaneous and interlocking systems. For this reason, they frequently operate with and through each other to produce social inequality. Not only are many existing inequalities exacerbated by the structural transformation of the economy, but the combined effects of class, race, and gender are producing new kinds of subordination and exclusion throughout society and especially the workplace. For example, the removal of manufacturing jobs has severely increased African American and Latino job loss. Just how much of this is due to class and how much is due to race remains an important question.

In contrast, many women of color have found their work opportunities expanded, albeit in poorly paid marginal work settings in service and high tech jobs. Does such growth offer traditionally oppressed race/gender groups new mobility opportunities or does the expansion of new kinds of work reproduce existing forms of inequality? The impact of economic restructuring on race and gender varies considerably. In some cases, it generates no jobs at all and displaces minority women workers. In other cases, minority women benefit by the creation of new jobs. Third world immigrant women provide the bulk of the high tech productive labor force in Silicon Valley (Hossfeld 1994, 1997). Yet a growing "underclass" in high tech consists of low-paid recent immigrant women from Mexico, Vietnam, Korea, and the Philippines. These examples reveal new labor systems as well as new forms of racial control based on class, race, and gender.

References

CNBC. 1996. "Business News" (April 8).

Coontz, Stephanie. 1997. *The Way We Really Are: Coming to Terms with America's Changing Families*. New York: Basic Books.

Dunn, Dana. 1996. "Gender and Earnings." In Paula J. Dubeck and Kathryn Borman (eds.), *Women and Work: A Handbook* (New York: Garland Publishing), pp. 61–63.

Folbre, Nancy. 1995. *The New Field Guide to the U.S. Economy*, New York: New Press.

Gordon, Marcy. 1997. "CEOs Get Layoff-Linked Rewards," Associated Press (May 2).

Herz, Diane E., and Barbara H. Wootton. 1996. "Women in the Workforce: An Overview." In Cynthia Costello and Barbara Kivimae Krimgold (eds.), *The American Woman, 1996–1997* (New York: W. W. Norton), pp. 44–78.

Hossfeld, Karen J. 1994. "Hiring Immigrant Women: Silicon Valley's 'Simple Formula.'" In Maxine Baca Zinn and Bonnie Thornton Dill (eds.), *Women of Color in U.S. Society* (Philadelphia: Temple University Press), pp. 65–93.

Hossfeld, Karen J. 1997. "'Their Logic Against Them': Contradictions in Sex, Race, and Class in Silicon Valley." In Maxine Baca Zinn, Pierrette Hondagneu-Sotelo, and Michael A. Messner (eds.), *Through the Prism of Difference* (Boston: Allyn & Bacon), pp. 388–400.

Kuttner, Robert. 1995. "The Fruits of Our Labor," *Washington Post National Weekly Edition* (September 6–12): 5.

Sklar, Holly. 1993. "Young and Guilty by Stereotype," *Z Magazine* 6 (July/August): 52–61.

Sklar, Holly. 1995. "The Snake Oil of Scapegoating," *Z Magazine* 8 (May): 49–56.

Sum, Andrew, Neal Fogg, and Robert Taggert. 1996. "The Economics of Despair," *The American Prospect*, Number 27 (July/August): 83–88.

Thurm, Scott. 1995. "Hispanic Employment Slips," *Denver Post* (January 27): E1,E5.

Timmer, Doug A., D. Stanley Eitzen, and Kathryn D. Talley. 1994. *Paths to Homelessness: Extreme Poverty and the Urban Housing Crisis.* Boulder, CO: Westview Press.

Uchitelle, Louis. 1996. "1995 Was Good for Companies, and Better for a Lot of C.E.O.'s," *New York Times* (March 29), A1.

U.S. Bureau of the Census. 1995. "Population Profile of the United States 1995," *Current Population Reports*, Series P23-189. Washington, D.C.: U.S. Government Printing Office.

Wilson, William Julius. 1996. *When Work Disappears: The World of the New Urban Poor.* New York: Alfred A. Knopf.

SHORTCHANGED: *Restructuring* Women's Work

23

Teresa Amott

The way women experience . . . economic crisis depends on where they are located in the occupational hierarchy. One concept that will help us understand this hierarchy is *occupational segregation*, which can be by gender—most jobs are held by either men *or* women and few are truly integrated. To take one extreme example, in 1990, 82 percent of architects were male; 95 percent of typists were female. Occupational segregation can also be by race-ethnicity, although this is sometimes more difficult to detect—since racial-ethnic workers are a minority, they rarely dominate a job category numerically, although they may be in the majority at a particular workplace or in a geographical region. At the national level, we have to look for evidence of occupational segregation by race-ethnicity by examining whether a particular group is over- or underrepresented relative to its percent of the total workforce. For instance, if African American women make up 5.1 percent of the total work-

From: Teresa Amott, *Caught in the Crisis: Women and the U.S. Economy Today* (New York: Monthly Review Press, 1993), pp. 49–82, 148–50. Copyright by Teresa Amott. Reprinted by permission of Monthly Review Foundation.

force, they are underrepresented in an occupation if they hold less than 5.1 percent of the jobs.[1] Law would be such an occupation: African American women made up only 1.8 percent of all lawyers in 1990. In contrast, they are overrepresented in licensed practical nursing, where they hold 16.9 percent of the jobs.

A second concept that will be helpful in our examination of women's situation during the economic crisis is *labor market segmentation*, a term that refers to the division of jobs into categories with distinct working conditions. Economists generally distinguish two such categories, which they call the *primary* and *secondary* sectors. The first includes high-wage jobs that provide good benefits, job security, and opportunities for advancement. The upper level of this sector includes elite jobs that require long years of training and certification and offer autonomy on the job and a chance to advance up the corporate ladder. Access to upper level jobs is by way of family connections, wealth, talent, education, and government programs (like the GI bill, which guaranteed higher education to veterans returning from World War II). The lower level includes those manufacturing jobs that offer relatively high wages and job security (as a result of unionization), but do not require advanced training or degrees. The fact that unionized workers are part of this lower level is the result of the capital-labor accord . . . through which employers offered some unionized workers better pay and working conditions in exchange for labor peace. In both levels of the primary sector, job turnover is relatively low because it is more difficult for employers to replace these workers. Both the upper and lower levels of the primary sector were for many years the preserve of white men, with women (mostly white women) confined to small niches, such as schoolteaching and nursing.

The secondary sector includes low-wage jobs with few fringe benefits and little opportunity for advancement. Here too, there is a predominantly white-collar upper level (which includes sales and clerical workers), where working conditions, pay, and benefits are better than in the blue-collar lower level (private household, laborer, and most service jobs). Turnover is high in both levels of this sector because these workers have relatively few marketable skills and are easily replaced. For decades, the majority of women of all racial-ethnic groups, along with most men of color, were found in the secondary sector. Mobility between the primary and secondary sectors is limited: no career ladder connects jobs in the secondary sector to jobs in the primary sector.

While most jobs fall into these two sectors . . . during the [1980s] a third sector began to grow rapidly. This is known as the *informal sector*, or the underground economy. This name is not entirely accurate, however, since these activities do not make up a separate, distinct *economy* but are linked in many ways to the formal, above-ground sectors. Journalist and economist Philip

Mattera believes that economic activity can be lined up along a continuum of formality and regulation.[2] At one end there is formal, regulated, and measured activity, where laws are observed, taxes are paid, inspections are frequent, and the participants report their activities to the relevant government entities. At the other end is work "off the books," where regulations are not enforced, participants do not report their activities, and taxes are evaded. Many economic activities exist somewhere in the middle. As the economic crisis deepens, many large corporations, whose own jobs are in the primary sector, have subcontracted some of their work out to underground firms that hire undocumented workers and escape health and safety, minimum wage, and environmental regulations. For example, in El Paso, Texas, only half of all garment industry workers earn adequate wages in union shops.[3] The other half work in sweatshops that contract work from big-name brands, such as Calvin Klein and Jordache, pay the minimum wage, and sometimes fail to pay anything at all. In 1990, the Labor Department fined one contractor for owing $30,000 in back wages and forced him to pay up. Fortunately for the employees, there is some justice in this world: the International Ladies' Garment Workers Union (ILGWU) then won a contract at his shops. According to Mattera,

> operating a business off the books—i.e., without any state regulation or union involvement—is the logical conclusion of the restructuring process. It represents the ultimate goal of the profit-maximizing entrepreneur: proverbial *free* enterprise. . . . The type of restructuring that has taken place makes it possible for firms that cannot or do not want to go underground to take advantage of unprotected labor nonetheless.[4]

Another reason for the growth in the informal sector is that worsening wages and conditions in the two formal sectors lead people to seek additional work "off the books" to supplement their shrinking incomes and inadequate welfare or social security benefits. Thus the numbers of people suffering what Mattera calls "the nightmarish working conditions of unregulated capitalism" grow rapidly. Women and men of color, particularly immigrants, are those most likely to be found in the informal sector. . . .

CAPITAL FLIGHT: A FLIGHT TO WOMEN?

. . . Over the past twenty years, many U.S. corporations shifted manufacturing jobs overseas. The creation of this "global assembly line" became a crucial component of the corporate strategy to cut costs. In their new locations, these companies hired women workers at minimal wages, both in the Third World and in such countries as Ireland. Poorly paid as these jobs were, they were attractive to the thousands of women who were moving from impoverished rural villages into the cities in search of a better life for their families.

But in the United States, millions of workers lost their jobs as the result of capital flight or corporate downsizing. When workers lose their jobs because their plants or businesses close down or move, or their positions or shifts are abolished, it is called worker *displacement.* Over 5 million workers were displaced between 1979 and 1983, and another 4 million between 1985 and 1989.[5] In both periods, women were slightly *less* likely to lose their jobs than men of the same racial-ethnic group. Women in secondary sector factory jobs were hit hardest, primarily because they lacked union protection and the education and skills to find better jobs. (In 1989, over 35 percent of the women in manufacturing operative jobs had less than a high school education.)[6]

The overall result was that even though women lost jobs to capital flight and corporate downsizing, they did so at a slower rate than men. In fact, the share of manufacturing jobs going to women *rose* between 1970 and 1990. Women, in other words, claimed a growing share of a shrinking pie. Sociologist Joan Smith studied this growing tendency to replace male workers with women (as well as a parallel tendency to replace white workers with African Americans and Latinos). In her research on heavy manufacturing industries such as steel and automobiles, Smith found that employers hired men and/or white workers only in those areas of manufacturing where profits were high, jobs were being created, and there was substantial investment in new plant and equipment. In contrast,

> in sectors where profits were slipping, the obvious search for less expensive workers led to the use of Black and women workers as a substitute for white workers and for men. . . . Close to 70 percent of women in these sectors and well over two-fifths of Blacks held their jobs as either substitutes or replacements for whites or men.[7]

While manufacturing jobs were feminizing, the rapidly expanding service sector was also hiring women in larger and larger numbers—both women entering the labor market for the first time and women displaced from manufacturing. . . .

A large part of this service sector growth took place in what were already predominantly female jobs, such as nurses' aides, child care workers, or hotel chambermaids (jobs that men would not take), as employers took advantage of the availability of a growing pool of women workers who were excluded from male-dominated jobs.[8] Chris Tilly argues that these sectors were able to grow so rapidly during the 1980s precisely *because* they were able to use low-wage, part-time labor.[9] In other words, if no women workers had been available, the jobs would not have been filled by men; instead, service employment would not have grown as rapidly.

The availability of service sector jobs helped hold the average official unemployment rates for women below those for men. However, the overall figure for women masks important differences by race-ethnicity. . . .[10] In addition, the official unemployment rate doesn't tell the whole story. If we construct a measure of underemployment, we get a different sense of the relative hardships faced by women and men. The underemployed are those who are working part-time but would prefer full-time work, those who are so discouraged that they have given up looking for work, and those who want a job but can't work because home responsibilities—such as caring for children or aged parents—or other reasons keep them out of the labor force. If we look at underemployment rather than unemployment, the rankings change: in contrast to the official unemployment rates, the underemployment rates are higher for women than for men. . . .

UNION-BUSTING

. . . A second corporate strategy to bolster profitability was to attack labor unions. As the assault took its toll, union membership and union representation declined until by 1990 only 14 percent of women and 22 percent of men were represented by a union—compared to 18 percent of women and 28 percent of men only six years earlier. Even though women are less likely to be represented by unions than men, the *drop* in unionization was not as severe for women, largely because they are concentrated in the service sector where there were some organizing victories. In addition, most women do not hold jobs in factories, the area most vulnerable to union-busting. The drop in unionization was highest among Latinas, who are over-represented in manufacturing, followed closely by African American women. Still, African American women have the highest rate of union representation among women (22 percent) because so many work in the public sector.[11]

The drop in unionized jobs is dangerous for women for several reasons. The most obvious is that unionized women earn an average weekly wage that is 1.3 times that of nonunionized women. The gap is especially large in the service sector, where unionized workers earn 1.8 times as much as nonunionized workers. . . .[12]

The loss of unionized jobs also hurts women in nonunion jobs because of what economists call a "spillover" effect from union to nonunion firms in the area of wages and benefits. While it is difficult to estimate the extent of this spillover, Harvard economists Richard Freeman and James Medoff suggest that it raises wages in blue collar jobs in large nonunion firms (those with lower level primary sector jobs) by anywhere from 10 to 20 percent, and also improves benefits and working conditions.[13] This happens for two reasons: (1) nonunion firms must compete for labor with the union firms and therefore have to meet unionized rates, and (2) some firms will keep wages higher than necessary in order to keep unions out. Thus when union workers lose wages and benefits, these spillover effects diminish, lowering wages and benefits in nonunion jobs as well.

Higher wages are only part of what women achieve when there is a strong labor movement. Unionization has an important spillover effect in the political as well as economic arena. For instance, support from the labor movement is responsible for the passage of most of the major safety net legislation in the United States, including the minimum wage and the Social Security Act. When the labor movement is weakened, important items on labor's agenda—including national health care, national child care, improved enforcement of job safety and health regulations, and broadened unemployment insurance coverage—all become more difficult to achieve, even though they address the needs of *all* workers. Further evidence of the critical role played by the labor movement comes from Europe, where family policies (parental leave, child allowances, and national day care) were all enacted with the backing, and sometimes at the initiative, of the labor movement.

WOMEN AND THE RESTRUCTURING OF WORK

. . . Another component of restructuring, and one that particularly affects women, is the use of homework. While homework can provide incomes for women who are unable to locate affordable child care or who live in rural areas, far from other employment, it also exposes women to intense exploitation. Both men and women homeworkers typically earn much less than those who do the same work outside the home: according to the federal Office of Technology Assessment, the poverty risk for homeworkers is nearly double that for other workers. In addition, working conditions in the home can be dangerous. In semiconductor manufacturing homework, for instance, workers are exposed to hazardous substances that can also contaminate residential sewage systems.[14]

Many homeworkers are undocumented immigrants who work out of their homes in order to escape detection by the Immigration and Naturalization

Service (INS). During the early 1970s, the majority of undocumented immigrants were male, but since then women have begun to arrive in ever larger numbers. . . .

WOMEN'S RESPONSES TO RESTRUCTURING

As wages fell and employers pushed more and more work into the secondary and informal sectors, women responded with a variety of individual strategies. Some started small businesses. Others sought "nontraditional" jobs in areas formerly dominated by men, hoping to earn a man's wage. And a growing number took on multiple jobs, "moonlighting" in a desperate effort to make a living wage out of two, or even three, different jobs. Each of these individual strategies . . . held some promise, but all failed to deliver substantial gains except to the lucky few. . . .

. . . Self-employment did not solve women's economic problems. The vast majority of these businesses remained small in scale: although they made up nearly 33 percent of all the businesses in the United States, they earned only 14 percent of the receipts. Nearly 40 percent had total receipts of less than $5,000 a year; only 10 percent had any employees.[15] Even the most successful women entrepreneurs faced difficulties finding affordable health insurance and pension coverage, which they had to buy for themselves out of their profits. Despite conservative rhetoric about the glories of entrepreneurship, self-employment has not proved to be a cure-all for women's economic troubles. . . .

As the number of poverty-level jobs has increased, more and more women have been forced to turn to moonlighting to boost their incomes. . . . Many more women were moonlighting because of economic hardship—to meet regular household expenses or pay off debts—not to save for something special, get experience, or help out a family member or friend. African American women and Latinas were the most likely to report that they were moonlighting to meet *regular* household expenses, while white women were more likely to report saving for the future or other reasons—a difference caused by the lower average income of women of color and the greater likelihood that they were raising children on their own.

But moonlighting is ultimately limited by the number of hours in the day. . . . As the crisis wears on and more and more women become fully employed, more families turn to children to help out. According to recent estimates, at least 4 million children under the age of 19 are employed legally, while at least 2 million more work "off the books," and the number is growing.[16]

NEW JOBS FOR WOMEN

. . . [S]ome women [have] made inroads into traditionally male jobs in the highly paid primary sector, and these gains are likely to be maintained. In addition, graduate degrees show how women are increasingly willing to prepare themselves for male-dominated occupations. . . . There has been an occupational "trickle down" effect, as white women improved their occupational status by moving into male-dominated professions such as law and medicine, while African American women moved into the *female-dominated* jobs, such as social work and teaching, vacated by white women. There is some evidence that the improvement for white women was related to federal civil rights legislation, particularly the requirement that firms receiving federal contracts comply with affirmative action guidelines.[17]

The movement of women into highly skilled blue-collar work, such as construction and auto-making, was sharply limited by the very slow growth in those jobs. Not coincidentally, male resistance to letting women enter these trades stiffened during the 1980s, when layoffs were common and union jobs were under attack. Moreover, all women lost ground in secondary sector manufacturing jobs, such as machine operators and laborers. Latinas were particularly hard hit, losing ground in most manufacturing jobs. . . .

UP THE DOWN ESCALATOR

If the postwar economic boom had continued into the 1970s and 1980s, women's economic status today would be substantially improved. The crisis produced some gains for women, but many of these evaporate on close inspection. The wage gap narrowed, but partly because men's wages fell. The gap between men's and women's rates of unionization and access to fringe benefits fell, but again partly because men's rates fell. More women entered the workforce, but they also worked longer hours than ever before, held multiple jobs, and sought work in the informal economy in order to maintain their standard of living. Finally, women's gains were not evenly distributed: highly educated women moved even further ahead of their less-educated counterparts.

All this took place against a backdrop of rising family responsibilities. . . . The most serious stress faced by married women . . . was associated with the reduced standard of living their families faced as a result of the cutbacks: "Their continued need to reduce what their wages can buy for their families means that the conflicts they feel between work and family life intensify.

Earning less makes it feel harder and harder for them to continue to work and take care of their families."[18]

For these women, who depended on manufacturing jobs, it is not surprising that the economic crisis took a heavy toll. What is surprising is that they experienced it most acutely in the home rather than on the job. Work for them was not a career, a satisfying route to self-actualization, but a fate they accepted in order to provide for their families. When their earnings fell, it was their work at home—the work of marketing, cooking, cleaning, caring—that became more difficult. An increasing number were the sole support of their households. Others found that their household's standard of living could only be maintained if they took on one—or more—paid jobs in addition to their homemaking. They were caught between shrinking incomes and growing responsibilities. . . .

NOTES

1. Unpublished data, Bureau of Labor Statistics.

2. Philip Mattera, *Off the Books: The Rise of the Underground Economy* (New York: St. Martin's Press, 1985), p. 38.

3. Colatosti, Camille, "A Job Without a Future," *Dollars and Sense* (May 1992), p. 10.

4. Philip Mattera, *Prosperity Lost* (Reading, MA: Addison-Wesley, 1992), pp. 34–35.

5. For instance, between 1985 and 1989 the displacement rate—the number of workers displaced for every 1,000 workers employed—was 6.3 for white women compared to 6.7 for white men; 6.1 for African American women compared to 7.3 for African American men; and 8.3 for Latinas compared to 9.0 for Latinos. See U.S. Department of Labor, Bureau of Labor Statistics, *Displaced Workers, 1985–89*, June 1991, Bulletin 2382, Table 4.

6. U.S. Bureau of the Census, *Statistical Abstract of the United States 1991*, Table 656, p. 400. African American women factory operatives are better educated than whites: only 29 percent lack a high school degree, compared to 36 percent of whites.

7. Joan Smith, "Impact of the Reagan Years: Race, Gender, and the Economic Restructuring," *First Annual Women's Policy Research Conference Proceedings* (Washington, D.C.: Institute for Women's Policy Research, 1989), p. 20.

8. Because the number of jobs held by women increased more than the number held by men, the percent of service sector jobs held by women increased from 43 percent to 52 percent between 1970 and 1990.

9 Polly Callaghan and Heidi Hartmann, *Contingent Work* (Washington, D.C.: Economic Policy Institute, 1991), p. 24.

10. For Latinas, on the other hand, women's unemployment was higher than men's. While it is difficult to pinpoint the reason for this, it may be because of the relatively rapid growth of Latina participation in the labor force. Although Latinas have the lowest labor force participation rate of the three groups of women, their participation

rates are growing the most rapidly. It may be that this growth in the Latina workforce outstripped job creation in the secondary sector, which had traditionally hired Latinas, while discrimination still barred their way in the primary sector—resulting in high unemployment. See National Council of La Raza, *State of Hispanic America 1991*, p. 26.

11. Paula Ries and Anne J. Stone, eds., *The American Woman 1992–93: A Status Report* (New York, W. W. Norton and Co., 1991), p. 369.

12. U.S. Bureau of Labor Statistics, *Employment and Earnings*, January 1992, Tables 59–60.

13. Richard Freeman and James Medoff, *What Do Unions Do?* (New York: Basic Books, 1981), p. 153.

14. Virginia DuRivage and David Jacobs, "Home-Based Work: Labor's Choices," in *Homework: Historical and Contemporary Perspectives on Paid Labor at Home*, ed. Eileen Boris and Cynthia R. Daniels (Urbana: University of Illinois Press, 1989), p. 259.

15. Ries and Stone, *The American Woman*, pp. 347–48.

16. Gina Kolata, "More Children Are Employed, Often Perilously," *New York Times*, 21 June 1992, p. 1.

17. Barbara Bergman, *The Economic Emergence of Women* (New York: Basic Books, 1986), p. 147.

18. Ellen Israel Rosen, *Bitter Choices: Blue Collar Women In and Out of Work* (Chicago: University of Chicago Press, 1987), p. 164.

THE GAP BETWEEN STRIVING AND ACHIEVING: *The Case of Asian American Women*

24

Deborah Woo

Much academic research on Asian Americans tends to underscore their success, a success which is attributed almost always to a cultural emphasis on education, hard work, and thrift. Less familiar is the story of potential not fully

From: Asian Women United of California (eds.) *Making Waves: An Anthology of Writings by and About Asian American Women* (Boston: Beacon Press, 1989), pp. 185–94. Copyright © 1989 by Asian Women United. Reprinted by permission of Beacon Press.

realized. For example, despite the appearance of being successful and highly educated, Asian American women do not necessarily gain the kind of recognition or rewards they deserve.

The story of unfulfilled dreams remains unwritten for many Asian Americans. It is specifically this story about the gap between striving and achieving that I am concerned with here. Conventional wisdom obscures the discrepancy by looking primarily at whether society is adequately rewarding individuals. By comparing how minorities as disadvantaged groups are doing relative to each other, the tendency is to view Asian Americans as a "model minority." This practice programs us to ignore structural barriers and inequities and to insist that any problems are simply due to different cultural values or failure of individual effort.

Myths about the Asian American community derive from many sources. All ethnic groups develop their own cultural myths. Sometimes, however, they create myths out of historical necessity, as a matter of subterfuge and survival. Chinese Americans, for example, were motivated to create new myths because institutional opportunities were closed off to them. Succeeding in America meant they had to invent fake aspects of an "Oriental culture," which became the beginning of the Chinatown tourist industry.

What has been referred to as the "model minority myth," however, essentially originated from without. The idea that Asian Americans have been a successful group has been a popular news media theme for the last twenty years. It has become a basis for cutbacks in governmental support for all ethnic minorities—for Asian Americans because they apparently are already successful as a group; for other ethnic minorities because they are presumably not working as hard as Asian Americans or they would not need assistance. Critics of this view argue that the portrayal of Asian Americans as socially and economically successful ignores fundamental inequities. That is, the question "Why have Asians been successful vis-à-vis other minorities?" has been asked at the expense of another equally important question: "What has kept Asians from *fully* reaping the fruits of their education and hard work?"

The achievements of Asian Americans are part reality, part myth. Part of the reality is that a highly visible group of Asian Americans are college-educated, occupationally well-situated, and earning relatively high incomes. The myth, however, is that hard work reaps commensurate rewards. This essay documents the gap between the level of education and subsequent occupational or income gains.

THE ROOTS AND CONTOURS OF THE "MODEL MINORITY" CONCEPT

Since World War II, social researchers and news media personnel have been quick to assert that Asian Americans excel over other ethnic groups in terms of earnings, education, and occupation. Asian Americans are said to save more, study more, work more, and so achieve more. The reason given: a cultural emphasis on education and hard work. Implicit in this view is a social judgment and moral injunction: if Asian Americans can make it on their own, why can't other minorities?

While the story of Asian American women workers is only beginning to be pieced together, the success theme is already being sung. The image prevails that despite cultural and racial oppression, they are somehow rapidly assimilating into the mainstream. As workers, they participate in the labor force at rates higher than all others, including Anglo women. Those Asian American women who pursue higher education surpass other women, and even men, in this respect. Moreover, they have acquired a reputation for not only being conscientious and industrious but docile, compliant, and uncomplaining as well.

In the last few decades American women in general have been demanding "equal pay for equal work," the legitimation of housework as work that needs to be recompensed, and greater representation in the professional fields. These demands, however, have not usually come from Asian American women. From the perspective of those in power, this reluctance to complain is another feature of the "model minority." But for those who seek to uncover employment abuses, the unwillingness to talk about problems on the job is itself a problem. The garment industry, for example, is a major area of exploitation, yet it is also one that is difficult to investigate and control. In a 1983 report on the Concentrated Employment Program of the California Department of Industrial Relations, it was noted:

> The major problem for investigators in San Francisco is that the Chinese community is very close-knit, and employers and employees cooperate in refusing to speak to investigators. In two years of enforcing the Garment Registration Act, the CEP has never received a complaint from an Asian employee. The few complaints received have been from Anglo or Latin workers.[1]

While many have argued vociferously either for or against the model minority concept, Asian Americans in general have been ambivalent in this regard. Asian Americans experience pride in achievement born of hard work and self-sacrifice, but at the same time, they resist the implication that all is well. Data provided here indicate that Asian Americans have not been successful in terms of benefitting fully, (i.e., monetarily), from their education. It is a myth that Asian Americans have proven the American Dream. How does this myth develop?

The Working Consumer: Income and Cost of Living

One striking feature about Asian Americans is that they are geographically concentrated in areas where both income and cost of living are very high. . . . [Immigration] has not only produced dramatic increases, especially in the Filipino and Chinese populations, but has also continued the overwhelming tendency for these groups to concentrate in the same geographical areas, especially those in California.[2] Interestingly enough, the very existence of large Asian communities in the West has stimulated among more recent refugee populations what is now officially referred to as "secondary migration," that is, the movement of refugees away from their sponsoring communities (usually places where there was no sizeable Asian population prior to their own arrival) to those areas where there are well-established Asian communities.[3]

This residential pattern means that while Asian Americans may earn more by living in high-income areas, they also pay more as consumers. The additional earning power gained from living in San Francisco or Los Angeles, say, is absorbed by the high cost of living in such cities. National income averages which compare the income of Asian American women with that of the more broadly dispersed Anglo women systematically distort the picture. Indeed, if we compare women within the same area, Asian American women are frequently less well-off than Anglo American females, and the difference between women pales when compared with Anglo males, whose mean income is much higher than that of any group of women.[4]

When we consider the large immigrant Asian population and the language barriers that restrict women to menial or entry-level jobs, we are talking about a group that not only earns minimum wage or less, but one whose purchasing power is substantially undermined by living in metropolitan areas of states where the cost of living is unusually high.

Another striking pattern about Asian American female employment is the high rate of labor force participation. Asian American women are more likely than Anglo American women to work full time and year round. The model minority interpretation tends to assume that mere high labor force participation is a sign of successful employment. One important factor motivating minority women to enter the work force, however, is the need to supplement family resources. For Anglo American women some of the necessity for working is partly offset by the fact that they often share in the higher incomes of Anglo males, who tend not only to earn more than all other groups but, as noted earlier, also tend to receive higher returns on their education. Moreover, once regional variation is adjusted for, Filipino and Chinese Americans had a median annual income equivalent to black males in four mainland SMSAs—Chicago, Los Angeles/Long Beach, New York, San Francisco/Oakland.[5] Census statistics point to the relatively lower earning capacity of Asian males compared to Anglo males, suggesting that Asian American women enter the work force to help compensate for this inequality. Thus, the mere fact of high employment must be read cautiously and analyzed within a larger context.

The Different Faces of Immigration

Over the last decade immigration has expanded the Chinese population by 85.3 percent, making it the largest Asian group in the country at 806,027, and has swelled the Filipino population by 125.8 percent, making it the second largest at 774,640.* Hence at present the majority of Chinese American and Filipino American women are foreign-born. In addition the Asian American "success story" is misleading in part because of a select group of these immigrants: foreign-educated professionals.

Since 1965 U.S. immigration laws have given priority to seven categories of individuals. Two of the seven allow admittance of people with special occupational skills or services needed in the United States. Four categories facilitate family reunification, and the last applies only to refugees. While occupation is estimated to account for no more than 20 percent of all visas, professionals are not precluded from entering under other preference categories. Yet this select group is frequently offered as evidence of the upward mobility possible in America when Asian Americans who are born and raised in the United States are far less likely to reach the doctoral level in their education. Over two-thirds of Asians with doctorates in the United States are trained and educated abroad.[6]

Also overlooked in some analyses is a great deal of downward mobility among the foreign-born. For example, while foreign-educated health professionals are given preferential status for entry into this country, restrictive licensing requirements deny them the opportunity to practice or utilize their special skills. They are told that their educational credentials, experience, and certifications are inadequate. Consequently, for many the only alternatives are

---

* Editors' note: In 1990, the Chinese American population was 1,645,000, and the Filipino American population was 1,407,000—still the two largest Asian American groups.

menial labor or unemployment.[7] Other highly educated immigrants become owner/managers of Asian businesses, which also suggests downward mobility and an inability to find jobs in their field of expertise.

"Professional" Obscures More Than It Reveals

Another major reason for the perception of "model minority" is that the census categories implying success, "professional-managerial" or "executive, administrative, managerial," frequently camouflage important inconsistencies with this image of success. As managers, Asian Americans, usually male, are concentrated in certain occupations. They tend to be self-employed in small-scale wholesale and retail trade and manufacturing. They are rarely buyers, sales managers, administrators, or salaried managers in large-scale retail trade, communications, or public utilities. Among foreign-born Asian women, executive-managerial status is limited primarily to auditors and accountants.[8]

In general, Asian American women with a college education are concentrated in narrow and select, usually less prestigious, rungs of the "professional-managerial" class. In 1970, 27 percent of native-born Japanese women were either elementary or secondary school teachers. Registered nurses made up the next largest group. Foreign-born Filipino women found this to be their single most important area of employment, with 19 percent being nurses. They were least represented in the more prestigious professions—physicians, judges, dentists, law professors, and lawyers.[9] In 1980 foreign-born Asian women with four or more years of college were most likely to find jobs in administrative support or clerical occupations.

Self-Help Through "Taking Care of One's Own"

Much of what is considered ideal or model behavior in American society is based on Anglo-Saxon, Protestant values. Chief among them is an ethic of individual self-help, of doing without outside assistance or governmental support. On the other hand, Asian Americans have historically relied to a large extent on family or community resources. Their tightly knit communities tend to be fairly closed to the outside world, even when under economic hardship. Many below the poverty level do not receive any form of public assistance.[10] Even if we include social security benefits as a form of supplementary income, the proportion of Asian Americans who use them is again very low, much lower than that for Anglo Americans.[11] Asian American families, in fact, are more likely than Anglo American families to bear economic hardships on their own.

While Asian Americans appear to have been self-sufficient as communities, we need to ask, at what personal cost? Moreover, have they as a group reaped rewards commensurate with their efforts? The following section presents data which document that while Asian American women may be motivated to achieve through education, monetary returns for them are less than for other groups.

THE NATURE OF INEQUALITY

The decision to use white males as the predominant reference group within the United States is a politically charged issue. When women raise and push the issue of "comparable worth," of "equal pay for equal work," they argue that women frequently do work equivalent to men's, but are paid far less for it. . . .

Another way of thinking about comparable worth is not to focus only on what individuals do on the job, but on what they bring to the job as well. Because formal education is one measure of merit in American society and because it is most frequently perceived as the means to upward mobility, we would expect greater education to have greater payoffs.

Asian American women tend to be extraordinarily successful in terms of attaining higher education. Filipino American women have the highest college completion rate of all women and graduate at a rate 50 percent greater than that of majority males. Chinese American and Japanese American women follow closely behind, exceeding both the majority male and female rate of college completion.[12] Higher levels of education, however, bring lower returns for Asian American women than they do for other groups.

While education enhances earnings capability, the return on education for Asian American women is not as great as that for other women, and is well below parity with white males. . . .

The fact that Asian American women do not reap the income benefits one might expect given their high levels of educational achievement raises questions about the reasons for such inequality. To what extent is this discrepancy based on outright discrimination? On self-imposed limitations related to cultural modesty? The absence of certain social or interpersonal skills required for upper managerial positions? Or institutional factors beyond their control? It is beyond the scope of this paper to address such concerns. However, the fact of inequality is itself noteworthy and poorly appreciated.

In general, Asian American women usually are overrepresented in clerical or administrative support jobs. While there is a somewhat greater tendency for foreign-born college-educated Asian women to find clerical-related jobs, both native- and foreign-born women have learned that clerical work is the area where they are most easily employed. . . . In addition Asian American women tend to be over-represented as cashiers, file clerks, office machine operators, and typists. They are less likely to get jobs as secretaries or receptionists. The former occupations not only carry less prestige but generally have "little or no decision-making authority, low mobility, and low public contact." [13]

In short, education may improve one's chances for success, but it cannot promise the American Dream. For Asian American women education seems to serve less as an opportunity for upward mobility than as a protection against jobs as service or assembly workers, or as machine operatives—all areas where foreign-born Asian women are far more likely to find themselves.

CONCLUSION

In this essay I have attempted to direct our attention to the gap between achievement and reward, specifically the failure to reward monetarily those who have demonstrated competence. Asian American women, like Asian American men, have been touted as "model minorities," praised for their outstanding achievements. The concept of model minority, however, obscures the fact that one's accomplishments are not adequately recognized in terms of commensurate income or choice of occupation. By focusing on the achievements of one minority in relation to another, our attention is diverted from larger institutional and historical factors which influence a group's success. Each ethnic group has a different history, and a simplistic method of modeling which assumes the experience of all immigrants is the same ignores the sociostructural context in which a certain kind of achievement occurred. For example, World War II enabled many Asian Americans who were technically trained and highly educated to move into lucrative war-related industries. [14] More recently, Korean immigrants during the 1960s were able to capitalize on the fast-growing demand for wigs in the United States. It was not simply cultural ingenuity or individual hard work which made them successful in this enterprise, but the fact that Korean immigrants were in the unique position of being able to import cheap hair products from their mother country. [15]

Just as there are structural opportunities, so there are structural barriers. However, the persistent emphasis in American society on individual effort deflects attention away from such barriers and creates self-doubt among those who have not "made it." The myth that Asian Americans have succeeded as a group, when in actuality there are serious discrepancies between effort and achievement, and between achievement and reward, adds still further to this self-doubt.

While others have also pointed out the myth of the model minority, I want to add that myths do have social functions. It would be a mistake to dismiss the model minority concept as merely a myth. Asian Americans are—however inappropriately—thrust into the role of being models for other minorities.

A closer look at the images associated with Asians as a model minority group suggests competing or contradictory themes. One image is that Asian Americans exemplify a competitive spirit, enabling them to overcome structural barriers through perseverance and ingenuity. On the other hand, they are also seen as complacent, content with their social lot, and expecting little in the way of outside help. A third image is that Asian Americans are experts at assimilation, demonstrating that this society still functions as a melting pot. Their values are sometimes equated with white, middle-class, Protestant values of hard work, determination, and thrift. Opposing this image, however, is still another, namely that Asian Americans have succeeded because they possess cultural values unique to them as a group—their family-centeredness and long tradition of reverence for scholarly achievement, for example.

Perhaps, then, this is why so many readily accept the myth, whose tenacity is due to its being vague and broad enough to appeal to a variety of different groups. Yet to the extent that the myth is based on misconceptions, we are called upon to reexamine it more closely in an effort to narrow the gap between striving and achieving.[16]

NOTES

1. Ted Bell, "Quiet Loyalty Keeps Shops Running," *Sacramento Bee*, 11 February 1985.

2. U.S. Bureau of the Census, *Race of the Population by States* (Washington, D.C., 1980). According to the census, 40 percent of all Chinese in America live in California, as well as 46 percent of all Filipinos, and 37 percent of all Japanese. New York ranks second for the number of Chinese residing there, and Hawaii is the second most populated state for Filipinos and Japanese.

3. Tricia Knoll, *Becoming Americans: Asian Sojourners, Immigrants, and Refugees in the Western United States* (Portland, Oreg.: Coast to Coast Books, 1982), 152.

4. U.S. Commission on Civil Rights, *Social Indicators of Equality for Minorities and Women* (Washington, D.C., 1978), 24, 50, 54, 58, 62.

5. David M. Moulton, "The Socioeconomic Status of Asian American Families in Five Major SMSAs" (Paper prepared for the Conference of Pacific and Asian American

Families and HEW-related Issues, San Francisco, 1978). No comparative data were available on blacks for the fifth SMSA, Honolulu.

6. James E. Blackwell, *Mainstreaming Outsiders* (New York: General Hall, Inc., 1981), 306; and Commission on Civil Rights, *Social Indicators*, 9.

7. California Advisory Committee, "A Dream Unfulfilled: Korean and Pilipino Health Professionals in California" (Report prepared for submission to U.S. Commission on Civil Rights, May 1975), iii.

8. See Amado Y. Cabezas, "A View of Poor Linkages Between Education, Occupation and Earnings for Asian Americans" (Paper presented at the Third National Forum on Education and Work, San Francisco, 1977), 17; and Census of Population, PUS, 1980.

9. Census of the Population, PUS, 1970, 1980.

10. A 1977 report on California families showed that an average of 9.3 percent of Japanese, Chinese, and Filipino families were below the poverty level, but that only 5.4 percent of these families received public assistance. The corresponding figures for Anglos were 6.3 percent and 5.9 percent. From Harold T. Yee, "The General Level of Well-Being of Asian Americans" (Paper presented to U.S. government officials in partial response to Justice Department amicus).

11. Moulton, "Socioeconomic Status," 70–71.

12. Commission on Civil Rights, *Social Indicators*, 54.

13. Bob H. Suzuki, "Education and the Socialization of Asian Americans: A Revisionist Analysis of the 'Model Minority' Thesis," *Amerasia Journal* 4:2 (1977): 43. See also Fong and Cabezas, "Economic and Employment Status," 48–49; and Commission on Civil Rights, *Social Indicators*, 97–98.

14. U.S. Commission on Civil Rights, "Education Issues" in *Civil Rights Issues of Asian and Pacific Americans: Myths and Realities* (Washington, D.C., 1979), 370–376. This material was presented by Ling-chi Wang, University of California, Berkeley.

15. Illsoo Kim, *New Urban Immigrants: The Korean Community in New York* (Princeton, N.J.: Princeton University Press, 1981).

16. For further discussion of the model minority myth and interpretation of census data, see Deborah Woo, "The Socioeconomic Status of Asian American Women in the Labor Force: An Alternative View," *Sociological Perspectives* 28:3 (July 1985): 307–338.

THE LATINO POPULATION: *The Importance of Economic Restructuring*

25

Joan Moore and Raquel Pinderhughes

. . . American minorities have been incorporated into the general social fabric in a variety of ways. Just as Chicago's black ghettos reflect a history of slavery, Jim Crow legislation, and struggles for civil and economic rights, so the nation's Latino barrios reflect a history of conquest, immigration, and a struggle to maintain cultural identity.

In 1990 there were some 22 million Latinos residing in the United States, approximately 9 percent of the total population. Of these, 61 percent were Mexican in origin, 12 percent Puerto Rican, and 5 percent Cuban. These three groups were the largest, yet 13 percent of Latinos were of Central and South American origin and another 9 percent were classified as "other Hispanic." Latinos were among the fastest-growing segments of the American population, increasing by 7.6 million, or 53 percent, between 1980 and 1990. There are predictions that Latinos will outnumber blacks by the twenty-first century. If Latino immigration and fertility continue at their current rate, there will be over 54 million Latinos in the United States by the year 2020.

This is an old population: as early as the sixteenth century, Spanish explorers settled what is now the American Southwest. In 1848, Spanish and Mexican settlers who lived in that region became United States citizens as a result of the Mexican-American War. Although the aftermath of conquest left a small elite population, the precarious position of the masses combined with the peculiarities of southwestern economic development to lay the foundation for poverty in the current period (see Barrera 1979; Moore and Pachon 1985).

In addition to those Mexicans who were incorporated into the United States after the Treaty of Guadalupe Hidalgo, Mexicans have continually crossed the border into the United States, where they have been used as a source of cheap labor by U.S. employers. The volume of immigration from Mexico has been highly dependent on fluctuations in certain segments of the U.S. economy. This dependence became glaringly obvious earlier in this century. During the Great Depression of the 1930s state and local governments

From: Joan Moore and Raquel Pinderhughes (eds.), *In the Barrios: Latinos and the Underclass Debate* (New York: Russell Sage Foundation, 1993), pp. xvi–xxix. © 1993 Russell Sage Foundation. Reprinted by permission of Russell Sage Foundation.

"repatriated" hundreds of thousands of unemployed Mexicans, and just a few years later World War II labor shortages reversed the process as Mexican contract-laborers (*braceros*) were eagerly sought. A little later, in the 1950s, massive deportations recurred when "operation Wetback" repatriated hundreds of thousands of Mexicans. Once again, in the 1980s, hundreds of thousands crossed the border to work in the United States, despite increasingly restrictive legislation.

High levels of immigration and high fertility mean that the Mexican-origin population is quite young—on the average, 9.5 years younger than the non-Latino population—and the typical household is large, with 3.8 persons, as compared with 2.6 persons in non-Latino households (U.S. Bureau of the Census 1991). Heavy immigration, problems in schooling, and industrial changes in the Southwest combine to constrain advancement. The occupational structure remains relatively steady, and though there is a growing middle class, there is also a growing number of very poor people.

The incorporation of Puerto Ricans into the United States began in 1898, when the United States took possession of Puerto Rico and Cuba during the Spanish-American War. Although Cuba gained its independence in 1902, Puerto Rico became a commonwealth of the United States in 1952. Thus Puerto Rican citizens are also citizens of the United States. The colonial relationship strongly influenced the structure of the Puerto Rican economy and the migration of Puerto Ricans to the mainland. As a result of the U.S. invasion, the island's economy was transformed from a diversified, subsistence economy, which emphasized tobacco, cattle, coffee, and sugar, to a one-crop sugar economy, of which more than 60 percent was controlled by absentee U.S. owners (Steward 1956). The constriction of the sugar economy in the 1920s resulted in high unemployment and widespread poverty, and propelled the first wave of Puerto Rican migration to the United States (Rodriguez 1989).

Puerto Rican migration to the mainland took place in roughly three periods. The first, 1900–1945, was marked by the arrival of rural migrants forced to leave the island to find work after some of these economic transformations. Many migrants directly responded to U.S. companies who valued Puerto Rican citizenship status and experience in agriculture and recruited Puerto Rican laborers for agriculture and industry in the United States (Morales 1986; Maldonado-Denis 1972). Almost all settled in New York City, most working in low-skilled occupations.

The second period, 1946–1964, is known as the "great migration" because it was during this period that the greatest number of Puerto Ricans migrated. This movement reflected factors that included the search for work, artificially low fares between the island and New York arranged by the island government, labor recruitment, and the emergence of Puerto Rican settlements on the mainland. Though Puerto Ricans were still relegated to low-wage jobs, they were employed in large numbers.

The period after 1965 has been characterized by a fluctuating pattern of net migration as well as greater dispersion to parts of the United States away from New York City (Rodriguez 1989). It is known as the "revolving-door migration": during most of this period the heavy flow from the island to the mainland was balanced by equally substantial flows in the opposite direction. However, since 1980 the net outflows from Puerto Rico have rivaled those experienced in the 1950s.

Over the past three decades the economic status of Puerto Ricans dropped precipitously. By 1990, 38 percent of all Puerto Rican families were below the poverty line. A growing proportion of these families were concentrated in poor urban neighborhoods located in declining industrial centers in the Northeast and Midwest, which experienced massive economic restructuring and diminished employment opportunities for those with less education and weaker skills. The rising poverty rate has also been linked to a dramatic increase in female-headed households. Recent studies show that the majority of recent migrants were not previously employed on the island. Many were single women who migrated with their young children (Falcon and Gurak 1991). Currently, Puerto Ricans are the most economically disadvantaged group of all Latinos. As a group they are poorer than African Americans.

Unlike other Latino migrants, who entered the United States as subordinate workers and were viewed as sources of cheap labor, the first large waves of Cuban refugees were educated middle- and upper-class professionals. Arriving in large numbers after Castro's 1959 revolution, Cubans were welcomed by the federal government as bona fide political refugees fleeing communism and were assisted in ways that significantly contributed to their economic well-being. Cubans had access to job-training programs and placement services, housing subsidies, English-language programs, and small-business loans. Federal and state assistance contributed to the growth of a vigorous enclave economy (with Cubans owning many of the businesses and hiring fellow Cubans) and also to the emergence of Miami as a center for Latin American trade. Cubans have the highest family income of all Latino groups. Nevertheless, in 1990, 16.9 percent of the Cuban population lived below the poverty line.

In recent years large numbers of Salvadorans and Guatemalans have come to the United States in search of refuge from political repression. But unlike Cubans, few have been recognized by the U.S. government as bona fide refugees. Their settlement and position in the labor market have been influenced by their undocumented (illegal) status. Dominicans have also come in large numbers to East Coast cities, many also arriving as undocumented workers.

Working for the lowest wages and minimum job security, undocumented workers are among the poorest in the nation.

Despite their long history and large numbers, Latinos have been an "invisible minority" in the United States. Until recently, few social scientists and policy analysts concerned with understanding stratification and social problems in the United States have noticed them. Because they were almost exclusively concerned with relations between blacks and whites, social scientists were primarily concerned with generating demographic information on the nation's black and white populations, providing almost no information on other groups. Consequently, it has been difficult, sometimes impossible, to obtain accurate data about Latinos.

Latinos began to be considered an important minority group when census figures showed a huge increase in the population. By 1980 there were significant Latino communities in almost every metropolitan area in the nation. As a group, Latinos have low education, low family incomes, and are more clustered in low-paid, less-skilled occupations. Most Latinos live in cities, and poverty has become an increasing problem. On the whole, Latinos are more likely to live in poverty than the general U.S. population: poverty is widespread for all Latino subgroups except Cubans. They were affected by structural factors that influenced the socioeconomic status of all U.S. workers. In 1990, 28 percent were poor as compared with 13 percent of African Americans (U.S. Bureau of the Census 1991). Puerto Ricans were particularly likely to be poor. . . .

. . . The earliest evidence of a new and important economic change appeared in the 1970s. Jobs seemed to be relocating: they declined massively in some formerly prosperous parts of the country and grew quickly in other, formerly peripheral regions—especially the South and West (Perry and Watkins 1977). It became obvious that the nation as a whole was losing "good" manufacturing jobs as production became internationalized (AFL-CIO Industrial Union Department 1986; Bluestone and Harrison 1982). In the 1990s, white-collar employment began to be restructured as well.

By the late 1980s there was consensus that the geographical shift in the location of job growth was a manifestation of a second and more important aspect of economic restructuring—the shift from a manufacturing to a service economy, and of the increasing globalization of the economy. This was a major transformation, and it became obvious that traditional manufacturing was not going to revive. Jobs continued to be created in the new service and information economy, but many were disproportionately at either the high or the low end of the wage and salary distributions, and many of the new firms functioned without the internal differentiation that might permit workers to move up within the company.

Rustbelt manufacturing decline and Sunbelt growth have come to epito-mize what economic restructuring means. But in reality things are a lot more subtle, a lot more complex, and demand a more elaborate conceptualization, especially as these trends affect Latino poverty. Elements of a more complex model are being developed by a number of researchers, but as of this writing none is yet adequate to understand the shifts that are evident in the cities represented in this volume. Several of these deserve particular emphasis.

First, there is the "Rustbelt in the Sunbelt" phenomenon. Some research-ers have argued that deindustrialization has been limited to the Rustbelt, and that the causal chain adduced by Wilson therefore does not apply outside that region. But the fact is that many Sunbelt cities developed manufacturing in-dustries, particularly during and after World War II. Thus Rustbelt-style economic restructuring—deindustrialization, in particular—has also affected them deeply. In the late 1970s and early 1980s cities like Los Angeles experi-enced a major wave of plant closings that put a fair number of Latinos out of work (Morales 1985; Soja, Morales, and Wolff 1983).

Second, there has been significant reindustrialization and many new jobs in many of these cities, a trend that is easily overlooked. Most of the expanding low-wage service and manufacturing industries, like electronics and garment manufacturing, employ Latinos (McCarthy and Valdez 1986; Muller and Espenshade 1986), and some depend almost completely on immigrant labor working at minimum wage (Fernandez-Kelly and Sassen 1991). In short, nei-ther the Rustbelt nor the Sunbelt has seen uniform economic restructuring.

Third, Latinos are affected by the "global cities" phenomenon, particu-larly evident in New York and Chicago. This term refers to a particular mix of new jobs and populations and an expansion of both high- and low-paid service jobs (see Sassen-Koob 1984). When large multinational corporations central-ize their service functions, upper-level service jobs expand. The growing cor-porate elite want more restaurants, more entertainment, more clothing, and more care for their homes and children, but these new consumer services usu-ally pay low wages and offer only temporary and part-time work. The new service workers in turn generate their own demand for low-cost goods and services. Many of them are Latino immigrants and they create what Sassen calls a "Third World city . . . located in dense groupings spread all over the city": this new "city" also provides new jobs (1989, p. 70).

Los Angeles . . . has experienced many of these patterns. The loss of manufacturing jobs has been far less visible than in New York or Chicago, for although traditional manufacturing declined, until the 1990s high-tech manu-facturing did not. Moreover, Los Angeles' international financial and trade functions flourished (Soja 1987). The real difference between Los Angeles on the one hand and New York and Chicago on the other was that more poor people in Los Angeles seemed to be working. In all three cities internationalization had similar consequences for the *structure* of jobs for the poor. More of the immigrants pouring into Los Angeles were finding jobs, while the poor residents of New York and Chicago were not.

Fourth, even though the deindustrialization framework remains of overarching importance in understanding variations in the urban context of Latino poverty, we must also understand that economic restructuring shows many different faces. It is different in economically specialized cities. Houston, for example, has been called "the oil capital of the world," and most of the devastating economic shifts in that city were due to "crisis and reorganization in the world oil-gas industry" (Hill and Feagin 1987, p. 174). Miami is another special case. The economic changes that have swept Miami have little to do with deindustrialization, or with Europe or the Pacific Rim, and much to do with the overpowering influence of its Cuban population, its important "enclave economy," and its "Latino Rim" functions (see Portes and Stepick 1993).

Finally, economic change has a different effect in peripheral areas. Both Albuquerque and Tucson are regional centers in an economically peripheral area. Historically, these two cities served the ranches, farms, and mines of their desert hinterlands. Since World War II, both became military centers, with substantial high-tech defense industrialization. Both cities are accustomed to having a large, poor Latino population, whose poverty is rarely viewed as a crisis. In Tucson, for example, unemployment for Mexican Americans has been low, and there is stable year-round income. But both cities remain marginal to the national economy, and this means that the fate of their poor depends more on local factors.

Laredo has many features in common with other cities along the Texas border, with its substantial military installations, and agricultural and tourist functions. All of these cities have been affected by general swings in the American and Texan economy. These border communities have long been the poorest in the nation, and their largely Mexican American populations have suffered even more from recent economic downturns. They are peripheral to the U.S. economy, but the important point is that their economic well-being is intimately tied to the Mexican economy. They were devastated by the collapse of the peso in the 1980s. They are also more involved than most American cities in international trade in illicit goods, and poverty in Laredo has been deeply affected by smuggling. Though Texas has a long history of discrimination against Mexican Americans, race is not an issue within Laredo itself, where most of the population—elite as well as poor—is of Mexican descent. This fact is of particular importance in evaluating the underclass debate.

THE INFORMAL AND ILLICIT ECONOMIES

The growth of an informal economy is part and parcel of late twentieth-century economic restructuring. Particularly in the global cities, a variety of "informal" economic activities proliferates—activities that are small-scale, informally organized, and largely outside government regulations (cf. Portes, Castells, and Benton 1989). Some low-wage reindustrialization, for example, makes use of new arrangements in well-established industries (like home work in the garment industry, as seamstresses take their work home with them). Small-scale individual activities such as street vending and "handyman" house repairs and alterations affect communities in peripheral as well as global cities. . . .

And, finally, there are illicit activities—most notoriously, a burgeoning drug market. There is not much doubt that the new poverty in the United States has often been accompanied by a resurgence of illicit economic activities (see Fagan, forthcoming, for details on five cities). . . . Most . . . Latino communities . . . have been able to contain or encapsulate such activities so that they do not dominate neighborhood life. But in most of them there is also little doubt that illicit economic activities form an "expanded industry". They rarely provide more than a pittance for the average worker: but for a very small fraction of barrio households they are part of the battery of survival strategies. . . .

IMMIGRATION

Immigration—both international and from Puerto Rico—is of major significance for poor Latino communities in almost every city in every region of the country. Further, there is every reason to believe that immigration will continue to be important.

First, it has important economic consequences. Immigration is a central feature of the economic life of global cities: for example, Los Angeles has been called the "capital of the Third World" because of its huge Latino and Asian immigration (Rieff 1991). . . . Those cities most bound to world trends (New York, Los Angeles, Chicago, Houston, and Miami) experienced massive Latino immigration in the 1980s. In . . . Los Angeles, Houston, and Miami . . . , immigration is a major factor in the labor market, and the residents of the "second settlement" Puerto Rican communities . . . in New York and Chicago operate within a context of both racial and ethnic change and of increased Latino immigration. The restructured economy provides marginal jobs for immigrant workers, and wage scales seem to drop for native-born Latinos in areas where immigration is high. This is a more complicated scenario than the simple loss of jobs accompanying Rustbelt deindustrialization. Immigrants are ineligible for most government benefits, are usually highly motivated, and are driven to take even the poorest-paying jobs. They are also more vulnerable to labor-market swings.

These may be construed as rather negative consequences, but in addition, immigrants have been a constructive force in many cities. For example, [there is] the economic vitality of immigrant-serving businesses [and] the revival of language and of traditional social controls, the strengthening of networks, and the emergence of new community institutions. . . .

References

AFL-CIO Industrial Union Department 1986. *The Polarization of America*. Washington, D.C.: AFL-CIO Industrial Union Department.

Barrera, Mario 1979. *Race and Class in the Southwest*. Notre Dame, IN: University of Notre Dame Press.

Bluestone, Barry, and Bennett Harrison 1982. *The Deindustrialization of America*. New York: Basic Books.

Falcon, Luis, and Douglas Gurak 1991. "Features of the Hispanic Underclass: Puerto Ricans and Dominicans in New York." Unpublished manuscript.

Fernandez-Kelly, Patricia, and Saskia Sassen 1991. "A Collaborative Study of Hispanic Women in the Garment and Electronics Industries: Executive Summary." New York: New York University, Center for Latin American and Caribbean Studies.

Hill, Richard Child, and Joe R. Feagin 1987. "Detroit and Houston: Two Cities in Global Perspective." In Michael Peter Smith and Joe R. Feagin, eds. In *The Capitalist City*, pp. 155–177. New York: Basil Blackwell.

Maldonado-Denis, Manuel 1972. *Puerto Rico: A Sociohistoric Interpretation*. New York: Random House.

McCarthy, Kevin, and R. B. Valdez 1986. *Current and Future Effects of Mexican Immigration in California*. Santa Monica, CA: Rand Corporation.

Moore, Joan, and Harry Pachon 1985. *Hispanics in the United States*. Englewood Cliffs, NJ: Prentice Hall.

Morales, Julio 1986. *Puerto Rican Poverty and Migration: We Just Had to Try Elsewhere*. New York: Praeger.

Morales, Rebecca 1985. "Transitional Labor: Undocumented Workers in the Los Angeles Automobile Industry." *International Migration Review* 17: 570–596.

Muller, Thomas, and Thomas J. Espenshade 1986. *The Fourth Wave*. Washington, D.C.: Urban Institute Press.

Perry, David, and Alfred Watkins 1977. *The Rise of the Sunbelt Cities*. Beverly Hills, CA: Sage.

Portes, Alejandro, Manuel Castells, and Lauren A. Benton 1989. *The Informal Economy.* Baltimore: Johns Hopkins University Press.

Portes, Alejandro, and Juan Clark 1987. "Mariel Refugees: Six Years After." *Migration World Magazine* 15:14–18.

Portes, Alejandro, and Alex Stepick 1993. *City on the Edge: The Transformation of Miami.* Berkeley: University of California Press.

Rieff, David 1987. *Going to Miami: Exiles, Tourists, and Refugees in the New America.* Boston: Little Brown.

Rodriguez, Clara 1989. *Puerto Ricans: Born in the U.S.A.* Boston: Unwin Hyman.

Sassen, Saskia 1989. "New Trends in the Sociospatial Organization of the New York City Economy." In Robert Beauregard, ed. *Economic Restructuring and Political Response.* Newbury Park, CA.

Sassen-Koob, Saskia 1984. "The New Labor Demand in Global Cities." In Michael Smith, ed. *Cities in Transformation.* Beverly Hills, CA: Sage.

Soja, Edward 1987. "Economic Restructuring and the Internationalization of the Los Angeles Region." In Michael Peter Smith and Joe R. Feagin, eds. *The Capitalist City,* pp. 178–198. New York: Basil Blackwell.

Soja, Edward W., Rebecca Morales, and G. Wolff 1983. "Urban Restructuring: An Analysis of Social and Spatial Change in Los Angeles." *Economic Geography* 59:195–230.

Stevens-Arroyo, Antonio M. 1974. *The Political Philosophy of Pedro Abizu Campos: Its Theory and Practice.* Ibero American Language and Area Center. New York: New York University Press.

Steward, Julian H. 1956. *The People of Puerto Rico.* Urbana, IL: University of Illinois Press.

U.S. Bureau of the Census 1991. *The Hispanic Population in the United States: March 1991.* Current Population Reports, Series P-20, No. 455. Washington, D.C.: U.S. Government Printing Office.

WORKING POOR, WORKING HARD

26

Katherine Newman

Conservatives insist that poor adults got where they are because they haven't the brains to do better, lack the moral fiber to restrain their sexual urges, or have succumbed to the easy out-of-state support that, we are told, puts people

From: *The Nation* 19 (July 29/August 5, 1996): 20–23. Reprinted by permission.

on the federal payroll for having children out of wedlock. What fails to register in the national imagination is the fact that the vast majority of poor people do work for a living. They hold the jobs that no one else really wants: the ones that pay the minimum wage, try the strength and patience of anyone who has ever tried to hold them, and subject their incumbents to a lingering stigma. Hamburger flippers, bedmakers, bedpan cleaners—these are the people Jesse Jackson once called attention to when he tried, in vain it seems, to elicit sympathy or at least recognition for the country's working poor.

There are approximately 30 million people in the United States who fill this bill. They are as far from the shiftless stereotype as one can imagine. Their full-time, year-round earnings are so meager that despite their best efforts they can't afford decent housing, diets, health care or child care. Apart from the Earned Income Tax Credit—perhaps the most important antipoverty program of the past twenty years, now threatened with dismemberment—we've devoted precious little attention to support of the working poor. We have been content to leave them to endure their struggle to survive, even though our best investment in poverty programs might well be to make sure the working poor can stay at work.

It is not surprising that we ignore the working poor. They do not impinge on the middle class; they are not poised to riot; and they are usually too busy trying to make ends meet to argue very loudly for a greater share of the public purse. Theirs is an invisible social problem, but a big one nonetheless. For the debilitating conditions that impinge on the working poor—substandard housing, crumbling schools, inaccessible health care—are hardly different from those that surround their non-working counterparts.

Indeed, for many these difficulties are measurably worse because the working poor lack access to many government supports like subsidized housing and medical care. Other benefits, like food stamps, are cut off at absurdly low levels and therefore unavailable to people who earn the minimum wage but work a forty-hour week. There isn't much of a safety net spread below the working poor, even though their struggle to survive can be as desperate as that of any family reliant on public support. We have put a great deal of energy into debating appropriate policy for the latter; we seem not to know the former exists.

But exist they do. In central Harlem—where I've been doing research on low-wage workers for the past three years—67 percent of households include at least one full-time worker. This, in a neighborhood where over 40 percent of the population exists below the poverty line and 29 percent receive public assistance, is the one number we never hear on the nightly news. The working people of central Harlem often find themselves subject to unstable hours—cut back with no notice because of a business downturn—or seasonal layoffs that hit their pocketbooks hard. Those who are part-time workers, most of whom are involuntary part-timers who would jump at the chance for a full-time position, are denied access to the unemployment insurance system. Hence, when the layoffs come or the hours shrink, there is no backstop save what is left of the welfare system.

Working mothers everywhere are faced with the problems of balancing the demands of a job and the responsibilities of child care. Poor mothers in Harlem who are working often find that their wages are so low that the best they can afford in the way of child care is of questionable quality and reliability. Even more problematic may be the arrangements for the care of the elderly, whose working adult children may be responsible for their care: there is no money to hire home aides or to make use of nursing homes, the strategies middle-aged middle-class families turn to (though at great cost).

Employers are less sympathetic, or less flexible, when faced with a store clerk who can't make it in today than they are when a high-priced accountant has to stay home. Vacation pay and sick pay are unknown benefits in the low-wage world. Instead, the working poor are deemed replaceable, and with good reason: there are indeed hundreds of people lining up to take their jobs, particularly in inner-city communities like Harlem, where we found the ratio of jobseekers to successful jobholders is 14 to 1.

Both child care and elder care obligations wreak havoc with the ability of a worker—especially a mother—to maintain her job when catastrophes or even garden-variety problems (the 5-year-old comes down with chicken pox) strike. The frequency with which these problems occur among poor families—working or not—is much higher than among those who are better off. Chronic asthma rates have doubled in the United States between 1980 and 1993, with over 5 million children presently suffering from this dangerous disease. Children in the ghetto are vastly overrepresented among asthma victims; medical researchers are unsure about the reasons. But like diabetes, tuberculosis and other chronic diseases, these are problems of poverty that are rising at alarming rates among ghetto residents. The working poor cannot earn their way out of crowded housing, exposure to toxic substances, inadequate diets and low birth weight babies—all conditions that contribute to disease. These health problems also set the stage for employment instability as parents struggle to cope with the endless rounds of hospitalization and doctors' visits that a chronic asthmatic requires. Many a working mother has found herself forced back to the welfare rolls because she can no longer manage these demands within the strictures of a low-wage job.

In recent years a new term has cropped up to describe ghetto dwellers: the urban underclass. Coined by the journalist Ken Auletta in an influential article in *The New Yorker*, the idea took on greater salience with the publication of William Julius Wilson's landmark volume *The Truly Disadvantaged* (see Adolph Reed Jr., "The Liberal Technocrat," February 6, 1988). The underclass literature paints a picture of the inner city as a public policy nightmare—a racially and economically segregated conglomeration of the welfare-dependent, criminals, broken families and children without stable roots or any understanding of how the employment world functions. Working people are erased from view; they are, presumably, living in better neighborhoods. What's wrong with this picture? There are, of course, millions of poor people on welfare in our ghettos. New York City, for example, has over 1 million people on A.F.D.C. But my research in Harlem shows that no Berlin wall separates the welfare recipient from the working poor. More often than not, both kinds of poor people live under a single roof.

This will not surprise anyone who has ever tried to live on either an A.F.D.C. family budget or the proceeds of a "McJob." Neither will really float a family: women on welfare cannot possibly pay the rent, the food bill, utilities and other basic necessities (let alone any luxuries) on the stipend provided by A.F.D.C. They have to find additional resources, and most of them do: they work off the books, they receive "in kind" assistance from family members or friends, which has to be hidden from view. The working poor are in the same situation: they cannot pay the rent on their apartments out of the minimum wage. They frequently rely on another family member's subsidized housing and Medicare. America's poorest workers are often intertwined with relatives and close friends who are on welfare. The combination of the two income streams makes it possible to manage the basic necessities of life at a very low level.

Descriptions of inner-city communities that stress a dramatic separation between people who are gainfully employed and those who are on welfare fail to grasp this fundamental connection. This matters because any reforms or cuts the legislative enacts, aiming at welfare recipients, will have a profound effect on the working poor. Any move to rearrange the lives of those on welfare will ricochet back on the working poor. At the end of the day the nation may not be much better off in reducing the costs of welfare: we may simply push one group out and find another waiting on the doorstep. Credible policy begins by understanding how deeply intertwined the two kinds of poverty, working and non-, really are.

America's middle class used to seem remote from the daily problems of the poor. But the massive layoffs of the 1990s have left many a former middle manager scrambling for a job at K Mart where work conditions share some of the features that make low-wage jobs so problematic. Some analysts would see an optimistic aspect to this situation, hoping that the jarring experience of economic insecurity among the relatively advantaged would generate the sympathy, the political will to reconsider the role of government in providing jobs, guaranteed income, health care and the other essentials in life. Most of us who have studied the shaken middle class believe we are very far from that kind of political reawakening. What middle-class insecurity seems to be fostering at the moment is a conservatism designed to punish.

Budget cuts will punish the poor who had the bad taste to get born into the wrong families at the wrong time in our economic and political history. But they will punish the rest of us too: the students who will be shut out of public universities, the elderly who will lose their home health aid (and their daughters who will have to forgo their jobs to care for them instead), the teachers who will be missing from classrooms that have already been stripped of books and pencils, and most of all the children who will inherit a labor market that is worsening steadily and a state sector that refuses to step in. And of course, the draconian legislation on the docket now will punish the very people who are among the strongest subscribers to the work ethic, the working poor and their families.

Families

OUR MOTHERS' GRIEF: *Racial-Ethnic* *Women and the Maintenance of Families*

27

Bonnie Thornton Dill

REPRODUCTIVE LABOR[1] FOR WHITE WOMEN IN EARLY AMERICA

In eighteenth- and nineteenth-century America, the lives of white[2] women in the United States were circumscribed within a legal and social system based on patriarchal authority. This authority took two forms: public and private. The social, legal, and economic position of women in this society was controlled through the private aspects of patriarchy and defined in terms of their relationship to families headed by men. The society was structured to confine white wives to reproductive labor within the domestic sphere. At the same time, the formation, preservation, and protection of families among white settlers was seen as crucial to the growth and development of American society. Building, maintaining, and supporting families was a concern of the State and of those organizations that prefigured the State. Thus, while white women had few legal rights as women, they were protected through public forms of patriarchy that acknowledged and supported their family roles of wives, mothers, and daughters because they were vital instruments for building American society.

The groundwork for public support of women's family roles was laid during the colonial period. As early as 1619, the London Company began

From: *Journal of Family History* 13 (1988): 415–31. Reprinted by permission.

planning for the importation of single women into the colonies to marry colonists, form families, and provide for a permanent settlement. The objective was to make the men "more settled and less moveable . . . instability would breed a dissolution, and so an overthrow of the Plantation" (cited in Spruill 1972, p. 8).

In accordance with this recognition of the importance of families, the London Company provided the economic basis necessary for the development of the family as a viable and essential institution within the nascent social structure of the colonies. Shares of land were allotted for both husbands and wives in recognition of the fact that "in a new plantation it is not known whether men or women be the most necessary" (cited in Spruill 1972, p. 9).

This pattern of providing an economic base designed to attract, promote, and maintain families was followed in the other colonial settlements. Lord Baltimore of Maryland ". . . offered to each adventurer a hundred acres for himself, a hundred for his wife, fifty for each child, a hundred for each man servant, and sixty for a woman servant. Women heads of families were treated just as men" (Spruill 1972, p. 11).

In Georgia, which appealed to poorer classes for settlers more than did Virginia or Maryland, ". . . among the advantages they offered men to emigrate was the gainful employment of their wives and children" (Spruill 1972, p. 16).

In colonial America, white women were seen as vital contributors to the stabilization and growth of society. They were therefore accorded some legal and economic recognition through a patriarchal family structure.

> While colonial life remained hard, . . . American women married earlier [than European women], were less restricted by dowries, and often had legal protection for themselves and their children in antenuptial contracts (Kennedy 1979, p. 7).

Throughout the colonial period, women's reproductive labor in the family was an integral part of the daily operation of small-scale family farms or artisan's shops. According to Kessler-Harris (1981), a gender-based division of labor was common, but not rigid. The participation of women in work that was essential to family survival reinforced the importance of their contributions to both the protection of the family and the growth of society.

Between the end of the eighteenth and mid-nineteenth century, what is labeled the "modern American family" developed. The growth of industrialization and an urban middle class, along with the accumulation of agrarian wealth among Southern planters, had two results that are particularly pertinent to this discussion. First, class differentiation increased and sharpened, and with it, distinctions in the content and nature of women's family lives.

Second, the organization of industrial labor resulted in the separation of home and family and the assignment to women of a separate sphere of activity focused on childcare and home maintenance. Whereas men's activities became increasingly focused upon the industrial competitive sphere of work, "women's activities were increasingly confined to the care of children, the nurturing of the husband, and the physical maintenance of the home" (Degler 1980, p. 26).

This separate sphere of domesticity and piety became both an ideal for all white women as well as a source of important distinctions between them. As Matthaei (1982) points out, tied to the notion of wife as homemaker is a definition of masculinity in which the husband's successful role performance was measured by his ability to keep his wife in the homemaker role. The entry of white women into the labor force came to be linked with the husband's assumed inability to fulfill his provider role.

For wealthy and middle-class women, the growth of the domestic sphere offered a potential for creative development as homemakers and mothers. Given ample financial support from their husband's earnings, some of these women were able to concentrate their energies on the development and elaboration of the more intangible elements of this separate sphere. They were also able to hire other women to perform the daily tasks such as cleaning, laundry, cooking, and ironing. Kessler-Harris cautions, however, that the separation of productive labor from the home did not seriously diminish the amount of physical drudgery associated with housework, even for middle-class women.

> It did relegate the continuing hard work to second place, transforming the public image of the household by the 1820s and 1830s from a place where productive labor was performed to one whose main goals were the preservation of virtue and morality. . . . Many of the "well-run" homes of the pre-Civil War period seem to have been the dwelling of overworked women. Short of household help, without modern conveniences, and frequently pregnant, these women complained bitterly of their harsh existence (Kessler-Harris 1981, p. 39).

In effect, household labor was transformed from economic productivity done by members of the family group to home maintenance; childcare and moral uplift done by an isolated woman who perhaps supervised some servants.

Working-class white women experienced this same transformation but their families' acceptance of the domestic code meant that their labor in the home intensified. Given the meager earnings of working-class men, working-class families had to develop alternative strategies to both survive and keep the wives at home. The result was that working-class women's reproductive labor increased to fill the gap between family need and family income. Women increased their own production of household goods through things such as canning and sewing; and by developing other sources of income, including boarders and homework. A final and very important source of other income was wages earned by the participation of sons and daughters in the labor force. In fact, Matthaei argues that "the domestic homemaking of married women was supported by the labors of their daughters" (1982, p. 130).

The question arises: Why did white working-class families sacrifice other aspects of this nineteenth-century notion of family, such as privacy and the protection of children, to keep wives as homemakers within the home? Zaretsky (1978) provides a possible answer.

> The Victorian emphasis on the sanctity of the family and on the autonomy of women within the family marked an advance for women of all classes over the interdependent but male dominated subsistence farm of the 18th century . . . most of women's adult life was taken up with childrearing. As a result, a special respect for her place within the home, and particularly for her childrearing activities was appreciated by working-class women (p. 211).

Another way in which white women's family roles were socially acknowledged and protected was through the existence of a separate sphere for women. The code of domesticity, attainable for affluent women, became an ideal toward which nonaffluent women aspired. Notwithstanding the personal constraints placed on women's development, the notion of separate spheres promoted the growth and stability of family life among the white middle class and became the basis for working-class men's efforts to achieve a family wage, so that they could keep their wives at home. Also, women gained a distinct sphere of authority and expertise that yielded them special recognition.

During the eighteenth and nineteenth centuries, American society accorded considerable importance to the development and sustenance of European immigrant families. As primary laborers in the reproduction and maintenance of family life, women were acknowledged and accorded the privileges and protections deemed socially appropriate to their family roles. This argument acknowledges the fact that the family structure denied these women many rights and privileges and seriously constrained their individual growth and development. Because women gained social recognition primarily through their membership in families, their personal rights were few and privileges were subject to the will of the male head of the household. Nevertheless, the recognition of women's reproductive labor as an essential building block of the family, combined with a view of the family as the cornerstone of the nation, distinguished the experiences of the white, dominant culture from those of racial ethnics.

Thus, in its founding, American society initiated legal, economic, and social practices designed to promote the growth of family life among European colonists. The reception colonial families found in the United States contrasts sharply with the lack of attention given to the families of racial-ethnics. Although the presence of racial-ethnics was equally as important for the growth of the nation, their political, economic, legal, and social status was quite different.

REPRODUCTIVE LABOR AMONG RACIAL-ETHNICS IN EARLY AMERICA

Unlike white women, racial-ethnic women experienced the oppressions of a patriarchal society but were denied the protections and buffering of a patriarchal family. Their families suffered as a direct result of the organization of the labor systems in which they participated.

Racial-ethnics were brought to this country to meet the need for a cheap and exploitable labor force. Little attention was given to their family and community life except as it related to their economic productivity. Labor, and not the existence or maintenance of families, was the critical aspect of their role in building the nation. Thus they were denied the social structural supports necessary to make *their* families a vital element in the social order. Family membership was not a key means of access to participation in the wider society. The lack of social, legal, and economic support for racial-ethnic families intensified and extended women's reproductive labor, created tensions and strains in family relationships, and set the stage for a variety of creative and adaptive forms of resistance.

AFRICAN-AMERICAN SLAVES

Among students of slavery, there has been considerable debate over the relative "harshness" of American slavery, and the degree to which slaves were permitted or encouraged to form families. It is generally acknowledged that many slaveowners found it economically advantageous to encourage family formation as a way of reproducing and perpetuating the slave labor force. This became increasingly true after 1807 when the importation of African slaves was explicitly prohibited. The existence of these families and many aspects of their functioning, however, were directly controlled by the master. In other words, slaves married and formed families but these groupings were completely subject to the master's decision to let them remain intact. One study has estimated that about 32% of all recorded slave marriages were disrupted by sale, about 45% by death of a spouse, about 10 percent by choice, with the remaining 13 percent not disrupted at all (Blassingame 1972, pp. 90–92). African slaves thus quickly learned that they had a limited degree of control over the formation and maintenance of their marriages and could not be assured of keeping their children with them. The threat of disruption was perhaps the most direct and pervasive cultural assault[3] on families that slaves encountered. Yet there were a number of other aspects of the slave system which reinforced the precariousness of slave family life.

In contrast to some African traditions and the Euro-American patterns of the period, slave men were not the main provider or authority figure in the family. The mother-child tie was basic and of greatest interest to the slave-owner because it was critical in the reproduction of the labor force.

In addition to the lack of authority and economic autonomy experienced by the husband-father in the slave family, use of the rape of women slaves as a weapon of terror and control further undermined the integrity of the slave family.

> It would be a mistake to regard the institutionalized pattern of rape during slavery as an expression of white men's sexual urges, otherwise stifled by the specter of the white womanhood's chastity . . . Rape was a weapon of domination, a weapon of repression, whose covert goal was to extinguish slave women's will to resist, and in the process, to demoralize their men (Davis 1981, pp. 23–24).

The slave family, therefore, was at the heart of a peculiar tension in the master-slave relationship. On the one hand, slaveowners sought to encourage familial ties among slaves because, as Matthaei (1982) states: ". . . these provided the basis of the development of the slave into a self-conscious socialized human being" (p. 81). They also hoped and believed that this socialization process would help children learn to accept their place in society as slaves. Yet the master's need to control and intervene in the familial life of the slaves is indicative of the other side of this tension. Family ties had the potential for becoming a competing and more potent source of allegiance than the slave-master himself. Also, kin were as likely to socialize children in forms of resistance as in acts of compliance.

It was within this context of surveillance, assault, and ambivalence that slave women's reproductive labor took place. She and her menfolk had the task of preserving the human and family ties that could ultimately give them a reason for living. They had to socialize their children to believe in the possibility of a life in which they were not enslaved. The slave woman's labor on behalf of the family was, as Davis (1971) has pointed out, the only labor the slave engaged in that could not be directly appropriated by the slaveowner for his own profit. Yet, its indirect appropriation, as labor crucial to the reproduction of the slaveowner's labor force, was the source of strong ambivalence for many slave women. Whereas some mothers murdered their babies to keep them from being slaves, many sought within the family sphere a degree of autonomy and creativity denied them in other realms of the society. The maintenance of a distinct African-American culture is testimony to the ways in which slaves maintained a degree of cultural autonomy and resisted the creation of a slave family that only served the needs of the master.

Gutman (1976) provides evidence of the ways in which slaves expressed a unique Afro-American culture through their family practices. He provides data on naming patterns and kinship ties among slaves that flies in the face of the dominant ideology of the period. That ideology argued that slaves were immoral and had little concern for or appreciation of family life.

Yet Gutman demonstrated that within a system which denied the father authority over his family, slave boys were frequently named after their fathers, and many children were named after blood relatives as a way of maintaining family ties. Gutman also suggested that after emancipation a number of slaves took the names of former owners in order to reestablish family ties that had been disrupted earlier. On plantation after plantation, Gutman found considerable evidence of the building and maintenance of extensive kinship ties among slaves. In instances where slave families had been disrupted, slaves in new communities reconstituted the kinds of family and kin ties that came to characterize black family life throughout the South. These patterns included, but were not limited to, a belief in the importance of marriage as a long-term commitment, rules of exogamy that included marriage between first cousins, and acceptance of women who had children outside of marriage. Kinship networks were an important source of resistance to the organization of labor that treated the individual slave, and not the family, as the unit of labor (Caulfield 1974).

Another interesting indicator of the slaves' maintenance of some degree of cultural autonomy has been pointed out by Wright (1981) in her discussion of slave housing. Until the early 1800s, slaves were often permitted to build their housing according to their own design and taste. During that period, housing built in an African style was quite common in the slave quarters. By 1830, however, slaveowners had begun to control the design and arrangement of slave housing and had introduced a degree of conformity and regularity to it that left little room for the slave's personalization of the home. Nevertheless, slaves did use some of their own techniques in construction and often hid it from their masters.

> Even the floors, which usually consisted of only tamped earth, were evidence of a hidden African tradition: slaves cooked clay over a fire, mixing in ox blood or cow dung, and then poured it in place to make hard dirt floors almost

like asphalt . . . In slave houses, in contrast to other crafts, these signs of skill
and tradition would then be covered over (Wright 1981, p. 48).

Housing is important in discussions of family because its design reflects
sociocultural attitudes about family life. The housing that slaveowners provided for their slaves reflected a view of Black family life consistent with the
stereotypes of the period. While the existence of slave families was acknowledged, it certainly was not nurtured. Thus, cabins were crowded, often con-
taining more than one family, and there were no provisions for privacy. Slaves
had to create their own.

> Slave couples hung up old clothes or quilts to establish boundaries; others
> built more substantial partitions from scrap wood. Parents sought to establish
> sexual privacy from children. A few ex-slaves described modified trundle beds
> designed to hide parental lovemaking . . . Even in one room cabins, sexual seg-
> regation was carefully organized (Wright 1981, p. 50).

Perhaps most critical in developing an understanding of slave women's reproductive labor is the gender-based division of labor in the domestic sphere.
The organization of slave labor enforced considerable equality among men
and women. The ways in which equality in the labor force was translated into
the family sphere is somewhat speculative. Davis (1981), for example, suggests
that egalitarianism between males and females was a direct result of slavery
when she says:

> Within the confines of their family and community life, therefore, Black
> people managed to accomplish a magnificent feat. They transformed that nega-
> tive equality which emanated from the equal oppression they suffered as slaves
> into a positive quality: the egalitarianism characterizing their social relations
> (p. 18).

It is likely, however, that this transformation was far less direct than Davis
implies. We know, for example, that slave women experienced what has recently been called the "double day" before most other women in this society.
Slave narratives (Jones 1985; White 1985; Blassingame 1977) reveal that
women had primary responsibility for their family's domestic chores. They
cooked (although on some plantations meals were prepared for all of the
slaves), sewed, cared for their children, and cleaned house, all after completing
a full day of labor for the master. Blassingame (1972) and others have pointed
out that slave men engaged in hunting, trapping, perhaps some gardening, and
furniture making as ways of contributing to the maintenance of their families.
Clearly, a gender-based division of labor did exist within the family and it
appears that women bore the larger share of the burden for housekeeping and
child care.

By contrast to white families of the period, however, the division of labor in the domestic sphere was neither reinforced in the relationship of slave women to work nor in the social institutions of the slave community. The gender-based division of labor among the slaves existed within a social system that treated men and women as almost equal, independent units of labor.[4] Thus Matthaei (1982) is probably correct in concluding that

> Whereas . . . the white homemaker interacted with the public sphere through her husband, and had her work life determined by him, the enslaved Afro-American homemaker was directly subordinated to and determined by her owner . . . The equal enslavement of husband and wife gave the slave marriage a curious kind of equality, an equality of oppression (p. 94).

Black men were denied the male resources of a patriarchal society and therefore were unable to turn gender distinctions into female subordination, even if that had been their desire. Black women, on the other hand, were denied support and protection for their roles as mothers and wives and thus had to modify and structure those roles around the demands of their labor. Thus, reproductive labor for slave women was intensified in several ways: by the demands of slave labor that forced them into the double-day of work; by the desire and need to maintain family ties in the face of a system that gave them only limited recognition; by the stresses of building a family with men who were denied the standard social privileges of manhood; and by the struggle to raise children who could survive in a hostile environment.

This intensification of reproductive labor made networks of kin and quasi-kin important instruments in carrying out the reproductive tasks of the slave community. Given an African cultural heritage where kinship ties formed the basis of social relations, it is not at all surprising that African American slaves developed an extensive system of kinship ties and obligations (Gutman 1976; Sudarkasa 1981). Research on Black families in slavery provides considerable documentation of participation of extended kin in childrearing, childbirth, and other domestic, social, and economic activities (Gutman 1976; Blassingame 1972; Genovese 1974).

After slavery, these ties continued to be an important factor linking individual household units in a variety of domestic activities. While kinship ties were also important among native-born whites and European immigrants, Gutman (1976) has suggested that these ties

> were comparatively more important to Afro-Americans than to lower-class native white and immigrant Americans, the result of their distinctive low economic status, a condition that denied them the advantages of an extensive associational life beyond the kin group and the advantages and disadvantages resulting from mobility opportunities (p. 213).

His argument is reaffirmed by research on Afro-American families after slavery (Shimkin et al. 1978; Aschenbrenner 1975; Davis 1981; Stack 1974). Sudarkasa (1981) takes this argument one step further and links this pattern to the African cultural heritage:

> historical realities require that the derivation of this aspect of Black family organization be traced to its African antecedents. Such a view does not deny the adaptive significance of consanguineal (kin) networks. In fact, it helps to clarify why these networks had the flexibility they had and why, they, rather than conjugal relationships came to be the stabilizing factor in Black families (p. 49).

With individual households, the gender-based division of labor experienced some important shifts during emancipation. In their first real opportunity to establish family life beyond the controls and constraints imposed by a slavemaster, family life among Black sharecroppers changed radically. Most women, at least those who were wives and daughters of able-bodied men, withdrew from field labor and concentrated on their domestic duties in the home. Husbands took primary responsibility for the fieldwork and for relations with the owners, such as signing contracts on behalf of the family. Black women were severely criticized by whites for removing themselves from field labor because they were seen to be aspiring to a model of womanhood that was considered inappropriate for them. This reorganization of female labor, however, represented an attempt on the part of Blacks to protect women from some of the abuses of the slave system and to thus secure their family life. It was more likely a response to the particular set of circumstances that the newly freed slaves faced than a reaction to the lives of their former masters. Jones (1985) argues that these patterns were "particularly significant" because at a time when industrial development was introducing a labor system that divided male and female labor, the freed Black family was establishing a pattern of joint work and complementary tasks between males and females that was reminiscent of the preindustrial American families. Unfortunately, these former slaves had to do this without the institutional supports that white farm families had in the midst of a sharecropping system that deprived them of economic independence.

CHINESE SOJOURNERS

An increase in the African slave population was a desired goal. Therefore, Africans were permitted and even encouraged at times to form families subject to the authority and whim of the master. By sharp contrast, Chinese people were explicitly denied the right to form families in the United States through both law and social practice. Although male laborers began coming to the United States in sizable numbers in the middle of the nineteenth century, it was more than a century before an appreciable number of children of Chinese parents were born in America. Tom, a respondent in Nee and Nee's (1973) book, *Longtime Californ'* says: "One thing about Chinese men in America was you had to be either a merchant or a big gambler, have lot of side money to have a family here. A working man, an ordinary man, just can't!" (p. 80).

Working in the United States was a means of gaining support for one's family with an end of obtaining sufficient capital to return to China and purchase land. The practice of sojourning was reinforced by laws preventing Chinese laborers from becoming citizens, and by restrictions on their entry into this country. Chinese laborers who arrived before 1882 could not bring their wives and were prevented by law from marrying whites. Thus, it is likely that the number of Chinese-American families might have been negligible had it not been for two things: the San Francisco earthquake and fire in 1906, which destroyed all municipal records; and the ingenuity and persistence of the Chinese people who used the opportunity created by the earthquake to increase their numbers in the United States. Since relatives of citizens were permitted entry, American-born Chinese (real and claimed) would visit China, report the birth of a son, and thus create an entry slot. Years later the slot could be used by a relative or purchased. The purchasers were called "paper sons." Paper sons became a major mechanism for increasing the Chinese population, but it was a slow process and the sojourner community remained predominantly male for decades.

The high concentration of males in the Chinese community before 1920 resulted in a split-household form of family. As Glenn observes:

> In the split-household family, production is separated from other functions and is carried out by a member living far from the rest of the household. The rest—consumption, reproduction and socialization—are carried out by the wife and other relatives from the home village. . . . The split-household form makes possible maximum exploitation of the workers . . . The labor of prime-age male workers can be bought relatively cheaply, since the cost of reproduction and family maintenance is borne partially by unpaid subsistence work of women and old people in the home village (Glenn 1981, pp. 14–15).

The women who were in the United States during this period consisted of a small number who were wives and daughters of merchants and a larger percentage who were prostitutes. Hirata (1979) has suggested that Chinese prostitution was an important element in helping to maintain the split-household family. In conjunction with laws prohibiting intermarriage, Chinese prostitution helped men avoid long-term relationships with women in the United

States and ensured that the bulk of their meager earnings would continue to support the family at home.

The reproductive labor of Chinese women, therefore, took on two dimensions primarily because of the split-household family form. Wives who remained in China were forced to raise children and care for in-laws on the meager remittances of their sojourning husband. Although we know few details about their lives, it is clear that the everyday work of bearing and maintaining children and a household fell entirely on their shoulders. Those women who immigrated and worked as prostitutes performed the more nurturant aspects of reproductive labor, that is, providing emotional and sexual companionship for men who were far from home. Yet their role as prostitute was more likely a means of supporting their families at home in China than a chosen vocation.

The Chinese family system during the nineteenth century was a patriarchal one wherein girls had little value. In fact, they were considered only temporary members of their father's family because when they married, they became members of their husband's families. They also had little social value: girls were sold by some poor parents to work as prostitutes, concubines, or servants. This saved the family the expense of raising them, and their earnings also became a source of family income. For most girls, however, marriages were arranged and families sought useful connections through this process.

With the development of a sojourning pattern in the United States, some Chinese women in those regions of China where this pattern was more prevalent would be sold to become prostitutes in the United States. Most, however, were married off to men whom they saw only once or twice in the 20- or 30-year period during which he was sojourning in the United States. Her status as wife ensured that a portion of the meager wages he earned would be returned to his family in China. This arrangement required considerable sacrifice and adjustment on the part of wives who remained in China and those who joined their husbands after a long separation.

Kingston (1977) tells the story of the unhappy meeting of her aunt, Moon Orchid, with her husband from whom she had been separated for 30 years.

> For thirty years she had been receiving money from him from America. But she had never told him that she wanted to come to the United States. She waited for him to suggest it, but he never did (p. 144).

His response to her when she arrived unexpectedly was to say:

> "Look at her. She'd never fit into an American household. I have important American guests who come inside my house to eat." He turned to Moon

Orchid, "You can't talk to them. You can barely talk to me." Moon Orchid was so ashamed, she held her hands over her face. She wished she could also hide her dappled hands (p. 178).

Despite these handicaps, Chinese people collaborated to establish the opportunity to form families and settle in the United States. In some cases it took as long as three generations for a child to be born on United States soil.

In one typical history, related by a 21 year old college student, great-grandfather arrived in the States in the 1890s as a "paper son" and worked for about 20 years as a laborer. He then sent for the grandfather, who worked alongside greatgrandfather in a small business for several years. Greatgrandfather subsequently returned to China, leaving grandfather to run the business and send remittance. In the 1940s, grandfather sent for father; up to this point, none of the wives had left China. Finally, in the late 1950s father returned to China and brought his wife back with him. Thus, after nearly 70 years, the first child was born in the United States (Glenn 1981, p. 14).

CHICANOS

Africans were uprooted from their native lands and encouraged to have families in order to increase the slave labor force. Chinese people were immigrant laborers whose "permanent" presence in the country was denied. By contrast, Mexican-Americans were colonized and their traditional family life was disrupted by war and the imposition of a new set of laws and conditions of labor. The hardships faced by Chicano families, therefore, were the result of the United States colonization of the indigenous Mexican population, accompanied by the beginnings of industrial development in the region. The treaty of Guadalupe Hidalgo, signed in 1848, granted American citizenship to Mexicans living in what is now called the Southwest. The American takeover, however, resulted in the gradual displacement of Mexicans from the land and their incorporation into a colonial labor force (Barrera 1979). In addition, Mexicans who immigrated into the United States after 1848 were also absorbed into the labor force.

Whether natives of Northern Mexico (which became the United States after 1848) or immigrants from Southern Mexico, Chicanos were a largely peasant population whose lives were defined by a feudal economy and a daily struggle on the land for economic survival. Patriarchal families were important instruments of community life and nuclear family units were linked together through an elaborate system of kinship and godparenting. Traditional life was characterized by hard work and a fairly distinct pattern of sex-role segregation.

> Most Mexican women were valued for their household qualities, men by their ability to work and to provide for a family. Children were taught to get up early, to contribute to the family's labor to prepare themselves for adult life . . . Such a life demanded discipline, authority, deference—values that cemented the working of a family surrounded and shaped by the requirements of Mexico's distinctive historical pattern of agricultural development, especially its pervasive debt peonage (Saragoza 1983, p. 8).

As the primary caretakers of hearth and home in a rural environment, *Las Chicanas* labor made a vital and important contribution to family survival. A description of women's reproductive labor in the early twentieth century can be used to gain insight into the work of the nineteenth-century rural women.

> For country women, work was seldom a salaried job. More often it was the work of growing and preparing food, of making adobes and plastering houses with mud, or making their children's clothes for school and teaching them the hymns and prayers of the church, or delivering babies and treating sicknesses with herbs and patience. In almost every town there were one or two women who, in addition to working in their own homes, served other families in the community as *curanderas* (healers), *parteras* (midwives), and schoolteachers (Elasser 1980, p. 10).

Although some scholars have argued that family rituals and community life showed little change before World War I (Saragoza 1983), the American conquest of Mexican lands, the introduction of a new system of labor, the loss of Mexican-owned land through the inability to document ownership, plus the transient nature of most of the jobs in which Chicanos were employed, resulted in the gradual erosion of this pastoral way of life. Families were uprooted as the economic basis for family life changed. Some immigrated from Mexico in search of a better standard of living and worked in the mines and railroads. Others who were native to the Southwest faced a job market that no longer required their skills and moved into mining, railroad, and agricultural labor in search of a means of earning a living. According to Camarillo (1979), the influx of Anglo[5] capital into the pastoral economy of Santa Barbara rendered obsolete the skills of many Chicano males who had worked as ranchhands and farmers prior to the urbanization of that economy. While some women and children accompanied their husbands to the railroad and mine camps, they often did so despite prohibitions against it. Initially many of these camps discouraged or prohibited family settlement.

The American period (post-1848) was characterized by considerable transiency for the Chicano population. Its impact on families is seen in the growth of female-headed households, which was reflected in the data as early as 1860. Griswold del Castillo (1979) found a sharp increase in female-headed households in Los Angeles, from a low of 13 percent in 1844 to 31% in 1880. Camarillo (1979, p. 120) documents a similar increase in Santa Barbara from 15% in 1844 to 30 percent by 1880. These increases appear to be due not so much to divorce, which was infrequent in this Catholic population, but to widowhood and temporary abandonment in search of work. Given the hazardous nature of work in the mines and railroad camps, the death of a husband, father or son who was laboring in these sites was not uncommon. Griswold del Castillo (1979) reports a higher death rate among men than women in Los Angeles. The rise in female-headed households, therefore, reflects the instabilities and insecurities introduced into women's lives as a result of the changing social organization of work.

One outcome, the increasing participation of women and children in the labor force was primarily a response to economic factors that required the modification of traditional values. According to Louisa Vigil, who was born in 1890:

> The women didn't work at that time. The man was supposed to marry that girl and take [care] of her . . . Your grandpa never did let me work for nobody. He always had to work, and we never did have really bad times (Elasser 1980, p. 14).

Señora Vigil's comments are reinforced in Garcia's (1980) study of El Paso. In the 393 households he examined in the 1900 census, he found 17.1% of the women to be employed. The majority of this group were daughters, mothers with no husbands, and single women. In the cases of Los Angeles and Santa Barbara, where there were even greater work opportunities for women than in El Paso, wives who were heads of household worked in seasonal and part-time jobs and lived from the earnings of children and relatives in an effort to maintain traditional female roles.

Slowly, entire families were encouraged to go to railroad workcamps and were eventually incorporated into the agricultural labor market. This was a response both to the extremely low wages paid to Chicano laborers and to the preferences of employers who saw family labor as a way of stabilizing the workforce. For Chicanos, engaging all family members in agricultural work was a means of increasing their earnings to a level close to subsistence for the entire group and of keeping the family unit together. Camarillo (1979, p. 93) provides a picture of the interplay of work, family, and migration in the Santa Barbara area in the following observation:

> The time of year when women and children were employed in the fruit cannery and participated in the almond and olive harvests coincided with the seasons when the men were most likely to be engaged in seasonal migratory work. There were seasons, however, especially in the early summer when the entire

family migrated from the city to pick fruit. This type of family seasonal harvest was evident in Santa Barbara by the 1890s. As walnuts replaced almonds and as the fruit industry expanded, Chicano family labor became essential.

This arrangement, while bringing families together, did not decrease the hardships that Chicanas had to confront in raising their families. We may infer something about the rigors of that life from Jesse Lopez de la Cruz's description of the workday of migrant farm laborers in the 1940s. Work conditions in the 1890s were as difficult, if not worse.

> We always went where the women and men were going to work, because if it were just the men working it wasn't worth going out there because we wouldn't earn enough to support a family . . . We would start around 6:30 a.m. and work for four or five hours, then walk home and eat and rest until about three-thirty in the afternoon when it cooled off. We would go back and work until we couldn't see. Then I'd clean up the kitchen. I was doing the housework and working out in the fields and taking care of two children (quoted in Goldman 1981, pp. 119–120).

In the towns, women's reproductive labor was intensified by the congested and unsanitary conditions of the *barrios* in which they lived. Garcia (1980) described the following conditions in El Paso:

> Mexican women had to haul water for washing and cooking from the river or public water pipes. To feed their families, they had to spend time marketing, often in Ciudad Juarez across the border, as well as long, hot hours cooking meals and coping with the burden of desert sand both inside and outside their homes. Besides the problem of raising children, unsanitary living conditions forced Mexican mothers to deal with disease and illness in their families. Diphtheria, tuberculosis, typhus and influenza were never too far away. Some diseases could be directly traced to inferior city services . . . As a result, Mexican mothers had to devote much energy to caring for sick children, many of whom died (pp. 320–321).

While the extended family has remained an important element of Chicano life, it was eroded in the American period in several ways. Griswold del Castillo (1979), for example, points out that in 1845 about 71 percent of Angelenos lived in extended families and that by 1880, fewer than half did. This decrease in extended families appears to be a response to the changed economic conditions and to the instabilities generated by the new sociopolitical structure. Additionally, the imposition of American law and custom ignored and ultimately undermined some aspects of the extended family. The extended family in traditional Mexican life consisted of an important set of familial, religious, and community obligations. Women, while valued primarily for their domesticity, had certain legal and property rights that acknowledged the importance of their work, their families of origin and their children. In California, for example:

> Equal ownership of property between husband and wife had been one of the mainstays of the Spanish and Mexican family systems. Community-property laws were written into the civil codes with the intention of strengthening the economic controls of the wife and her relatives. The American government incorporated these Mexican laws into the state constitution, but later court decisions interpreted these statutes so as to undermine the wife's economic rights. In 1861, the legislature passed a law that allowed the deceased wife's property to revert to her husband. Previously it had been inherited by her children and relatives if she died without a will (Griswold del Castillo 1979, p. 69).

The impact of this and other similar court rulings was to "strengthen the property rights of the husband at the expense of his wife and children" (Griswold del Castillo 1979, p. 69).

In the face of the legal, social, and economic changes that occurred during the American period, Chicanas were forced to cope with a series of dislocations in traditional life. They were caught between conflicting pressures to maintain traditional women's roles and family customs and the need to participate in the economic support of their families by working outside the home. During this period the preservation of some traditional customs became an important force for resisting complete disarray.

According to Saragoza (1983), transiency, the effects of racism and segregation, and proximity to Mexico aided in the maintenance of traditional family practices. Garcia has suggested that women were the guardians of Mexican cultural traditions within the family. He cites the work of anthropologist Manuel Gamio, who identified the retention of many Mexican customs among Chicanos in settlements around the United States in the early 1900s.

> These included folklore, songs and ballads, birthday celebrations, saints' day, baptism, weddings, and funerals in the traditional style. Because of poverty, a lack of physicians in the barrios, and adherence to traditional customs, Mexicans continued to use medicinal herbs. Gamio also identified the maintenance of a number of oral traditions, and Mexican style cooking (Garcia 1980, p. 322).

Of vital importance to the integrity of traditional culture was the perpetuation of the Spanish language. Factors that aided in the maintenance of other aspects of Mexican culture also helped in sustaining the language. However, entry into English-language public schools introduced the children and their families to systematic efforts to erase their native tongue. Griswold del Castillo reports that in the early 1880s there was considerable pressure against the speaker of Spanish in the public school. He also found that some Chicano parents responded to this kind of discrimination by helping support independent bilingual schools. These efforts, however, were short-lived.

Another key factor in conserving Chicano culture was the extended family network, particularly the system of *compadrazgo* or godparenting. Although the full extent of the impact of the American period on the Chicano extended family is not known, it is generally acknowledged that this family system, though lacking many legal and social sanctions, played an important role in the preservation of the Mexican community (Camarillo 1979, p. 13). In Mexican society, godparents were an important way of linking family and community through respected friends or authorities. Named at the important rites of passage in a child's life, such as birth, confirmation, first communion, and marriage, *compadrazgo* created a moral obligation for godparents to act as guardians, to provide financial assistance in times of need, and to substitute in case of the death of a parent. Camarillo (1979) points out that in traditional society these bonds cut across class and racial lines.

> The rites of baptism established kinship networks between rich and poor—between Spanish, mestizo and Indian—and often carried with them political loyalty and economic-occupational ties. The leading California patriarchs in the pueblo played important roles in the compadrazgo network. They sponsored dozens of children for their workers or poor relatives. The kindness of the *padrino* and *madrina* was repaid with respect and support from the *pobladores* (pp. 12–13).

The extended family network—which included godparents—expanded the support groups for women who were widowed or temporarily abandoned and for those who were in seasonal, part-, or full-time work. It suggests, therefore, the potential for an exchange of services among poor people whose income did not provide the basis for family subsistence. Griswold del Castillo (1980) argues that family organization influenced literacy rates and socioeconomic mobility among Chicanos in Los Angeles between 1850 and 1880. His data suggest that children in extended families (defined as those with at least one relative living in a nuclear family household) had higher literacy rates than those in nuclear families. He also argues that those in larger families fared better economically, and experienced less downward mobility. The data here are too limited to generalize to the Chicano experience as a whole but they do reinforce the actual and potential importance of this family form to the continued cultural autonomy of the Chicano community.

CONCLUSION: OUR MOTHERS' GRIEF

Reproductive labor for Afro-American, Chinese-American, and Mexican-American women in the nineteenth century centered on the struggle to maintain family units in the face of a variety of cultural assaults. Treated primarily as individual units of labor rather than as members of family groups, these women labored to maintain, sustain, stabilize, and reproduce their families while working in both the public (productive) and private (reproductive) spheres. Thus, the concept of reproductive labor, when applied to women of color, must be modified to account for the fact that labor in the productive sphere was required to achieve even minimal levels of family subsistence. Long after industrialization had begun to reshape family roles among middle-class white families, driving white women into a cult of domesticity, women of color were coping with an extended day. This day included subsistence labor outside the family and domestic labor within the family. For slaves, domestics, migrant farm laborers, seasonal factory-workers, and prostitutes, the distinctions between labor that reproduced family life and which economically sustained it were minimized. The expanded workday was one of the primary ways in which reproductive labor increased.

Racial-ethnic families were sustained and maintained in the face of various forms of disruption. Yet racial-ethnic women and their families paid a high price in the process. High rates of infant mortality, a shortened life span, the early onset of crippling and debilitating disease provided some insight into the costs of survival.

The poor quality of housing and the neglect of communities further increased reproductive labor. Not only did racial-ethnic women work hard outside the home for a mere subsistence, they worked very hard inside the home to achieve even minimal standards of privacy and cleanliness. They were continually faced with disease and illness that directly resulted from the absence of basic sanitation. The fact that some African women murdered their children to prevent them from becoming slaves is an indication of the emotional strain associated with bearing and raising children while participating in the colonial labor system.

We have uncovered little information about the use of birth control, the prevalence of infanticide, or the motivations that may have generated these or other behaviors. We can surmise, however, that no matter how much children were accepted, loved, or valued among any of these groups of people, their futures in a colonial labor system were a source of grief for their mothers. For those children who were born, the task of keeping them alive, of helping them to understand and participate in a system that exploited them, and the challenge of maintaining a measure—no matter how small—of cultural integrity, intensified reproductive labor.

Being a racial-ethnic woman in nineteenth-century American society meant having extra work both inside and outside the home. It meant having a contradictory relationship to the norms and values about women that were being generated in the dominant white culture. As pointed out earlier, the notion of separate spheres of male and female labor had contradictory outcomes for the nineteenth-century whites. It was the basis for the confinement of women to the household and for much of the protective legislation that subsequently developed. At the same time, it sustained white families by providing social acknowledgment and support to women in the performance of their family roles. For racial-ethnic women, however, the notion of separate spheres served to reinforce their subordinate status and became, in effect, another assault. As they increased their work outside the home, they were forced into a productive labor sphere that was organized for men and "desperate" women who were so unfortunate or immoral that they could not confine their work to the domestic sphere. In the productive sphere, racial-ethnic women faced exploitative jobs and depressed wages. In the reproductive sphere, however, they were denied the opportunity to embrace the dominant ideological definition of "good" wife or mother. In essence, they were faced with a double-bind situation, one that required their participation in the labor force to sustain family life but damned them as women, wives, and mothers because they did not confine their labor to the home. Thus, the conflict between ideology and reality in the lives of racial-ethnic women during the nineteenth century sets the stage for stereotypes, issues of self-esteem, and conflicts around gender-role prescriptions that surface more fully in the twentieth century. Further, the tensions and conflicts that characterized their lives during this period provided the impulse for community activism to jointly address the inequities, which they and their children and families faced.

ACKNOWLEDGMENTS

The research in this study is the result of the author's participation in a larger collaborative project examining family, community, and work lives of racial-ethnic women in the United States. The author is deeply indebted to the scholarship and creativity of members of the group in the development of this study. Appreciation is extended to Elizabeth Higginbotham, Cheryl Townsend Gilkes, Evelyn Nakano Glenn, and Ruth Zambrana (members of the original working group), and to the Ford Foundation for a grant that supported in part the work of this study.

NOTES

1. The term *reproductive labor* is used to refer to all of the work of women in the home. This includes but is not limited to: the buying and preparation of food and clothing, provision of emotional support and nurturance for all family members, bearing children, and planning, organizing, and carrying out a wide variety of tasks associated with their socialization. All of these activities are necessary for the growth of patriarchal capitalism because they maintain, sustain, stabilize, and *reproduce* (both biologically and socially) the labor force.

2. The term *white* is a global construct used to characterize peoples of European descent who migrated to and helped colonize America. In the seventeenth century, most of these immigrants were from the British Isles. However, during the time period covered by this article, European immigrants became increasingly diverse. It is a limitation of this article that time and space does not permit a fuller discussion of the variations in the white European immigrant experience. For the purposes of the argument made herein and of the contrast it seeks to draw between the experiences of mainstream (European) cultural groups and that of racial/ethnic minorities, the differences among European settlers are joined and the broad similarities emphasized.

3. Cultural assaults, according to Caulfield (1974), are benign and systematic attacks on the institutions and forms of social organization that are fundamental to the maintenance and flourishing of a group's culture.

4. Recent research suggests that there were some tasks that were primarily assigned to males and some others to females. Whereas some gender-role distinctions with regard to work may have existed on some plantations, it is clear that slave women were not exempt from strenuous physical labor.

5. This term is used to refer to white Americans of European ancestry.

References

Aschenbrenner, Joyce. 1975. *Lifelines: Black Families in Chicago.* New York, NY: Holt, Rinehart, and Winston.

Barrera, Mario. 1979. *Race and Class in the Southwest.* South Bend, IN: Notre Dame University Press.

Blassingame, John. 1972. *The Slave Community: Plantation Life in the Antebellum South.* New York: Oxford University Press.

———. 1977. *Slave Testimony: Two Centuries of Letters, Speeches, Interviews, and Autobiographies.* Baton Rouge, LA: Louisiana State University Press.

Camarillo, Albert. 1979. *Chicanos in a Changing Society.* Cambridge, MA: Harvard University Press.

Caulfield, Mina Davis. 1974. "Imperialism, the Family, and Cultures of Resistance." *Socialist Review* 4(2)(October): 67–85.

Davis, Angela. 1971. "The Black Woman's Role in the Community of Slaves." *Black Scholar* 3(4)(December): 2–15.

———. 1981. *Women, Race and Class.* New York: Random House.

Degler, Carl. 1980. *At Odds.* New York: Oxford University Press.

Elasser, Nan Kyle MacKenzie, and Yvonne Tixier Y. Vigil. 1980. *Las Mujeres.* New York: The Feminist Press.

Garcia, Mario T. 1980. "The Chicano in American History: The Mexican Women of El Paso, 1880–1920—A Case Study." *Pacific Historical Review* 49(2)(May): 315–358.

Genovese, Eugene D. and Elinor Miller, eds. 1974. *Plantation, Town, and County: Essays on the Local History of American Slave Society.* Urbana: University of Illinois Press.

Glenn, Evelyn Nakano. 1981. "Family Strategies of Chinese-Americans: An Institutional Analysis." Paper presented at the Society for the Study of Social Problems Annual Meetings.

Goldman, Marion S. 1981. *Gold Diggers and Silver Miners.* Ann Arbor: The University of Michigan Press.

Griswold del Castillo, Richard. 1979. *The Los Angeles Barrio: 1850–1890.* Los Angeles: The University of California Press.

Gutman, Herbert. 1976. *The Black Family in Slavery and Freedom: 1750–1925.* New York: Pantheon.

Hirata, Lucie Cheng. 1979. "Free, Indentured, Enslaved: Chinese Prostitutes in Nineteenth-Century America." *Signs* 5 (Autumn): 3–29.

Jones, Jacqueline. 1985. *Labor of Love, Labor of Sorrow.* New York: Basic Books.

Kennedy, Susan Estabrook. 1979. *If All We Did Was to Weep at Home: A History of White Working-Class Women in America.* Bloomington: Indiana University Press.

Kessler-Harris, Alice. 1981. *Women Have Always Worked.* Old Westbury: The Feminist Press.

———. 1982. *Out to Work.* New York: Oxford University Press.

Kingston, Maxine Hong. 1977. *The Woman Warrior.* Vintage Books.

Matthaei, Julie. 1982. *An Economic History of Women in America.* New York: Schocken Books.

Nee, Victor G., and Brett de Bary Nee. 1973. *Longtime Californ'.* New York: Pantheon Books.

Saragoza, Alex M. 1983. "The Conceptualization of the History of the Chicano Family: Work, Family, and Migration in Chicanos." Research Proceedings of the Symposium on Chicano Research and Public Policy. Stanford, CA: Stanford University, Center for Chicano Research.

Shimkin, Demetri, E. M. Shimkin, and D. A. Frate, eds. 1978. *The Extended Family in Black Societies.* The Hague: Mouton.

Spruill, Julia Cherry. 1972. *Women's Life and Work in the Southern Colonies.* New York: W. W. Norton and Company (first published in 1938, University of North Carolina Press).

Stack, Carol S. 1974. *All Our Kin: Strategies for Survival in a Black Community.* New York: Harper and Row.

Sudarkasa, Niara. 1981. "Interpreting the African Heritage in Afro-American Family Organization." Pp. 37–53 in *Black Families,* edited by Harriette Pipes McAdoo. Beverly Hills, CA: Sage Publications.

White, Deborah Gray. 1985. *Ar'n't I a Woman?: Female Slaves in the Plantation South.* New York: W. W. Norton.

Wright, Gwendolyn. 1981. *Building the Dream: A Social History of Housing in America.* New York: Pantheon Books.

Zaretsky, Eli. 1978. "The Effects of the Economic Crisis on the Family." Pp. 209–218 in *U.S. Capitalism in Crisis,* edited by Crisis Reader Editorial Collective. New York: Union of Radical Political Economists.

THE DIVERSITY OF AMERICAN FAMILIES   28

Eleanor Palo Stoller and Rose Campbell Gibson

Our understanding of family, perhaps more than any other institution, is clouded by myth. Masking diversity among contemporary U.S. families is an ideology that describes the American family as "one man who goes off to earn the bacon, one woman who waits at home to fry it, and roughly 2.5 children poised to eat it" (Cole 1986, p. 11). This image of family-dominated U.S. culture during the decades following World War II, a period during which the majority of older people today experienced at least part of their child-rearing years. As described by Betty Friedan (1964) in *The Feminine Mystique,* this ideology told women that caring for their home and family was the path to fulfillment and gratification (Andersen 1993). The sociologist Talcott Parsons (1955) described the American family as an isolated, nuclear unit in which the husband, wife, and their dependent children live geographically and economically independent from other relatives. Families were viewed as safe havens from the competitiveness of the workplace, and both popular images and social scientific treatises emphasized the split between the public sphere of employment and the private sphere of home.

From: *Worlds of Difference: Inequality in the Aging Experience,* ed. by Eleanor Palo Stoller and Rose Campbell Gibson (Thousand Oaks, CA: Pine Forge Press, 1997), pp. 213–18. Reprinted by permission.

We will explore the consequences of this idealized image of the American family in our efforts to understand the diversity of family experiences among older people today. Our exploration of older families will ask the following questions:

1. How does the myth of the American family mask the diversity of families in which today's cohorts of elderly people grew up and grew old?
2. How do the interlocking hierarchies based on gender, race, and class structure the family experiences of today's older Americans?
3. How do definitions of family based on the experiences of the dominant groups in society bias research on other older families?

HOW THE MYTH OF THE AMERICAN FAMILY MASKS THE DIVERSITY OF FAMILIES IN WHICH TODAY'S COHORTS OF ELDERLY PEOPLE GREW UP AND GREW OLD

There are a number of problems with the ideological view of the nuclear family as a key to understanding the family experiences of older people today. First, not all older people married and not all married people had children. Of people 65 and older today, 5% never married, 6% divorced or separated, and 20% do not have adult children. Since the 1950s, the percentage of never-married older people has declined, and the percentage of divorced elderly people has begun to increase (Myers 1990, p. 39). Marriage and divorce rates also vary by gender and ethnicity. Older Native American and Asian American women are slightly less likely to have remained single throughout their lives than are other older people. Rates of divorce and separation for elderly African Americans, Hispanic Americans, Native Americans, and Asian Americans are higher than rates for elderly whites.

Second, the idealized view of the family glorified the role of the full-time housewife, wife, and mother. Married women who did not enter the paid labor force and worked full-time at home often found that their lives did not correspond to the image fostered by popular culture. Many full-time mothers were troubled by the isolation of their daily lives, the repetitiveness of household tasks, and the invisibility of their accomplishments (Oakley 1974). Betty Friedan (1964) labeled the vague, unspoken anxiety experienced by the full-time housewives in her study as "the problem that has no name." According to popular ideology, these women should have been among the most happy and fulfilled. Their families came closest to the idealized Amercan family. Yet Friedan's (1964) subjects shared doubts and dissatisfactions. Moreover, they felt responsible for this unease, because the ideology of the feminine mystique shifted the blame to women themselves if they were unsatisfied with their lives. Women were taught that they

> could desire no greater destiny than to glory in their femininity. . . . If a woman had a problem in the 1950s and 1960s, she knew that something must be wrong with her marriage and with herself. . . . What kind of woman was she if she did not feel this mysterious fulfillment waxing the kitchen floor? (p. 19)

Not all women who opted for full-time work at home over paid work in the labor market were dissatisfied with their situation. Caring for a home and family offered opportunities for creativity and satisfaction. Because work at home provides freedom from supervision and some flexibility in organizing tasks, many women preferred it to the more alienating conditions they would have encountered in the paid positions available to them (Andersen 1993). Nevertheless, as the 1960s came to a close, the vague discontent described by Friedan began to crystallize into a feminist analysis that defined motherhood as a serious obstacle to women's fulfillment. This strain of feminist thought reflects the race and class position of its creators. In gathering data for *The Feminine Mystique,* Friedan interviewed her former classmates at Smith College, and her results reflect the experiences of these white, middle-class, college-educated women. bell hooks (1984) describes the bias of an analysis developed from the standpoint of this particular group:

> [They] argued that motherhood was a serious obstacle to women's liberation, a trap confining women to the home, keeping them tied to cleaning, cooking, and child care. . . . Had black women voiced their views on motherhood, [motherhood] would not have been named a serious obstacle to our freedom as women. Racism, availability of jobs, lack of skills or education, and a number of other issues would have been at the top of the list—but not motherhood. Black women would not have said motherhood prevented us from entering the world of paid work, because we have always worked. (p. 133)

. . . [N]ot all of today's older women were able to choose whether or not they would stay out of the paid labor force when their children were young, despite the popular and scholarly rhetoric advocating full-time motherhood. Women rearing children alone because of single motherhood, divorce, or widowhood had little choice but to combine the nurturing roles assigned to mothers with the breadwinning responsibilities supposedly reserved for fathers. Further, poor and working-class two-parent families often needed two incomes to survive. Women of color were more likely than white women to find themselves in this situation, as racial oppression denied many sufficient economic resources to maintain nuclear family households (Collins 1990). Elderly African American women, for example, typically combined child rearing with paid work at a time when employment opportunities for women and for people of color were even more restricted than they are today.

The rigid gender-based division of labor that characterized popular constructions of the American family of the 1950s did not apply across all combinations of race and class. For example, the strict gender-role segregation within the domestic sphere applied more to white families than to African American families. African American couples have traditionally been more flexible in allocating family obligations, with parents sharing household tasks and child rearing as well as responsibility for earning a living (McAdoo 1986). Upper-class women, most of whom were white, retained personal responsibility for managing children and household but rarely did the actual work themselves (Ostrander 1984). Rather, they supervised the work of paid domestic workers, who were usually poor or working-class women and often women of color (Cole 1986).

Upper-class women also had more resources with which to exercise their family responsibilities than less affluent women. Susan Ostrander's (1984) wealthy subjects wanted their children "to develop to their fullest potential. . . . to be 'the best'" (p. 75). They expressed concerns about their children's personal happiness and their future success. Most important, they had resources to facilitate these goals for their children. For example, if their children encountered academic difficulties or adjustment problems, the parents often sent them to private schools. Mrs. Miles, a woman now in her seventies whose husband had been chairman of the board of a major bank, describes her decision to enroll her teenage daughter in a private boarding school: "She was a very shy child . . . so I thought . . . going away to school would be good for her, would give her some confidence and put her on her feet" (p. 85). As Ostrander explains, "Upper class children . . . are not allowed to fail academically or personally. This gives them strong advantages" (p. 84). Parents from other social classes share many of the same dreams and concerns for their children expressed by the women Ostrander interviewed, but they do not share the same resources for helping their children realize these dreams.

Another aspect of the mythic American family that clouds our understanding of the family experiences of today's elderly people is the assumption that the nuclear family is isolated from other relatives. Contrary to this image of independent nuclear households, women have traditionally maintained family relationships across households through visiting, writing letters, organizing holiday gatherings, and remembering birthdays and anniversaries— activities that have been described as "kinship work" (Rosenthal 1985). The work of "kinkeeping" can produce complex networks of female kin who entrust confidences, pool resources, and share social activities. Furthermore, as social gerontologists have demonstrated, women are the primary caregivers for the frail elderly people in their families.

Strong female-centered networks have also linked families and households among African Americans. Mothering responsibilities have traditionally been shared among women in African American communities. Boundaries distinguishing biological mothers from other women who care for children are less rigid within these communities, and women often feel a sense of obligation for *our children*, a term that includes all of the children in their community. Patricia Hill Collins (1991) explains that "African American communities recognized that vesting one person with full responsibility for mothering a child may not be wise or possible. As a result, 'othermothers,' women who assist bloodmothers by sharing mothering responsibilities, traditionally have been a central part of the institution of Black motherhood" (p. 47). When children are orphaned, when parents are ill or at work, or when biological mothers are too young to care for their children alone, other women in the community take on child care responsibilities, sometimes temporarily, but other times permanently.

These women-centered networks of bloodmothers and othermothers within African American communities have often been described as a reaction to the legacy of slavery and to generations of poverty and oppression. More recently, scholars have challenged this interpretation. The impact of racial oppression on African American families must be acknowledged, but it is also important to recognize the ways in which today's elderly African Americans evolved new definitions of family from their everyday experiences and cultural legacies. Networks of community-based child care, for example, provided supervision for children while biological parents worked to provide economic support. Alternative definitions of motherhood, emphasizing both emotional support and physical provision of care, were adapted from West African culture, which had been retained as a culture of resistance since slavery (Sudarkasa 1981). Building on this cultural heritage, African American women devised strategies for ensuring their children's physical survival within a racist society (Collins 1991). . . .

Other examples also challenge the myth of self-reliance of isolated, nuclear households. Coontz (1992) points out that working-class and ethnic subcommunities "evolved mutual aid in finding jobs, surviving tough times, and pooling money for recreation" (p. 71). Immigrant groups developed an infrastructure of churches, temperance societies, workers' associations, fraternal orders, and cooperatives that provided instrumental and emotional aid beyond the confines of the nuclear household. Godparenting created extra-familial bonds among Catholic populations, bonds that often cut across social class boundaries, and, as Coontz (1992) points out, "the notion of 'going for sisters' has long and still thriving roots in black communities" (p. 72). Extended networks of kin also characterize Latino families, in which child care responsibilities are often shared with older siblings, aunts, uncles, and grandparents. Ceremonial events frequently involve close friends linked to the family through *compadrazgo*, a network of kinlike ties among very close friends who exchange tangible assistance and social support (Heyck 1994).

The idealized image of the American family also assumes a heterosexual couple. Despite more recent attention to gay and lesbian rights, it is important to remember that today's cohorts of elderly gay men and lesbians grew up in a social environment in which they encountered strong pressures to hide their sexual orientation. Many kept their relationships secret to avoid discriminatory treatment by their families, their employers, and their communities (D'Augelli & Hart 1987). Although marriage between gay and lesbian partners is not legally sanctioned, many couples establish long-term relationships. . . . Gay and lesbian elders often establish networks of friends who, like family members, exchange emotional and instrumental support (Kimmel 1992). Lipman (1986) reported that elderly gay men and lesbians tend to have more friends than heterosexuals of similar age. Most of these friends are of the same gender, and many are also gay or lesbian.

Generalizing the image of the American family to all couples masks the complexity of family experiences and resources with which older Americans face the challenges of aging. Uncritically accepting dominant stereotypes of older families is like wearing blinders. Our understanding is limited because we don't think to ask questions about dimensions of social reality that we overlook. We forget to ask how people who never married develop other relationships to meet their needs for companionship in old age. We forget to ask how gay and lesbian couples activate a support network to care for them when they are ill. We forget to ask how more fluid definitions of family and more flexible divisions of labor both within and beyond nuclear family households influence strategies used by older African Americans in coping with disability in old age. These oversights are not intentional. They reflect limited views of social reality from particular standpoints. . . .

References

Andersen, M. L. 1993. *Thinking About Women: Sociological and Feminist Perspectives on Sex and Gender*, 3d ed. New York: Macmillan.

Cole, J. 1986. *All American Women: Lives That Divide, Ties That Bind*. New York: Free Press.

Collins, P. H. 1991. "The Meaning of Motherhood in Black Culture and Black Mother-Daughter Relationships." In P. Bell-Scott, B. Guy-Sheftall, J. Jones Royster, J. Sims-Wood, M. DeCosta-Willis, and L. Fultz, eds., *Double Stitch: Black Women Write about Mothers and Daughters* (Boston: Beacon Press), pp. 42–60.

Collins, P. H. 1990. *Black Feminist Thought: Knowledge, Consciousness, and the Politics of Empowerment.* Boston: Unwin Hyman.

D'Augelli, A., and Hart, M. M. 1987. "Gay Women, Men and Families in Rural Settings: Toward the Development of Helping Communities." *American Journal of Community Psychology* 15 (1): 79–93.

Heyck, D. L. D. 1994. *Barrios and Borderlands: Cultures of Latins and Latinas in the United States.* New York: Routledge.

Kimmel, D. C. 1992. "The Families of Older Gay Men and Lesbians." *Generations* 16 (Summer): 37–38.

McAdoo, H. 1986. "Societal Stress: The Black Family." In J. Cole, ed., *All American Women: Lines That Divide, Ties That Bind* (New York: Free Press), pp. 187–97.

Myers, G. 1990. "Demography of Aging." In R. Binstock and L. K. George, eds., *Handbook of Aging and the Social Sciences*, 3d ed. (San Diego, CA: Academic Press), pp. 19–45.

Oakley, A. 1974. *The Sociology of Housework.* Oxford: Martin Robertson.

Ostrander, S. 1984. *Women of the Upper Class* (Philadelphia: Temple University Press), pp. 111–139.

Rosenthal, C. J. 1985. "Kinkeeping in the Familial Division of Labor." *Journal of Marriage and the Family* 47: 965–74.

Sudarkasa, N. 1981. "Interpreting the African Heritage in Afro-American Family Organizations." In H. McAdoo, ed., *Black Families* (Beverly Hills, CA: Sage), pp. 37–53.

ARE "ABSENT FATHERS" REALLY ABSENT? *Urban Adolescent Girls Speak Out About Their Fathers*

29

Niobe Way and Helena Stauber

In the midst of a lunch break during one of our writing sessions for this paper, we ran into a friend who asked us what we were writing about. When we told her we were writing about urban girls' relationships with their fathers, she immediately inquired, "Do these girls even have fathers in their lives?" Her question reflected one of the most pervasive beliefs about the family relation-

From: Bonnie J. Ross Leadbeater and Niobe Way, *Urban Girls: Resisting Stereotypes, Creating Identities*, (New York: New York University, 1996), pp. 132–48. Reprinted by permission.

ships of low-income, inner-city children and adolescents, particularly those who are black or Hispanic. Quite in contrast to this stereotype of father absence in the inner city, however, our studies with low-income urban adolescent girls suggest that these fathers are typically very much present in their daughters' lives. Although many of the girls in our studies told us that their fathers did not live with them, the majority offered accounts of complex, multilayered relationships with their fathers. In this paper we describe the stories of father-daughter relationships told to us by a group of urban adolescent girls.

. . . Although some researchers have examined middle-class adolescent girls' perceptions of their fathers (see Youniss and Smollar 1985), there is a dearth of studies that examine how low-income or ethnic minority adolescent girls experience this relationship. Reasons for the shortage may lie with the seemingly implicit assumption in the research literature that those fathers who do not live with their daughters have no relationships with their daughters. Studies do exist, however, that show that fathers from low-income communities who do not live with their children often continue to play an active role in their children's lives (see Earl and Lohmann 1978; Liebow 1967; Rivara, Sweeney, and Henderson 1986; Stack 1974).

The purpose of our study is to address this void in the research literature by exploring urban adolescent girls' perceptions of their relationships with their fathers. . . . Our intent is not to produce findings that can be generalized to larger populations of adolescent girls, but rather to listen to the subtleties and nuances in the narratives of a small group of urban adolescent girls.

THE STUDY

Method

Sample.   The sample is comprised of 45 adolescent girls who attended an inner-city public high school in a large northeastern city. The girls ranged in age from fifteen to nineteen and identified themselves as African American (22), Puerto Rican (5), Dominican (4), Irish or Italian American (5), West Indian (7), Syrian (1), and Native American (1). The adolescents all came from poor or working-class families (social class was inferred from the parents' educational backgrounds and current occupations). The sample was drawn from classrooms with students who represent the range of academic abilities. All of the students who volunteered to participate in the study (95% of those who were told about the research project) were included.[1]

Procedure.   The participants were interviewed when they were in the ninth (N = 12), tenth (N = 8), eleventh (N = 14), or twelfth (N = 11) grade. Twelve of the girls (randomly selected from the forty-five girls) were also involved in a three-year longitudinal study (see Way 1995; Way in press). The interviewers—psychology doctoral students at the time of the study—included an African American middle-class man, two white middle-class women (the authors), and one white man from a working-class family. The interview lasted from one and a half to two hours and included questions concerning the girls' relationships with their parents or with the people who raised them (e.g., How would you describe your relationship with your father/mother? What do you like/dislike about this relationship? What would you like to change about this relationship? How do you think it has changed over the years?). Although each interview included a standard set of initial questions, follow-up questions were open-ended in order to capture the adolescents' own ways of describing their relationships. All interviews were tape-recorded and transcribed by a professional transcriber. . . .

Demographic Findings.   Forty-two percent (N = 19) of the girls lived with their fathers (two lived only with their fathers); 15% (N = 7) did not live with their fathers but had frequent contact (daily or weekly) with them; 22% (N = 10) did not live with their fathers but had occasional contact with them (e.g., anywhere from once or twice a month to two or three times a year); and the fathers of 20% (N = 9) were either dead (N = 3) or entirely absent (N = 6). In sum, 36 of the adolescent girls had daily, weekly, or occasional contact with their fathers, and only 9 had completely absent fathers. We detected ethnic/racial differences within the group of girls who lived with their fathers: 77% of the Latina, 80% of the white, and only 24% of the black girls in the sample lived with their fathers (28% of these black girls were of West Indian origin, and the remainder were African American).

Most of the 19 girls who lived with their fathers reported that they had lived with their fathers all of their lives and that their parents had been married or together for twenty years or longer. Among the 17 girls who did not live with their fathers but had occasional or regular contact, 11 reported that their fathers lived in their neighborhood or in close proximity (e.g., in nearby cities), while 6 girls reported that their fathers lived at a considerable distance (e.g., in another state or country). Of the 6 girls who had no contact with their fathers but whose fathers were still alive, 5 girls stated that their fathers lived in a distant location, and only 1 stated that her father lived close by.

Qualitative Findings.   The girls in our sample who had contact with their biological fathers (N = 36) typically appeared to have one of three types of relationships with their fathers: conflicted and engaged (33%), warm and open (36%), or emotionally distant (25%). These three categories were created in response to our interview data—they are data-driven categories. They represent our attempts to capture the prevailing quality of the interpersonal contact between the girls and their fathers. Those girls who described conflictual and engaged relationships with their fathers were more likely to live with their fathers, whereas those who had distant relationships were more likely to live apart from their fathers. Girls who lived with their fathers were as likely to describe warm and open relationships with their fathers as were those who did not live with their fathers.

1. **Feeling Conflicted/Being Engaged.** When Melissa, an Irish American adolescent girl, is asked about her relationship with her father in her junior year, she reaches into her backpack and pulls out a poem she has recently written. . . . [The] poem provides a poignant illustration of the anger and attachment that may coincide within a father-daughter relationship and the tension and dialogue between these two conflicting emotions. . . .

Melissa, who recently chose to leave her mother and move in with her father, expresses through her poetry an admiration for her father ("he's filled with knowledge . . ."), a desire for his attention ("I hate being second in line"), and an impassioned anger toward him ("Why do I hate you so much?"). Despite, or perhaps because of, the intensity of her feelings, Melissa is ambivalent about revealing them directly to her father. She tells her interviewer that her father discovered her poem when he opened her mail (she had submitted this poem to a youth newspaper), and consequently she felt bad because she knew the poem would "hurt his feelings." At the same time, her choice to leave the letter out in the open, knowing that her father has a tendency to read her mail (which did not appear to bother her), suggests a competing wish to reveal her feelings to her father.

After discussing her poem, Melissa says to her interviewer that she gets along with her father but only "talk[s] to him . . . like I don't really speak to him that much." "Speaking," she implies, means openly discussing her thoughts and feelings. In her decision to move in with her father, in "talking" but not "speaking" to him, in her concern about hurting his feelings, in her assessment that she "gets along" with her father, and in her statements of simultaneous admiration and hatred, Melissa suggests a relationship that is characterized by strong and sometimes opposing emotions of ire, frustration, affection, and love. Melissa exhibits conflicted feelings about her father, and also appears to be deeply emotionally engaged with him.

Similar themes were heard from eleven of Melissa's peers, seven of whom lived with their fathers and four of whom did not. These girls spoke at length and with great emphasis about their anger or frustration and their feelings of being hurt by their fathers. Concurrently, however, they also spoke about wanting to connect with, admiring, and/or identifying with their fathers. These twelve girls expressed emotional attachment with their fathers in tandem with feelings of pain and frustration.

Marie, a Dominican student who lives with both her parents, says in her freshman year:

> Sometimes I feel like my father treats me like shit, you know . . . sometimes he treats me nice. . . . Ok, he gives me everything I want. He cares for me a lot and everything. It's not like he's a bad father . . . and . . . thing is, sometimes . . . he's not understanding. It's like he always—he always every time something happens he always thinks it's me, it was my fault that that happened. . . . He'a nice and everything but he doesn't think when he does things. . . . So I'm always like trying to defend myself. Every time they blame me for something, I try to talk back and they're always "you ain't got no respect. You always talk back."

Marie's words reflect a theme that was heard repeatedly among the girls in the sample—especially among those who lived with their fathers and who had conflicted and engaged relationships with their fathers. Eight of the twelve girls who had conflicted and engaged relationships (as well as five of the thirteen girls who had warm and open relationships) stated that they were outspoken with their fathers. When they experienced frustration with or felt that they were being treated unfairly or unkindly by their fathers, these girls expressed their feelings directly to their fathers in protest.

Marie, interviewed again in her sophomore year, reiterates the complexity and variability of her feelings for her father. On the one hand, she says that her father gives her anything she wants and is "sweet" and "there" for her, while on the other hand, she says that he's a "pain" because he is not understanding of her. . . .

. . . Rather than silencing herself with or distancing herself from her father when she is angry, Marie speaks back to her father's accusations.

Christine, an African American student interviewed in her sophomore year, also seems to be irritated with, embittered by, and engaged with her father. Christine's engagement, like Marie's, is suggested when she discusses her ability to speak her mind with her father:

> We always have discussions every once in a while. But then I don't agree with his view so. When he says something, you know, I try to challenge it. You know, find out why he thinks that way. And he'll get mad. 'Cause he doesn't like when

people try to challenge him. He wants to think that he's always right. And when you do tell him he's wrong, he gets mad and tries to tell you you're just trying to act grown [up].

Christine does not retreat when she disagrees with her father but challenges him directly even when he disapproves of her challenges. In her desire and ability to speak out, Christine appears actively engaged with her father. At a later point in her interview, Christine tells her interviewer the reason for her behavior:

I'm really the only person that'll stand up to [my father]. Talk to him word for word. So he really doesn't argue with me or try to you know. I can't really explain. . . . I get it from him. . . . He doesn't bug down . . . to nobody and so neither do I. . . . He doesn't like to listen, he wants everything to go his way. He likes to be his own boss. And I'm the same way . . . 'cause he always taught me never to let people take advantage of you, take control of you. You should always speak up, talk for yourself. And that's what I do. It just so happens I do it with him.

Stating her similarity to her father in their shared tendency to assert themselves, Christine suggests a bond and a sense of identification with her father. Five of the twelve girls who had conflicted and engaged relationships with their fathers and one of the thirteen girls who had warm and open relationships, identified with their fathers. These girls expressed feelings of frustration and anger toward their fathers—with whom they regularly fought—and also stated they were "just like" their fathers and appreciated the ways in which they were alike. These statements of identification and appreciation in the face of irritation and animosity contributed to our interpretation that an adolescent felt *conflicted* rather than exclusively angry or loving toward her father.

The coexistence of anger and identification is heard in the words of Tyiesha, an African American student, who says in her freshman year: "[My father and I] argue. We're too much alike. That's why we don't get along too well. . . . We always arguing—too much alike." In her junior year, Verlinda, also African American, says that she does not like her father's "attitude" or temper. However, she says that she thinks she gets her "attitude" and temper from her father: "Me and my father are pretty much the same. . . . My mother always tells me 'You're just like your father.'" Verlinda says at the end of her interview that sometimes she likes her "attitude," implying that occasionally she appreciates her father's influence. A similar theme is heard in the words of Gabriela, an Irish American adolescent, who says that her father "expects me to keep my mouth shut [when he gets angry] but I don't. I'm not gonna keep my mouth shut. . . . I ain't taking it from him. . . . Like he gets mad when I'm like this, but he gave me this." While Gabriela is angry at her father, she also realizes, much like Christine, that her father has taught her a skill she considers valuable—her ability to speak her mind.

In sum, twelve of the thirty-six adolescent girls in the sample (33%) who had contact with their fathers voiced conflicted feelings about their fathers in their expressions of outrage, hurt, appreciation, and attachment. They did not typically distance themselves from their fathers when they were angry or upset, but engaged with their father through arguments, disagreements, and confrontations—skills that they often said they had learned from their fathers. Their conflicted feelings were suggested by their statements of identification, their stories of speaking out, and their verbal shifts between expressions of appreciation, admiration, and warmth, on the one hand, and displeasure, frustration, and anger, on the other.

2. **Open and Warm Relationships.**   Mary, an Irish American student, says in her freshman year that her father is her role model because he "has so much patience with me. . . . He can cope with, you know to have a teenage daughter and everything." Mary's mother died when she was ten, and since that time, she says that she and her father have been close. Interviewed again in her sophomore year, Mary says that she is getting along with her father even better than she did the previous year because she listens to him more "where before I was like, 'You don't know anything.' . . . He listens to me and understands that, you know, that he can't always treat me like a baby." She says she is open and honest with her father and considers him to be her role model because he has "patience . . . and he has the energy and he's just—I don't know, he's great." In her junior year, Mary says that she enjoys her relationship with her father because she can sit down and tell him "everything . . . frustrating problems like my boyfriend . . . and school. . . . Anything to do with that I'd go to my father." She says that out of all her relationships, she is closest to her father. Her perspectives on her father over the three years that she is interviewed suggest a consistent theme of warmth and openness.

Thirteen of the thirty-six girls (36%) who had contact with their fathers described their relationships with their fathers as being warm, open, and connected (six lived with their fathers and seven did not). Sonia, a Puerto Rican student who lives with both of her parents, says in her freshman year that she feels closer to her father than to her mother—a common feeling among those who appeared to have warm and open relationships with their fathers. Sonia says that she feels that she can more freely speak with her father and that he is more understanding of her than is her mother. She explains, "He shows that he cares more than my mother at times. . . . He's just always there for us, you know.

He's like a good father. He never hit us, or, you know, or really argue with us. You know, we play around." In her sophomore year, Sonia's reflections about her relationship with her father suggest a continuity and a stability to her positive feelings as she says that her relationship is "wonderful.... We've always had a good relationship." She always reiterates in this interview that she would turn to her father more than to her mother if she wanted to discuss personal concerns and that she has become more open with him about her personal life over time. She repeats that he has "always been there for [me]."

Yolanda, a Puerto Rican senior who lives with both of her parents, also says that she feels closer to her father than to her mother:

> *Like when you're concerned about anything, will you turn to your mother or will you turn to your friends first?*
> I feel more comfortable talking to my father.
> *Oh really?*
> Yes 'cause he's just for some reason, he won't say—he'll just be like "um-hm." . . . And my mother, if I tell her, she'll give me like a whole hour speech about . . . and I'm like "Forget it. Thanks anyway." But I feel more comfortable talking with my father about things. . . . He's more easy to talk to.

Resisting the interviewer's assumption that she would choose between her mother and her friends, Yolanda states with emphasis that it is her father to whom she prefers to turn in times of need.

Souad, a Syrian adolescent, conveys a similar theme in her junior year:

> My father has, you know, too much knowledge and he has an open mind, he's different from other parents, Arabic parents. Because Arabic people are very closed sometimes, especially with dealing with their children, like they are stubborn, when they say anything, you know the children should do it. My father is not like that. We can argue with each other. We can always, if I have any problems, I can speak with him . . . I can speak with him. . . . My mother is a little bit more, she's nervous and she sometimes, I can speak with her because she is like my friend, you know, but not like my father. My father can understand more than my mother why we do that, or I can argue with my father more than my mother. . . . My relationship with my father is easier.

Unlike some of her peers who have conflictual relationships with their fathers, Souad seems to experience her arguments with her father as one of the more positive aspects of their relationship: their disagreements seem to enhance, rather than compromise, their sense of connection. Souad, who also lives with both her parents, says that when she grows up, she would like to have her father's "open mind" and her mother's "patience."

Florence, an African American girl whose parents have been separated since she was six years old, says when interviewed in her freshman year that she sees her father "all the time . . . every week." She describes their relationship as communicative:

> We get along good. He's like—I can talk to him about—I, I think I could talk to him more than I could talk to my mother. . . . Since he's not with me all the time, he like wants to know everything I'm doing so he's always like "What are you doing?" and—so he gets to know more about what I do than my mother does. And my mother's there so she thinks she sees everything. Sometimes she asks. But my father asks all the time.

Interviewed again in her junior year, Florence brings more detail and insight to her reflections about her father's role in her development:

> I think my father had a lot to do with [my maturity and independence] . . . because he would call me—even though we didn't live together—he would just preach to me every day. School this, school that. And [as I was growing up] we had a good relationship, me and my father. And like we—he didn't live with me but yet it was almost like he was there 'cause he kind of like raised me from a distance. You know, like things [my mother] didn't tell me, he told me. You know, he talks to me about sex, about boyfriends, about school. Things that she never sat down and told me. So it was almost as if he was like my mother and she was like the father.

Directly challenging the idea that a father who is not at home with his daughter has little influence, Florence describes the important role her father has played in her life. She also appreciates his absence:

> I keep telling him "In a way I'm glad you didn't live with me,"—he probably could have taught me more things if he was there. But then again I wouldn't have got a chance to go out and have so much freedom. I don't know. In a way freedom made me kind of smart, you know. I can't explain it. . . . I know I'm glad he didn't [live with me] 'cause he might have held me back from a lot of things and you know like going to parties, being with my friends and stuff.

Florence suggests that she recognizes both the losses and gains of not living with her father during her childhood and adolescence. She implies that her mother's and father's different styles of parenting have complemented each other to promote her growth and her sense of responsibility. Her father's relative distance by virtue of living in a different home seems to have balanced rather than jeopardized their relationship and contributed to her appreciation of the relationship.

Chloe, interviewed in her sophomore year, also portrays a warm, involved relationship with her father, who lives near her and visits her daily. Unlike Florence, Chloe reports that her father lived in Trinidad when she was growing up, and that she had little contact with him until recently. Now, she says that "me and my father get along great." She goes on to describe how her father assumes the role of mediator in her often contentious relationship with her mother:

> Like we talk about things, like you know, when I'm staying with my mother and we're having conflicts and you know, he always gets involved, because my mother makes him get involved. . . . When I talk to her, nothing gets through to her, you know, nothing gets through. So he tries to talk to me, and when he talks to me we have an open mind. I listen to his side, and he listens to mine and the reasons I did whatever I did, you know, and I like that.

For Chloe, as for Florence, not living with her father does not seem to endanger her perception of him as available and involved.

These thirteen girls whom we have described as having warm and open relationships with their fathers appeared consistently willing to discuss their thoughts and feelings with their fathers and commonly felt more willing to speak openly with their fathers than with their mothers. Some did not have to live with their fathers to maintain such warmth and openness and, in fact, for a few of the girls, the physical distance between themselves and their fathers seemed to enhance the warmth and openness in their relationships. Although these warm and open relationships involved conflicts or disagreements at times, the conflicts did not appear to stem from or lead to conflicted feelings about the relationships per se; these thirteen girls appeared unambivalent in their warm feelings for their fathers.

3. **Disengaged.**   Nine of the thirty-six girls who had contact with their fathers (25%) portrayed emotionally distant relationships with them. At times these girls spoke about their fathers in a language that seemed as sparse and as disconnected affectively as the relationships they described. Denise, an African American senior, recently moved back with her mother after living with her father and stepmother in another city. She says that while she is currently having "a little bit of trouble" with her father, she generally gets along with both her parents because "I can get along with anybody." Her descriptions of her relationships with both her father and her mother are notably reserved and are characterized neither by adversity nor by warmth. Emotional distance from both of her parents is captured in her statement that "my parents aren't important figures in my life really."

Affective distance is also heard in the words of Patrice, an African American junior whose father has been separated from her mother for seven years and lives in her community:

> *What's your relationship like with your dad?*
> I dunno.
> *It sounds like it's mixed.*
> We're not close.
> *You're not close. Like how often will you see him about?*
> Twice a month.
> *Twice a month. And when you see each other, what will you guys do usually?*
> Nothing.
> *Nothing?*
> No. Just talk.
> *Just talk. So you won't share things with him that are personal at all.*
> Nope.
> *It sounds like you're somewhat angry at him. Is that a right reading or no?*
> No. Not really.
> *Not really? Do you feel like he makes any effort to be close to you?*
> No.
> *Do you think that you try to be close to him, or not really?*
> No. Not any more.
> *Not any more. What happened that you started to not make the effort any more?*
> I didn't want to.

Patrice makes no attempt to communicate her feelings directly to her father. While anger might be discerned in her terse description of their relationship (she was more talkative in other parts of the interview), Patrice seems primarily resigned to her disconnected relationship with her father and communicates little interest in changing it.

In the narratives of Cerise, an African American sophomore whose father lives out of state, anger is clearly evident. Cerise's contact with her father is currently "like off and on. Now that I live with my aunt, he calls occasionally. Comes up here like Christmas and stuff." Her description of her history with her father suggests that disillusionment and emotional distance predated their current "off and on" contact. She explains, "When I lived with my mother, I never really talked to him. And like when he did come over, I'd get confused 'cause I didn't want to call him Daddy or whatever." Cerise expresses disappointment in her current relationships both with her mother, whom she describes as "inconsistent," and her father, who "makes a lot of empty promises." Cerise also reveals that although her relationship with her mother has shifted from a closer to a more distant one, she has never felt close to her father.

Delane is an African American freshman whose father moved out of the family home two years earlier. She estimates the frequency of her current contact with him as "a few times a month," and the quality of their contact as "all right." She claims that she does not like to be with her father because she does not like to "be with a man. I don't care if he is my father." She says she does not understand the origins of this discomfort and claims to have "always felt this way." . . .

. . . In her sophomore year Delane says she sees her father every week and their relationship is "fine . . . but I still wouldn't tell him a lot of things, like I would . . . [with] my mother." In her junior year Delane says she still sees her father every week. When asked if she will confide in her father, she says:

> I don't talk to males about any problem that I have. Rather keep it to myself. I feel that fathers aren't—some people can go to their fathers and talk to them about things and others don't. I just don't have that kind of relationship where I can—I don't like—I don't feel—I feel—feel that I should talk to my mother more so than my father. The only thing I talk about to him is like school . . . , car, jobs, something like that. The basics.

Although resignation, disappointment, discomfort, or anger appears to have contributed to and resulted from the distant relationships described by Patrice, Cerise, and Delane, other girls described distant relationships with their fathers that seemed to have less negative origins and consequences. These girls described their relationships with their fathers as "fine" but not close. Some claimed that the reason for their emotionally distant relationships with their fathers was that they rarely saw their fathers, because of their fathers' work schedules or because they lived far away. When asked about her relationship with her father, Bettina, an African-American senior, says: "Well, I don't talk to him all the time. But it's better than with my mother. I never argue with him. He gives me anything I want whenever I see him. But he has his own family now so I don't really go—I don't go down to see him. . . . I see him only when he comes up here." Bettina's relationship with her father does not seem particularly close, angry, or engaged; she seems at once appreciative of and emotionally distant from her father.

Whether or not these nine girls saw their fathers occasionally or regularly, the girls did not seem actively involved in these relationships. They did not confide in or argue with their fathers, and their response to anger and frustration with their fathers was typically to avoid direct confrontation. In contrast to their peers who had conflicted and engaged or warm and open relationships, these girls did not express admiration, connection, or identification with their fathers. These disengaged relationships were more common among those who did not live with their fathers ($N = 6$) than among those who did ($N = 3$).

DISCUSSION

Whereas nine of the girls in this sample had no contact with their fathers and had not had any contact for many years (three of the fathers had died), twenty-six had daily or weekly contact (nineteen lived with their fathers), and ten had occasional contact with their fathers. The father-daughter relationships described by those girls who had daily, weekly, or occasional contact were varied. Some involved simultaneous feelings of attachment and anger, and others primarily involved feelings of affection and understanding. A smaller percentage of girls who had contact with their fathers appeared to have emotionally distant relationships with them. . . .

Sixty-nine percent of the adolescent girls who had contact with their fathers described relationships in which there was much affect and communication—either in the form of conflicts or in the sharing of thoughts and feelings, or in both. While many of the adolescent girls suggested feeling misunderstood by and angry with their fathers, just as many girls spoke about feeling closer to their fathers than to their mothers, being able to express themselves more to their fathers than to their mothers, and identifying with and admiring their fathers. Given the previous findings on primarily white middle-class populations, these current findings suggest that there may be important differences across social class, race, and ethnicity in the quality of relationships between adolescent girls and their fathers. It also suggests that girls' relationships with their fathers may be changing as we move into an era of more involved fathers. Future studies should explore these possibilities.

Our data suggest that fathers who do not live with their daughters, as well as those who do, often play critical roles in their daughters' lives. Far from being absent, most of the fathers appeared actively engaged with their daughters. Continued qualitative exploration of father-daughter relationships, particularly among the under-studied population of urban adolescent girls and focused especially on girls' own perspectives, is essential to a growing understanding of adolescent female development.

NOTES

1. This study is part of a larger, cross-sectional study of ninety urban adolescents (forty-five girls and forty-five boys), which focused on the socioemotional correlates of

risk-taking behavior. This larger project was funded by the National Institute of Drug Abuse (principal investigator, Perry London).

References

Earl, L., and N. Lohmann. 1978. Absent fathers and black male children. *Social Work* 23 (5): 413–15.

Liebow, E. 1967. *Tally's corner.* Boston: Little, Brown.

Rivara, F., P. Sweeney, and B. Henderson. 1986. Black teenage fathers: What happens when the child is born? *Pediatrics* 78 (1): 151–58.

Stack, C. 1974. *All our kin: Strategies for survival in a black community.* New York: Harper and Row.

Youniss, J., and J. Smollar. 1985. *Adolescent relations with mothers, fathers, and friends.* Chicago: University of Chicago Press.

UNDOCUMENTED LATINAS: **30**
The New "Employable Mothers"

Grace Chang

The nomination of Zoe Baird for U.S. Attorney General in 1993 forced a confession that provoked a public uproar. Baird admitted to employing two undocumented Peruvian immigrants, as a baby-sitter and a driver, in clear violation of current immigration law prohibiting the hiring of "illegal" aliens. Responses to Baird's disclosure indicate her "crime" is a pervasive phenomenon.[1] Deborah Sontag reported in the *New York Times* that two-career, middle-class families employing so-called illegal immigrants to do child care and domestic work is so common that employment agencies routinely recommend undocumented immigrants to their clients. As the director of one Manhattan nanny agency said, "It's just a reality of life that without the illegal girls, there wouldn't be any nannies, and the mommies would have to stay home and mind their own kids."[2] Another agency's director said bluntly, "It

From: Evelyn Nakano Glenn, Grace Chang, and Linda Rennie Forcey, eds., *Mothering: Ideology, Experience, and Agency* (New York: Routledge, 1994), pp. 259–85. Reprinted by permission.

all comes down to money . . . the reason that people hire immigrants without papers is that they're looking to save. If they want legal, they can get it, but it costs." [3] According to a survey of eighteen New York agencies, "illegal" workers earned as little as 175 dollars a week and "legal" workers as much as six hundred dollars. [4]

Thus the uproar surrounding Zoe Baird was not so much a response to the discovery that some people flouted the law by employing undocumented workers. This was hardly news. Rather, the public outcry was a reflection of resentment that this practice was so easily accessible to the more privileged classes while others, that is, working-class, "working" mothers, struggled to find any child care. As one critic of Baird commented, "I don't think it's fair. I raised my kids while I was working. I worked days. My husband worked nights at the post office. Our in-laws filled in when they had to." [5] Another woman pointed out "Average working mothers don't make nearly what she makes, and yet we are obligated to follow the law." [6]

What was conspicuously absent from most of the commentary on the Baird controversy was concern for the plight of the undocumented workers themselves. Ironically, two other news stories involving immigrant women working in private households appeared in a California newspaper at the same time Zoe Baird's situation was making headlines across the nation; yet these stories did not receive comparable attention. The first of these involved Claudia Garate, who immigrated from Chile at the age of nineteen in order to take a job as an au pair for a professional couple. Ms. Garate testified before the state Labor Commissioner in Sonoma County that she slept on the floor and worked on call twenty-four hours a day, seven days a week as a maid, baby-sitter, cook and gardener for fifty dollars a month. Garate's employers held on to her visa and passport, and withheld her pay for thirteen months, claiming they would deposit it in a bank account for her. The second case involved Maria de Jesus Ramos Hernandez, who left her three children in Mexico to work as a housekeeper in California. Once here, her employer repeatedly raped her, telling her that he had paid her way here and would have her jailed if she did not submit to him. [7]

Evidence indicates that while Garate's and Hernandez's cases may have been extreme, abuses of undocumented women working in private households were not uncommon. Lina Avidan, program director for the San Francisco-based Coalition for Immigrant and Refugee Rights and Services (CIRRS), said "I have clients who work . . . seven days a week, doing child care from 6 a.m. to 10 p.m. [for] $200 a month. Clearly, they are working in the homes of the wealthy and they're not even getting minimum wage." [8] Spokeswomen for Mujeres Unidas y Activas, a San Francisco-based advocacy group for Latina immigrants, said they had heard countless reports from Latinas working as domestics who endure conditions approaching slavery or indentured servitude.[9] These statements were echoed by undocumented household workers in New York. For example, Dorothea Grant, a Jamaican woman who received a green card after working as a nanny for seven years, explained why American-born workers rarely apply for nanny jobs: "These days, most Americans see it as some kind of slavery."[10] Others reported that because their employers agreed to sponsor them for legal residency they felt like indentured servants in the interim, sometimes waiting up to ten years. One woman from Guyana who had applied for residency six years before was working on call round the clock as a housekeeper for her sponsoring family.[11]

Taken together, these accounts indicate that middle-class households often make exploitative use of immigrant women to do child care and domestic work. They also suggest that the advances of many middle-class, white women in the work force have been largely predicated on the exploitation of poor, immigrant women. While middle- and upper-class women entrust their children and homes to undocumented immigrant women, the immigrant women often must leave their own children in order to work. Some leave their children with family in their home countries, hoping to earn enough to return or send money back to them. Thus, middle- and upper-class women are readily able to find "affordable" care for their children at the expense of poor immigrant women and their children. The employment of undocumented women in dead-end, low-wage, temporary service jobs—often under exploitative conditions—makes it possible for middle- and upper-class women to pursue salaried jobs, and not have to contend with the "second shift" when they come home.

A predictable outgrowth of the Baird controversy has been the proposal that the existing law, the Immigration Reform and Control Act (IRCA) of 1986, be changed so that household employers are exempted from the prohibition against hiring "illegal" immigrants, or that household workers are given special visas. If the law were changed to meet this "popular demand," it would only serve to perpetuate—and authorize by law—the exploitation of thousands of undocumented immigrants. Certainly there is much historical precedent for government-sanctioned exploitation of immigrants as cheap laborers in the U.S. For example, in response to the wartime demands of Southwestern agricultural employers for laborers, the Bracero Program was instituted in 1942, allowing for the importation of millions of Mexicans as temporary workers. These workers were bound to fixed low wages, and obliged to stay on certain farms until they were to be returned to Mexico at the end of their contractual periods. Following this model, agribusiness lobbyists succeeded in getting provisions for agricultural "guest workers" written into IRCA, enabling growers to continue to draw on immigrants as a superexploitable labor pool. The current proposals raise the specter of a counterpart to these agricultural "guest workers" in private household work: "disposable nannies" who may be dumped once babies become older or newer immigrants can be found who are willing to work for even lower wages. . . .

THE IMMIGRATION REFORM AND CONTROL ACT OF 1986: A COMPROMISE

The IRCA emerged in 1986 after nearly a decade of debate in Congress and in the public domain about what impact immigration, particularly "illegal" immigration, had on the U.S. economy. The Act had two main objectives that were possibly contradictory: to stem the tide of illegal immigration, and to provide rights and the chance to legalize their status to those undocumented immigrants who had already lived and/or worked in the country. Unable to reconcile these conflicting impulses, Congress incorporated a number of provisions into the law as concessions to various interest groups. First, to discourage illegal immigration, the law established employer sanctions against those who knowingly employed "illegal" immigrants. Second, to provide rights and protections to undocumented persons, the amnesty program offered those who could prove they had lived in the country "illegally" since at least 1982 the chance to apply for temporary resident status. Finally, in response to the concerns of growers about how the law might affect the availability of agricultural labor, Congress created three special classes of those who could enter the country or gain residency as agricultural workers.

Some of the most heated debate surrounding IRCA centered around the issue of whether immigrants contribute to or deplete the public coffers. This debate led lawmakers to include in IRCA two provisions, the public charge exclusion and the five-year bar, governing whether those perceived as potentially welfare-dependent should be able to gain residency, and what entitlements "legalized persons" should be allowed to receive. Before examining these provisions more closely, it will be useful to look at some of the dimensions of this debate, in order to understand the political context in which these provisions were formulated and implemented. In the past, most public views and scholarly discussions on the "costs and benefits" of immigrants have emphasized the charge that male migrant laborers steal jobs from "native" workers. In the last decade, however, this concern has been drowned out by cries that immigrants impose a heavy welfare burden on "natives." A 1986 CBS/ *New York Times* poll found that forty-seven percent of Americans believed that "most immigrants wind up on welfare." [12] In a review of studies on the economic impacts of immigration to the U.S., Annie Nakao reported for the *San Francisco Examiner*, "What is generally accepted is that immigrants do not take jobs from natives. . . ."[13] The abundance of studies examining how immigrants affect the U.S. economy disagree on many points, but most recent works seem to agree that Americans should be more worried about protecting public revenues than their jobs.

Thus, a new twist in anti-immigrant rhetoric has emerged, with a focus on immigrants as economic welfare burdens. For example, Governor Pete Wilson deployed this rhetoric in marketing his proposals for slashing social service funds in California's 1992 budget. In his administration's report "The Growing Taxpayer Squeeze," Wilson called immigrants "tax receivers" and identified the "rising foreign immigrant population" as a major cause of growing tax expenditures for welfare, Medi-Cal and public schools.[14] The new emphasis on the alleged depletion of public revenues by immigrants signals an implicit shift in the main target of anti-immigrant attacks. Men as job stealers are no longer seen as the "immigrant problem." Instead, immigrant women as idle, welfare-dependent mothers and inordinate breeders of dependents are seen as the great menace. Thus, a legislative analyst on Governor Wilson's staff reported that Latinas have an AFDC dependency rate twenty-three percent higher than the rate for all other women.[15] Such "findings" are almost always coupled with statements about higher birth rates among immigrant women, and the threat they pose to controlling population growth.

Perhaps this new strategy, identifying immigrant women (and particularly Latinas) as the major threat to American public resources, reflects a growing awareness of changes that have occurred in the composition and nature of Mexican migration to the U.S. in the last two decades. Wayne Cornelius, of the Center for U.S.-Mexican Studies, reports that, beginning in the 1970s and through the 1980s, there was a shift in Mexican migration from that dominated by "lone male" (single or unaccompanied by dependents), seasonally employed, and highly mobile, migrant laborers to a "de facto" permanent Mexican immigrant population including more women, children and entire families.[16] The change consists of more migration by whole families, more family reunification, and more migration by single women. Cornelius explains that Mexico's crisis has driven more women to migrate to the United States, where there is "an abundance of new employment opportunities for which women are the preferred labor source," including child care, cleaning and laundry work.

Cornelius's analysis of U.S. Census Bureau (1988) data suggests that, as a result of this expanded female migration, females may now represent the majority of "settled" undocumented Mexican immigrants. In her study of undocumented Mexican immigrant communities, Pierette Hondagneu reports that it is the women who advocate and mobilize families toward permanent settlement in the U.S. Thus, she theorizes, U.S. xenophobia has come to focus on women because they are perceived as the leaders of this threatening demographic shift.[17]

Heightened awareness of these new demographics contributed to hysteria about protecting public revenues and guarding against the growth of a population of welfare dependents. These concerns undoubtedly influenced the inclusion of two provisions of IRCA, the public charge exclusion and the five-year bar, to restrict aliens' access to social services and public benefits. Clearly, these restrictions were formulated with the goal of limiting welfare expenditures. In executing IRCA, the INS went even further, utilizing an interpretation of the law that effectively denied amnesty to those seen as potential welfare abusers, that is, undocumented women with children. . . .

CONCLUSION

Some feminists have proposed that subsidies to women with children should be expanded in the recognition that full-time mothering is work and should be properly rewarded. Wendy Sarvasy, for example, has called for us to recapture some of the original principles behind the Mothers' Pension program: that mothers be seen as civil servants and provided with pensions as compensation for their services in nurturing future citizens.[18] Such proposals repeat the original flaw of the Mothers' Pension program: limiting support to an elite group of women by defining "deserving" mothers as full-time mothers, while few women actually find full-time mothering viable. Under the current racial division of reproductive labor, some women cannot stay at home with their own children while they mother other people's children and keep other people's homes. Ironically, the assumption of these reproducive functions by women of color and immigrant women for white middle-class or professional women allows the latter group *not* to choose full-time mothering, opting for careers and other pursuits which may be more rewarding economically or personally. Thus, proposals to reward full-time mothering offer nothing to most women of color, for whom this occupation is rarely an option.

Furthermore, women of color may not view full-time mothering as the ideal. Historically, women of color have had to work, even while raising small children, either to supplement inadequate wages garnered by their men, or to provide for families in the absence of male providers. In response, communities of color have often constructed alternatives to dominant society's model of the family in which men are providers and women primarily dependents and consumers. For example, Carol Stack and Linda Burton report that male, female, old, and young members of low-income African-American families negotiated shared caretaking responsibilities, enabling women to earn wages during early childbearing years.[19] Similarly, the Mexicana mothers in Denise Segura's study viewed employment as compatible with mothering, as it enabled them to contribute toward the collective good of the family.[20]

Thus, proposals to reform the welfare system through revaluing the work of full-time mothering fail to address the needs of women of color and further marginalize them in their struggles to provide for their families. A more radical proposition—and one which might begin to address the plight of women of color who are poor working mothers—would be to recognize and reward women for the services they provide through both their productive and reproductive labors. . . . This would necessitate a demystification of these women as welfare-dependents, and a recognition that they are working mothers, often heads of household or at least significant providers for their families. More importantly, such a demystification might compel the state to recognize the many ways in which these women benefit American capital and society at large—through paid (but grossly undercompensated) productive labor, through reproductive labor for others, and through the reproductive functions they perform within their own families. The fact that they too, in raising their own children, provide a service in nurturing future adult citizens should not be obscured by public ignorance casting their children as somehow less worthy.

In return for all of these contributions, these women should at least be afforded access to citizenship. This would mean that they would not jeopardize their chances to gain legal status in seeking public assistance, including AFDC, for their children, many of whom are American citizens by birth and therefore fully entitled to these benefits. The professed goal of AFDC is to give temporary support to poor mothers so that ultimately they may be able to provide for their families through wage earning. If this is indeed the objective of AFDC, then denying amnesty to undocumented women who have received aid for their children contradicts this purpose by undermining these women's abilities to increase their employment options and earning powers. As we have seen, government policies have been utilized to handicap rather than support undocumented mothers. These practices facilitate U.S. employers' ability to extract cheap labor from these women and at the same time allow the state to evade responsibility for the welfare of citizen children. Undocumented women have been forced to choose between aid for their children or the possibility to gain legal status. Those who "choose" aid for their children condemn themselves to remain in an underclass, unrecognized as "productive citizens," yet functioning as perhaps the ultimate servants of our society.

NOTES

1. The *San Francisco Chronicle* reported that, although no precise figures exist, "experts believe a large percentage of the estimated 3 million undocumented workers now residing in the United States are employed in child-care and domestic work." "Hiring of Aliens Is a Widespread Practice," *San Francisco Chronicle* (January 15, 1993), p. A6.

2. Deborah Sontag, "Increasingly, Two-Career Family Means Illegal Immigrant Help," *New York Times* (January 24, 1993), p. A-1.

3. Ibid., p. A-13.

4. Ibid.

5. Felicity Barringer, "What Many Say About Baird: What She Did Wasn't Right," *New York Times* (January 22, 1993), p. A1.

6. Ibid., p. A10.

7. Carla Marinucci, "Immigrant Abuse: 'Slavery—Pure and Simple,'" *San Francisco Examiner* (January 10, 1993), pp. A-1, A-8.

8. Carla Marinucci, pp. A1, A8.

9. Carla Marinucci, "Silence Shields Abuse of Immigrant Women," *San Francisco Examiner* (January 11, 1993), pp. A-1, A-10.

10. Sontag, p. A13.

11. Ibid.

12. CBS/*New York Times* poll, July 14, 1986, cited in Julian Simon, *The Economic Consequences of Immigration* (Cambridge: Basil Blackwell, Inc., 1989), p. 105.

13. Annie Nakao, "Assessing the Cost of Immigration," *San Francisco Examiner* (December 1, 1991), pp. B-1, B-3.

14. Terri Lobdell and Lewis Butler, "Tending Our Future Together," *California Perspectives* (November, 1991), pp. 31–41; and California Department of Finance, "California's Growing Taxpayer Squeeze," *California Perspectives* (November, 1991), p. 4.

15. Nakao, p. B-3.

16. Wayne Cornelius, "From Sojourners to Settlers: The Changing Profile of Mexican Migration to the U.S." (San Diego: Center for U.S.-Mexican Studies, University of California, San Diego, August 15, 1990), p. 17; published in Jorge Bustamante, Raul Hinojosa, and Clark Reynolds, eds., *U.S.-Mexico Relations: Labor Market Interdependence* (Stanford, CA: Stanford University Press, 1991).

17. Pierrette M. Hondagneu, "Gender and the Politics of Mexican Undocumented Immigrant Settlement," Ph.D. Dissertation, University of California (1990), p. 249.

18. Wendy Sarvasy, "Reagan and Low-Income Mothers: A Feminist Recasting of the Debate," in M. K. Brown, eds., *Remaking the Welfare State: Retrenchment and Social Policy in America and Europe* (Philadelphia: Temple University Press, 1988), pp. 253–276.

19. Carol Stack and Linda Burton, "Kinscripts: Reflections on Family, Generation, and Culture," in E. Nakano Glenn, G. Chang, and L. Rennie Forcey, eds., *Mothering: Ideology, Experience, and Agency* (New York: Routledge, 1993).

20. Denise Segura, "Working at Motherhood: Chicana and Mexican Immigrant Mothers and Employment," in E. Nakano Glenn, G. Chang, and L. Rennie Forcey, eds., *Mothering: Ideology, Experience, and Agency* (New York: Routledge, 1993).

MIGRATION AND VIETNAMESE AMERICAN WOMEN: *Remaking Ethnicity*

31

Nazli Kibria

VIETNAMESE AMERICANS AND THE RISE IN WOMEN'S POWER

My research on the adaptive strategies of a community of Vietnamese refugees in Philadelphia revealed some of the ways in which women and men struggled and clashed with each other in efforts to shape the social organization of family and community life. From 1983 to 1985, I gathered information on family life and gender relations through participant observation in household and community settings, as well as in-depth interviews with women and men in the ethnic community.

The Vietnamese of the study were recent immigrants who had arrived in the United States during the late 1970s and early 1980s. Most were from urban, middle-class backgrounds in southern Vietnam. At the time of the study, over 30 percent of the adult men in the households of study were unemployed.

From: Maxine Baca Zinn and Bonnie Thornton Dill, eds. *Women of Color in U.S. Society* (Philadelphia: Temple University Press, 1994), pp. 247–61. Reprinted by permission.

Of the men who were employed, over half worked in low-paying, unskilled jobs in the urban service sector or in factories located in the outlying areas of the city. Women tended to work periodically, occupying jobs in the informal economic sector as well as in the urban service economy. Eight of the twelve households had members who collected public assistance. Both the family economy and informal community exchange networks were important means by which the households dealt with economic scarcities. Family and community were of tremendous economic salience to the group, as they were important resources for survival in the face of a rather inhospitable economic and social environment.

As suggested by the high rate of the men's unemployment, settlement in the United States had generated some shifts in power in favor of the women in the group. Traditional Vietnamese family and gender relations were modeled on Confucian principals, which placed women in subordination to men in every aspect of life. A key aspect of the social and economic oppression of women in traditional Vietnamese life was the patrilineal extended household. Its organization dictated that women married at a young age, following which they entered the household of their husband's father. This structure ensured the concentration of economic resources in the hands of men and men's control of women through the isolation of women from their families of origin.[1]

It is important to note the deep-seated changes in traditional family and gender structures in Vietnam during this century. War and urbanization eroded the structure of the patrilineal extended household. While unemployment was high in the cities, men from middle-class backgrounds were able to take advantage of the expansion of middle-level positions in the government bureaucracy and army. Such occupational opportunities were more limited for women: the women study participants indicated that they engaged in seasonal and informal income-generating activities or worked in low-level jobs in the growing war-generated service sector in the cities. The transition from rural to urban life had generated a shift in the basis of men's control over economic and social resources. However, families relied on men's income to maintain a middle-class standard of living. Thus women remained in a position of economic subordination to men, a situation that served to sustain the ideals of the traditional family system and men's authority in the family. Restrictions on women's sexuality were important for middle-class families who sought to distinguish themselves from the lower social strata. My data suggest that families were especially conscious of the need to distance themselves from poorer "fallen" women who had become associated with the prostitution generated by the American military presence.

Within the Vietnamese American community of study, I found several conditions that were working to undermine the bases on which male authority had rested in Vietnam. Most important, for the Vietnamese men, the move to the United States had involved a profound loss of social economic status. Whereas in pre-1975 Vietnam the men held middle-class occupations, in the United States they had access to largely unskilled, low-status, and low-paying jobs. Also, because of their difficulties with English and their racial-ethnic status, the men found themselves disadvantaged within social arenas of the dominant society. Compounding these problems was the dearth of strong formal ethnic organizations in the community that could have served as a vehicle for the men's political assertion into the dominant society.

As a result of these losses, the comparative access of men and women to the resources of the dominant society had to some extent become equalized. In contrast to the experiences of the men, migration had not significantly altered the position of the women in the economy. As in Vietnam, the women tended to work sporadically, sometimes in family businesses or, more commonly, in temporary jobs in the informal and service sector economies of the city. However, the economic contributions of women to the family budget had risen in proportion to those of the men. I have suggested that in modern, urban South Vietnam the force and legitimacy of male authority had rested heavily on the ability of men to ensure a middle-class status and standard of living for their families. In the United States, the ability of men to fulfull this expectation had been eroded. Among the men, there was widespread concern about the consequences of this situation for their status in the family, as is revealed by the words of a former lieutenant of the South Vietnamese army: "In Vietnam, the man earns and everyone depends on him. In most families, one or two men could provide for the whole family. Here the man finds he can never make enough money to take care of the family. His wife has to work, his children have to work, and so they look at him in a different way. The man isn't strong anymore, like he was in Vietnam."

Such changes had opened up the possibilities for a renegotiation of gender relations, and were the cause of considerable conflict between men and women in the family and community. The shifts in power had also enhanced the ability of women to construct and channel familial and ethnic resources in ways that they chose. Previously I suggested that the changes in the balance of power between men and women generated by migration are crucial to understanding the manner and degree to which immigrant family and community reveal themselves to be gender contested. How, then, did the fairly drastic shift in the gender balance of power among the Vietnamese Americans reflect itself in the ability of the men and women in this group to influence family and community life? In the following section, I describe some of the ways in which gender interests and conflict shaped family and community life for the Vietnamese Americans.

FAMILY AND ETHNICITY AS GENDER CONTESTED

One of the most intriguing and important strategies of Vietnamese American adaptation that I observed was the rebuilding of kinship networks. Family ties had undergone tremendous disruption in the process of escape from Vietnam and resettlement in the United States. Despite this, the households of the group tended to be large and extended. The process by which this occurred was one in which the study participants actively worked to reconstruct family networks by building kin relationships. In order for this to take place, the criteria for inclusion in the family had become extremely flexible. Thus close friends were often incorporated into family groups as fictive kin. Also, relationships with relatives who were distant or vaguely known in Vietnam were elevated in importance. Perhaps most important for women, the somewhat greater significance traditionally accorded the husband's kin had receded in importance.[2] Given the scarcity of relatives in the United States, such distinctions were considered a luxury, and the demands of life made the rebuilding of family a valuable, if not a necessary, step in the process of adaptation to the dominant society.

While important for the group as a whole, the reconstruction of kinship as it took place had some special advantages for women. One consequence of the more varied and inclusive nature of the kinship network was that women were rarely surrounded exclusively by the husband's relatives and/or friends. As a result, they were often able to turn to close fictive kin and perhaps members of their families of origin for support during conflicts with men in the family. Another condition that enhanced the power of married women in the family was that few had to deal with the competing authority of their mother-in-law in the household, because elderly women have not been among those likely to leave Vietnam.

The reconstruction of kinship thus had important advantages for women, particularly as it moved the Vietnamese perhaps even further from the ideal model of the patrilineal extended household that it had been in the past. But women were not simply passive beneficiaries of the family rebuilding process. Rather, they played an active part in family reconstruction, attempting to shape family boundaries in ways that were to their advantage. I found women playing a vital part in creating fictive kin by forging close ties. And women were often important, if not central, "gatekeepers" to the family group and household. Thus the women helped to decide such matters as whether the marriage of a particular family member was a positive event and could be taken as an opportunity to expand kinship networks. At other times the women passed judgment on current or potential family members, as to whether they had demonstrated enough commitment to such important familial obligations as the sharing of economic and social resources with kin.

Although women undoubtedly played an important part in family reconstruction, their control over decisions about family membership was by no means exclusive or absolute. In fact, the question of who was legitimately included in the family group was often a source of tension within families, particularly between men and women. The frequency of disputes over this issue stemmed in part from the fluidity and subsequent uncertainty about family boundaries, as well as the great pressures often placed on individuals to subordinate their needs to those of the family collective. Beyond this, I also suggest that disputes over boundaries arose from the fundamental underlying gender divisions in the family. That is, the different interests of women and men in the family spurred efforts to shape the family in ways that were of particular advantage to them. For the reasons I have previously discussed, the Vietnamese American women had greater influence and opportunity in the shaping of family in the United States than they had in the past. The women tended to use this influence to construct family groups that extended their power in the family.

In one case that I observed, considerable tension developed between a couple named Nguyet and Phong concerning the sponsorship[3] of Nguyet's nephew and his family from a refugee camp in Southeast Asia. Nguyet and Phong had been together with their three children (two from Nguyet's previous marriage) for about seven years, since they had met in a refugee camp in Thailand. Phong remained married to a woman who was still living in Vietnam with his children, a fact that was the source of some stress for Nguyet and Phong. The issue of the nephew's sponsorship seemed to exacerbate tensions in the relationship. Phong did not want to undertake the sponsorship because of the potentially heavy financial obligations it entailed. He also confessed that he was worried that Nguyet would leave him after the nephew's arrival, a threat often made by Nguyet during their quarrels. Finally, he talked of how Nguyet's relationship with the nephew was too distant to justify the sponsorship. Nguyet had never even met the nephew, who was the son of a first cousin rather than of a sibling.

Confirming some of Phong's fears, Nguyet saw the presence of the nephew and his family as a potentially important source of support for herself. She spoke of how she had none of "my family" in the country, in comparison with Phong, whose sister lived in the city. She agreed that she did not know much about her nephew, but nonetheless felt that his presence would ease her sense of isolation and also would provide a source of aid if her relationship with Phong deteriorated. Eventually she proceeded with the sponsorship, but only after a lengthy dispute with Phong.

While the issue of sponsorship posed questions about kinship in an especially sharp manner, there were other circumstances under which women and men clashed over family boundaries. When kin connections could not be questioned (for example, in the case of a sibling), what came under dispute was the commitment of the particular person involved to familial norms and obligations. One of my woman respondents fought bitterly with her older brother about whether their male cousin should live with them. Her brother objected to the cousin's presence in the household on the grounds that he had not responded to their request for a loan of money two years ago. The woman respondent wanted to overlook this breach of conduct because of her extremely close relationship with the cousin, who had been her "best friend" in Vietnam.

Regardless of the particular circumstances, gender conflict seemed an important part of the family reconstruction process. Women and men shared an interest in creating and maintaining a family group that was large and cohesive enough to provide economic and social support. However, their responses to the family reconstruction process were framed by their differing interests, as men and women, within the family. Men and women attempted to channel family membership in ways that were to their advantage, such that their control over the resources of the family group was enhanced.

Gender divisions and conflicts also entered into the community life of the group. The social networks of the Vietnamese American women were central to the dynamics and organization of the ethnic community. They served to organize and regulate exchange between households. While "hanging out" at informal social gatherings, I observed women exchanging information, money, goods, food, and tasks such as child care and cooking. Given the precarious economic situation of the group, these exchanges played an important role in ensuring the economic survival and stability of the households. The women's centrality to these social networks gave them the power not only to regulate household exchange but also to act as agents of social control in the community in a more general sense. I found that women, through the censure of gossip and the threat of ostracism, played an important part in defining community norms. In short, the relative rise in power that had accrued to the Vietnamese American women as a result of migration expressed itself in their considerable influence over the organization and dynamics of the ethnic community. Like kinship, community life was a negotiated arena, one over which women and men struggled to gain control.

The gender-contested quality of ethnic forms was also apparent in the efforts of women to reinterpret traditional Vietnamese familial ideologies on their own terms. In general, the Vietnamese American women continued to espouse and support traditional ideologies of gender relations as important ideals. For example, when asked during interviews to describe the "best" or ideal roles of men and women in the family, most of my respondents talked of a clear division of roles in which women assumed primary responsibility for maintaining the home and taking care of the children, and men for the economic support of the family. Most felt that household decisions should be made jointly, although the opinion of the man was seen to carry more weight. About half of those interviewed felt that a wife should almost always obey her husband. Even more widespread were beliefs in the importance of restrictions on female (but not male) sexuality before marriage.

While women often professed such beliefs, their relationship to traditional ideologies was active rather than passive and inflexible. In other words, the women tended to emphasize certain aspects of the traditional familial ideology over others. In particular, they emphasized parental authority and the obligation of men to sacrifice individual needs and concerns in order to fulfill the needs of the family, traditional precepts they valued and hoped to preserve in the United States. The women's selective approach to Vietnamese "tradition" emerged most clearly in situations of conflict between men and women in the family. In such disputes, women selectively used the traditional ideologies to protect themselves and to legitimate their actions and demands (Kibria 1990). Thus, husbands who were beating their wives were attacked by other women in the community on the grounds that they (the husbands) were inadequate breadwinners. The women focused not on the husband's treatment of his wife but on his failure to fulfill his family caretaker role. Through this selective emphasis, the women managed to condemn the delinquent husband without appearing to depart from "tradition." In short, for the Vietnamese American women, migration had resulted in a greater ability to shape family and community life.

CONCLUSION

For immigrant women, ethnic ties and institutions may be both a source of resistance and support, and of patriarchal oppression. Through an acknowledgment of this duality we can arrive at a fuller understanding of immigrant women's lives: one that captures the multifaceted constraints as well as the resistances that are offered by immigrant women to the oppressive forces in their lives. In patterns similar to those noted by studies of other racial-ethnic groups (Stack 1976; Baca Zinn 1975), the Vietnamese Americans presented in this [article] relied on family and community for survival and resistance. Their marginal status made the preservation of these institutions an important priority.

Like other racial-ethnic women, the ability of the Vietnamese American women to shape ethnicity was constrained by their social-structural location in the dominant society. These women saw the traditional family system as key to their cultural autonomy and economic security in American society. Migration may have equalized the economic resources of the men and women, but it had not expanded the economic opportunities of the women enough to make independence from men an attractive economic reality. The Vietnamese American women, as is true for other women of color, were especially constrained in their efforts to "negotiate" family and community in that they faced triple disadvantages (the combination of social class, racial-ethnic, and gender statuses) in their dealings with the dominant society.

Recognition of the role of ethnic institutions in facilitating immigrant adaptation and resistance is essential. However, it is equally important to not lose sight of gender divisions and conflicts, and the ways in which these influence the construction of ethnic institutions. Feminist scholars have begun to explore the diverse ways in which immigrant women manipulate family and community to enhance their own power, albeit in ways that are deeply constrained by the web of multiple oppressions that surround them (Andezian 1986; Bhachu 1986; Kibria 1990). Such work begins to suggest the complexity of immigrant women's relationship to ethnic structures, which is informed by both strength and oppression.

NOTES

1. Some scholars stress the fact that the reality of women's lives was far different from that suggested by these Confucian ideals. Women in traditional Vietnam also had a relatively favorable economic position in comparison with Chinese women due to Vietnamese women's rights of inheritance as well as their involvement in commercial activities (see Hickey 1964; Keyes 1977). Despite these qualifications, there is little to suggest that the economic and social subordination of women was not a fundamental reality in Vietnam.

2. Hy Van Luong (1984) has noted the importance of two models of kinship in Vietnamese life, one that is patrilineal-oriented and another in which bilateral kin are of significance. Thus the flexible, encompassing conceptions of family that I found among the group were not entirely new, but had their roots in Vietnamese life; however, they had acquired greater significance in the context of the United States.

3. Refugee resettlement in the United States involves a system of sponsorship by family members or other interested parties who agree to assume part of the responsibility for taking care of those sponsored for a period of time after their arrival.

References

Andezian, Sossie. 1986. "Women's Roles in Organizing Symbolic Life: Algerian Female Immigrants in France." Pp. 254–266 in *International Migration: The Female Experience*, edited by R. J. Simon and C. B. Brettell. Totowa, N.J.: Rowman and Allenheld.

Baca Zinn, Maxine. 1975. "Political Familism: Toward Sex Role Equality in Chicano Families." *Aztlan* 6, no. 1: 13–26.

Bhachu, Parminder K. 1986. "Work, Dowry and Marriage Among East African Sikh Women in the U.K." Pp. 241–254 in *International Migration: The Female Experience*, edited by R. J. Simon and C. B. Brettell. Totowa, N.J.: Rowman and Allenheld.

Hickey, Gerald C. 1964. *Village in Vietnam*. New Haven: Yale University Press.

Keyes, Charles F. 1977. *The Golden Peninsula*. New York: Macmillan.

Kibria, Nazli. 1990. "Power, Patriarchy and Gender Conflict in the Vietnamese Immigrant Community." *Gender and Society* 4, no. 1 (March): 9–24.

Luong, Hy Van. 1984. "'Brother' and 'Uncle': An Analysis of Rules, Structural Contradictions and Meaning in Vietnamese Kinship." *American Anthropologist* 86, no. 2: 290–313.

Stack, Carol. 1974. *All Our Kin*. New York: Harper & Row.

Education

CAN EDUCATION ELIMINATE RACE, CLASS, AND GENDER INEQUALITY?

32

Roslyn Arlin Mickelson and Stephen Samuel Smith

INTRODUCTION

Parents, politicians, and educational policy makers share the belief that a "good education" is *the* meal ticket. It will unlock the door to economic opportunity and thus enable disadvantaged groups or individuals to improve their lot dramatically.[1] This belief is one of the assumptions that has long been part of the American Dream. According to the putative dominant ideology, the United States is basically a meritocracy in which hard work and individual effort are rewarded, especially in financial terms.[2] Related to this central belief are a series of culturally enshrined misconceptions about poverty and wealth. The central one is that poverty and wealth are the result of individual inadequacies or strengths rather than the results of the distributive mechanisms of the capitalist economy. A second misconception is the belief that everyone is the master of her or his own fate. The dominant ideology assumes that American society is open and competitive, a place where an individual's status depends on talent and motivation, not inherited position, connections, or privileges linked to ascriptive characteristics like gender or race. To compete fairly, everyone must have access to education free of the fetters of family background, gender, and race. Since the middle of this century, the reform policies of the federal government have been designed, at least officially, to enhance individuals' opportunities to acquire education. The question we will explore in this essay is whether expanding educational opportunity is enough to reduce the inequalities of race, social class, and gender which continue to characterize U.S. society.

We begin by discussing some of the major educational policies and pro-

---

grams of the past forty-five years that sought to reduce social inequality through expanding equality of educational opportunity. This discussion highlights the success and failures of programs such as school desegregation, compensatory education, Title IX, and job training. We then focus on the barriers these programs face in actually reducing social inequality. Our point is that inequality is so deeply rooted in the structure and operation of the U.S. political economy, that, at best, educational reforms can play only a limited role in ameliorating such inequality. In fact, there is considerable evidence that indicates that, for poor and many minority children, education helps legitimate, if not actually reproduce, significant aspects of social inequality in their lives. Finally, we speculate about education's potential role in individual and social transformation.

First, it is necessary to distinguish among equality, equality of opportunity, and equality of educational opportunity. The term *equality* has been the subject of extensive scholarly and political debate, much of which is beyond the scope of this essay. Most Americans reject equality of life conditions as a goal, because it would require a fundamental transformation of our basic economic and political institutions, a scenario most are unwilling to accept. As Ralph Waldo Emerson put it, "The genius of our country has worked out our true policy—opportunity."

The distinction between equality of opportunity and equality of outcome is important. Through this country's history, equality has most typically been understood in the former way. Rather than a call for the equal distribution of money, property, or many other social goods, the concern over equality has been with equal opportunity in pursuit of these goods. In the words of Jennifer Hochschild, "So long as we live in a democratic capitalist society—that is, so long as we maintain the formal promise of political and social equality while encouraging the practice of economic inequality—we need the idea of equal opportunity to bridge that otherwise unacceptable contradiction."[3] To use a current metaphor: If life is a game, the playing field must be level; if life is a race, the starting line must be in the same place for everyone. For the playing field to be level, many believe education is crucial because it gives individuals the wherewithal to compete in the allegedly meritocratic system. In America, then, *equality* is really understood to mean *equality of opportunity*, which itself hinges on *equality of educational opportunity*.

THE SPOTTY RECORD OF FEDERAL EDUCATIONAL REFORMS

In the past forty-five years, a series of educational reforms initiated at the national level has been introduced into local school systems. All of the reforms aimed to move education closer to the ideal of equality of educational opportunity. Here we discuss several of these reforms, and how the concept of equality of educational opportunity has evolved. Given the importance of race and racism in U.S. history, many of the federal education policies during this period attempted to redress the most egregious forms of inequality based on race.

School Desegregation

Although American society has long claimed to be based on equality of opportunity, the history of race relations suggests the opposite. Perhaps the most influential early discussion of this disparity was Gunnar Myrdal's *An American Dilemma*, published in 1944. The book vividly exposed the contradictions between the ethos of freedom, justice, equality of opportunity and the actual experiences of African Americans in the United States.[4]

The links among desegregation, expanded educational opportunity, and the larger issue of equality of opportunity are very clear from the history of the desegregation movement. This movement, whose first phase culminated in the 1954 *Brown* decision outlawing *de jure* segregation in school, was the first orchestrated attempt in U.S. history to directly address inequality of educational opportunity. The NAACP strategically chose school segregation to be the camel's nose under the tent of the Jim Crow (segregated) society. That one of the nation's foremost civil rights organizations saw the attack on segregated schools as the opening salvo in the battle against society-wide inequality is a powerful example of the American belief that education has a pivotal role in promoting equality of opportunity.

Has desegregation succeeded? This is really three questions: First, to what extent are the nation's schools desegregated? Second, have desegregation efforts enhanced students' academic outcomes? Third, what are the long term outcomes of desegregated educational experiences?

Since 1954, progress toward the desegregation of the nation's public schools has been uneven and limited. Blacks experienced little progress in desegregation until the mid-1960s when, in response to the civil rights movement, a series of federal laws, executive actions, and judicial decisions resulted in significant gains, especially in the South. Progress continued until 1988, when the effects of a series of federal court decisions and various local and national political developments precipitated marked trends toward the resegregation of black students. Nationally, in 1994–1995, 33 percent of Black students attended majority White schools compared with the approximately 37 percent who attended majority White schools for much of the 1980s.[5]

Historically, Latinos were relatively less segregated than African Americans. However, from the mid-1960s to the mid-1990s there was a steady increase in the percentage of the Latino students who attended segregated schools. As a result, education for Latinos is now more segregated than it is for Blacks.

Given the long history of legalized segregation in the south, it is ironic that the South's school systems are now generally the country's most *deseg-regated*, while those in the northeast are the most intensely segregated. How-ever, even desegregated schools are often resegregated at the classroom level by tracking or ability grouping. There is a strong relationship between race and social class, and racial isolation is often an outgrowth of residential seg-regation and socioeconomic background.

Has desegregation helped to equalize educational outcomes? A better question might be which desegregation programs under what circumstances accomplish which goals? Evidence from recent desegregation research sug-gests that, overall, children benefit academically and socially from well-run programs. Black students enjoy modest academic gains, while the academic achievement of White children is not hurt, and in some cases is helped, by desegregation. In school systems which have undergone desegregation efforts, the racial gap in educational outcomes has generally been reduced, but not eliminated.

More important than short-term academic gains are the long-term con-sequence of desegregation for Black students. Compared to those who at-tended racially isolated schools, Black adults who experienced desegregated education as children are more likely to attend multiracial colleges and gradu-ate from them, work in higher-status jobs, live in integrated neighborhoods, assess their abilities more realistically when choosing an occupation, and to report interracial friendships.[6]

Despite these modest, but positive, outcomes, in the last decade of the twentieth century, most American children attend schools segregated by race, ethnicity, and social class. Consequently, forty-five years of official federal in-terventions aimed at achieving equality of educational opportunity through school desegregation have only made small steps toward achieving that goal; children from different race and class backgrounds continue to receive segre-gated and, in many respects, unequal educations.

The Coleman Report

Largely because evidence introduced in the 1954 *Brown* case showed that re-sources in segregated Black and White schools were grossly unequal, Con-gress mandated in 1964 a national study of the "lack of availability of equality of educational opportunity for individuals due to race, color, religious, or na-tional origin in public schools." The authors of the subsequent study, James Coleman and his associates, expected to find glaring disparities in educational resources available to African-American and White students and that these differences would explain the substantial achievement differences between majority and minority students.

Instead, the Coleman Report, released in 1966, produced some very unexpected findings which became the underpinning for many subsequent educational policies and programs. The researchers found that twelve years after *Brown*, most Americans still attended segregated schools but that the characteristics of Black and White schools (e.g., facilities, books, labs, teacher experience, and expenditures) were surprisingly similar. Apparently, segregated Southern districts had upgraded Black educational facilities in the wake of the *Brown* decision. Coleman and his colleagues also found that variations in school resources had relatively little to do with the variations in students' school performance. Instead, they found that family background influenced an individual's school achievement more than any other factor, including school characteristics.[7] Subsequent research has provided a better understanding of when, where, and how resources and school characteristics influence student outcomes.

However, at the time, the Coleman Report had dramatic and long-lasting effects. The report tended to deflect attention away from how schools operated and instead focused public policy upon poor and minority children and their families as the ultimate sources of unequal school outcomes. Numerous observers concluded incorrectly that schools had little to do with Black-White educational differences because they paid insufficient attention to another of the report's findings that implicated schools in inequality of educational outcomes. That finding showed that African-American and White achievement differences increased with every year of schooling. That is, the achievement gap between Black and White first graders was much smaller than the gap between twelfth graders. This finding suggested that, at best, schools reinforce the disadvantages of race and class and, at worst, are themselves a major source of educational inequality.

Although published over thirty years ago, the Coleman Report remains one of the most important and controversial pieces of social research ever completed in the United States. One of its many lasting results was a redefinition of the concept of equality of educational opportunity. The report made it clear that putting greater resources into schools, in and of itself, was not necessarily associated with greater student achievement or with eliminating the racial gap in school outcomes. In other words, resource levels alone were no longer considered a satisfactory measure of equality of educational opportunity. As a result, the Coleman Report helped reconceptualize the notion of equality of educational opportunity. It became a matter of equality of educational outcomes, measured in terms of academic achievement (performance) and attainment (amount of schooling completed) irrespective of race, gender, and socioeconomic background. This goal has yet to be reached in the United States.

Compensatory Education

A second outcome of the Coleman Report was widespread support for compensatory education. Policy makers interpreted the finding that family background was the strongest predictor of students' achievement as evidence of "cultural deprivation" among poor and minority families. This interpretation gave impetus to an education movement designed to compensate for the alleged cultural deficiencies of families that were neither middle class nor White, so that when so-called disadvantaged children came to school, they could compete without the handicaps of their background.

Beginning with the passage of the Elementary and Secondary Education Act in 1965, a series of educational programs offered low-income and underachieving children developmental preschool followed by a host of individualized programs in math, reading, and language arts once they arrived in elementary school. Examples of compensatory education programs include:

- Early childhood education such as Head Start
- Follow Through, where Head Start children, now in elementary school, continue to receive special programs
- Title I (formerly called Chapter I), which provides language arts and math programs plus food, medicine, and clothing to needy children in primary schools
- Guidance and counseling in secondary schools

Compensatory education programs have had a controversial history. Critics from the left charge that the underlying premise of compensatory education—that poor and minority families are deficient relative to middle-class White families—is racist and elitist. Critics on the right argue that compensatory education is a waste of time and money because the lower achievement scores of minority and poor children are due to their inferior intelligence. Policy critics charge that it is impossible to judge the effectiveness of compensatory education programs unless they are fully funded and implemented so that all eligible children receive services. Since the inception of compensatory education programs, less than half of eligible students have received services.

Despite criticism from such diverse quarters, the compensatory education movement survived the past thirty years. The Head Start program, for example, is currently embraced by a wide range of Americans who consider it a cost-effective strategy to help poor children do better in school. A growing body of research demonstrates the existence of both cognitive and social benefits from early childhood education for low-income and minority children. However, the achievement gaps between minority and White, and between working- and middle-class, children remain. Furthermore, evaluations of Title I and Follow Through have been unable to demonstrate unambiguous benefits. One must conclude that compensatory education, like desegregated education, has neither leveled the playing field nor eliminated racial or social class inequality in educational outcomes.

Human Capital Theory and Workforce Education Programs

The widely held belief that a good education is *the* meal ticket to reducing inequality receives its most sophisticated exposition in human capital theory, which holds that greater levels of education are investments in human beings' productive capacities. People who are poor, according to this theory, have had inadequate investments in their education. Over many decades, numerous education and training programs have been implemented, but the school-to-work transition remains problematic for many noncollege-bound youth. During the last third of this century, specific programs linked to anti-poverty efforts were implemented and eventually scrapped for failing to provide low-income youth with what they needed. While the Comprehensive Education and Training Act (CETA) and its successor, the Job Training and Partnership Act (JTPA), two of the best known programs, gave skill training to low-income youth, the jobs needed to employ them were simply not there. Moreover, JTPA graduates were no more likely to obtain a job than those without such training.

A variety of workforce education programs exist today in America's schools. These are attempts to provide broad-based technical skills to students, although they have not completely displaced the more traditional vocational education programs. However, current programs are of questionable value for reducing inequality for a number of reasons. First, there is a striking absence of bridges from schools to workplaces. Other than tapping into their personal networks, youth usually do not know alternative options for obtaining jobs. Schools rarely have outreach programs, and employers who hire entry-level workers only occasionally check grades and transcripts. The federal government's initiative for community-based one-stop job centers, where potential workers and employers can match their availabilities via computerized databases, may prove useful in the future by providing the necessary bridges and pathways from school to work. But they are only in their infancy.

A second reason to question the utility of these programs is that the human capital approach views the problem of inequality as a lack of worker skill, not a paucity of well-paying jobs. The poor often have a great deal of skills and education. What they lack are well-paying jobs in which to invest their skills. In fact, many studies indicate there is an adequate match between skill requirements of current jobs and those possessed by the workforce. Rather, the complaints of most employers are that entry-level new hires lack a strong work ethic, a problem human capital theory does not address.

The third reason to question the utility of workforce development education for reducing inequality has to do with the changing nature of the post-industrial economy. There is little certain knowledge of what the restructuring and globalization of the economy will mean for future workers. It is entirely possible that workforce education will prepare people for jobs which have been relocated to Mexico, Thailand, or India. Certainly, there is already evidence of such a trend. And the relocation of these jobs has more to do with the cost of labor in these nations than with the education or skill levels of Americans. It is not surprising, then, that educational reforms based on human capital theory have not, and cannot, substantially narrow race, class, and gender differences in equality of opportunity in this society.

Title IX

Title IX of the 1972 Higher Education Act is the primary federal law prohibiting sex discrimination in education. It states, "No person in the United States shall, on the basis of sex, be excluded from participation in, be denied benefit of, or be subjected to discrimination under any program or activity receiving Federal financial assistance." Until Title IX's passage, gender inequality in educational opportunity received minimal legislative attention. The act mandates gender equality of treatment in admission, courses, financial aid, counseling services, employment, and athletics.

The effect of Title IX upon college athletics has been especially controversial. While women constitute 53 percent of undergraduates, they are only 37 percent of college athletes. This is undoubtedly due to the complex interaction between institutional practices and gender-role socialization over the life course. Certainly, the fact that the vast majority of colleges spend much more money on recruiting and scholarships for male athletes contributes to the disparities.

In spring 1997, the U.S. Supreme Court refused to review a lower court's ruling in *Brown v. Cohen* that states in essence that Title IX requires universities to provide equal athletic opportunities for male and female students regardless of cost. Courts have generally upheld the following three-pronged test for compliance: (1) the percentage of athletes who are females should reflect the percentage of students who are female; (2) there must be a continuous record of expanding athletic opportunities for females athletes; and, (3) schools must accommodate the athletic interests and abilities of female students. As of the ruling, very few universities were in compliance with the law.

Gender discrimination exists in other areas of education where it takes a variety of forms. For example, in K–12 education official curricular materials frequently feature a preponderance of male characters. Male and female characters typically exhibit traditional gender roles. Vocational education at the high school and college level remains gender-segregated to some degree.

School administrators at all levels are overwhelmingly male although most teachers in elementary and secondary schools are female. In higher education, the situation is more complex. Faculty women in academia are found disproportionately in the lower ranks, are less likely to be promoted, and continue to earn less than their male colleagues.

Like the laws and policies aimed at eliminating race differences in school processes and outcomes, those designed to eliminate gender differences in educational opportunities have, at best, only narrowed them. Access to educational opportunity in the United States remains unequal for people of different gender, race, ethnic, and socioeconomic backgrounds.

EQUALITY OF EDUCATIONAL OPPORTUNITY AND EQUALITY OF INCOME

Despite the failures of these many programs to eliminate the inequality of educational opportunity over the past 45 years, there is one indicator of substantial progress: measured in median years, the gap in educational attainment between Blacks and Whites, and between males and females, has all but disappeared. In 1997, the median educational attainment of most groups was slightly more than twelve years. In the 1940s, by contrast, White males and females had a median educational attainment of just under nine years, African American males about five years, and African American women about six.

However, the main goal of educational reform is not merely to give all groups the opportunity to receive the same quality and quantity of education. According to the dominant ideology, the ultimate goal of these reforms is to provide equal educational opportunity in order to facilitate equal access to jobs, housing, and various other aspects of the American dream. It thus becomes crucial to examine whether the virtual elimination of the gap in educational attainment has been accompanied by a comparable decrease in other measures of inequality.

Of the various ways inequality can be measured, income is one of the most useful. Much of a person's social standing and access to the good things in life depends on his or her income.[8] Unfortunately, the dramatic progress in narrowing the gap in educational attainment has not been matched by a comparable narrowing of the gap in income inequality. Median individual earnings by race and gender indicate that White men still earn significantly more than any other group. Black men trail White men, and all women earn significantly less than all men. Even when occupation, experience, and level of education are controlled, women earn less than men, and Black men earn less than White men. It is only Black and White women with comparable educational credentials in similar jobs who earn about the same.

The discrepancy between the near elimination of the gap in median educational attainment and the ongoing gaps in median income is further evidence that addressing the inequality of educational opportunity is woefully insufficient for addressing broader sources of inequality throughout society.

This discrepancy can be explained by the nature of the U.S. political economy. The main cause of income inequality is the structure and operation of U.S. capitalism, a set of institutions which scarcely have been affected by the educational reforms discussed earlier. Greater equality of educational opportunity has not led to a corresponding decrease in income inequality because educational reforms do not create more good-paying jobs, affect gender-segregated and racially segmented occupational strutures, or limit the mobility of capital either between regions of the country or between the United States and other countries. For example, no matter how good an education White working-class or minority youth may receive, it does nothing to alter the fact that thousands of relatively good paying manufacturing jobs have left northern inner cities for northern suburbs, the sunbelt, or foreign countries.

Many argue that numerous service jobs remain or that new manufacturing positions have been created in the wake of this capital flight. But these pay less than the departed manufacturing jobs, are often part-time or temporary, and frequently do not provide benefits. Even middle-class youth are beginning to fear the nature of the jobs which await them once they complete their formal education. Without changes in the structure and operation of the capitalist economy, educational reforms alone cannot markedly improve the social and economic position of disadvantaged groups. This is the primary reason that educational reforms do little to affect the gross social inequalities that inspired them in the first place.

BEYOND ATTAINMENT: THE PERSISTENCE OF EDUCATIONAL INEQUALITY

Educational reforms have not led to greater overall equality for several additional reasons. While race and gender gaps in educational attainment have narrowed considerably, educational achievement remains highly differentiated by social class, gender, and race. Many aspects of school processes and curricular content are deeply connected to race, social class, and gender inequality. But gross measures of educational outputs, such as median years of schooling completed, mask these indicators of inequality.

Not all educational experiences are alike. Four years of public high school in Beverly Hills are quite different from four years in an inner-city school. Family background, race, and gender have a great deal to do with whether a person goes to college and which institution of higher education she or he attends. The more privileged the background, the more likely a person is to attend an elite private university.

For example, according to Jacobs, women trail men slightly in representation in high status institutions of higher education because women are less likely to attend engineering programs and are more likely to be part-time students (who are themselves more likely to attend lower status institutions such as community colleges). Gender segregation in fields of study remains marked, with women less likely than men to study in scientific and mathematical fields. Furthermore, there is substantial race and ethnic segregation between institutions of higher education. Asian-Americans and Latinos are more segregated from Whites than are African-Americans. Whites and Asian-Americans are more likely to attend higher status universities than are African-Americans and Latinos.[9]

These patterns of race and gender segregation in higher education have direct implications for gender and race gaps in occupational and income attainment. Math and science degree recipients are more likely to obtain more lucrative jobs. A degree from a state college is not as competitive as one from an elite private university. Part of the advantage of attending more prestigious schools comes from the social networks to which a person has access and can join.

Another example of persistent inequality of educational opportunity is credential inflation. Even though women, minorities, and members of the working class now obtain higher levels of education than they did before, members of more privileged social groups gain even higher levels of education. At the same time, the educational requirements for the best jobs (those with the highest salaries, benefits, agreeable working conditions, autonomy, responsibility) are growing. Those with the most education from the best schools tend to be the top candidates for the best jobs. Because people from more privileged backgrounds are almost always in a better position to gain these desirable educational credentials, members of the working class, women, and minorities are still at a competitive disadvantage. Due to the dynamics of credential inflation, educational requirements previously necessary for the better jobs and now within the reach of many dispossessed groups are inadequate and insufficient in today's labor market. The credential inflation process keeps the already privileged one step (educational credential) ahead of the rest of the job seekers.[10]

One additional aspect of the persistent inequalities in educational opportunities concerns what sociologists of education call the hidden curriculum. This concept refers to two separate but related processes. The first is that the content and process of education differ for children according to their race, gender, and class. The second is that these differences reflect and thus help reproduce the inequalities based on race, gender, and class that characterize U.S. society as a whole.

One aspect of the hidden curriculum is the formal curriculum's ideological content. Anyon's work on U.S. history texts demonstrates that children from more privileged backgrounds are more likely to be exposed to rich, sophisticated, and complex materials than are their working-class counterparts.[11] Another aspect of the hidden curriculum concerns the social organization of the school and the classroom. Some hidden curriculum theorists suggest that tracking, ability grouping, and conventional teacher-centered classroom interactions contribute to the reproduction of the social relations of production at the workplace. Lower-track classrooms are disproportionately filled with working-class and minority students. Students in lower tracks are more likely than those in higher tracks to be assigned repetitive exercises with low levels of cognitive challenge. Lower-track students are likely to work individually and to lack classroom experience with problem solving or other independent, creative activities. Such activities are more conducive to preparing students for working-class jobs than for professional and managerial positions. Correspondence principle theorists argue that educational experiences from preschool to high school are designed to differentially prepare students for their ultimate positions in the work force, and that a student's placement in various school programs is strongly related to her or his race and class origin. Critics charge that the correspondence principle has been applied in too deterministic and mechanical a fashion. Evidence abounds of student resistance to class, gender, and race differentiated education.[12] This is undoubtedly why so many students drop out or graduate from high school with minimal levels of literacy and formal skills. Nonetheless, hidden curriculum theory offers a compelling contribution to explanations of how and why school processes and outcomes are so markedly different according to the race, gender, and social class of students.

CONCLUSION

In this [essay] we have argued that educational reforms alone cannot reduce inequality. Nevertheless, education remains important to any struggle to reduce inequality. Moreover, education is more than a meal ticket; it is intrinsically worthwhile and crucially important for the survival of democratic society. Many of the programs discussed in this essay contribute to the enhancement of individuals' cognitive growth and thus promote important nonsexist, nonracist attitudes and practices. Many of these programs also make schools somewhat more humane places for adults and children. Furthermore, education, even reformist liberal education, contains the seeds of individual and social transformation. Those of us committed to the struggle against inequality cannot be paralyzed by the structural barriers that make it impossible for education to eliminate inequality. We must look upon the schools as arenas of struggle against race, gender, and social class inequality.

NOTES

1. This essay draws on an article by Roslyn Arlin Mickelson that appeared as "Education and the Struggle Against Race, Class and Gender Inequality," *Humanity and Society* 11(4) (1987): 440–64.

2. Ascertaining whether a set of beliefs constitutes the dominant ideology in a particular society involves a host of difficult theoretical and empirical questions. For this reason we use the term *putative dominant ideology*. For discussion of these questions, see Nicholas Abercrombie et al., *The Dominant Ideology Thesis* (London: George Allen & Unwin, 1980); James C. Scott, *Weapons of the Weak* (New Haven: Yale University Press, 1985); Stephen Samuel Smith, "Political Acquiescence and Beliefs About State Coercion" (unpublished Ph.D. dissertation, Stanford University, 1990).

3. Jennifer Hochschild, "The Double-Edged Sword of Equal Educational Opportunity." Paper presented at the meeting of the American Education Research Association, Washington, D.C., April 22, 1987.

4. Gunnar Myrdal, *An American Dilemma: The Negro Problem and Modern Democracy* (New York: Harper & Row, 1944).

5. Gary Orfield, Mark D. Bachmeier, David R. James, and Tamela Eitle, "Deepening Segregation in American Public Schools" (Cambridge, MA: Harvard Project on School Desegregation, 1997).

6. Amy Stuart Wells and Robert L. Crain, "Perpetuation Theory and the Long-Term Effects of School Desegregation," *Review of Educational Research* 64(4) (1994): 531–55.

7. J. S. Coleman et al., *Equality of Educational Opportunity* (Washington, D.C.: Government Printing Office, 1966).

8. To be sure, income does not measure class-based inequality, but there is a positive correlation between income and class. Income has the additional advantage of being easily quantifiable. Were we to use another measure of inequality, e.g., wealth, the disjuncture between it and increases in educational attainment would be even larger. Although the distribution of wealth in U.S. society has remained fairly stable since the Depression, the gap between rich and poor increased in the 1980s and 1990s. Although accurate data are difficult to obtain, a 1992 study by the Federal Reserve found that in 1989 the top one-half of one percent of households held 29 percent of the wealth held by all households.

9. Jerry A. Jacobs, "Gender and Race Segregation Between and Within Colleges" (paper presented at the Eastern Sociological Society, Boston, MA, April 1996).

10. Randall Collins, *The Credential Society* (New York: Academic Press, 1979).

11. Jean Anyon, "Social Class and the Hidden Curriculum of Work," *Journal of Education* 162(1) (1980): 67–92; Jean Anyon, "Social Class and School Knowledge," *Curriculum Inquiry* 10 (1981): 3–42.

12. Samuel Bowles and Herbert Gintis, *Schooling in Capitalist America: Educational Reform and the Contradictions of Economic Life* (New York: Basic Books, 1976); Roslyn Arlin Mickelson, "The Case of the Missing Brackets: Teachers and Social Reproduction," *Journal of Education* 169(2) (1987): 78–88.

CANTO, LOCURA Y POESIA

33

Olivia Castellano

I am a walking contradiction. I have no Ph.D. yet I'm a full professor of English at a state university. By all definitions and designs I should not even have made it to college. I am the second of five children of a Southern Pacific Railroad worker with a fifth-grade education and a woman who dropped out of the second grade to help raise ten siblings—while her mother worked ten hours a day cleaning houses and doing laundry for rich Texan ranchers.

In Comstock, the Tex-Mex border town about fifteen miles from the Rio Grande where I spent the first twelve years of my life, I saw the despair that poverty and hopelessness had etched in the faces of young Chicano men who, like my father, walked back and forth on the dusty path between Comstock and the Southern Pacific Railroad station. They would set out every day on rail carts to repair the railroad. The women of Comstock fared no better. Most married early. I had seen them in their kitchens toiling at a stove, with one baby propped on one hip and two toddlers tugging at their skirts. Or they followed their working mothers' route, cleaning houses and doing laundry for rich Texan ranchers who paid them a pittance. I decided very early that this was not the future I wanted.

In 1958 my father, tired of seeing his days fade into each other without promise, moved us to California where we became farmworkers in the San Jose area (then a major agricultural center). I saw the same futile look in the faces of young Chicanos and Chicanas working beside my family. Those faces already lined so young with sadness made me deadly serious about my books and my education.

At a young age—between eleven and fourteen—I began my intellectual and spiritual rebellion against my parents and society. I fell in love with books and created space of my own where I could dare to dream. Yet in school I remained shy and introverted, terrified of my white, male professors. In my adolescence I rebelled against my mother's insistence that Mexican girls should marry young, as she did at eighteen. I told her that I didn't care if my cousins Alicia and Anita were getting married and having babies early. "I was put on this earth to make books, not babies!" I announced and ran into my room.

From: *Women's Review of Books* 7 (5) (February 1990). Reprinted by permission of the author.

Books were my obsession. I wanted to read everything that I was not supposed to. By fourteen I was already getting to know the Marquis de Sade, Rimbaud, Lautréamont, Whitman, Dostoyevsky, Marx. I came by these writers serendipitously. To get from home to Sacramento High School I had to walk through one of the toughest neighborhoods in the city, Oak Park. There were men hanging out with liquor in brown paper bags, playing dice, shooting craps and calling from cars: "Hey, baby, get in here with me!" I'd run into a little library called Oak Park Library, which turned out to have a little bit of everything. I would walk around and stare at the shelves, killing time till the shifty-eyed men would go away.

The librarians knew and tolerated me with skepticism: "Are you sure you're going to read the Marquis de Sade? Do your parents know you're checking out this material? What are you doing with the *Communist Manifesto?*" One librarian even forbade me to check the books out, so I'd sit reading in the library for hours on end. Later, at sixteen or seventeen, I was allowed to check anything and everything out.

So it was that I came to grapple with tough language and ideas. These books were hot! Yet I also was obsessed with wanting to be pretty, mysterious, silent and sexy. I wanted to have long curly hair, red lips and long red nails; to wear black tight dresses and high heels. I wanted desperately to look like the sensuous femmes fatales of the Mexican cinema—María Féliz, one of the most beautiful and famous of Mexico's screen goddesses, and Libertad Lamarque, the smoky-voiced, green-eyed Argentinian singer. These were the women I admired when my mother and I went to the movies together. So these were my "outward" models. My "inward" models, the voices of the intellect that spoke to me when I shut the door to my room, were, as you have gathered, a writer of erotica, two mad surrealists, a crazy Romantic, an epileptic literary genius and a radical socialist.

I needed to sabotage society in a major, intellectually radical way. I needed to be a warrior who would catch everyone off guard. But to be a warrior, you must never let your opponent figure you out. When the bullets of racism and sexism are flying at you, you must be very clever in deciding how you want to live. I knew that everything around me—school, teachers, television, friends, men, even my own parents, who in their internalized racism and self-hatred didn't really believe I'd amount to much though they hoped like hell that life would prove them wrong—everything was against me, and I understood this fully.

To protect myself I fell in love with language—all of it, poems, stories, novels, plays, songs, biographies, "cuentos" or little vignettes, movies—all manifestations of spoken and written language. I fell in love with ideas, with essays by writers like Bacon or Montaigne. I began my serious reading crusade around age eleven, when I was already convinced that books were central to my life. Only through them and through songs, I felt, would I be free to structure some kind of future for myself.

I wanted to prove to anyone who cared to ask (though by now I was convinced no one gave a damn) that I, the daughter of a laborer-farmworker, could dare to be somebody. Try to imagine what it is like to be always full of rage—rage at everything: at white teachers who could never even pronounce my name (I was often called anything from "Odilia" to "Otilia" to "Estela"); rage at those teachers who asked me point-blank, "But how did you get to be so smart? You are Mexican, aren't you?"; rage at my eleventh-grade English teacher who said to me in front of the class, "You stick to essay writing; never try to write a poem again because a poet you are not!" (This, after I had worked for two diligent weeks on an imitation of "La Belle Dame Sans Merci"! Now I can laugh. Then it was pitiful.)

From age thirteen I was also angry at boys who hounded me for dates. When I'd reject them they'd yell, "So what do you plan to do for the rest of your life, fuck a book?" I was angry at my Chicana classmates in high school who, perhaps jealous of my high grades, would say, "What are you trying to do, be like the whites?" I regret to say that I was also angry at my parents, exasperated by their docility, their limited expectations of me. Oh, I knew they were proud; but sometimes, in their own misdirected rage (maybe afraid of my little successes), they would make painful comments. "Te vas a volver loca con esos jodidos libros" ("You'll go nuts with those fucking books") was my mother's frequent statement. Or the even more sickening, "Esta nunca se va a casar." ("Give up on this one; she'll never get married.") This was the tenor of my adolescent years. When nothing on either side of the two cultures, Mexican or Anglo-American, affirms your existence, that is how rage is shaped.

While I managed to escape at least from the obvious entrapments—a teen pregnancy, a destructive early marriage—I did not escape years of being told I wasn't quite right, that because of my ethnicity and gender I was somehow defective, incomplete. Those years left wounds on my self-esteem, wounds so deep that even armed with my books and stolen knowledge I could not entirely escape deep feelings of unworthiness.

By the time I graduated from high school and managed to get a little scholarship to California State University in Sacramento, where I now teach (in 1962 it was called Sacramento State College), I had become very unassertive, immensely shy. I was afraid to look unfeminine if I raised my hand in class, afraid to seem ridiculous if I asked a "bad" question and all eyes turned on me. A deeper part of me was afraid that my rage might rear its ugly head and I would be considered "an angry Mexican accusing everybody of racism."

I was painfully concerned with my physical appearance: wasn't I supposed to look beautiful like Féliz and Lamarque? Yet while I wanted to look pretty for the boys, the thought of having sex terrified me. What if I got pregnant, had to quit college and couldn't read my books any more? The more I feared boys, the more I made myself attractive for them and the more they would make advances, the more I rejected them.

The constant tension sapped my energy and distracted me from my creative journeys into language. Oh, I would write little things (poems, sketches for stories, journal entries), but I was afraid to show them to anyone. Besides, no one knew I was writing them. I was so frightened by my white, male professors, especially in the English department—they looked so arrogant and were so ungiving of their knowledge—that I didn't have the nerve to major in English, though it was the major I really wanted.

Instead, I chose to major in French. The "Parisiens" and "Québecois" in the French department faculty admired my French accent: "Mademoiselle, êtes-vous certaine que vous n'êtes pas parisienne?" they would ask. In short, they cared. They engaged me in dialogue, asked why I preferred to study French instead of Spanish. ("I already know Spanish," I'd say.) French became my adopted language. I could play with it, sing songs in it and sound exotic. It complemented my Spanish; besides, I didn't have to worry about speaking English with my heavy Spanish accent and risk being ridiculed. At one point, my spoken French was better than my oral Spanish; my written French has remained better than my written Spanish.

At 23, armed with a secondary school teaching credential and B.A. in French with an English minor, I became a high school teacher of French and English. Soon after that I began to work for a school district where the majority of the students were Chicanos and Blacks from families on welfare and/ or from households run by women.

After two years of high school teaching, I returned to Cal State at Sacramento for the Master's degree. Professionally and artistically, it was the best decision I have ever made. The Master's program to which I applied was a pilot program in its second year at CSUS. Called the Mexican American Experienced Teachers' Fellowship, it was run by a team of anthropology professors, central among whom was Professor Steven Arvizu. The program was designed to turn us into "agents of cultural change." It was 1969 and the program was one of the first federally funded (Title V) ones to address Mexican American students' needs by re-educating their teachers.

My interests were literary, but all twenty of us "fellows" had to get an M.A. in social anthropology, since this experiment took the "anthropologizing education" approach. We studied social dynamics, psycholinguistics, history of Mexico, history of the American Southwest, community activism and confrontational strategies and the nature of the Chicano movement. The courses were eye-openers. I had never heard the terms Chicano, biculturalism, marginality, assimilation, Chicanismo, protest art. I had never heard of Cesar Chavez and the farmworkers nor of Luis Valdez and the Teatro Campesino. I had never studied the nature of racism and identity. The theme of the program was that culture is a powerful tool for learning, self-expression, solidarity and positive change. Exploring it can help Chicano students understand their bicultural circumstances.

The program brought me face to face with nineteen other Chicano men and women, all experienced public school teachers like myself, with backgrounds like mine. The program challenged every aspect of my life. Through group counseling, group encounter, classroom interaction, course content and community involvement I was allowed to express my rage and to examine it in the company of peers who had a similar anger. Most of our instructors, moreover, were Chicano or white professors sensitive to Chicanos. For the first time, at 25, I had found my role models. I vowed to do for other students what these people had done for me.

Eighteen years of teaching primarily white women students, Chicanos and Blacks at California State University, Sacramento, have led me to see myself less as a teacher and more as a cultural worker, struggling against society to undo the damage of years of abuse. I continue to see myself as a warrior empowered by my rage. Racism and sexism leave two clear-cut scars on my students; internalized self-hatred and fear of their own creative passion, in my view the two most serious obstacles in the classroom. Confronting this two-headed monster has made me razor-sharp. Given their tragic personal stories, the hope in my students' eyes reconfirms daily the incredible beauty, the tenacity of the human spirit.

Teaching white women students (ages 30–45) is no different from working with Chicano and Black students (both men and women): you have to bring about changes in the way they view themselves, their abilities, their right to get educated and their relation to a world that has systematically oppressed them simply for being who they are. You have to help them channel and understand the seething rage they carry deep inside, a rage which, left unexpressed, can make them turn against each other and, more sadly, against themselves.

I teach four courses per semester: English 109G, Writing for Proficiency for Bilingual/Bidialectal Students (a course taken mainly by Chicano and Black students, ages 19–24); English 115A, Pedagogy/Language Arts for Prospective Elementary School Teachers (a course taken mainly by women aged 25–45, 50 percent white, 50 percent Chicano); English 180G, Chicano Literature, an advanced studies General Education course for non-English majors (taken by excellent students, aged 24–45, about 40 percent white, 40 percent Chicano, 20 percent Black/Vietnamese/Filipino/South American).

The fourth course is English 1, Basic Language Skills, a pre-freshman composition course taken primarily by Black and Chicano freshmen, male and female, aged 18–22, who score too low on the English Placement Test to be placed in "regular" Freshman Composition.

Mine is a teaching load that, in my younger days at CSUS, used to drive me close to insanity from physical, mental and spiritual exhaustion—spiritual from having internalized my students' pain. Perhaps not fully empowered myself, not fully emplumed in the feathers of my own creativity (to borrow the wonderful "emplumada" metaphor coined by Lorna Dee Cervantes, the brilliant Chicana poet), I allowed their rage to become part of mine. This kind of rage can kill you. And so through years of working with these kinds of students I have learned to make my spirit strong with "canto, locura y poesia" (song, madness, and poetry). Judging from my students' progress, the songs have worked.

Truly, it takes a conjurer, a magus with all her teaching cards up her sleeve, to deal with the fragmented souls that show up in my classes. Among the Chicanos and Blacks I get ex-offenders (mostly men but occasionally a woman who has done time), orphans, single women heads of household, high school dropouts who took years to complete their Graduation Equivalency Diploma.

I get women who have been raped and/or who have been sexually abused either by a father figure or by male relatives—Sylvia Tracey, for example, a 30-year-old Chicana feminist, mother of two, whose parents pressured her to marry her (white) rapist and who is going through divorce after ten years of marriage. I get women who have been battered. And, of course, I get the young Chicano and Black little yuppies who don't believe the world existed before 1970, who know nothing about the sixties' history of struggle and student protest, who—in the case of the Chicanos—feel ashamed that their parents speak English with an accent or were once farmworkers. I get Chicanos, Blacks and white women, especially, who are ashamed of their writing skills, who have never once been told that they could succeed in school.

Annetta Jones is typical. A 45-year-old Black woman, who single-handedly raised three children, all college-educated and successful, she is still married to a man who served ten years in prison for being a "hit man." She visited him faithfully in prison and underwent all kinds of humiliation at the hands of correctional officers—even granting them sexual favors just to be allowed to have conjugal visits. When her husband completed his time he fell in love with a young woman from Chicago, where he now lives.

Among my white women students (ranging in age from 25 to 40, though occasionally I get a 45-year-old woman who wants to be an elementary or high school teacher and "help out young kids so they won't have to go through what I went through"—their exact words) I get women who are either divorced or divorcing; rarely do I get a "happily" married woman. This is especially true of the white women who take my Chicano literature and my credential-pedagogy classes. Take Lynne Trebeck, for instance, a white woman about 40 years old who runs a farm. When she entered the university her husband objected, so she divorced him! They continue to live in the same house (he refused to leave), "but now he has no control over me," she told me triumphantly midway through the semester. She has two sons, fifteen and eighteen years old; as a young woman she did jail time as the accomplice of a convicted drug dealer.

Every semester I get two or three white lesbian feminists. This semester there was Vivianne Rose, about 40, in my Chicano literature class. Apparently sensing too much conservatism in the students, and knowing that she wanted to be an elementary school teacher, she chose to conceal her sexual orientation. On the first day of class she wore Levi pants, a baggy sweat shirt, white tennis shoes and a beige baseball cap. By the end of the first week she had switched to ultrafeminine dresses and skirts, brightly colored blouses, nylons and medium-heeled black shoes, not to mention lipstick and eye makeup. When she spoke in class she occasionally made references to "my husband who is Native American." She and Sylvia Tracey became very close friends. Halfway through the course they informed me that "Shit, it's about time we tell her." (This, from Sylvia.) "Oh hell, why not," Vivianne said; "my 'husband' is a woman." The woman *is* Native American; Vivianne Rose lived on a reservation for years and taught young Native American children to read and write. She speaks "Res" talk (reservation speech) and has adopted her "husband's" last name.

Among my white women students there are also divorced women who are raising two to four children, usually between the ages of eight and seventeen. Sometimes I get older widowed white women who are taking classes for their own enjoyment, not for a degree. These women also tell stories of torment: rapes, beatings, verbal and emotional harassment from their men. On occasion I get women who have done jail time, usually for taking the rap for drug-connected boyfriends. I rarely get a married woman, but when I do there is pain: "My husband doesn't really want me in school." "My husband doesn't really care what I do in college as long as I take care of his needs and the kids' needs." "My husband doesn't really know what I'm studying—he has never asked and I've never told him."

Most of the white women as well as the minority students come to the university under special programs. There is the "Educational Opportunity Program" for students who do not meet all university entrance requirements or whose grade point average is simply too low for regular admission. There is the "Student Affirmative Action Program" for students who need special counseling and tutoring to bring their academic skills up to par or deal with emotional trauma. There is the "College Assistance Migrant Program" for students whose parents are migrant farmworkers in the agricultural areas surrounding Sacramento. There is a wonderful program called PASAR for older women students entering the university for the first time or returning after a multiple-year absence. The Women's Resource Center also provides small grants and scholarships for re-entry women. A large number of my students (both white and minority women) come severely handicapped in their basic language, math and science skills; many have never used a computer. It is not uncommon (especially among Chicanos and Blacks) to get an incoming student who scores at the fifth- and sixth-grade reading levels.

The task is herculean, the rewards spiritually fulfilling. I would not have it any other way. Every day is a lesson in humility and audacity. That my students have endured nothing but obstacles and putdowns, yet still have the courage and strength to seek a college education, humbles me. They are, like me, walking paradoxes. They have won against all the odds (their very presence on campus attests to that). Yet really they haven't won: they carry a deeply ingrained sense of inferiority, a firm conviction that they are not worthy of success.

This is my challenge: I embrace it wholeheartedly. There is no place I'd rather be, no profession more noble. Sure, I sometimes have doubts: every day something new, sad, even tragic comes up. Just as I was typing this article, for instance, Vicky, one of the white students in my Chicano literature class, called in tears, barely able to talk. "Professor, I can't possibly turn in my paper to your mailbox by four o'clock," she cried. "Everything in my house is falling apart! My husband just fought with my oldest daughter [from a previous marriage], has thrown her out of the house. He's running up and down the street, yelling and threatening to leave us. And I'm sitting here trying to write your paper! I'm going crazy. I feel like walking away from it all!" I took an hour from writing this article to help her contain herself. By the end of our conversation, I had her laughing. I also put her in touch with a counselor friend of mine and gave her a two-day extension for her final paper. And naturally I was one more hour late with my own writing!

I teach in a totally non-traditional way. I use every trick in the book: lots of positive reinforcement, both oral and written; lots of one-on-one conferences. I network women with each other, refer them to professor friends who can help them; connect them to graduate students and/or former students who are already pursuing careers. In the classroom I force my students to come up in front of their classmates, explain concepts or read their essays aloud. I create panels representing opposing viewpoints and hold debates—lots of oral participation, role-playing, reading their own texts. Their own writing and opinions become part of the course. On exams I ask them questions about their classmates' presentations. I meet with individual students in local coffeehouses or taverns: it's much easier to talk about personal pain over coffee or a beer or a glass of wine than in my office. My students, for the most part, do not have a network of support away from the university. There are no supportive husbands, lovers (except on rare occasions, as with my lesbian students), no relatives saying, "Yes, you can do it."

Is it any wonder that when these students come to me they have a deep sense of personal shame about everything—poor skills, being older students? They are also very angry, not only at themselves but at the schools for having victimized them; at poor, uninspired teaching; at their parents for not having had high enough expectations of them or (in the case of the women) for having allowed them to marry so young. Sylvia, my Chicana feminist student, put it best when I was pointing out incomplete sentences in her essay: "Where the hell was I when all this was being taught in high school? And why didn't anybody give a damn that I wasn't learning it?"

I never teach content for the first two weeks of any of my courses. I talk about anger, sexism, racism and the sixties—a time when people believed in something larger than themselves. We dialogue—about prisons and why so many Chicano and Black young men are behind bars in California; why people fear differences; why they are so homophobic. I give my students a chance to talk about their anger ("coraje" in Spanish). I often read them the poem by my friend and colleague Jose Montoya, called "Eslipping and Esliding," where he talks about "locura" (craziness) and says that with a little locura, a little eslipping and esliding, we can survive the madness that surrounds us. We laugh at ourselves, sharing our tragic, tattered pasts; we undo everything and let the anger out. "I know why so many of you are afraid of doing well," I say. "You've been told you can't do it, and you're so pissed off about it, you can't concentrate." Courage takes pure concentration. By the end of these initial two or three weeks we have become friends and defined our mutual respect. Only then do we enter the course content.

I am not good at endings; I prefer to celebrate beginnings. The struggle continues and the success stories abound. Students come back, year after year, to say "Thank you." Usually I pull these visitors into the classroom: "Tell my class that they can do it. Tell them how you did it!" They start talking and can't stop. "Look, Olivia, when I first came into your class," said Sylvia, "I couldn't even put a fucking sentence together. And now look at me, three years later I'm even writing poetry!"

REMINISCENCE OF A POST-INTEGRATION KID:

34

Or, Where Have We Come Since Then?

Gaye Williams

I want to dedicate this to the young woman who sat in the University of Tennessee dining hall while a table of seven white kids—who would share her microbiology class an hour later—stood to leave as she sat down. They stood to prove that they were too good to sit next to a Black woman. However, one of them (her lab partner) could later laugh and joke with her by the safety of the microscope where no one could see them, as if nothing had happened. Sharing space were the young man and the woman, who was smarter but darker, and who knew what it was to be hurt.

Post-segregation. I asked my mother why, when I got around to attending schools, everything had calmed down. "After the whites realized they had lost," she said, "they mostly gave up fighting desegregation." So, no mobs lined up to keep me out of North Roebuck School's third grade class, or my Black teacher from her job teaching all those white kids. I even won Top Scholar of my class.

Someone asked me when I first began to identify as a feminist. I realized then that I was outraged about the unequal treatment of women in relation to men long before I understood what racism meant. Before being asked the question, I had not thought about the fact that I lacked a racial consciousness growing up in Birmingham, Alabama, in the sixties. I thought some more.

The absence of a racial consciousness was a result of my being a post-integration kid. I came up through a conspiracy to silence the Black movement, which was not a topic in the white schools I attended. Even though one of my babysitters was among the girls killed in the 16th Street Baptist Church bombing, everything was fine, or so the silence suggested. There were no bombs, no riots, no state troopers taking me to school. There was also no one to reinforce the radio's message that Black was beautiful, just James Brown shouting as I got ready for school "Say It Loud, I'm Black and I'm Proud!"

But JB's message did not sink in. In elementary school, I thought the white kids would not know I was Black if I did not tell them, although I am not

From: *Sage: A Scholarly Journal on Black Women* 1 (Spring 1984): 20–21. Reprinted by permission.

at all light enough to pass. I made myself invisible and pretended I was someone else, for protection from the self-hatred that I had learned from somewhere. The fact that my imaginary self was always male, and more often white than Black, says something about the effects of curriculum, climate, and the availability of suitable role models to my sense of self-validation. Very little in my pre-college schooling gave me an alternative to hating my Blackness and femaleness, although at the time I am sure I would not have spoken of it that way. I retreated into identities I felt were safer, more attractive than my own.

Was that all the fault of curriculum, climate, lack of suitable role models? Maybe not, but I spent the great bulk of my pre-college time either in school or studying, with very little free time left over. When I was not studying, I was acting, and the classes and plays were little different in content from my school lessons.

Very early in my education, the lines were drawn between the kids singled out as smart and the others. I think about the church schools I attended—no rods spared nor children spoiled there. Being "good" and "smart" were my best defenses. From there I went to public elementary school for second grade, where I wandered outside of class to the school library and the reading specialist's lab. I was a favored child, along with my friend Eric. We were taught at home by our school teacher/administrator families. We both advanced beyond the other kids and were treated something like a prince and princess—"golden" children.

I thought I was rich, with my light skin and proper talk, because I was set apart by the teacher, attractive to boys, and picked on by at least one girl. I knew, even while she made me cry, that Debra was as much attracted to me as she was envious or curious. That was the year when the teacher performed a wedding ceremony for Eric and me in front of all the others.

The chance to develop my talent also set me apart from other Black kids. I was not a singer, coming along in the bosom of a Black church choir, but acting in plays that had Black characters only because their actors happened to be Black. After a successful audition, I went to the Alabama School of the Fine Arts in drama. The number of Black kids in the school slowly grew, but most of them were musicians. Even among the Black actors, I was still set apart by my lack of a Black or southern accent.

The other side of being a golden child marked for success was the feeling of being separated from my peers. I was held up for praise and vulnerable to disdain. One problem of post-integration kids is that the divisions between us of class, sex, and color have become more virulent. These differences have always been present in the Black community, and their resulting hierarchy a divisive problem. But this distancing effect of the combination of light skin, brains and opportunity became much greater post-integration because the possibility of admission to white society became much more probable. With post-integration came more options.

I came from a family that sent its children, daughters included, to the best schools the Black community had to offer, at whatever sacrifice. They considered education the necessary ticket to advancement. By the time I was ready for college, even the Ivy League was within reach. My admission was bought by sit-ins, marches, riots, freedom rides, and legislation, but I had only the vaguest understanding of the importance of these occurrences.

My mother once said that she wondered whether she had done the best thing sending me to mostly white private schools. Which is more important—that kids have the best opportunity to develop their minds in language, art and science, or that their souls and spirits benefit from the best the Black tradition has to offer, so that they face the world's hostility feeling part of a strong community and heritage? She could not ask nor answer then the question of why we should have to choose between the two.

It was in college that I began to gain a full sense of what it meant to me to be born Black and female. Knowing that the Ivy League probably had little idea of what a young Black girl/woman from the South needed or wanted so far away from home, I had the sense to build a community of friends around me very soon. Convinced that I was not prepared to be in this prestigious university and that I was going to have a difficult time academically (even though I never before had trouble in school), I surrounded myself with people and activities to help see me through. First, I found upperclass women students who were intrigued, I think, by my boldness and always ready with advice and comfort. They were my survival, along with special teachers, the Boston women's community, and the Black children and their families I met in the Intergenerational Outreach Saturday Educational Program.

I met older Black kids who had a strong racial and feminist consciousness. But they had also felt separated from most Blacks for the same reasons I had. These folks were discovering together how to bridge gaps, knowing that these were the most important lessons to learn. I grew from a base of self-love to alliances and friendships with students who were Chicano, Puerto Rican, African, Asian and American Indian, and the gay folks of all ethnicities.

The most important thing about going to college when I did and where I did was that I began to broaden my idea of education. Thinking about what education has meant to me seems almost the same as thinking about what being young means. There is the education that happens inside classrooms or other clearly designated learning places. But it is difficult, and perhaps undesirable, to separate the learning I did there from the constant process of figuring out what I was doing in this world. The point to consider is how the learning in everyday life and that of school fit together. Then I can begin to answer questions of how my racial/feminist consciousness was shaped.

I went to a lot of conferences and events in the Boston area while in school. The most significant one was the 1981 Women and Law Conference. Meeting feminist women of color there had a catalytic effect on my education, helping to bring everything together. Afterwards, I began demanding that my course work in school complement the questions I was framing and hearing outside the classroom. I insisted that this expensive schooling give me something that would help me understand the meaning of my life as a Black woman. From the example of the women I met there and at subsequent events, I gained the inspiration and courage necessary to run successfully for president of the Harvard-Radcliffe Black Students Association, and learned to bring the force that women's work embodied onto the campus and into the lives of the other students.

I have not liked feeling cut off from people, especially from Black folks. It has been my good fortune to make friendships that showed me how silly and immaterial were the barriers which I thought kept me apart. With the help of my friends of all races I have come to insist that the post-integration decades be just that: a time that allows me to "integrate" and to use all the parts of myself. I insist on it being a time that does not demand that I be either Black **or** female, but recognizes that I am both, **all** of the time. From there I can go on to the business of helping make this world fit for all of us to live in.

Of course, my education does not end with the close of classes. The beat, I know, will keep going on.

TITLE IX: *Does Help for Women Come at the Expense of African Americans?*

35

Craig T. Greenlee

Gender equity has created an intriguing set of circumstances in the world of college athletics.

On the one hand, Title IX, the federal law which forbids sex discrimination in educational institutions receiving federal funds, has opened the window of opportunity for scores of female athletes.

From: *Black Issues in Higher Education* (April 17, 1997): 24–26. Reprinted by permission.

The NCAA women's basketball tournament offers ample proof. The Women's Final Four has attracted an average of almost 50,000 fans over the past two years.

And there are other examples. Soccer has blossomed as a premier women's sport in America. Colleges and universities are a major part of the feeder system that produced players for the 1996 Olympic gold-medal winning U.S. soccer team. Women's gymnastics and swimming are also on the rise as collegiate sports which feature top-caliber competition and widespread fan support.

But there is a down side. While there are now more women sports programs on the collegiate scene, critics say that in general, women have benefitted at the expense of men's sports.

IT'S ALL ABOUT PROPORTIONALITY

In order for schools to comply with Title IX, schools have to provide opportunities for female athletes that are in line with the percentage of females on that campus. Put another way, if a school's student body is 55 percent women, 55 percent of its total athletic offerings must be geared toward women. The law doesn't mandate that schools treat men's and women's sports identically, but it does say that the benefits for both should be comparable. Few schools have yet met this test, according to recent surveys, but pressure to comply may increase after a landmark Title IX case against Brown University works its way through the Supreme Court.

For many schools, adhering to Title IX means cutting men's sports to provide funding for women. In many instances, schools have had to eliminate some men's sports or reduce—sometimes dramatically—the number of scholarships and coaches in those sports.

"If you increase opportunities for one group, I'm not so sure that you don't wind up denying another group," says Alex Wood, head football coach at James Madison University and vice-president of the Black Coaches Association. "And because there's only so much money available to operate a college sports program, somebody will inevitably get the short end of the stick."

Football has become a main target for Title IX advocates because it eats up a large chunk of the athletic budget. The sport is expensive because of the large roster sizes (80–100 players), equipment, and recruiting costs.

Title IX supporters assert that schools can reduce football scholarships and still maintain a competitive program. They point to National Football League (NFL) teams which have roster limits of forty-five players. When compared to the eighty-five scholarship limit that the major college football programs have, they ask, "If the pros prosper with forty-five, why can't the colleges?" Decreasing the number of football scholarships, Title IX proponents explain, will free up sufficient money to finance women's sports.

RACE V. GENDER

This is where race and gender wind up on a collision course.

"The race versus gender issue is very real," says Wood. "In football, a large number of the players are Black. So when you start cutting scholarships, you not only take away the opportunity to play, you take away the opportunity to go to school. Playing football is the only way that a lot of Black players get to go to college at all."

Black males aren't the only ones to feel the pinch. Black women, ironically, are also caught in the crunch.

As a group, Black women have not benefitted from Title IX because the expansion in women's athletics involves sports where Black female participation is minimal. It is estimated that approximately 97 percent of the 4,000 Black female collegiate athletes participate in basketball or track and field. However, the so-called "emerging" or non-traditional sports—gymnastics, swimming, crew, lacrosse and soccer, to name a few—are the ones that many schools are opting to add to help meet Title IX guidelines.

"Women of color are hurt because they don't participate in those sports where all the expansion is taking place," says Dee Todd, assistant commissioner of the Atlantic Coast Conference. "Women of color have a double protected status [because of race and gender], but they're still left out. Most play basketball or run track. You'll see a handful in volleyball, softball and soccer, but that's about it. As a result, Title IX doesn't do a whole lot for women of color."

While Blacks don't participate in the non-traditional sports in large numbers, Todd says there is one sport that many colleges and Black athletes have yet to look at as an alternative—team handball.

"I can't say why more schools aren't playing team handball," says Todd. "You don't need a lot in terms of facilities, all you need is a wall. But the people who can do well in this sport are athletes who've played basketball and volleyball—sports that require good hand-to-eye coordination."

BROADENING ATHLETIC HORIZONS

In the long run, Black athletes—male and female—will have to broaden their athletic horizons if they want to earn college athletic scholarships. In other words, Blacks will have to begin taking up sports other than football, basketball and track because there won't be any expansion in those sports.

Todd feels Blacks can be steered toward other sports if they're exposed at an early age.

"I talk to youngsters all the time and I tell them if they want a college scholarship, get a golf club and learn how to play, or take swim lessons, or get into youth soccer. There's no reason why Black youngsters can't do well in those sports. It's all a matter of exposure."

There are no easy solutions in the athletic competition between race and gender. In too many instances, it seems that the two are always in direct conflict. But even when they are not, problems can arise. For example, a school might add field hockey to its sports menu to comply with Title IX, then discover that there is not sufficient interest among the students to maintain it. In that scenario, the sport was added strictly because of Title IX, not because the students wanted it.

In the short-term, however, Wood contends that schools can individually do the right thing by choosing to allocate their athletic resources in a fair manner among men and women's sports.

"Everybody should have the opportunity to play and have a good experience in doing so," Wood says. "The same kinds of opportunities should be provided for everybody, and nobody should feel like they're getting second-class treatment. Each school has to look at its own situation and make a decision based on what their individual needs are. Schools have to look at what their constituents want."

The State and Social Policy

THE FIRST AMERICANS: *American Indians*

36

C. Matthew Snipp

By the end of the nineteenth century, many observers predicted that American Indians were destined for extinction. Within a few generations, disease, warfare, famine, and outright genocide had reduced their numbers from millions to less than 250,000 in 1890. Once a self-governing, self-sufficient people, American Indians were forced to give up their homes and their land, and to subordinate themselves to an alien culture. The forced resettlement to reservation lands or the Indian Territory (now Oklahoma) frequently meant a life of destitution, hunger, and complete dependency on the federal government for material needs.

Today, American Indians are more numerous than they have been for several centuries. While still one of the most destitute groups in American society, tribes have more autonomy and are now more self-sufficient than at any time since the last century. In cities, modern pan-Indian organizations have been successful in making the presence of American Indians known to the larger community, and have mobilized to meet the needs of their people (Cornell 1988; Nagel 1986; Weibel-Orlando 1991). In many rural areas, American Indians and especially tribal governments have become increasingly more important and increasingly more visible by virtue of their growing political and economic power. The balance of this [reading] is devoted to explaining their unique place in American society.

From: Silvia Pedraza and Rubén G. Rumbaut, eds., *Origins and Destinies: Immigration, Race, and Ethnicity in America*. (Belmont, CA: Wadsworth, 1996), pp. 390–403. Reprinted by permission.

THE INCORPORATION OF AMERICAN INDIANS

The current political and economic status of American Indians is the result of the process by which they were incorporated into Euro-American society (Hall 1989). This amounts to a long history of efforts aimed at subordinating an otherwise self-governing and self-sufficient people that eventually culminated in widespread economic dependency. The role of the U.S. government in this process can be seen in the five major historical periods of federal Indian relations: removal, assimilation, the Indian New Deal, termination and relocation, and self-determination.

Removal

In the early nineteenth century, the population of the United States expanded rapidly at the same time that the federal government increased its political and military capabilities. The character of Indian-American relations changed after the War of 1812. The federal government increasingly pressured tribes settled east of the Appalachian Mountains to move west to the territory acquired in the Louisiana purchase. Numerous treaties were negotiated by which the tribes relinquished most of their land and eventually were forced to move west.

Initially the federal government used bargaining and negotiation to accomplish removal, but many tribes resisted (Prucha 1984). However, the election of Andrew Jackson by a frontier constituency signaled the beginning of more forceful measures to accomplish removal. In 1830 Congress passed the Indian Removal Act, which mandated the eventual removal of the eastern tribes to points west of the Mississippi River, in an area which was to become the Indian Territory and is now the state of Oklahoma. Dozens of tribes were forcibly removed from the eastern half of the United States to the Indian Territory and newly created reservations in the west, a long process ridden with conflict and bloodshed.

As the nation expanded beyond the Mississippi River, tribes of the plains, southwest, and west coast were forcibly settled and quarantined on isolated reservations. This was accompanied by the so-called Indian Wars—a bloody chapter in the history of Indian-White relations (Prucha 1984; Utley 1984). This period in American history is especially remarkable because the U.S. government was responsible for what is unquestionably one of the largest forced migrations in history.

The actual process of removal spanned more than a half-century and affected nearly every tribe east of the Mississippi River. Removal often meant extreme hardships for American Indians, and in some cases this hardship reached legendary proportions. For example, the Cherokee removal has become known as the "Trail of Tears." In 1838, nearly 17,000 Cherokees were ordered to leave their homes and assemble in military stockades (Thornton 1987, p. 117). The march to the Indian Territory began in October and continued through the winter months. As many as 8,000 Cherokees died from cold weather and diseases such as influenza (Thornton 1987, p. 118).

According to William Hagan (1979), removal also caused the Creeks to suffer dearly as their society underwent a profound disintegration. The contractors who forcibly removed them from their homes refused to do anything for "the large number who had nothing but a cotton garment to protect them from the sleet storms and no shoes between them and the frozen ground of the last stages of their hegira. About half of the Creek nation did not survive the migration and the difficult early years in the West" (Hagan 1979, p. 77–81). In the West, a band of Nez Perce men, women, and children, under the leadership of Chief Joseph, resisted resettlement in 1877. Heavily outnumbered, they were pursued by cavalry troops from the Wallowa valley in eastern Oregon and finally captured in Montana near the Canadian border. Although the Nez Perce were eventually captured and moved to the Indian Territory, and later to Idaho, their resistance to resettlement has been described by one historian as "one of the great military movements in history" (Prucha 1984, p. 541).

Assimilation

Near the end of the nineteenth century, the goal of isolating American Indians on reservations and the Indian Territory was finally achieved. The Indian population also was near extinction. Their numbers had declined steadily throughout the nineteenth century, leading most observers to predict their disappearance (Hoxie 1984). Reformers urged the federal government to adopt measures that would humanely ease American Indians into extinction. The federal government responded by creating boarding schools and the allotment acts—both were intended to "civilize" and assimilate American Indians into American society by Christianizing them, educating them, introducing them to private property, and making them into farmers. American Indian boarding schools sought to accomplish this task by indoctrinating Indian children with the belief that tribal culture was an inferior relic of the past and that Euro-American culture was vastly superior and preferable. Indian children were forbidden to wear their native attire, to eat their native foods, to speak their native language, or to practice their traditional religion. Instead, they were issued Euro-American clothes, and expected to speak English and become Christians. Indian children who did not relinquish their culture were punished by school authorities. The curriculum of these schools taught vocational arts along with "civilization" courses.

The impact of allotment policies is still evident today. The 1887 General Allotment Act (the Dawes Severalty Act) and subsequent legislation mandated that tribal lands were to be allotted to individual American Indians in fee simple title, and the surplus lands left over from allotment were to be sold on the open market. Indians who received allotted tribal lands also received citizenship, farm implements, and encouragement from Indian agents to adopt farming as a livelihood (Hoxie 1984, Prucha 1984).

For a variety of reasons, Indian lands were not completely liquidated by allotment, many Indians did not receive allotments, and relatively few changed their lifestyles to become farmers. Nonetheless, the allotment era was a disaster because a significant number of allotees eventually lost their land. Through tax foreclosures, real estate fraud, and their own need for cash, many American Indians lost what for most of them was their last remaining asset (Hoxie 1984).

Allotment took a heavy toll on Indian lands. It caused about 90 million acres of Indian land to be lost, approximately two-thirds of the land that had belonged to tribes in 1887 (O'Brien 1989). This created another problem that continues to vex many reservations: "checkerboarding." Reservations that were subjected to allotment are typically a crazy quilt composed of tribal lands, privately owned "fee" land, and trust land belonging to individual Indian families. Checkerboarding presents reservation officials with enormous administrative problems when trying to develop land use management plans, zoning ordinances, or economic development projects that require the construction of physical infrastructure such as roads or bridges.

The Indian New Deal

The Indian New Deal was short-lived but profoundly important. Implemented in the early 1930s along with the other New Deal programs of the Roosevelt administration, the Indian New Deal was important for at least three reasons. First, signaling the end of the disastrous allotment era as well as a new respect for American Indian tribal culture, the Indian New Deal repudiated allotment as a policy. Instead of continuing its futile efforts to detribalize American Indians, the federal government acknowledged that tribal culture was worthy of respect. Much of this change was due to John Collier, a long-time Indian rights advocate appointed by Franklin Roosevelt to serve as Commissioner of Indian Affairs (Prucha 1984).

Like other New Deal policies, the Indian New Deal also offered some relief from the Great Depression and brought essential infrastructure development to many reservations, such as projects to control soil erosion and to build hydroelectric dams, roads, and other public facilities. These projects created jobs in New Deal programs such as the Civilian Conservation Corps and the Works Progress Administration.

An especially important and enduring legacy of the Indian New Deal was the passage of the Indian Reorganization Act (IRA) of 1934. Until then, Indian self-government had been forbidden by law. This act allowed tribal governments, for the first time in decades, to reconstitute themselves for the purpose of overseeing their own affairs on the reservation. Critics charge that this law imposed an alien form of government, representative democracy, on traditional tribal authority. On some reservations, this has been an on-going source of conflict (O'Brien 1989). Some reservations rejected the IRA for this reason, but now have tribal governments authorized under different legislation.

Termination and Relocation

After World War II, the federal government moved to terminate its long-standing relationship with Indian tribes by settling the tribes' outstanding legal claims, by terminating the special status of reservations, and by helping reservation Indians relocate to urban areas (Fixico 1986). The Indian Claims Commission was a special tribunal created in 1946 to hasten the settlement of legal claims that tribes had brought against the federal government. In fact, the Indian Claims Commission became bogged down with prolonged cases, and in 1978 the commission was dissolved by Congress. At that time, there were 133 claims still unresolved out of an original 617 that were first heard by the commission three decades earlier (Fixico 1986, p. 186). The unresolved claims that were still pending were transferred to the Federal Court of Claims.

Congress also moved to terminate the federal government's relationship with Indian tribes. House Concurrent Resolution (HCR) 108, passed in 1953, called for steps that eventually would abolish all reservations and abolish all special programs serving American Indians. It also established a priority list of reservations slated for immediate termination. However, this bill and subsequent attempts to abolish reservations were vigorously opposed by Indian advocacy groups such as the National Congress of American Indians. Only two reservations were actually terminated, the Klamath in Oregon and the Menominee in Wisconsin. The Menominee reservation regained its trust status in 1975 and the Klamath reservation was restored in 1986.

The Bureau of Indian Affairs (BIA) also encouraged reservation Indians to relocate and seek work in urban job markets. This was prompted partly by the desperate economic prospects on most reservations, and partly because of the federal government's desire to "get out of the Indian business." The BIA's relocation programs aided reservation Indians in moving to designated cities, such as Los Angeles and Chicago, where they also assisted them in finding housing and employment. Between 1952 and 1972, the BIA relocated more than 100,000 American Indians (Sorkin 1978). However, many Indians returned to their reservations (Fixico 1986). For some American Indians, the return to the reservation was only temporary; for example, during periods when seasonal employment such as construction work was hard to find.

Self-Determination

Many of the policies enacted during the termination and relocation era were steadfastly opposed by American Indian leaders and their supporters. As these programs became stalled, critics attacked them for being harmful, ineffective, or both. By the mid-1960s, these policies had very little serious support. Perhaps inspired by the gains of the Civil Rights movement, American Indian leaders and their supporters made "self-determination" the first priority on their political agendas. For these activists, self-determination meant that Indian people would have the autonomy to control their own affairs, free from the paternalism of the federal government.

The idea of self-determination was well received by members of Congress sympathetic to American Indians. It also was consistent with the "New Federalism" of the Nixon administration. Thus, the policies of termination and relocation were repudiated in a process that culminated in 1975 with the passage of the American Indian Self-Determination and Education Assistance Act, a profound shift in federal Indian policy. For the first time since this nation's founding, American Indians were authorized to oversee the affairs of their own communities, free of federal intervention. In practice, the Self-Determination Act established measures that would allow tribal governments to assume a larger role in reservation administration of programs for welfare assistance, housing, job training, education, natural resource conservation, and the maintenance of reservation roads and bridges (Snipp and Summers 1991). Some reservations also have their own police forces and game wardens, and can issue licenses and levy taxes. The Onondaga tribe in upstate New York have taken their sovereignty one step further by issuing passports that are internationally recognized. Yet there is a great deal of variability in terms of how much autonomy tribes have over reservation affairs. Some tribes, especially those on large and well-organized reservations have nearly complete control over their reservations, while smaller reservations with limited resources often depend heavily on BIA services. . . .

CONCLUSION

Though small in number, American Indians have an enduring place in American society. Growing numbers of American Indians occupy reservation and other trust lands, and equally important has been the revitalization of tribal governments. Tribal governments now have a larger role in reservation affairs than ever in the past. Another significant development has been the urbanization of American Indians. Since 1950, the proportion of American Indians in cities has grown rapidly. These American Indians have in common with reservation Indians many of the same problems and disadvantages, but they also face other challenges unique to city life.

The challenges facing tribal governments are daunting. American Indians are among the poorest groups in the nation. Reservation Indians have substantial needs for improved housing, adequate health care, educational opportunities, and employment, as well as developing and maintaining reservation infrastructure. In the face of declining federal assistance, tribal governments are assuming an ever-larger burden. On a handful of reservations, tribal governments have assumed completely the tasks once performed by the BIA.

As tribes have taken greater responsibility for their communities, they also have struggled with the problems of raising revenues and providing economic opportunities for their people. Reservation land bases provide many reservations with resources for development. However, these resources are not always abundant, much less unlimited, and they have not always been well managed. It will be yet another challenge for tribes to explore ways of efficiently managing their existing resources. Legal challenges also face tribes seeking to exploit unconventional resources such as gambling revenues. Their success depends on many complicated legal and political contingencies.

Urban American Indians have few of the resources found on reservations, and they face other difficult problems. Preserving their culture and identity is an especially pressing concern. However, urban Indians have successfully adapted to city environments in ways that preserve valued customs and activities—powwows, for example, are an important event in all cities where there is a large Indian community. In addition, pan-Indianism has helped urban Indians set aside tribal differences and forge alliances for the betterment of urban Indian communities.

These alliances are essential, because unlike reservation Indians, urban American Indians do not have their own form of self-government. Tribal governments do not have jurisdiction over urban Indians. For this reason, urban Indians must depend on other strategies for ensuring that the needs of their community are met, especially for those new to city life. Coping with the transition to urban life poses a multitude of difficult challenges for many American Indians. Some succumb to these problems, especially the hardships of unemployment, economic deprivation, and related maladies such as substance abuse, crime, and violence. But most successfully overcome these difficulties, often with help from other members of the urban Indian community.

Perhaps the greatest strength of American Indians has been their ability to find creative ways for dealing with adversity, whether in cities or on reservations. In the past, this quality enabled them to survive centuries of oppression and persecution. Today this is reflected in the practice of cultural traditions that Indian people are proud to embrace. The resilience of American Indians is an abiding quality that will no doubt ensure that they will remain part of the ethnic mosaic of American society throughout the twenty-first century and beyond.

References

Cornell, Stephen. 1988. *The Return of the Native: American Indian Political Resurgence.* New York: Oxford University Press.

Hall, Thomas D. 1989. *Social Change in the Southwest, 1350–1880.* Lawrence, KS: University Press of Kansas.

Nagel, Joanne. 1989. "American Indian Repertoires of Contention." Paper presented at the annual meeting of the American Sociological Association, San Francisco, CA.

Weibel-Orlando, Joan. 1991. *Indian Country, L.A.* Urbana, IL: University of Illinois Press.

THOUGHTS ON CLASS, RACE, AND PRISON

37

Alan Berkman and Tim Blunk

Having been locked up as political prisoners for the past five years, we have found a lot in several recent articles which was familiar and thought provoking. Certain descriptions of ping-ponging from jail to jail could be a log of some of our own trips conducted by the Bureau of Prisons (BoP), and we can verify accounts of how prisoners typically support one another. Perhaps most importantly, we strongly agree with the view—best expressed in an article by Sam Day published in the October 13, 1989 issue of *Isthmus*—that the threat of prison should not deter political activists from acting. We are concerned,

From: Alan Berkman and Tim Blunk, "Thoughts on Class, Race, and Prison," in *Cages of Steel: The Politics of Imprisonment in the United States*, eds. Ward Churchill and J. J. Vanderwell (Washington, D.C.: Maisonnueve Press, 1992), pp. 190–93. Reprinted by permission.

though, that in so vividly portraying his own experience and utilizing it to demystify prison life, he may have played into some common misrepresentations of life in U.S. prisons.

First, a word about our own experiences. We've each spent time in prison for a variety of politically motivated acts. One of us (Alan Berkman) spent two years in a number of county jails in the Philadelphia area. The other (Tim Blunk) spent a year in New York's Riker's Island. We've both been in a number of federal prisons, including the BoP's most repressive facility at Marion, Illinois. For the past eighteen months, we've been together in the equivalent of a county jail in Washington, D.C. where we and four women activists face federal charges of resisting U.S. war crimes through "violent and illegal means" (we call this the Resistance Conspiracy Case). Tim is presently serving a fifty-eight year sentence; Alan, twelve years. Having spent considerable time in the whole range of penal institutions, from county jails where most prisoners are pre-trial to Marion, where the average sentence is forty years, we've been struck by how directly prisons reflect the social and economic realities of society as a whole.

We live in a country where large numbers of people, particularly young Afro-American, Latino and Native American women and men, have been written off by society. Government leaders look at the human devastation caused by their policies and declare war on the victims (the "war on drugs"). The racism and malignant neglect that permeate the schools, the labor market, the welfare system and social services of the Third World and poor white neighborhoods of our cities bear their inevitable fruit in the prisons. In prison, though, even the usual facade of "fairness" and "justice" by which such inequity is usually accompanied is dropped.

There are almost a million people in U.S. jails, state and federal prisons. More than ninety percent of these prisoners are in county jails and state prisons. D.C. jail, for example, is filled to capacity with poor, mis-educated, profoundly alienated young Afro-American men and women. It's a warehouse. There are no programs, no contact visiting, no privacy, and most importantly, no justice.

Particularly for political activists who are white and middle class, prison can be one of the few places where we look at America from the bottom up. We can understand poverty differently when we live with young people who've never had $20 in their pocket and who are constantly bombarded with the Gucci ads, the BMWs on *Miami Vice*, and the idea of free trips to Rio on *The Price is Right*. In a consumer society, if you've got nothing, you're considered to be nothing, and the frustration of this reality leads people into crime and drugs. You can see the deadness and despair in the eyes of too many of the people around you as they realize the American Dream is not for them now, and probably wasn't from the day they were born into the ghetto, the barrio or the reservation. And you'll get a feeling in your gut about the rage racism and poverty generate in the hearts of the oppressed. In fact, if you allow yourself, you'll learn quite a lot about the feelings that accompany that word, "oppressed."

What we've come to understand ever more deeply from our experience as prisoners is that the essence of oppression lies in dehumanization and disrespect. Being oppressed isn't just being a bit poorer or having a rougher life than your oppressor; it means being treated as if you're less of a human being than they are. For instance, we've consistently experienced the fact that you can be getting along okay with the guards and prison officials when they suddenly do something which is not only totally outrageous, but often a complete violation of their own supposed rule structure. You call them on it, and their response is to look at you—*through* you would be more accurate—as if you just weren't there. It doesn't matter a bit whether you're right or wrong. They deny that you have the right even to hold an opinion on the matter. It cuts at the very core of your sense of humanity and self-worth. It is infuriating and degrading.

Self-worth. Dignity. Self-respect. These are the feelings prison consciously or unconsciously is designed to destroy. These are the same feelings that racism and economic inequality in society work to destroy in the poor and people of color. This is why we believe Sam and others err when they reduce prison life to "three hots [meals] and a cot" and other minor indignities. It's like reducing chattel slavery to hard work, bad pay and poor living conditions. It leaves out the truth of the human dimension of oppression. And it leaves out the human reaction to prison conditions: rebellion. Prison "riots" are fundamentally slave rebellions.

Over the past few weeks, there have been one major and two minor prison uprisings in the Pennsylvania prison system. At Camp Hill, where prisoners destroyed much of the prison, the precipitating event seems trivial: the prison cancelled the semi-annual family day picnics where prisoners' relatives can bring home-cooked food and eat with their loved ones out on the yard. What's the big deal about two afternoons per year? Well, at one level, when you don't have much, even a *little* more deprivation hurts rather badly, especially when that deprivation concerns those precious bits of time you can spend with those closest to you and when you manage to feel most fully human. At a deeper level, the arbitrariness of the prison officials' decision in Pennsylvania gave the prisoners a clear message that they had nothing at all coming to them: *everything* is a "privilege," *nothing* is a right. The oppressor giveth, the oppressor taketh away. The oppressed have nothing to say in the matter, one way or the other. The oppressed—the prisoner in this instance—is a cipher, a statistic, nothing more than an animal whose basic needs of calories, shelter and waste disposal must (usually) be met. No more.

The "senseless violence" of burning buildings and smashing equipment which marked the Camp Hill rebellion—and which has marked prison uprisings from Attica to Atlanta, New Mexico, Los Angeles, and elsewhere—is, perhaps paradoxically, the most basic assertion of human dignity and self-respect. "We do have a say," the prisoners are insisting. "We will exercise our rights as human beings one way or another, either as reasonable human beings in a reasonable situation, or this way if you insist the situation remain unreasonable. But either way you can ultimately do nothing to provide our demonstration of the fact that *we are people*." It is a very human response to a very inhuman situation, the same impulse which led to the great ghetto uprisings of Watts, Detroit and Newark during the '60s, and which has led to similar phenomena in places like Miami during the '80s.

Political activists who come to jail and are able to keep their eyes and hearts open will learn an enormous amount, both about themselves and about the world they live in. How do we deal with our own race/class biases to build principled relationships with those we live with? If we're white, how do we respond when the Aryan Brotherhood tells us they'll kick our ass if we eat with Afro-American, Latino or American Indian prisoners in the *de facto* segregated mess hall? How do we struggle against the pervasive and intense sexism in men's prisons? How do we remain caring and giving people without allowing ourselves to be taken advantage of and disrespected? How do we decide when to directly confront the power of the officials, and when not to? When do we get involved in other people's beefs, whether among prisoners or between prisoners and guards?

There's a lot to learn in prison, although it's certainly not the only, or even necessarily the best, place to learn it. Whatever else may be said, time spent behind bars—no more or less than any other sort of time—need never be "wasted." We agree with Sam and others when they say that fear of prison should never be paralyzing. While locked up, you'll more than likely get in shape, have the opportunity to read many of the books you always meant to plow through but never managed to get around to, learn to write letters which say more than just "howdy-do." Most importantly, you'll finally have the chance to identify with the oppressed on a gut level which is typically absent from the life experiences of most Euro-Americans. You'll come to understand power relationships in a new and far more meaningful way. And we're willing to bet that you'll come out more committed than ever to the need for fundamental social change.

IT'S A FAMILY AFFAIR: *Women, Poverty, and Welfare*   **38**

Randy Albelda and Chris Tilly

Hating poor women for being poor is all the rage—literally. Radio talk show hosts, conservative think tanks, and many elected officials bash poor single mothers for being too "lazy," too "dependent," and too fertile. Poor mothers are blamed for almost every imaginable economic and social ill under the sun. Largely based on anecdotal information, mythical characterizations, and a recognition that the welfare system just isn't alleviating poverty, legislatures across the land and the federal government are proposing and passing draconian welfare "reform" measures.

It is true that current welfare policies do not work well—but not for the reasons usually presented. Welfare "reform" refuses to address the real issues facing single-mother families, and is heavily permeated by myths.

Aid to Families with Dependent Children (AFDC), the government income transfer program for poor non-elder families in the United States, serves only about 5 percent of the population at any given time, with over 90 percent of those receiving AFDC benefits being single mothers and their children. In 1993, 14 million people (two-thirds of them children) in the United States received AFDC. That same year, just under 40 million people were poor. Despite garnering a lion's share of political discussion, AFDC receives a minuscule amount of funding: it accounts for less than 1 percent of the federal budget and less than 3 percent of the state budgets.

Single mothers work. Not only do they do the unpaid work of raising children, they also average the same number of hours in the paid labor force as other mothers do—about 1,000 hours a year (a full-time, year-round job is about 2,000 hours a year). [1] And while close to 80 percent of all AFDC recipients are off in two years, over half of those return at some later point—usually because their wages in the jobs that got them off welfare just didn't match the cost of health care and childcare needed so they could keep the jobs. In fact, most AFDC recipients "cycle" between relying on families, work, and AFDC

From: Diane Dujon and Ann Withorn, eds., *For Crying Out Loud: Women's Poverty in the United States*. (Boston: South End Press, 1996), pp. 79–85. Reprinted by permission.

benefits to get or keep their families afloat.[2] That means that, for many single mothers, AFDC serves the same function as unemployment insurance does for higher-paid, full-time workers.

And, contrary to a highly volatile stereotype, welfare mothers, on average, have fewer kids than other mothers. And once on AFDC, they are less likely to have another child.

POVERTY AND THE "TRIPLE WHAMMY"

Poverty is a persistent problem in the United States. People without access to income are poor. In the United States, most people get access to income by living in a family with one or more wage-earners (either themselves or others). Income from ownership (rent, dividends, interest, and profits) provides only a few families with a large source of income. Government assistance is limited—with elders getting the bulk of it. So wages account for about 80 percent of all income generated in the United States. Not surprisingly, people whose labor market activity is limited, or who face discrimination, are the people most at risk for poverty. Children, people of color, and single mothers are most likely to be poor.

In 1993, 46 percent of single-mother families in the United States were living in poverty, but only 9 percent of two-adult families with chidren were poor.[3]

Why are so many single-mother families poor? Are they lazy, do they lack initiative, or are they just unlucky? The answer to all of these is a resounding "No." Single-mother families have a very hard time generating enough income to keep themselves above the poverty line for a remarkably straightforward reason: one female adult supports the family—and one female adult usually does not earn enough to provide both childcare expenses and adequate earnings.

To spell it out, single mothers face a "triple whammy." First, like all women, when they do paid work they often face low wages—far lower than men with comparable education and experience. In 1992, the median income (the midpoint) for all women who worked full-time was $13,677. That means that about 40 percent of all working women (regardless of their marital status) would not have made enough to support a family of three above the poverty line. Even when women work year-round full-time, they make 70 percent of what men do.

Second, like all mothers, single mothers must juggle paid and unpaid work. Taking care of healthy, and sometimes sick, children, and knowing where they are when at work, requires time and flexibility that few full-time jobs afford. All mothers are more likely to earn less and work less than other women workers because of it.

Finally, *unlike* married mothers, many single mothers must juggle earning income and taking care of children without the help of another adult. Single-mother families have only one adult to send into the labor market. And that same adult must also make sure children get through their day.

The deck is stacked—but not just for single mothers. All women with children face a job market which has little sympathy for their caregiving responsibilities and at the same time places no economic value on their time spent at home. The economic activity of raising children is one that no society can do without. In our society, we do not recognize it as work worth paying mothers for. For a married mother, this contradiction is the "double day." For a single mother, the contradiction frequently results in poverty for her and her children.

DENYING THE REAL PROBLEMS

The lack of affordable childcare, the large number of jobs that fail to pay living wages, and the lack of job flexibility are the real problems that face all mothers (and increasingly everyone). For single mothers, these problems compound into crisis.

But instead of tackling these problems head on, politicians and pundits attack AFDC. Why? One reason is that non-AFDC families themselves are becoming more desperate, and resent the limited assistance that welfare provides to the worst-off. With men's wages falling over the last 30 years, fewer and fewer families can get by with only one wage earner. The government is not providing help for many low-income families who are struggling but are still above the AFDC eligibility threshold. This family "speed-up" has helped contribute to the idea that if both parents in a two-parent household can work (in order to be poor), then all AFDC recipients should have to work too.

Instead of facing the real problems, debates about welfare reform are dominated by three dead ends. First, politicians argue that single mothers must be made to work in the paid labor market. But, most single mothers already work as much as they can. Studies confirm that AFDC recipients already do cycle in and out of the labor force. Further, as surveys indicate, mothers receiving AFDC would like to work. The issue is not whether or not to work, but whether paid work is available, how much it pays, and how to balance work and childcare.

Second, there is a notion of replacing the social responsibilities of government assistance with individual "family" responsibilities: make men pay child support, demand behavioral changes of AFDC recipients, or even pressure single women to get married. While child support can help, for most single mothers it offers a poor substitute for reliable government assistance. Penalizing women and their children for ascribed behaviors (such as having more children to collect welfare) that are supported by anecdotes but not facts is at best mean-spirited.

Third, there is an expectation that people only need support for a limited amount of time—many states and some versions of federal welfare reform limit families to 24 months of aid over some period of time (from a lifetime to five years). Yet, limiting the amount of time women receive AFDC will not reduce or limit the need for support. Children do not grow up in 24 months, nor will many women with few skills and little education necessarily become job ready. But more important, many women who do leave AFDC for the workplace will not make enough to pay for childcare or the health insurance they need to go to work.

In short, welfare "reform" that means less spending and no labor market supports will do little beyond making poor women's lives more miserable.

NOTES

1. These data, and others throughout the paper, were calculated by the authors using current population survey tapes.

2. Five recent studies have looked at welfare dynamics and all come to these conclusions. LaDonna Pavetti, "The Dynamics of Welfare and Work: Exploring the Process by Which Young Women Work Their Way Off Welfare," paper presented at the APPAM Annual Research Conference, 1992; Kathleen Harris, "Work and Welfare Among Single Mothers in Poverty," *American Journal of Sociology*, vol. 99 (2), September 1993, pp. 317–52; Roberta Spalter-Roth, Beverly Burr, Heidi Hartmann, and Lois Shaw, "Welfare That Works: The Working Lives of AFDC Recipients," Institute for Women's Policy Research, 1995; Rebecca Blank and Patricia Ruggles, "Short-Term Recidivism Among Public Assistance Recipients," *American Economic Review*, vol. 84 (2), May 1994; pp. 49–53; and Peter David Brandon, "Vulnerability to Future Dependence Among Former AFDC Mothers," Institute for Research on Poverty discussion paper DP1005-95, University of Wisconsin, Madison, Wis., 1995.

3. U.S. Department of Commerce, Census Bureau, "Income, Poverty and Valuation of Noncash Benefits," *Current Population Reports*, 1995, pp. 60–188, p. D-22.

AID TO DEPENDENT CORPORATIONS: **39**
Exposing Federal Handouts to the Wealthy

Chuck Collins

In 1992 rancher J. R. Simplot of Grandview, Idaho paid the U.S. government $87,000 for grazing rights on federal lands, about one-quarter the rate charged by private landowners. Simplot's implicit subsidy from U.S. taxpayers, $261,000, would have covered the welfare costs of about 60 poor families. With a net worth exceeding $500 million, it's hard to argue that Simplot needed the money.

Since 1987, American Barrick Resources Corporation has pocketed $8.75 billion by extracting gold from a Nevada mine owned by the U.S. government. But Barrick has paid only minimal rent to the Department of the Interior. In 1992 Barrick's founder was rewarded for his business acumen with a $32 million annual salary.

Such discounts are only one form of corporate welfare, dubbed "wealthfare" by some activists, that U.S. taxpayers fund. At a time when Congress is attempting to slash or eliminate the meager benefits received by the poor, we are spending far more to subsidize wealthy corporations and individuals. Wealthfare comes in five main varieties: discounted user fees for public resources; direct grants; corporate tax reductions and loopholes; giveaways of publicly funded research and development (R&D) to private profit-making companies; and tax breaks for wealthy individuals.

Within the Clinton administration Secretary of Labor Robert Reich and Budget Director Alice Rivlin have attacked "welfare for the rich." Armed with a study from the Progressive Policy Institute, the Democratic Leadership Council's think tank, Reich floated the notion that over $200 billion in corporate welfare could be trimmed over the next five years. In a sign of the problems with our two-party system, Clinton discouraged Reich from taking this campaign further, for fear of alienating big Democratic Party funders.

From: *Dollars & Sense* 1 (May/June 1995): 15–17, 40. Reprinted by permission.

TAX AVOIDANCE

The largest, yet most invisible, part of wealthfare is tax breaks for corporations and wealthy individuals. The federal Office of Management and Budget (OMB) estimates that these credits, deductions, and exemptions, called "tax expenditures," will cost $440 billion in fiscal 1996. This compares, for example, to the $16 billion annual federal cost of child support programs.

Due both to lower basic tax rates and to myriad loopholes, corporate taxes fell from one-third of total federal revenues in 1953 to less than 10% today (see "Disappearing Corporate Taxes," *Dollars & Sense,* July 1994). Were corporations paying as much tax now as they did in the 1950s, the government would take in another $250 billion a year—more than the entire budget deficit.

The tax code is riddled with tax breaks for the natural resource, construction, corporate agri-business, and financial industries. Some serve legitimate purposes, or did at one time. Others have been distorted to create tax shelters and perpetuate bad business practices. During the 1993 budget battle, New Jersey Senator Bill Bradley attacked the "loophole writing" industry in Washington, where inserting a single sentence into the tax laws can save millions, even billions, in taxes for a corporate client.

Depreciation on equipment and buildings, for example, is a legitimate business expense. But the "accelerated depreciation" rule allows corporations to take this deduction far faster than their assets are wearing out. This simply lets businesses make billions of dollars in untaxed profits. One estimate is that this loophole will cost $164 billion over the next five years.

One particularly generous tax break is the foreign tax credit, which allows U.S.-based multinational corporations to deduct from their U.S. taxes the income taxes they pay to other nations. Donald Bartlett and James Steele, authors of *America: Who Really Pays the Taxes,* say that by 1990 this writeoff was worth $25 billion a year.

While in many cases this credit is a valid method of preventing double taxation on profits earned overseas, the oil companies have used it to avoid most of their U.S. tax obligations. Until 1950, Saudi Arabia had no income tax, but charged royalties on all oil taken from their wells. Such royalties are a payment for use of a natural resource. They are a standard business expense, payable *before* a corporation calculates the profits on which it will pay taxes.

These royalties were a major cost to ARAMCO, the oil consortium operating there (consisting of Exxon, Mobil, Chevron, and Texaco). But since royalties are not income taxes, they could not be used to reduce Exxon and friends' tax bills back home.

When King Saud decided to increase the royalty payments, ARAMCO convinced him to institute a corporate income tax and to substitute this for the royalties. The tax was a sham, since it applied only to ARAMCO, not to any other business in Saudi Arabia's relatively primitive economy. The result was that the oil companies avoided hundreds of millions of dollars in their American taxes. Eventually the other oil-producing nations, including Kuwait, Iraq, and Nigeria, followed suit, at huge cost to the U.S. Treasury.

In contrast to the ARAMCO problem, many corporate executive salaries should not be counted as deductible expenses. These salaries and bonuses are often so large today that they constitute disguised profits. Twenty years ago the average top executive made 34 times the wages of the firm's lowest paid workers. Today the ratio is 140 to 1. The Hospital Corporation of America, for example, paid its chairman $127 million in 1992—$61,000 an hour! In 1993 the Clinton administration capped the deductibility of salaries at $1 million, but the law has several loopholes that allow for easy evasion.

CHICKEN McNUGGETS
AND OTHER VITAL MATTERS

Taxes are but one form of wealthfare. Subsidized use of public resources, as with J. R. Simplot's grazing and American Barrick's mining, is also widespread. Barrick's profit-making was allowed by the General Mining Law of 1872. Just last year the government finally put a one-year moratorium on this resource-raiding.

In a manner similar to the mining situation, the U.S. Forest Service under-charges timber companies for the logs they take from publicly owned land. The Forest Service also builds roads and other infrastructure needed by the timber industry, investing $140 million last year.

Many corporations also receive direct payments from the federal government. The libertarian Cato Institute argues that every cabinet department "has become a conduit for government funding of private industry. Within some cabinet agencies, such as the U.S. Department of Agriculture and the Department of Commerce, almost every spending program underwrites private business."

Agriculture subsidies typically flow in greater quantities the larger is the recipient firm. Of the $1.4 billion in annual sugar price supports, for example, 40% of the money goes to the largest 1% of firms, with the largest ones receiving more than $1 million each.

The Agriculture Department also spends $110 million a year to help U.S. companies advertise abroad. In 1992 Sunkist Growers got $10 million, Gallo

Wines $4.5 million, M&M/Mars $1.1 million, McDonalds $466,000 to promote Chicken McNuggets, and American Legend Fur Coats $1.2 million.

The Progressive Policy Institute estimates that taxpayers could save $114 billion over five years by eliminating or restricting such direct subsidies. Farm subsidies, for example, could be limited to only small farmers.

The government also pays for scientific research and development, then allows the benefits to be reaped by private firms. This occurs commonly in medical research. One product, the anti-cancer drug Taxol, cost the U.S. government $32 million to develop as part of a joint venture with private industry. But in the end the government gave its share to Bristol-Myers Squibb, which now charges cancer patients almost $1,000 for a three-week supply of the drug.

WHO IS ENTITLED?

Beyond corporate subsidies, the government also spends far more than necessary to help support the lifestyles of wealthy individuals. This largess pertains to several of the most expensive and popular "entitlements" in the federal budget, such as Social Security, Medicare, and the deductibility of interest on home mortgages. As the current budget-cutting moves in Congress demonstrate, such universal programs have much greater political strength than do programs targeted solely at low-income households.

While this broad appeal is essential to maintain, billions of dollars could be saved by restricting the degree to which the wealthy benefit from universal programs. If Social Security and Medicare payments were denied to just the richest 3% of households this would reduce federal spending by $30 to $40 billion a year—more than the total federal cost of food stamps.

Similarly, mortgage interest is currently deductible up to $1 million per home, justifying the term "mansion subsidy" for its use by the rich. The government could continue allowing everyone to use this deduction, but limit it to $250,000 per home. This would affect only the wealthiest 5% of Americans, but would save taxpayers $10 billion a year.

Progressive organizations have mounted a renewed focus on the myriad handouts to the corporate and individual rich. One effort is the Green Scissors coalition, an unusual alliance of environmental groups such as Friends of the Earth, and conservative taxcutters, such as the National Taxpayers Union. Last January Green Scissors proposed cutting $33 billion over the next ten years in subsidies that they contend are wasteful and environmentally damaging. These include boondoggle water projects, public land subsidies, highways, foreign aid projects, and agricultural programs.

Another new organization, Share the Wealth, is a coalition of labor, religious, and economic justice organizations. It recently launched the "Campaign for Wealth-Fare Reform," whose initial proposal targets over $35 billion in annual subsidies that benefit the wealthiest 3% of the population. The campaign rejects the term "corporate welfare" because it reinforces punitive anti-welfare sentiments. Welfare is something a humane society guarantees to people facing poverty, unemployment, low wages, and racism. "Wealthfare," in contrast, is the fees and subsidies extracted from the public by the wealthy and powerful—those who are least in need.

Today's Congress is not sympathetic to such arguments. But the blatant anti-poor, pro-corporate bias of the Republicans has already begun to awaken a dormant public consciousness. This will leave more openings, not less, for progressives to engage in public education around the true nature of government waste.

References

Bartlett, Donald, and James Steele. 1994. *America: Who Really Pays the Taxes?* New York: Simon & Schuster.

Crittendon, Anne. 1993. *Killing the Sacred Cows.* New York: Penguin Books.

Friends of the Earth. 1995. *Green Scissors Report: Cutting Wasteful and Environmentally Harmful Spending and Subsidies.* Washington, DC: Green Scissors Campaign of Citizens United to Terminate Subsidies.

Shapiro, Robert. 1994. *Cut-and-Invest to Compete and Win: A Budget Strategy for American Growth.* Washington, DC: Progressive Policy Institute.

Shields, Janice C. 1994. *Aid to Dependent Corporations.* Washington, DC: Essential Information.

THE BRUTALITY OF THE BUREAUCRACY   **40**

Theresa Funiciello

Poverty is the worst form of violence.

—Mahatma Gandhi

In our own time, attention to experience may signal that the greatest threat to due process principles is formal reason severed from the insights of passion. . . . In the bureaucratic welfare state of the late twentieth century, it may be that what we need most acutely is the passion that understands the pulse of life beneath the official version of events . . . the characteristic complaint of our time seems to be not that government provides no reasons, but that its reasons often seem remote from the human beings who must live with their consequences.

—Justice William J. Brennan, Jr.

Think of the worst experience you've ever had with a clerk in some government service job—motor vehicles, hospital, whatever—and add the life-threatening condition of impending starvation or homelessness to the waiting line, multiply the anxiety by an exponent of 10, and you have some idea of what it's like in a welfare center. You wait and wait, shuttling back and forth in various lines like cattle to the slaughter. You want to wring the workers' necks, but you don't dare talk back. The slightest remark can set your case back hours, days, weeks, or forever. Occasionally someone loses it and starts cursing at the top of her lungs. Then she's carted away by security guards. In the early days, I thought this meant she was getting served faster for having had the nerve to lay it all out. It wasn't long before it became evident that these were among the ones who would be arrested and sometimes beaten up. It's truly amazing that more welfare workers aren't killed; the torment so many of them inflict would break the patience of anyone whose life wasn't on the line. But that's always their ace in the hole. No check, no life.

Babies and other small children are squirming all over the place, and always at least one worker shoots verbal bullets if they cry or run around. Half the time they're hungry as well as bored, but the mother has run out of the food she brought, had none to begin with, or can't leave to get some for fear her number will be called while she's out tracking down an apple, candy bar, or quart of milk. If she yells at the kid, a worker yells at her. If she doesn't yell at the kid, a worker yells at her.

You run the same risk taking one of the kids or yourself to the bathroom, so "accidents" are common. In any event, the bathrooms in welfare centers are not to be believed. The stalls have no doors, and there's rarely any toilet paper. If they are ever cleaned, it must be on holidays; they are putrid with the stench of weeks-old rags piled up anywhere because there are usually no bins to put them in. You could yearn for the good life in a prison after walking into a welfare center bathroom.

Not all workers are ghouls, but they all have to contend with their own predicaments, too. What happens and what doesn't is in many ways the consequence of the draw—what worker you got on what day of the week. Most workers know little about the actual law governing eligibility for public assistance, and even those who start out with good intentions often get blown away by the ever-changing legal script. Each month, new regulations by the dozens are distributed. Most of the workers are so overwhelmed with the sheer volume of clients that only the truly stalwart keep up with the changes.

On the one side, the department is constantly trying to figure itself out, so as "to minimize waste and abuse." On the other, slews of lawsuits are constantly changing policies out of compliance with state or federal laws. And, because most lawmakers are oblivious of the Constitution as it pertains to poor people, it is not uncommon for laws that are unconstitutional to be passed, put into operation, challenged, and overturned. A welfare worker in Kingston, New York, once told one of our members, "When you sign up for welfare, that's when you sign all your rights away." Though this attitude is pervasive throughout the system, it is simply not true.

If you have the uncanny luck of getting a worker who knows the law, you also have to get that person on a good day—which makes all the difference between being treated decently and being treated inhumanely. Inhumanity, by the nature of the job, is the more common treatment. God forbid your worker had a fight with a family member that morning. If the worker has previously been on welfare herself, she might be more helpful, but it's more likely that she'll be more bitter about having to "work" at this low-paying, essentially dead-end job while someone else makes off with a welfare check. Still other workers have this attitude that the welfare is coming right out of their pockets—an outlook the hierarchy likes to cultivate. Some welfare centers even had a system of rewards for intake workers who denied the most applications or cut off the most recipients in a given period.

When you walk into a center, you have the right to get and file an application. To do that, you must first be deemed worthy of one by the persons whose job it is to hand them out and initiate the herding process. Often people never get past point one, because the worker *who is obligated by law* to give you that application, not to question or determine your eligibility, often refuses to do so. When that happens, people who are rejected at the door, who have no resources to turn to for advice, just fall through the cracks. They are included in no statistics, anywhere. For all practical intents and purposes, from the state's point of view, they do not exist unless they have submitted an application. At DWAC we used to get many people with this problem. There had to be thousands of others, because we were such a small, virtually unknown operation at the time.

Getting the application into the hands of an "intake worker" who will interview you and check that you have proper documentation—like birth certificates, leases, and social security numbers—is the next hurdle. We were convinced that some of the workers were sadists and would find any excuse to humiliate an applicant, like the one who raged that an applicant was using a verboten red pen. The applicant had made a quarter-inch mark on the form, not even completing the first letter of her name. "You can't use red ink," bellowed the application approver nearby. Okay. The applicant switched to black ink and continued. But over an hour later, as the applicant handed the worker the completed form in black ink, she was told that the entire application was no good because of the tiny red mark on it.

This worker, whose job was to look at applications to determine if they were completely filled out, simply refused to pass the form on to the intake worker. She insisted the applicant do the whole thing over again. She could have told her to get a new form the moment she spotted the red mark. She could have simply handed her another. She could have ignored the obviously irrelevant red mark. But this may have been the only part of her life in which she held the reins of power. She obviously liked it.

There is no rational, uniform method for qualifying for welfare. The attitude of the worker prevails on the one end and that of the state on the other, with rules of the federal government (subject to the politics of the moment) having some jurisdiction over all. At a minimum, there are fifty administrative systems for welfare, one in each state. That's then subdivided into whatever local categories the state chooses—like by county. In New York at the time, to qualify for welfare under the best of circumstances, you could have zero dollars in your possession, and if you had outside income (from a job or child support) it had to be less than the maximum you could have gotten on welfare if you had nothing. As a practical matter, if you walked in and said you had fifty dollars to your name, which the worker would know could not possibly last until your case was accepted, you could not qualify for welfare. If you had any asset whatsoever that the worker deemed sellable—like an old wedding ring—it had to be sold. *All* your resources had to be expended before you were eligible. One of the few things that improved for New Yorkers at least, during the Reagan administration, was the establishment of a one-thousand-dollar "resource limit," which meant an applicant could hold on to any asset deemed worth less than a thousand dollars and still qualify as long as all the other income requirements were met.

In states like Mississippi, income of more than $48 a month for a three-person family would, as a practical matter, disqualify them for welfare eligibility. In New York City, the maximum possible benefit for a three-person family at that time was $332. Job or other outside net income would have to be less than that.

When the intake worker gets through with you, the case supervisor normally has the final thumbs up or down. So all the variables that affected your outcome with the intake worker are once again operative with the supervisor. The case supervisor can demand that you see still other workers to approve this or that required document or answer on your seventeen-page application form. You can be sent to a dozen places before a determination is made. Pregnant women virtually ready to give birth must supply a doctor's verification of their pregnancies. Processing the application takes weeks, sometimes even months (in spite of the law, which says a determination must be made within thirty days). Even after you are determined eligible, you can wait days for the money to come through. Whether you are in heartrending, desperate, obvious need with a sick baby on your lap does not matter. People can fake heartrending, desperate, obvious need. That possibility guides the entire department as well as the local, state, and federal policies. For the poor there is no honor system like the one that guides paying taxes. No innocent until proven guilty like in the criminal justice system. No siree, Betsy.

More often than not, it is quite the opposite. There was a woman who had been raped and was being denied welfare for "refusing to comply" by naming the father. Father? Who the hell were they talking about? This young woman was leaving her infant home alone in a makeshift apartment in someone's attic while she worked a few hours a night in an all-night diner. She was taking home $41 a week. She called in tears. She simply couldn't make it on $41 a week.

If the welfare system worked the way it should, it would still not give people enough to live on. But securing and keeping what little it is possible to get would be difficult for an FBI agent. For instance, since the rape victim had only $41 a week for herself and her child, she was eligible for the difference between what she was paid on her job and the amount set for a welfare grant for two people. However, when asked who the father of her child was, she was in an impossible bind. He had told her neither his name nor his address during the rape, and she'd never seen him before in her life. Her inability to supply his name drove the case worker into a paroxysm of doubt. If she had known who the father was and he was not providing for the child, even if they were married she would still have been eligible for welfare.

Once the worker was dissatisfied with the answer to the name-that-father question, whole avenues of questions followed. Among them: had she reported the rape to the police? She hadn't. Now the worker felt justified in rejecting the application or postponing action until the woman went to the police to report a rape that had occurred over a year ago, securing proof of the report in the process. As you might imagine, this intrusion on their day-to-day work does not make the police happy and can and usually does evolve into another hassle.

Though the welfare is required to assist the applicant in obtaining documents that might be difficult for her to obtain on her own, this assistance is rarely given. Reluctantly, she went to the police, who resisted the paperwork. She called me from the police station, frustrated by another bureaucracy. I had to pressure them to give her a statement verifying that she made the report, which they thought was crazy, given the time lapse. Then it was back to the welfare department for the next round. We hoped she had satisfied them, but they wouldn't say. She was told that she would hear by mail.

This time they rejected her for lack of an address at which to contact a man whose name she did not know. And so on and on. In cases like these, the only way an applicant usually gets through the maze is by getting the aggressive intervention of an advocate or friend who knows the rules and how to enforce them or by going to court, which, given the dearth of lawyers who can or will take such a case, is rare.

TRACING THE POLITICS OF AFFIRMATIVE ACTION

41

Linda Faye Williams

The politics of civil rights have never been settled, but current tensions reveal that once again they have reached a particularly feverish pitch. At the eye of the political storm is the nation's affirmative action policy—a fixture on the American political horizon since the mid-1960s.

Affirmative action, born in an atmosphere in which the large political shock of the modern Civil Rights movement temporarily lowered political and institutional barriers to reform, is fast being transformed into the ultimate political wedge issue in the mid-1990s. Confusion, anxiety, and demagoguery pervade the debate over the issue and threaten to envelop U.S. society. . . . In short, in its three-decade-long history, affirmative action has never been more embattled than it is in the mid-1990s. But why have the politics of affirmative action become so explosive? To answer that question, one must analyze broader currents in the American political economy, particularly economic and technological transformation as well as the continuing uses of what has been called the "race card" in U.S. politics.

A LESSON FROM THE PAST

To begin, it is important to understand the political atmosphere in which affirmative action evolved in the first place. An appropriate starting point is the turbulent Civil Rights movement of the 1950s and 1960s and its major concomitant, the Civil Rights Act of 1964. This act ostensibly guaranteed blacks equal opportunity and became the backbone of efforts to eliminate discrimination throughout American society. By allowing private litigation, the act made every victim a monitor of civil rights and put enforcement potential in the hands of those with intimate knowledge of workplaces, educational institutions, and public accommodations.

From: George E. Curry, ed. *The Affirmative Action Debate* (Reading, MA: Addison-Wesley, 1996), pp. 24–57. Reprinted by permission.

On passage of the act, black individuals set out to prove that they had been discriminated against. They soon discovered problems. The difficulty in proving discrimination in an individual case is due in part to effective counterstrategies by those who discriminate. How, for instance, does a black applicant know whether a job really has "just been filled"? How can a black candidate prove that he would have gotten the job in a fair competition if the employer had not hired the son of a friend? What unemployed person looking for a job has the time and resources to assemble proof that she is the victim of racial bias? In short, the difficulty of proving biased intent when employers have the power to cover up such bias reduces the effectiveness of individual suits.[1] Thus, it soon became clear that something more than relying on individuals to show disparate treatment was needed if the United States was ever to actually live up to its new claim of equal opportunity for blacks.

Conceptually, then, the 1964 Civil Rights Act opened a second line of attack. Individuals could sue not only on the basis of individual discrimination, but also on that of systemic (adverse impact) discrimination. Moreover, if government was to generate real progress after centuries of employing its legislative, judicial, and executive powers to enforce the subjugation of blacks, the idea developed that it needed to do something "affirmative." President Lyndon B. Johnson drew the analogy of two racers in a track meet. The nation, Johnson concluded, could never expect two people in a race to have an equal chance of winning the race if one runner started at midpoint while the other began at the starting line. Something must be done to make the race fair. In antidiscrimination law, this meant taking measures that went beyond merely ceasing or avoiding discrimination; it meant taking measures that attempted to undo or compensate for the effects of past discrimination.[2]

Two executive orders enshrined these basic principles—the first by President John F. Kennedy, Executive Order 10925, and the second by President Lyndon Baines Johnson, Executive Order 11246. As Kennedy, who first used the term "affirmative action," saw it, the policy would provide those discriminated against a chance to demonstrate their skills and thus to break the preconceptions on which prejudicial barriers were based. Johnson's executive order began to flesh out what this actually meant in a way that could be implemented. As administered by the Office of Federal Contract Compliance Programs (OFCCP), Executive Order 11246 required all employers with federal contracts (today defined as those in excess of $50,000) to file written affirmative action plans with the government. The OFCCP, which regulates about a third of private employment, could intervene proactively as well to reduce barriers to employment of minorities and women. Plans could be either mandatory or voluntary, however, since many institutions might adopt voluntary plans to gain or retain federal contracts or to enhance their social and political status.

It was Richard M. Nixon's administration that added the requirement that all affirmative action plans must include minority and female hiring goals and timetables to which the contractor must commit its "good-faith" efforts. As then Undersecretary of Labor Laurence Silberman saw it, the notable absence of blacks from the workforce would not be remedied by vague employer promises to look for black applicants, but by setting a specific, reasonable numerical goal for hiring them.[3]

In today's world of a Republican-led attack on affirmative action, it may seem shocking that it was a Republican administration that added the now controversial goals and timetables. But in the political environment of the 1960s, it was not even particularly surprising. First, it should be remembered that before 1964, most poll data demonstrated that the two parties were not distinguished in the public eye vis-à-vis their commitment to civil rights. As Democratic pollster Stanley Greenberg has pointed out, both whites and blacks were about as likely to report that the Republicans were "the party most committed to black interests" as that the Democrats were.[4] In the presidential contest between Kennedy and Nixon in 1960, nearly one out of three blacks voted for Nixon. Playing the race card, albeit a successful strategy in the making, was not as firmly an established route for Republican politicians in the late 1960s and early 1970s as it became over time.

Indeed, affirmative action was a popular enough policy in both the Johnson and Nixon years for more and more groups to be added to its categories of protected classes: other people of color, women, and ultimately the disabled. In fact, by the mid-1970s affirmative action was anything but a race-specific policy; rather, its inclusion of white women meant that a substantial *majority* of Americans were covered by affirmative action.

The expansion of protected classes occurred so consensually in the late 1960s and early 1970s because there was little or no political price to pay at that time for supporting affirmative action. Plans for Progress, a cooperative program of more than 165 *Fortune* 500 corporations engaged in voluntary affirmative action in the recruitment of blacks and other people of color, strongly supported Kennedy's and Johnson's equal opportunity initiatives applicable to government contractors.[5] Polls of the day showed there was no white backlash to affirmative action policies.[6] Instead, at the birth of affirmative action there was widespread agreement throughout American society that the policy could fulfill the necessary function of upgrading the education and skills of a rapidly growing sector of the American labor force. As G. Williams Miller, then president of Textron Inc., put it: "The American economy cannot afford the burden of underdeveloped human resources. While the Civil Rights Act of 1964 sets the national standard, Plans for Progress remains an essential program to assure the affirmative action necessary to translate principle into reality." [7]

The basis for corporate support of affirmative action at its inception is instructive. In a period of tight labor markets, cheap energy supplies, easy access to credit on favorable terms, and presumptions of long-term national prosperity and a rapidly expanding economic pie, affirmative action was not controversial. As James Tobin, one of Kennedy's principal economic advisers, later explained, in the 1960s liberals believed that growth would unendingly provide new resources that could be devoted to the historically disadvantaged, thus expanding and upgrading opportunities for African-Americans, other people of color, and women—all without divisive conflicts over taxes, the size of the public sector, defense spending, and the distribution of income and wealth. [8]

In a nutshell, positive economic conditions encouraged a sanguine view of affirmative action. Then came the deficit spending produced by the war in Vietnam, the initial successes of the OPEC cartel at managing energy supplies, and the concomitant increase in prices and eventually in wage demands. All these combined to help produce a decade-long spiral of increasing prices and rising unemployment, called stagflation. Until these factors arose, however, most American corporations and citizens alike strongly backed affirmative action. Indeed, had the economy not gone into a tailspin, the backlash against affirmative action might never have developed.

THE ECONOMIC ROOTS
OF ANTI–AFFIRMATIVE ACTION POLITICS

The problem, of course, was that the "balanced growth" economy of the mid-1960s rapidly became a slow-growth one. Even when it "recovered," a steadily increasing proportion of Americans faced unemployment and underemployment. For instance, in the 1960s unemployment stood at an average of 4.8%. In the 1970s it rose again, to 6.2%. In the 1980s it averaged 7.3%. In the first four years of the 1990s, unemployment averaged 6.0%. Even allowing for short-term dips in the unemployment rate, the long-term trend is toward higher rates of unemployment. Add to this the fact that the new jobs being created are mostly in low-paying sectors (especially services and trade) and are often contingent employment. Moreover, wages have been steadily declining since 1974. Weighing these factors that have contributed to current economic conditions, one begins to get at the heart of the developments that undercut public support for affirmative action.

In particular, three key factors have created the economic basis for the political hostility to affirmative action. First, global economic change has produced job loss, especially in the relatively high-wage manufacturing sector, as a result of foreign competition and cheap labor markets abroad. Fewer and fewer decent employment opportunities helped transform the image of affirmative action from a distributive policy to a redistributive one. As white Americans began to think of affirmative action as a fixture in a zero-sum game, their views began to shift.

The second key factor, automation, has produced a much more formidable and perhaps intractable problem and hastened the growth of redistributive anxieties. The supposedly exciting new world of high-tech automated production has replaced human beings with intelligent machines in countless tasks, forcing millions of blue- and white-collar workers into unemployment lines or, for many, breadlines. While industrial workers have been the hardest hit to date, it is clear that automation and reengineering will ultimately replace a wide swath of service jobs as well. Even as the U.S. economy rebounded in the years since 1992, half a million additional clerical and technical jobs disappeared. Rapid advances in computer technology, including parallel processing and artificial intelligence, are likely to make many more white-collar workers redundant by the early decades of the next century.

According to Jeremy Rifkin, the all too few good jobs that are becoming available in the new high-tech global economy are in the knowledge sector: for example, openings for physicists, computer scientists, high-level technicians, molecular biologists, business consultants, lawyers, and accountants. The gap between those who have attained the educational levels required to hold these jobs and the growing number of those at the bottom who need jobs is so wide that only the very naive could believe that short-term retraining programs will adequately upgrade the performance of workers to match the kind of limited professional employment opportunities that are growing. More frightening, even if further education and retraining could be implemented on a mass scale, there might not be enough high-tech jobs available in the automated economy of the next century to absorb the vast numbers of dislocated workers.[9] Nobel laureate economist Wassily Leontief and his colleague Faye Duchin draw a scary comparison: "The role of humans as the most important factor of production is bound to diminish in the same way that the role of horses in agricultural production was first diminished and then eliminated by the introduction of tractors."[10] As good jobs become more scarce, whites—the group accustomed to occupying the best jobs—become less and less willing to share them with members of other groups.

The third key factor producing a bottoming out of opportunities for American workers and undercutting support for affirmative action was the economic policies of the Reagan-Bush era. As Kevin Phillips has argued persuasively, the Reagan administration's reduction of taxes for the rich, budgetary policies that amounted to shrinking domestic spending while increasing defense spending, deregulation, and monetary policies all benefited the top fifth of the population (and especially the top 5%) at the expense of everybody else. These policies were especially detrimental to members of the working class, who lost more through program cuts than they gained in tax cuts. Concomitantly, top-bottom income polarization deepened as the working and middle classes got poorer and the rich got richer. The big losers were workers in unionized and formerly regulated industries who found their wages cut and work environments changed.[11]

In a society where class has always been little more than a dirty secret, workers were poised to turn against each other on racial grounds rather than turn against capital. The politics of the Reagan-Bush era sought to guarantee this response. Working- and middle-class Americans were encouraged to divert their attention away from the global economic and technological roots of their problems and instead to affirmative action, other antidiscrimination policies, social programs for the poor, and immigration. Ronald Reagan in particular argued that racism was nearly a thing of the past and that most antidiscrimination measures, particularly affirmative action, were in effect reverse discrimination against white men.

NOTES

1. Gertrude Ezorsky, "Individual Candidate Remedies: Why They Won't Work," in *Moral Rights in the Workplace*, ed. Gertrude Ezorsky (Albany: State University of New York Press, 1987), 259–63.

2. Arthur Larson, "Affirmative Action," *The Guide to American Law* (New York: West, 1983).

3. Ezorsky, "Individual Candidate Remedies," 261.

4. Stanley Greenberg, *Middle Class Dreams: The Politics and Power of the New American Majority* (New York: Times Books, 1995), 41.

5. White House's Committee on Equal Employment Opportunity "Plans for Progress," first-year Report of the President's Committee on Equal Employment Opportunity (August 1964).

6. Andrew Hacker, *Two Nations: Black and White, Separate, Hostile, Unequal* (New York: Ballantine, 1995), 114.

7. Ibid.

8. James Tobin, *Politics for Prosperity* (Cambridge: MIT Press, 1987), 422.

9. Jeremy Rifkin, *The End of Work: The Decline of Mass Labor in the Production of Goods and Services* (New York: Putnam, 1994), 3–4.

10. Wassily Leontief and Faye Duchin, *The Future Impact of Automation on Workers* (New York: Oxford University Press, 1986).

11. Kevin Phillips, *The Politics of Rich and Poor: Wealth and the American Electorate in the Reagan Aftermath* (New York: Random House, 1990), chap. 4.

IV

Analyzing Social Issues

In the United States, public debates about issues such as multicultural education, abortion, school prayer, and welfare reform all reflect race, class, and gender arrangements in society. Race, class, and gender are systemic features of American society and do not change rapidly; however, the issues they generate change over time. Thinking inclusively about race, class, and gender can help us analyze contemporary social issues in at least three ways.

First, themes that emerge as social issues—affirmative action, immigration, and welfare reform, among others—are products of social institutions and institutional policies reflecting race, class, and gender arrangements. Since social institutions themselves are framed by race, class, and gender politics, social issues reflect these institutional politics. The organization of health care in the United States offers a compelling instance of how the politics of race, class, and gender operate. Depending on their placement in the race, class, and gender system, different groups have widely varying patterns of health and illness. Something as fundamental as who gets to live and who is left to die reflects the politics of race, class, and gender. Death rates for diabetes, for example—a disease exacerbated by poor nutrition and health care—are noticeably higher among African American, Latina, and Native American women than among Asian Americans and Whites. Access to health care also reflects the politics of race, class, and gender. People of color are much less likely than Whites to have health insurance. About 10 percent of Whites report they have no health coverage, compared to 17 percent of Asians, 20 percent of African Americans and Native Americans, and 33 percent of Latinos (O'Hare 1992: 17). Race, class, and gender also shape the contours of health research and health care delivery. For example, differences in funding for breast cancer research and heart disease research speak to the differential power of men and women in the health care system. Similarly, funding for infertility treatments for middle-class White couples is quite different from ongoing efforts to control the fertility of poor women via funding for sterilization.

Second, how social issues are presented and understood also operates within a race-, class-, and gender-inclusive framework. Media representations of homelessness provide an example of how dominant ideologies frame American understandings of important social issues. Increasing levels of underemployment coupled with a housing shortage of affordable, low-income housing pushed many people into poverty in the 1980s. The result was a large increase in the number of homeless people in the United States. But while the social issue of homelessness affected individuals and families across race, class, and gender groups, not all segments of the homeless population were seen as equally worthy of assistance. Building on longstanding notions of the deserving and undeserving poor, newly homeless families—especially women with dependent children—garnered more sympathy than groups long seen as part of the undeserving poor—namely, single men, persons with addictions, veterans, and the mentally ill. Moreover, racial politics intersected with class and gender to subdivide the homeless further: White families were seen as more deserving of aid than people of color, especially African Americans. Moreover, despite the centrality of ideologies of race, class, and gender in shaping understandings of homelessness, the importance of race, class, and gender to these understandings typically remains hidden. This is because the same social institutions that are organized through race, class, and gender simultaneously use ideologies to obscure how homelessness and other social issues reflect the politics of race, class, and gender.

Third, since current arrangements of race, class, and gender create social issues, an inclusive framework is required to effectively address such issues. It is not enough to critique existing social structures; we must also use our understanding of race, class, and gender to analyze existing social issues. For example, how might the United States redesign health care to make it equitable and responsive to all citizens? How might American communities provide safe, affordable housing so that everyone has a home? Not only do social institutions structure and legitimate existing race, class, and gender power relations, but individuals and groups use these same social institutions as sites of struggle. Thinking inclusively allows us to create alternative analyses of social issues and more effectively use our social locations to effect change.

As an example, consider the environment. Thinking narrowly about the environment leaves many people of color with the perception that the environmental movement is more concerned with protecting animals and other wildlife than with protecting people. In which types of communities do polluting industries locate? What is the least desirable land, and who is forced to live there? Some observers argue that communities with high concentrations of poor people and/or people of color incur more of the negative costs associated with industrial development than do more affluent communities. A more broadly defined environmental movement—such as the environmental justice movement, a national coalition of grass-roots organizations from all types of communities—recognizes that uses of natural resources reflect institutional decision making, and it takes into account all people and their relationship to the environment.

In Part IV, we include readings on three significant social issues that are currently contested in the United States: American national identity, sexuality, and violence. Although these three issues are not the only ones meriting attention, they remain prominent issues of public debate in the United States, and they can only be fully understood if they are located within a race, class, and gender analysis. Contemporary discussions about what constitutes American identity and culture, about the changing meaning of sexuality, and about the significance of violence are framed by implicit and explicit assumptions that stem from competing ideologies of race, class, and gender. How do race, class, and gender help our analysis of these important social issues? Conversely, how do these issues aid our understanding of the politics of race, class, and gender?

RETHINKING AMERICAN NATIONAL IDENTITY

Despite the ideology of the "melting pot," American national identity has been closely linked to a history of racial privilege. The term "American" is assumed to mean White; other types of "Americans," such as African Americans and Asian Americans, become distinguished from the "real" Americans by virtue of their race. "To be an 'all-American' is, by definition, *not* to be an Asian American, Pacific American, American Indian, Latino, Arab American, or African American," observes political theorist Manning Marable (Marable 1995: 364). More importantly, certain benefits are reserved for those deemed to be deserving "Americans."

This longstanding view of American national identity and of the social rewards that should accompany American citizenship is being challenged by the changing contours of the U.S. population. The overall percentage of the American population consisting of people of color is changing. Between 1900 and 1960, this percentage increased slightly, from 13.1 to 14.9 percent; between 1960 and 1990, it nearly doubled, to approximately 25 percent. In 1992, the combined population of Native Americans, African Americans, Latinos, and Asian Americans was estimated at 64.3 million; moreover, the composition of people of color in the United States is also undergoing substantial change. Just a few generations ago, African Americans were by far the largest "minority" group. Although still the most numerous, African Americans now account for less than half of all people of color, and their share continues to decline. Latino and Asian American populations are growing more rapidly than African American and Native American populations (O'Hare 1992: 9).

Longstanding perceptions of American national identity have also been challenged by profound global changes. You need only consult an atlas from the 1980s to get a sense of the widespread changes that characterize the late twentieth century. Political, economic, and social trends such as the dismantling of the Soviet Union, the increasing importance of Pacific Rim nations in global economic development, the rise of nationalisms of all sorts, and the erosion and collapse of socialism and communism have all affected the United States. The emergence of transnational corporations promises to dramatically change all societies, including the United States. Civil wars, rebellions, and economic hardships have led to shifts in national borders.

Changes of this magnitude both within and outside formal American borders raise fundamental questions about American national identity. How will America define itself as a nation-state in the changing global context? How will the increased visibility and vocality of people of color in the United States, in Latin America, in Africa, in Asia, and within the borders of former European colonial powers shape this process? Who will control the cultural representations and institutions that define American culture and transport it across the globe? How can such a diverse American population craft one country from so much difference created by race, class, and gender arrangements? What will it mean to be "American"?

Several themes emerge in rethinking American national identity via the lens of race, class, and gender. One is the issue of Americans who never "melted" into the melting pot. Many people think that race is like ethnicity and that the failure of people of color to shed their cultures and assimilate into the mainstream as White ethnic groups purportedly have done represents an unwillingness to embrace American national identity. But as Mary Waters points out in "Optional Ethnicities: For Whites Only?" this view seriously misreads the meaning of both race and ethnicity in shaping American national identity. Waters suggests that White Americans of European ancestry have a great deal of choice in terms of their ethnic identities and that ethnicity does not influence later-generation White ethnics' lives unless they want it to. Waters contrasts this individualistic *symbolic ethnic identity* among many Whites with a socially enforced and imposed racial identity among African Americans. Because race operates as a physical marker in the United States, intersections of race and ethnicity operate differently for Whites and for people of color.

For Native Americans, the seeming failure to assimilate into the dominant national culture takes on special meaning. Placing his analysis in a global human rights context, Ward Churchill discusses how American national identity has been constructed against the interests of Native Americans. He suggests that, while Native American symbolism and imagery is ubiquitous in the United States, the use of Native American mascots and the like for sports teams continues to harm Native Americans. Ironically, while the original "Americans" experienced and continue to experience genocidal policies, distorted ideas about Native American culture remain in use as symbols in constructing American national identity.

Current trends in African American culture can also be seen as a response to this theme of being an unmeltable racial-ethnic group within contemporary American national identity. Dominant ideologies about race, class, and gender and ideologies of resistance developed by subordinated groups gain institutional expression in language, the media, humor, and music. Institutional structures of racism have restricted Black access to wealth, power, and prestige; African Americans have lacked opportunities to assimilate, to "melt" into the American identity; yet, the continued racial segregation of African Americans has simultaneously provided community structures with their own outcomes. Many young Black people in particular see bleak futures for themselves, are warehoused in deteriorating neighborhoods and schools, and in response have created various cultures of resistance. The refusal to "melt" or assimilate is often accomplished by holding on to culture.

Ironically, while Native Americans and African Americans struggle to broaden definitions of "American" to include their experiences, Asians and Latinos migrate to the United States, in part, in search of an American identity. This leads to a second theme regarding American national identity—namely, the impact that immigration and changing population patterns are having on American culture. Many people of color see the United States as a land of opportunity, and their arrival and struggles to fit into existing structures of race, class, and gender reveal both the tenacity of these structures and how the structures are being changed. For example, Sucheng Chan's description of her experiences of immigration and of "becoming an American" reveals how American identity looks different to those choosing to migrate and to those lacking these choices. Recent migrations by people of color have fostered a corresponding discussion of the rights that accrue to citizens. What happens if citizens are defined as "White" and if English is the only "American" language? Which benefits should be reserved for American citizens, and which citizens should receive these benefits? In the United States, this process is profoundly affected by race, class, and gender.

Changing gender relations in the United States present yet another challenge to the process of rethinking American national identity. While the term "American" historically has meant "White," it has also been cast in gendered terms. Efforts to come to terms with contested American racial identity occur in conjunction with a renegotiated gender identity. People with the power to define reality create dominant ideologies reflecting their own interests, as Gloria Steinem reminds us in "If Men Could Menstruate." These ideologies shape all social institutions, especially ones that are central to reinforcing beliefs. Changes in gender politics resulting from feminist activism in the 1970s and 1980s have also fostered an ideological "backlash" against women that can take race- and ethnic-specific forms. In "From 'Kike' to 'JAP': How Misogyny, Anti-Semitism, and Racism Construct the 'Jewish American Princess,'" Evelyn Torton Beck demonstrates how the emergence of the stereotype of the so-called Jewish American Princess reflects the race, class, and gender politics characteristic of this particularly contentious period of American history.

While many social issues intersect with this overarching theme of rethinking American national identity, two social issues present particularly difficult challenges within this process. These challenges are raised by the changing nature of sexuality and the role of violence in American national identity and culture.

SEXUALITY

As a critical part of structures of race, class, and gender, sexuality in American culture has emerged as a central social issue in contemporary American life. Sexuality links systems of race, class, and gender, and can be seen as a dimension of each. For example, constructing images about Black sexuality was central to maintaining institutional racism. Similarly, beliefs about women's sexuality operate in structuring gender oppression. At the same time, sexuality operates as a system of oppression comparable to the systems of race, class, and gender. *Heterosexism*, the institutionalized structures and beliefs that define and enforce heterosexual behavior as the only natural and permissible form of sexual expression, can be seen as a system of oppression that affects everyone. These two meanings of sexuality are linked: manipulating sexualities in structuring systems of race, class, and gender occurs in conjunction with making assumptions of heterosexism and establishing definitions of so-called normal sexuality.

Currently, both meanings of sexuality—namely, sexuality as a dimension of race, class, and gender oppression and the rethinking of sexualities via challenges to heterosexism—are being contested in American culture. June Jordan refers to this situation as the "Politics of Sexuality" and posits that it "subsumes all of the different ways in which some of us seek to dictate to others of us what we should do, what we should desire, what we should dream about, and how we should behave ourselves, generally." For Jordan, race, class, gender, and sexuality are linked: challenging one necessarily involves challenging the rest. As she points out, "Freedom is indivisible, and either we are working for freedom or you are working for the sake of your self-interests and I am working for mine."

The interview by Amy Gluckman and Betsy Reed with longtime activist Barbara Smith explores race, class, and gender in the context of sexual oppressions. Challenging views that collapse race, class, gender, and sexuality into one another as interchangeable substitutes, Smith carefully identifies some of the ways in which race and sexuality, for example, are not similar. Smith notes that racial oppression is historically and structurally embedded in the founding of the United States, whereas lesbian and gay oppression was not. Differences in historical paths and varying structural arrangements should not blind us, however, to the similarities that link targets of racial oppression and homophobia. Smith notes that similar strategies are used to oppress people by race and by sexuality. As she succinctly puts it, "Our enemies are the same. That to me is the major thing that should be pulling us together."

In "The Beauty Myth," Naomi Wolf analyzes how sexuality is linked to the social-class system. As she points out, women's sexuality is used to sell products and is itself turned into a commodity for sale on the open market. Wolf emphasizes how the commodification and marketing of women's sexuality works to limit fully human, erotic love relationships among women and men. By keeping heterosexual men and women estranged and fearful concerning their heterosexual identities, businesses create a market for their products.

Dana Takagi's article problematizes the question of sexuality within Asian American groups as constructed under heterosexism and racism. Using the term *sexualities* to refer to the "variety of practices and identities that range from homoerotic to heterosexual desire," Takagi opens up the kind of conceptual space for discussion of sexualities called for by June Jordan. By focusing on issues shaping the construction of a lesbian and Asian identity, Takagi uses this particular social location to question how a broader conceptualization of sexuality might be constructed within the confines of racism and heterosexism.

In "Getting Off on Feminism," Jason Schultz takes on a similar question. Where Takagi focuses on lesbian sexuality and Asian American populations, Schultz questions White heterosexual male sexualities. He points to the narrow range of sexualities that are available to men, especially middle-class White men, under heterosexism. "Just as hetero women are often forced to choose between the images of the virgin and the whore, modern straight men are caught in a cultural tug-of-war between the Marlboro Man and the Wimp," Schultz observes. Rather than bemoan this dilemma, Schultz describes how he and his male friends tried to explore new ways of thinking about heterosexual masculinity.

VIOLENCE AND SOCIAL CONTROL

The escalating level of violence permeating the United States has emerged as another social issue that challenges the very terms of American national identity. Rather than being the spontaneous result of individual acts, violence represents a device of social control used to maintain systems of race, class, gender, and sexual inequality. Within social institutions organized around the politics of race, class, and gender, actual or threatened violence has long been essential in maintaining oppression. When people resist, violence is often used to keep them in their place.

Many of the current forms of violence are directly linked to inequality. For example, many people argue that pornography is a form of violence against women: it represents the link between commodified sexuality and violence, because both sell products. Joining sexuality and violence in this particular commodity fuels the social-class system's ongoing search for new consumer markets. The fact that women and children are used in this industry further reflects inequalities of gender and age.

The arming of America via gun sales and promotion offers another example of the marketing of violence within American culture and invites inquiry into the consequences that this practice can have. Unfortunately, violence often hurts the most vulnerable segments of society. For example, violence against children has reached alarming proportions. Homicide is the fourth leading cause of death for children 1 to 4 years old and the third leading cause of death for 5- to 14-year-olds. These data are heightened for African American children; homicide is the second leading cause of death for Black children of ages 1 to 14; a Black child dies from a gunshot wound every six hours (Robinson and Briggs 1993).

It is important to see violence not as acts of individual social deviance but as one outcome of how the politics of race, class, and gender permeate a range of social institutions. Violent acts, the threat of violence, and more generalized policies based on the use of force find organizational homes in designated institutions of social control—primarily the police, the military, and the criminal justice system. Violence simultaneously permeates a range of other social institutions. Some social institutions are known for their use of force, while others are less peaceful than they appear. Men's violence against women and children (whether physical, emotional, or sexual) in families of all social classes and racial/ethnic compositions contradicts the dominant belief that the home is a place of tranquility and love. Sexual harassment on the job as a dimension of social control belies the belief that men and women encounter a similar work environment. In the media, depicting rape as pleasurable to women and portraying violence against Native Americans, Asian Americans, and African Americans in numerous movies as justified contributes to a generalized belief system condoning violence. Thus, violence is simultaneously concentrated and diffuse.

One prominent theme in the readings on "Violence and Social Control" involves the links between individual acts of violence and more routinized, systemic violence. Lynchings of African American men can be seen as the random acts of unruly mobs; yet, the failure to arrest, prosecute, and convict those who did the lynching is powerful testament to public endorsement of these seemingly individual acts of violence. Rape appears to be a private act, but as Patricia Yancey Martin and Robert Hummer compellingly explore in "Fraternities and Rape on Campus," rapes occur as part of a generalized climate that condones violence against women. Although violence may be experienced individually, it occurs in specific organizational and institutional contexts. Race, class, and gender are fundamental determinants of organizational policies that foster social control. As Elijah Anderson's work points out, African American men's experience with the police exemplifies the differential treatment that individuals receive based on race, class, and gender.

Investigating violence and social control also reveals the interconnected nature of dominant ideologies of masculinity, the individualized and systemic forms of violence they justify, and the mechanisms of social control that support masculinity. "More Power Than We Want: Masculine Sexuality and Violence," by Bruce Kokopeli and George Lakey, investigates these links and elaborates on Martin's and Hummer's discussion of the connection between masculine identity and campus rape and on Hall's analysis of lynching and white masculinity. Although we include no articles on violence against gays and lesbians, homophobic violence and violence against women are both ways of enforcing interlocking belief systems concerning heterosexism and masculinity. Violence based on race, gender, and heterosexism are parts of one system of social control: all of these types of violence aim to reinforce systems of privilege.

In a sense, the escalating violence within American society is a curious outcome of these same conditions of inequality. Renegotiating a new American national identity and culture that no longer has the certainty of fixed race, class, and gender categories creates change, confusion, and often violence. When people are wronged, they may use violence as a weapon. Conflict that escalates into violence can characterize relationships among groups disadvantaged within existing race, class, and gender arrangements—sometimes with tragic consequences, as was the case with the Los Angeles riot in 1992. Sumi Cho's article "Korean Americans vs. African Americans: Conflict and Construction" examines some of the tensions among Black people and Koreans. She grounds her analysis in the politics of race, class, and gender, helping us rethink American national identity and culture.

Sexuality and violence are but two of the social issues that intertwine with the need to rethink American national identity in the context of globalization. How can the United States build a national community that acknowledges the differences within its population created by race, ethnicity, class, gender, sexuality, and citizenship status? This final and in some ways most central theme to rethinking American national identity requires avoiding two problematic stances to social change. On the one hand, structures where Whiteness, masculinity, and wealth set the standards by which everyone else is judged cannot continue. Although this way of building community—namely, placing one group in the center and requiring everyone else to try to be like them—can foster a certain sense of certainty within the American national community, it is unjust. Social issues such as oppressive sexualities and societal violence emerge in this context. On the other hand, the politics of self-interest, where each group lobbies only for itself and cares little about the American national community as a whole, is equally undesirable. Crafting new views of community that take race, class, and gender relations, among others, into account while aiming to construct an American national identity in which democracy is full realized is a vital task that promises to engage us all for some time.

References

Marable, Manning. 1995. "Beyond Racial Identity Politics: Toward a Liberation Theory for Multicultural Democracy." In *Race, Class, and Gender: An Anthology*, 2d ed., Eds. Margaret L. Anderson and Patricia Hill Collins. Belmont, CA: Wadsworth.

O'Hare, William P. 1992. "America's Minorities—The Demographics of Diversity." *Population Bulletin* 47 (4) (December). Washington, D.C.: Population Reference Bureau.

Robinson, Lori S., and Jimmie Briggs. 1993. "Kids and Violence." *Emerge* 5 (November): 44–50.

American Identity and Culture

OPTIONAL ETHNICITIES:

For Whites Only?

Mary C. Waters

42

What does it mean to talk about ethnicity as an option for an individual? To argue that an individual has some degree of choice in their ethnic identity flies in the face of the commonsense notion of ethnicity many of us believe in— that one's ethnic identity is a fixed characteristic, reflective of blood ties and given at birth. However, social scientists who study ethnicity have long concluded that while ethnicity is based on a *belief* in a common ancestry, ethnicity is primarily a *social* phenomenon, not a biological one (Alba 1985, 1990; Barth 1969; Weber [1921] 1968, p. 389). The belief that members of an ethnic group have that they share a common ancestry may not be a fact. There is a great deal of change in ethnic identities across generations through intermarriage, changing allegiances, and changing social categories. There is also a much larger amount of change in the identities of individuals over their lives than is commonly believed. While most people are aware of the phenomenon known as "passing"—people raised as one race who change at some point and claim a different race as their identity—there are similar life course changes in ethnicity that happen all the time and are not given the same degree of attention as "racial passing."

From: Silvia Pedraza and Rubén G. Rumbaut, eds. *Origins and Destinies: Immigration, Race and Ethnicity in America* (Belmont, CA: Wadsworth, 1996), pp. 444–54. Reprinted by permission.

White Americans of European ancestry can be described as having a great deal of choice in terms of their ethnic identities. The two major types of options White Americans can exercise are (1) the option of whether to claim any specific ancestry, or to just be "White" or American, [Lieberson (1985) called these people "unhyphenated Whites"] and (2) the choice of which of their European ancestries to choose to include in their description of their own identities. In both cases, the option of choosing how to present yourself on surveys and in everyday social interactions exists for Whites because of social changes and societal conditions that have created a great deal of social mobility, immigrant assimilation, and political and economic power for Whites in the United States. Specifically, the option of being able to not claim any ethnic identity exists for Whites of European background in the United States because they are the majority group—in terms of holding political and social power, as well as being a numerical majority. The option of choosing among different ethnicities in their family backgrounds exists because the degree of discrimination and social distance attached to specific European backgrounds has diminished over time. . . .

SYMBOLIC ETHNICITIES FOR WHITE AMERICANS

What do these ethnic identities mean to people and why do they cling to them rather than just abandoning the tie and calling themselves American? My own field research with suburban Whites in California and Pennsylvania found that later-generation descendants of European origin maintain what are called "symbolic ethnicities." Symbolic ethnicity is a term coined by Herbert Gans (1979) to refer to ethnicity that is individualistic in nature and without real social cost for the individual. These symbolic identifications are essentially leisure-time activities, rooted in nuclear family traditions and reinforced by the voluntary enjoyable aspects of being ethnic (Waters 1990). Richard Alba (1990) also found later-generation Whites in Albany, New York, who chose to keep a tie with an ethnic identity because of the enjoyable and voluntary aspects to those identities, along with the feelings of specialness they entailed. An example of symbolic ethnicity is individuals who identify as Irish, for example, on occasions such as Saint Patrick's Day, on family holidays, or for vacations. They do not usually belong to Irish American organizations, live in Irish neighborhoods, work in Irish jobs, or marry other Irish people. The symbolic meaning of being Irish American can be constructed by individuals from mass media images, family traditions, or other intermittent social activities. In other words, for later-generation White ethnics, ethnicity is not something that influences their lives unless they want it to. In the world of work and school and neighborhood, individuals do not have to admit to being ethnic unless they choose to. And for an increasing number of European-origin individuals whose parents and grandparents have intermarried, the ethnicity they claim is largely a matter of personal choice as they sort through all of the possible combinations of groups in their genealogies. . . .

RACE RELATIONS AND SYMBOLIC ETHNICITY

However much symbolic ethnicity is without cost for the individual, there is a cost associated with symbolic ethnicity for the society. That is because symbolic ethnicities of the type described here are confined to White Americans of European origin. Black Americans, Hispanic Americans, Asian Americans, and American Indians do not have the option of a symbolic ethnicity at present in the United States. For all of the ways in which ethnicity does not matter for White Americans, it does matter for non-Whites. Who your ancestors are does affect your choice of spouse, where you live, what job you have, who your friends are, and what your chances are for success in American society, if those ancestors happen not to be from Europe. The reality is that White ethnics have a lot more choice and room for maneuver than they themselves think they do. The situation is very different for members of racial minorities, whose lives are strongly influenced by their race or national origin regardless of how much they may choose not to identify themselves in terms of their ancestries.

When White Americans learn the stories of how their grandparents and great-grandparents triumphed in the United States over adversity, they are usually told in terms of their individual efforts and triumphs. The important role of labor unions and other organized political and economic actors in their social and economic successes are left out of the story in favor of a generational story of individual Americans rising up against communitarian, Old World intolerance, and New World resistance. As a result, the "individualized" voluntary, cultural view of ethnicity for Whites is what is remembered.

One important implication of these identities is that they tend to be very individualistic. There is a tendency to view valuing diversity in a pluralist environment as equating all groups. The symbolic ethnic tends to think that all groups are equal; everyone has a background that is their right to celebrate and pass on to their children. This leads to the conclusion that all identities are equal and all identities in some sense are interchangeable—"I'm Italian American, you're Polish American. I'm Irish American, you're African American." The important thing is to treat people as individuals and all equally. However, this assumption ignores the very big difference between an individualistic symbolic ethnic identity and a socially enforced and imposed racial identity.

My favorite example of how this type of thinking can lead to some severe misunderstandings between people of different backgrounds is from the *Dear Abby* advice column. A few years back a person wrote in who had asked an acquaintance of Asian background where his family was from. His acquaintance answered that this was a rude question and he would not reply. The bewildered White asked Abby why it was rude, since he thought it was a sign of respect to wonder where people were from, and he certainly would not mind anyone asking HIM about where his family was from. Abby asked her readers to write in to say whether it was rude to ask about a person's ethnic background. She reported that she got a large response, that most non-Whites thought it was a sign of disrespect, and Whites thought it was flattering:

> Dear Abby,
> I am 100 percent American and because I am of Asian ancestry I am often asked "What are you?" It's not the personal nature of this question that bothers me, it's the question itself. This query seems to question my very humanity. "What am I? Why I am a person like everyone else!"
>
> Signed, A REAL AMERICAN

> Dear Abby,
> Why do people resent being asked what they are? The Irish are so proud of being Irish, they tell you before you even ask. Tip O'Neill has never tried to hide his Irish ancestry.
>
> Signed, JIMMY.

In this exchange Jimmy cannot understand why Asians are not as happy to be asked about their ethnicity as he is, because he understands his ethnicity and theirs to be separate but equal. Everyone has to come from somewhere—his family from Ireland, another's family from Asia—each has a history and each should be proud of it. But the reason he cannot understand the perspective of the Asian American is that all ethnicities are not equal; all are not symbolic, costless, and voluntary. When White Americans equate their own symbolic ethnicities with the socially enforced identities of non-White Americans, they obscure the fact that the experiences of Whites and non-Whites have been qualitatively different in the United States and that the current identities of individuals partly reflect that unequal history.

In the next section I describe how relations between Black and White students on college campuses reflect some of these asymmetries in the understanding of what a racial or ethnic identity means. While I focus on Black and

White students in the following discussion, you should be aware that the myriad other groups in the United States—Mexican Americans, American Indians, Japanese Americans—all have some degree of social and individual influences on their identities, which reflect the group's social and economic history and present circumstance.

RELATIONS ON COLLEGE CAMPUSES

Both Black and White students face the task of developing their race and ethnic identities. Sociologists and psychologists note that at the time people leave home and begin to live independently from their parents, often ages eighteen to twenty-two, they report a heightened sense of racial and ethnic identity as they sort through how much of their beliefs and behaviors are idiosyncratic to their families and how much are shared with other people. It is not until one comes in close contact with many people who are different from oneself that individuals realize the ways in which their backgrounds may influence their individual personality. This involves coming into contact with people who are different in terms of their ethnicity, class, religion, region, and race. For White students, the ethnicity they claim is more often than not a symbolic one—with all of the voluntary, enjoyable, and intermittent characteristics I have described above.

Black students at the university are also developing identities through interactions with others who are different from them. Their identity development is more complicated than that of Whites because of the added element of racial discrimination and racism, along with the "ethnic" developments of finding others who share their background. Thus Black students have the positive attraction of being around other Black students who share some cultural elements, as well as the need to band together with other students in a reactive and oppositional way in the face of racist incidents on campus.

Colleges and universities across the country have been increasing diversity among their student bodies in the last few decades. This has led in many cases to strained relations among students from different racial and ethnic backgrounds. The 1980s and 1990s produced a great number of racial incidents and high racial tensions on campuses. While there were a number of racial incidents that were due to bigotry, unlawful behavior, and violent or vicious attacks, much of what happens among students on campuses involves a low level of tension and awkwardness in social interactions.

Many Black students experience racism personally for the first time on campus. The upper-middle-class students from White suburbs were often isolated enough that their presence was not threatening to racists in their high schools. Also, their class background was known by their residence and this may have prevented attacks being directed at them. Often Black students at the university who begin talking with other students and recognizing racial slights will remember incidents that happened to them earlier that they might not have thought were related to race.

Black college students across the country experience a sizeable number of incidents that are clearly the result of racism. Many of the most blatant ones that occur between students are the result of drinking. Sometimes late at night, drunken groups of White students coming home from parties will yell slurs at single Black students on the street. The other types of incidents that happen include being singled out for special treatment by employees, such as being followed when shopping at the campus bookstore, or going to the art museum with your class and the guard stops you and asks for your I.D. Others involve impersonal encounters on the street—being called a nigger by a truck driver while crossing the street, or seeing old ladies clutch their pocketbooks and shake in terror as you pass them on the street. For the most part these incidents are not specific to the university environment, they are the types of incidents middle-class Blacks face every day throughout American society, and they have been documented by sociologists (Feagin 1991).

In such a climate, however, with students experiencing these types of incidents and talking with each other about them, Black students do experience a tension and a feeling of being singled out. It is unfair that this is part of their college experience and not that of White students. Dealing with incidents like this, or the ever-present threat of such incidents, is an ongoing developmental task for Black students that takes energy, attention, and strength of character. It should be clearly understood that this is an asymmetry in the "college experience" for Black and White students. It is one of the unfair aspects of life that results from living in a society with ongoing racial prejudice and discrimination. It is also very understandable that it makes some students angry at the unfairness of it all, even if there is no one to blame specifically. It is also very troubling because, while most Whites do not create these incidents, some do, and it is never clear until you know someone well whether they are the type of person who could do something like this. So one of the reactions of Black students to these incidents is to band together.

In some sense then, as Blauner (1992) has argued, you can see Black students coming together on campus as both an "ethnic" pull of wanting to be together to share common experiences and community, and a "racial" push of banding together defensively because of perceived rejection and tension from Whites. In this way the ethnic identities of Black students are in some sense similar to, say, Korean students wanting to be together to share experiences. And it is an ethnicity that is generally much stronger than, say, Italian Americans. But for Koreans who come together there is generally a definition of themselves as "different from" Whites. For Blacks reacting to exclusion, there is a tendency for the coming together to involve both being "different from" but also "opposed to" Whites.

The anthropologist John Ogbu (1990) has documented the tendency of minorities in a variety of societies around the world, who have experienced severe blocked mobility for long periods of time, to develop such oppositional identities. An important component of having such an identity is to describe others of your group who do not join in the group solidarity as devaluing and denying their very core identity. This is why it is not common for successful Asians to be accused by others of "acting White" in the United States, but it is quite common for such a term to be used by Blacks and Latinos. The oppositional component of a Black identity also explains how Black people can question whether others are acting "Black enough." On campus, it explains some of the intense pressures felt by Black students who do not make their racial identity central and who choose to hang out primarily with non-Blacks. This pressure from the group, which is partly defining itself by not being White, is exacerbated by the fact that race is a physical marker in American society. No one immediately notices the Jewish students sitting together in the dining hall, or the one Jewish student sitting surrounded by non-Jews, or the Texan sitting with the Californians, but everyone notices the Black student who is or is not at the "Black table" in the cafeteria.

An example of the kinds of misunderstandings that can arise because of different understandings of the meanings and implications of symbolic versus oppositional identities concerns questions students ask one another in the dorms about personal appearances and customs. A very common type of interaction in the dorm concerns questions Whites ask Blacks about their hair. Because Whites tend to know little about Blacks, and Blacks know a lot about Whites, there is a general asymmetry in the level of curiosity people have about one another. Whites, as the numerical majority, have had little contact with Black culture; Blacks, especially those who are in college, have had to develop bicultural skills—knowledge about the social worlds of both Whites and Blacks. Miscommunication and hurt feelings about White students' questions about Black students' hair illustrate this point. One of the things that happens freshman year is that White students are around Black students as they fix their hair. White students are generally quite curious about Black students' hair—they have basic questions such as how often Blacks wash their hair, how they get it straightened or curled, what products they use on their hair, how they comb it, etc. Whites often wonder to themselves whether they should ask these questions. One thought experiment Whites perform is to ask themselves whether a particular question would upset them. Adopting the "do unto others" rule, they ask themselves, "If a Black person was curious about my hair would I get upset?" The answer usually is "No, I would be happy to tell them." Another example is an Italian American student wondering to herself, "Would I be upset if someone asked me about calamari?" The answer is no, so she asks her Black roommate about collard greens, and the roommate explodes with an angry response such as, "Do you think all Black people eat watermelon too?" Note that if this Italian American knew her friend was Trinidadian American and asked about peas and rice the situation would be more similar and would not necessarily ignite underlying tensions.

Like the debate in *Dear Abby*, these innocent questions are likely to lead to resentment. The issue of stereotypes about Black Americans and the assumption that all Blacks are alike and have the same stereotypical cultural traits has more power to hurt or offend a Black person than vice versa. The innocent questions about Black hair also bring up a number of asymmetries between the Black and White experience. Because Blacks tend to have more knowledge about Whites than vice versa, there is not an even exchange going on, the Black freshman is likely to have fewer basic questions about his White roommate than his White roommate has about him. Because of the differences historically in the group experiences of Blacks and Whites there are some connotations to Black hair that don't exist about White hair. (For instance, is straightening your hair a form of assimilation, do some people distinguish between women having "good hair" and "bad hair" in terms of beauty and how is that related to looking "White"?) Finally, even a Black freshman who cheerfully disregards or is unaware that there are these asymmetries will soon slam into another asymmetry if she willingly answers every innocent question asked of her. In a situation where Blacks make up only 10 percent of the student body, if every non-Black needs to be educated about hair, she will have to explain it to nine other students. As one Black student explained to me, after you've been asked a couple of times about something so personal you begin to feel like you are an attraction in a zoo, that you are at the university for the education of the White students.

INSTITUTIONAL RESPONSES

Our society asks a lot of young people. We ask young people to do something that no one else does as successfully on such a wide scale—that is to live together with people from very different backgrounds, to respect one another, to appreciate one another, and to enjoy and learn from one another. The successes that occur every day in this endeavor are many, and they are too often overlooked. However, the problems and tensions are also real, and they will not vanish on their own. We tend to see pluralism working in the United States in much the same way some people expect capitalism to work. If you put together people with various interests and abilities and resources, the "invisible hand" of capitalism is supposed to make all the parts work together in an economy for the common good.

There is much to be said for such a model—the invisible hand of the market can solve complicated problems of production and distribution better than any "visible hand" of a state plan. However, we have learned that unequal power relations among the actors in the capitalist marketplace, as well as "externalities" that the market cannot account for, such as long-term pollution, or collusion between corporations, or the exploitation of child labor, means that state regulation is often needed. Pluralism and the relations between groups are very similar. There is a lot to be said for the idea that bringing people who belong to different ethnic or racial groups together in institutions with no interference will have good consequences. Students from different backgrounds will make friends if they share a dorm room or corridor, and there is no need for the institution to do any more than provide the locale. But like capitalism, the invisible hand of pluralism does not do well when power relations and externalities are ignored. When you bring together individuals from groups that are differentially valued in the wider society and provide no guidance, there will be problems. In these cases the "invisible hand" of pluralist relations does not work, and tensions and disagreements can arise without any particular individual or group of individuals being "to blame." On college campuses in the 1990s some of the tensions between students are of this sort. They arise from honest misunderstandings, lack of a common background, and very different experiences of what race and ethnicity mean to the individual.

The implications of symbolic ethnicities for thinking about race relations are subtle but consequential. If your understanding of your own ethnicity and its relationship to society and politics is one of individual choice, it becomes harder to understand the need for programs like affirmative action, which recognize the ongoing need for group struggle and group recognition, in order to bring about social change. It also is hard for a White college student to understand the need that minority students feel to band together against discrimination. It also is easy, on the individual level, to expect everyone else to be able to turn their ethnicity on and off at will, the way you are able to, without understanding that ongoing discrimination and societal attention to minority status makes that impossible for individuals from minority groups to do. The paradox of symbolic ethnicity is that it depends upon the ultimate goal of a pluralist society, and at the same time makes it more difficult to achieve that ultimate goal. It is dependent upon the concept that all ethnicities mean the same thing, that enjoying the traditions of one's heritage is an option available to a group or an individual, but that such a heritage should not have any social costs associated with it.

As the Asian Americans who wrote to *Dear Abby* make clear, there are many societal issues and involuntary ascriptions associated with non-White identities. The developments necessary for this to change are not individual but societal in nature. Social mobility and declining racial and ethnic sensitivity are closely associated. The legacy and the present reality of discrimination on the basis of race or ethnicity must be overcome before the ideal of a pluralist society, where all heritages are treated equally and are equally available for individuals to choose or discard at will, is realized.

References

Alba, Richard D. 1985. *Italian Americans: Into the Twilight of Ethnicity*. Englewood Cliffs, NJ: Prentice-Hall.

———. 1990. *Ethnic Identity: The Transformation of White America*. New Haven: Yale University Press.

Barth, Frederick. 1969. *Ethnic Groups and Boundaries*. Boston: Little, Brown.

Blauner, Robert. 1992. "Talking Past Each Other: Black and White Languages of Race." *American Prospect* (Summer): 55–64.

Feagin, Joe R. 1991. "The Continuing Significance of Race: Anti-Black Discrimination in Public Places." *American Sociological Review* 56: 101–17.

Gans, Herbert. 1979. "Symbolic Ethnicity: The Future of Ethnic Groups and Cultures in America." *Ethnic and Racial Studies* 2: 1–20.

Ogbu, John. 1990. "Minority Status and Literacy in Comparative Perspective." *Daedalus* 119: 141–69.

Waters, Mary C. 1990. *Ethnic Options: Choosing Identities in America*. Berkeley: University of California Press.

Weber, Max. [1921]/1968. *Economy and Society: An Outline of Interpretive Sociology*, Eds. Guenther Roth and Claus Wittich, trans. Ephraim Fischoff. New York: Bedminister Press.

CRIMES AGAINST HUMANITY

43

Ward Churchill

If nifty little "pep" gestures like the "Indian Chant" and the "Tomahawk Chop" are just good clean fun, then let's spread the fun around, shall we?

During the past couple of seasons, there has been an increasing wave of controversy regarding the names of professional sports teams like the Atlanta "Braves," Cleveland "Indians," Washington "Redskins," and Kansas City "Chiefs." The issue extends to the names of college teams like Florida State University "Seminoles," University of Illinois "Fighting Illini," and so on, right on down to high school outfits like the Lamar (Colorado) "Savages." Also involved have been team adoption of "mascots," replete with feathers, buckskins, beads, spears, and "warpaint" (some fans have opted to adorn themselves in the same fashion), and nifty little "pep" gestures like the "Indian Chant" and "Tomahawk Chop."

A substantial number of American Indians have protested that use of native names, images, and symbols as sports team mascots and the like is, by definition, a virulently racist practice. Given the historical relationship between Indians and non-Indians during what has been called the "Conquest of America," American Indian Movement leader (and American Indian Anti-Defamation Council founder) Russell Means has compared the practice to contemporary Germans naming their soccer teams the "Jews," "Hebrews," and "Yids," while adorning their uniforms with grotesque caricatures of Jewish faces taken from the Nazis' anti-Semitic propaganda of the 1930s. Numerous demonstrations have occurred in conjunction with games—most notably during the November 15, 1992 match-up between the Chiefs and Redskins in Kansas City—by angry Indians and their supporters.

In response, a number of players—especially African Americans and other minority athletes—have been trotted out by professional team owners like Ted Turner, as well as university and public school officials, to announce that they mean not to insult but to honor native people. They have been joined by the television networks and most major newspapers, all of which have

From: Ward Churchill, "Crimes Against Humanity," *Z Magazine* 6 (March 1993): 43–47. Reprinted by permission of the author.

editorialized that Indian discomfort with the situation is "no big deal," insisting that the whole thing is just "good, clean fun." The country needs more such fun, they've argued, and "a few disgruntled Native Americans" have no right to undermine the nation's enjoyment of its leisure time by complaining. This is especially the case, some have argued, "in hard times like these." It has even been contended that Indian outrage at being systematically degraded—rather than the degradation itself—creates "a serious barrier to the sort of intergroup communication so necessary in a multicultural society such as ours."

Okay, let's communicate. We are frankly dubious that those advancing such positions really believe their own rhetoric, but, just for the sake of argument, let's accept the premise that they are sincere. If what they say is true, then isn't it time we spread such "inoffensiveness" and "good cheer" around among *all* groups so that *everybody* can participate *equally* in fostering the round of national laughs they call for? Sure it is—the country can't have too much fun or "intergroup involvement"—so the more, the merrier. Simple consistency demands that anyone who thinks the Tomahawk Chop is a swell pastime must be just as hearty in their endorsement of the following ideas—by the logic used to defend the defamation of American Indians—should help us all really start yukking it up.

First, as a counterpart to the Redskins, we need an NFL team called "Niggers" to honor Afro-Americans. Half-time festivities for fans might include a simulated stewing of the opposing coach in a large pot while players and cheerleaders dance around it, garbed in leopard skins and wearing fake bones in their noses. This concept obviously goes along with the kind of gaiety attending the Chop, but also with the actions of the Kansas City Chiefs, whose team members—prominently including black team members—lately appeared on a poster looking "fierce" and "savage" by way of wearing Indian regalia. Just a bit of harmless "morale boosting," says the Chiefs' front office. You bet.

So that the newly formed Niggers sports club won't end up too out of sync while expressing the "spirit" and "identity" of Afro-Americans in the above fashion, a baseball franchise—let's call this one the "Sambos"—should be formed. How about a basketball team called the "Spearchuckers"? A hockey team called the "Jungle Bunnies"? Maybe the "essence" of these teams could be depicted by images of tiny black faces adorned with huge pairs of lips. The players could appear on TV every week or so gnawing on chicken legs and spitting watermelon seeds at one another. Catchy, eh? Well, there's "nothing to be upset about," according to those who love wearing "war bonnets" to the Super Bowl or having "Chief Illiniwik" dance around the sports arenas of Urbana, Illinois.

And why stop there? There are plenty of other groups to include. Hispanics? They can be "represented" by the Galveston "Greasers" and San

Diego "Spics," at least until the Wisconsin "Wetbacks" and Baltimore "Bean-ers" get off the ground. Asian Americans? How about the "Slopes," "Dinks," "Gooks," and "Zipperheads?" Owners of the latter teams might get their logo ideas from editorial page cartoons printed in the nation's newspapers during World War II: slant-eyes, buck teeth, big glasses, but nothing racially insult-ing or derogatory, according to the editors and artists involved at the time. Indeed, this Second World War–vintage stuff can be seen as just another bar-rel of laughs, at least by what current editors say are their "local standards" concerning American Indians.

Let's see. Who's been left out? Teams like the Kansas City "Kikes," Han-over "Honkies," San Leandro "Shylocks," Daytona "Dagos," and Pittsburgh "Polacks" will fill a certain social void among white folk. Have a religious belief? Let's all go for the gusto and gear up the Milwaukee "Mackerel Snap-pers" and Hollywood "Holy Rollers." The Fighting Irish of Notre Dame can be rechristened the "Drunken Irish" or "Papist Pigs." Issues of gender and sexual preferences can be addressed through creation of teams like the St. Louis "Sluts," Boston "Bimbos," Detroit "Dykes," and the Fresno "Fags." How about the Gainesville "Gimps" and Richmond "Retards," so the physi-cally and mentally impaired won't be excluded from our fun and games?

Now, don't go getting "overly sensitive" out there. None of this is de-meaning or insulting, at least not when it's being done to Indians. Just ask the folks who are doing it, or their apologists like Andy Rooney in the national media. They'll tell you—as in fact they *have* been telling you—that there's been no harm done, regardless of what their victims think, feel, or say. The situation is exactly the same as when those with precisely the same mentality used to insist that Step 'n' Fetchit was okay, or Rochester on the Jack Benny Show, or Amos and Andy, Charlie Chan, the Frito Bandito, or any of the other cutsey symbols making up the lexicon of American racism. Have we commu-nicated yet?

Let's get just a little bit real here. The notion of "fun" embodied in rituals like the Tomahawk Chop must be understood for what it is. There's not a single non-Indian example used above which can be considered socially ac-ceptable in even the most marginal sense. The reasons are obvious enough. So why is it different where American Indians are concerned? One can only con-clude that, in contrast to the other groups at issue, Indians are (falsely) per-ceived as being too few, and therefore too weak, to defend themselves effec-tively against racist and otherwise offensive behavior.

Fortunately, there are some glimmers of hope. A few teams and their fans have gotten the message and have responded appropriately. Stanford Univer-sity, which opted to drop the name "Indians" from Stanford, has experi-enced no resulting drop-off in attendance. Meanwhile, the local newspaper in

Portland, Oregon, recently decided its long-standing editorial policy prohibiting use of racial epithets should include derogatory team names. The Redskins, for instance, are now referred to as "the Washington team," and will continue to be described in this way until the franchise adopts an inoffensive moniker (newspaper sales in Portland have suffered no decline as a result).

Such examples are to be applauded and encouraged. They stand as figurative beacons in the night, proving beyond all doubt that it is quite possible to indulge in the pleasure of athletics without accepting blatant racism into the bargain.

NUREMBERG PRECEDENTS

On October 16, 1946, a man named Julius Streicher mounted the steps of a gallows. Moments later he was dead, the sentence of an international tribunal composed of representatives of the United States, France, Great Britain, and the Soviet Union having been imposed. Streicher's body was then cremated, and—so horrendous were his crimes thought to have been—his ashes dumped into an unspecified German river so that "no one should ever know a particular place to go for reasons of mourning his memory."

Julius Streicher had been convicted at Nuremberg, Germany, of what were termed "Crimes Against Humanity." The lead prosecutor in his case—Justice Robert Jackson of the United States Supreme Court—had not argued that the defendant had killed anyone, nor that he had personally committed any especially violent act. Nor was it contended that Streicher had held any particularly important position in the German government during the period in which the so-called Third Reich had exterminated some 6,000,000 Jews, as well as several million Gypsies, Poles, Slavs, homosexuals, and other untermenschen (subhumans).

The sole offense for which the accused was ordered put to death was in having served as publisher/editor of a Bavarian tabloid entitled *Der Sturmer* during the early-to-mid 1930s, years before the Nazi genocide actually began. In this capacity, he had penned a long series of virulently anti-Semitic editorials and "news" stories, usually accompanied by cartoons and other images graphically depicting Jews in extraordinarily derogatory fashion. This, the prosecution asserted, had done much to "dehumanize" the targets of his distortion in the mind of the German public. In turn, such dehumanization had made it possible—or at least easier—for average Germans to later indulge in the outright liquidation of Jewish "vermin." The tribunal agreed, holding that Streicher was therefore complicit in genocide and deserving of death by hanging.

During his remarks to the Nuremberg tribunal, Justice Jackson observed that, in implementing its sentences, the participating powers were morally and legally binding themselves to adhere forever after to the same standards of conduct that were being applied to Streicher and the other Nazi leaders. In the alternative, he said, the victorious allies would have committed "pure murder" at Nuremberg—no different in substance from that carried out by those they presumed to judge—rather than establishing the "permanent benchmark for justice" which was intended.

Yet in the United States of Robert Jackson, the indigenous American Indian population had already been reduced, in a process which is ongoing to this day, from perhaps 12.5 million in the year 1500 to fewer than 250,000 by the beginning of the 20th century. This was accomplished, according to official sources, "largely through the cruelty of [Euro-American] settlers," and an informal but clear governmental policy which had made it an articulated goal to "exterminate these red vermin," or at least whole segments of them.

Bounties had been placed on the scalps of Indians—any Indians—in places as diverse as Georgia, Kentucky, Texas, the Dakotas, Oregon, and California, and had been maintained until resident Indian populations were decimated or disappeared altogether. Entire peoples such as the Cherokee had been reduced to half their size through a policy of forced removal from their homelands east of the Mississippi River to what were then considered less preferable areas in the West.

Others, such as the Navajo, suffered the same fate while under military guard for years on end. The United States Army had also perpetrated a long series of wholesale massacres of Indians at places like Horseshoe Bend, Bear River, Sand Creek, the Washita River, the Marias River, Camp Robinson, and Wounded Knee.

Through it all, hundreds of popular novels—each competing with the next to make Indians appear more grotesque, menacing, and inhuman—were sold in the tens of millions of copies in the U.S. Plainly, the Euro-American public was being conditioned to see Indians in such a way as to allow their eradication to continue. And continue it did until the Manifest Destiny of the U.S.—a direct precursor to what Hitler would subsequently call Lebensraumpolitik (the politics of living space)—was consummated.

By 1900, the national project of "clearing" Native Americans from their land and replacing them with "superior" Anglo-American settlers was complete: the indigenous population had been reduced by as much as 98 percent while approximately 97.5 percent of their original territory had "passed" to the invaders. The survivors had been concentrated, out of sight and mind of the public, on scattered "reservations," all of them under the self-assigned "plenary" (full) power of the federal government. There was, of course, no

Nuremberg-style tribunal passing judgment on those who had fostered such circumstances in North America. No U.S. official or private citizen was ever imprisoned—never mind hanged—for implementing or propagandizing what had been done. Nor had the process of genocide afflicting Indians been completed. Instead, it merely changed form.

Between the 1880s and the 1980s, nearly half of all Native American children were coercively transferred from their own families, communities, and cultures to those of the conquering society. This was done through compulsory attendance at remote boarding schools, often hundreds of miles from their homes, where native children were kept for years on end while being systematically "deculturated" (indoctrinated to think and act in the manner of Euro-Americans rather than Indians). It was also accomplished through a pervasive foster home and adoption program—including "blind" adoptions, where children would be permanently denied information as to who they were/are and where they'd come from—placing native youths in non-Indian homes.

The express purpose of all this was to facilitate a U.S. governmental policy to bring about the "assimilation" (dissolution) of indigenous societies. In other words, Indian cultures as such were to be caused to disappear. Such policy objectives are directly contrary to the United Nations 1948 Convention on Punishment and Prevention of the Crime of Genocide, an element of international law arising from the Nuremberg proceedings. The forced "transfer of the children" of a targeted "racial, ethnical, or religious group" is explicitly prohibited as a genocidal activity under the Convention's second article.

Article II of the Genocide Convention also expressly prohibits involuntary sterilization as a means of "preventing births among" a targeted population. Yet, in 1975, it was conceded by the U.S. government that its Indian Health Service (IHS), then a subpart of the Bureau of Indian Affairs (BIA), was even then conducting a secret program of involuntary sterilization that had affected approximately 40 percent of all Indian women. The program was allegedly discontinued, and the IHS was transferred to the Public Health Service, but no one was punished. In 1990, it came out that the IHS was inoculating Inuit children in Alaska with Hepatitis-B vaccine. The vaccine had already been banned by the World Health Organization as having a demonstrated correlation with the HIV-Syndrome which is itself correlated to AIDS. As this is written, a "field test" of Hepatitis-A vaccine, also HIV-correlated, is being conducted on Indian reservations in the northern plains region.

The Genocide Convention makes it a "crime against humanity" to create conditions leading to the destruction of an identifiable human group, as such.

Yet the BIA has utilized the government's plenary prerogatives to negotiate mineral leases "on behalf of" Indian peoples paying a fraction of standard royalty rates. The result has been "super profits" for a number of preferred U.S. corporations. Meanwhile, Indians, whose reservations ironically turned out to be in some of the most mineral-rich areas of North America, which makes us, the nominally wealthiest segment of the continent's population, live in dire poverty.

By the government's own data in the mid-1980s, Indians received the lowest annual and lifetime per capita incomes of any aggregate population group in the United States. Concomitantly, we suffer the highest rate of infant mortality, death by exposure and malnutrition, disease, and the like. Under such circumstances, alcoholism and other escapist forms of substance abuse are endemic in the Indian community, a situation which leads both to a general physical debilitation of the population and a catastrophic accident rate. Teen suicide among Indians is several times the national average.

The average life expectancy of a reservation-based Native American man is barely 45 years; women can expect to live less than three years longer.

Such itemizations could be continued at great length, including matters like the radioactive contamination of large portions of contemporary Indian Country, the forced relocation of traditional Navajos, and so on. But the point should be made: genocide, as defined in international law, is a continuing fact of day-to-day life (and death) for North America's native peoples. Yet there has been—and is—only the barest flicker of public concern about, or even consciousness of, this reality. Absent any serious expression of public outrage, no one is punished and the process continues.

A salient reason for public acquiescence before the ongoing holocaust in Native North America has been a continuation of the popular legacy, often through more effective media. Since 1925, Hollywood has released more than 2,000 films, many of them rerun frequently on television, portraying Indians as strange, perverted, ridiculous, and often dangerous things of the past. Moreover, we are habitually presented to mass audiences one-dimensionally, devoid of recognizable human motivations and emotions; Indians thus serve as props, little more. We have thus been thoroughly and systematically dehumanized.

Nor is this the extent of it. Everywhere, we are used as logos, as mascots, as jokes: "Big-Chief" writing tablets, "Red Man" chewing tobacco, "Winnebago" campers, "Navajo" and "Cherokee" and "Pontiac" and "Cadillac" pickups and automobiles. There are the Cleveland "Indians," the Kansas City "Chiefs," the Atlanta "Braves" and the Washington "Redskins" professional sports teams—not to mention those in thousands of colleges, high schools, and elementary schools across the country—each with their own degrading caricatures and parodies of Indians and/or things Indian. Pop fiction continues in the same vein, including an unending stream of New Age manuals purporting to expose the inner works of indigenous spirituality in everything from pseudo-philosophical to do-it-yourself styles. Blond yuppies from Beverly Hills amble about the country claiming to be reincarnated 17th century Cheyenne Ushamans ready to perform previously secret ceremonies.

In effect, a concerted, sustained, and in some ways accelerating effort has gone into making Indians unreal. It is thus of obvious importance that the American public begin to think about the implications of such things the next time they witness a gaggle of face-painted and war-bonneted buffoons doing the "Tomahawk Chop" at a baseball or football game. It is necessary that they think about the implications of the grade-school teacher adorning their child in turkey feathers to commemorate Thanksgiving. Think about the significance of John Wayne or Charlton Heston killing a dozen "savages" with a single bullet the next time a western comes on TV. Think about why Land-o-Lakes finds it appropriate to market its butter with the stereotyped image of an "Indian princess" on the wrapper. Think about what it means when non-Indian academics profess—as they often do—to "know more about Indians than Indians do themselves." Think about the significance of charlatans like Carlos Castaneda and Jamake Highwater and Mary Summer Rain and Lynn Andrews churning out "Indian" bestsellers, one after the other, while Indians typically can't get into print.

Think about the real situation of American Indians. Think about Julius Streicher. Remember Justice Jackson's admonition. Understand that the treatment of Indians in American popular culture is not "cute" or "amusing" or just "good, clean fun."

Know that it causes real pain and real suffering to real people. Know that it threatens our very survival. And know that this is just as much a crime against humanity as anything the Nazis ever did. It is likely that the indigenous people of the United States will never demand that those guilty of such criminal activity be punished for their deeds. But the least we have the right to expect—indeed, to demand—is that such practices finally be brought to a halt.

YOU'RE SHORT, BESIDES!

44

Sucheng Chan

When asked to write about being a physically handicapped Asian American woman, I considered it an insult. After all, my accomplishments are many, yet I was not asked to write about any of them. Is being handicapped the most salient feature about me? The fact that it might be in the eyes of others made me decide to write the essay as requested. I realized that the way I think about myself may differ considerably from the way others perceive me. And maybe that's what being physically handicapped is all about.

I was stricken simultaneously with pneumonia and polio at the age of four. Uncertain whether I had polio of the lungs, seven of the eight doctors who attended me—all practitioners of Western medicine—told my parents they should not feel optimistic about my survival. A Chinese fortune teller my mother consulted also gave a grim prognosis, but for an entirely different reason: I had been stricken because my name was offensive to the gods. My grandmother had named me "grandchild of wisdom," a name that the fortune teller said was too presumptuous for a girl. So he advised my parents to change my name to "chaste virgin." All these pessimistic predictions notwithstanding, I hung onto life, if only by a thread. For three years, my body was periodically pierced with electric shocks as the muscles of my legs atrophied. Before my illness, I had been an active, rambunctious, precocious, and very curious child. Being confined to bed was thus a mental agony as great as my physical pain. Living in war-torn China, I received little medical attention; physical therapy was unheard of. But I was determined to walk. So one day, when I was six or seven, I instructed my mother to set up two rows of chairs to face each other so that I could use them as I would parallel bars. I attempted to walk by holding my body up and moving it forward with my arms while dragging my legs along behind. Each time I fell, my mother gasped, but I badgered her until she let me try again. After four nonambulatory years, I finally walked once more by pressing my hands against my thighs so my knees wouldn't buckle.

From: Asian Women United of California (eds.), *Making Waves: An Anthology of Writings By and About Asian American Women* (Boston: Beacon Press, 1989), pp. 265–72. Copyright © 1989 by Asian Women United. Reprinted by permission of Beacon Press.

My father had been away from home during most of those years because of the war. When he returned, I had to confront the guilt he felt about my condition. In many East Asian cultures, there is a strong folk belief that a person's physical state in this life is a reflection of how morally or sinfully he or she lived in previous lives. Furthermore, because of the tendency to view the family as a single unit, it is believed that the fate of one member can be caused by the behavior of another. Some of my father's relatives told him that my illness had doubtless been caused by the wild carousing he did in his youth. A well-meaning but somewhat simple man, my father believed them.

Throughout my childhood, he sometimes apologized to me for having to suffer retribution for his former bad behavior. This upset me; it was bad enough that I had to deal with the anguish of not being able to walk, but to have to assuage his guilt as well was a real burden! In other ways, my father was very good to me. He took me out often, carrying me on his shoulders or back, to give me fresh air and sunshine. He did this until I was too large and heavy for him to carry. And ever since I can remember, he has told me that I am pretty.

After getting over her anxieties about my constant falls, my mother decided to send me to school. I had already learned to read some words of Chinese at the age of three by asking my parents to teach me the sounds and meaning of various characters in the daily newspaper. But between the ages of four and eight, I received no education since just staying alive was a full-time job. Much to her chagrin, my mother found no school in Shanghai, where we lived at the time, which would accept me as a student. Finally, as a last resort, she approached the American School which agreed to enroll me only if my family kept an *amah* (a servant who takes care of children) by my side at all times. The tuition at the school was twenty U.S. dollars per month—a huge sum of money during those years of runaway inflation in China—and payable only in U.S. dollars. My family afforded the high cost of tuition and the expense of employing a full-time *amah* for less than a year.

We left China as the Communist forces swept across the country in victory. We found an apartment in Hong Kong across the street from a school run by Seventh-Day Adventists. By that time I could walk a little, so the principal was persuaded to accept me. An *amah* now had to take care of me only during recess when my classmates might easily knock me over as they ran about the playground.

After a year and a half in Hong Kong, we moved to Malaysia, where my father's family had lived for four generations. There I learned to swim in the lovely warm waters of the tropics and fell in love with the sea. On land I was a cripple; in the ocean I could move with the grace of a fish. I liked the freedom of being in the water so much that many years later, when I was a graduate student in Hawaii, I became greatly enamored with a man just because he called me a "Polynesian water nymph."

As my overall health improved, my mother became less anxious about all aspects of my life. She did everything possible to enable me to lead as normal a life as possible. I remember how once some of her colleagues in the high school where she taught criticized her for letting me wear short skirts. They felt my legs should not be exposed to public view. My mother's response was, "All girls her age wear short skirts, so why shouldn't she?"

The years in Malaysia were the happiest of my childhood, even though I was constantly fending off children who ran after me calling, "*Baikah! Baikah!*" ("Cripple! Cripple!" in the Hokkien dialect commonly spoken in Malaysia). The taunts of children mattered little because I was a star pupil. I won one award after another for general scholarship as well as for art and public speaking. Whenever the school had important visitors, my teacher always called on me to recite in front of the class.

A significant event that marked me indelibly occurred when I was twelve. That year my school held a music recital and I was one of the students chosen to play the piano. I managed to get up the steps to the stage without any problem, but as I walked across the stage, I fell. Out of the audience, a voice said loudly and clearly, "Ayah! A *baikah* shouldn't be allowed to perform in public." I got up before anyone could get on stage to help me and, with tears streaming uncontrollably down my face, I rushed to the piano and began to play. Beethoven's "Für Elise" had never been played so fiendishly fast before or since, but I managed to finish the whole piece. That I managed to do so made me feel really strong. I never again feared ridicule.

In later years I was reminded of this experience from time to time. During my fourth year as an assistant professor at the University of California at Berkeley, I won a distinguished teaching award. Some weeks later I ran into a former professor who congratulated me enthusiastically. But I said to him, "You know what? I became a distinguished teacher by *limping* across the stage of Dwinelle 155!" (Dwinelle 155 is a large, cold classroom that most colleagues of mine hate to teach in.) I was rude not because I lacked graciousness but because this man, who had told me that my dissertation was the finest piece of work he had read in fifteen years, had nevertheless advised me to eschew a teaching career.

"Why?" I asked.

"Your leg . . ." he responded.

"What about my leg?" I said, puzzled.

"Well, how would you feel standing in front of a large lecture class?"

"If it makes any difference, I want you to know I've won a number of speech contests in my life, and I am not the least bit self-conscious about speaking in front of large audiences. . . . Look, why don't you write me a letter of recommendation to tell people how brilliant I am, and let *me* worry about my leg!"

This incident is worth recounting only because it illustrates a dilemma that handicapped persons face frequently: those who care about us sometimes get so protective that they unwittingly limit our growth. This former professor of mine had been one of my greatest supporters for two decades. Time after time, he had written glowing letters of recommendation on my behalf. He had spoken as he did because he thought he had my best interests at heart; he thought that if I got a desk job rather than one that required me to be a visible, public person, I would be spared the misery of being stared at.

Americans, for the most part, do not believe as Asians do that physically handicapped persons are morally flawed. But they are equally inept at interacting with those of us who are not able-bodied. Cultural differences in the perception and treatment of handicapped people are most clearly expressed by adults. Children, regardless of where they are, tend to be openly curious about people who do not look "normal." Adults in Asia have no hesitation in asking visibly handicapped people what is wrong with them, often expressing their sympathy with looks of pity, whereas adults in the United States try desperately to be polite by pretending not to notice.

One interesting response I often elicited from people in Asia but have never encountered in America is the attempt to link my physical condition to the state of my soul. Many a time while living and traveling in Asia people would ask me what religion I belonged to. I would tell them that my mother is a devout Buddhist, that my father was baptized a Catholic but has never practiced Catholicism, and that I am an agnostic. Upon hearing this, people would try strenuously to convert me to their religion so that whichever God they believed in could bless me. If I would only attend this church or that temple regularly, they urged, I would surely get cured. Catholics and Buddhists alike have pressed religious medallions into my palm, telling me if I would wear these, the relevant deity or saint would make me well. Once while visiting the tomb of Muhammad Ali Jinnah in Karachi, Pakistan, an old Muslim, after finishing his evening prayers, spotted me, gestured toward my legs, raised his arms heavenward, and began a new round of prayers, apparently on my behalf.

In the United States adults who try to act "civilized" towards handicapped people by pretending they don't notice anything unusual sometimes end up ignoring handicapped people completely. In the first few months I lived in this country, I was struck by the fact that whenever children asked me what was the matter with my leg, their adult companions would hurriedly shush them up, furtively look at me, mumble apologies, and rush their children away. After a few months of such encounters, I decided it was my responsibility to educate these people. So I would say to the flustered adults, "It's okay, let the kid ask." Turning to the child, I would say, "When I was a little girl, no bigger than you are, I became sick with something called polio. The muscles in my leg shrank up and I couldn't walk very well. You're much luckier than I am because now you can get a vaccine to make sure you never get my disease. So don't cry when your mommy takes you to get a polio vaccine, okay?" Some adults and their little companions I talked to this way were glad to be rescued from embarrassment; others thought I was strange.

Americans have another way of covering up their uneasiness: they become jovially patronizing. Sometimes when people spot my crutch, they ask if I've had a skiing accident. When I answer that unfortunately it is something less glamorous than that, they say, "I bet you *could* ski if you put your mind to it!" Alternately, at parties where people dance, men who ask me to dance with them get almost belligerent when I decline their invitation. They say, "Of course you can dance if you *want* to!" Some have given me pep talks about how if I would only develop the right mental attitude, I would have more fun in life.

Different cultural attitudes toward handicapped persons came out clearly during my wedding. My father-in-law, as solid a representative of middle America as could be found, had no qualms about objecting to the marriage on racial grounds, but he could bring himself to comment on my handicap only indirectly. He wondered why his son, who had dated numerous high school and college beauty queens, couldn't marry one of them instead of me. My mother-in-law, a devout Christian, did not share her husband's prejudices, but she worried aloud about whether I could have children. Some Chinese friends of my parents, on the other hand, said that I was lucky to have found such a noble man, one who would marry me despite my handicap. I, for my part, appeared in church in a white lace wedding dress I had designed and made myself—a miniskirt!

How Asian Americans treat me with respect to my handicap tells me a great deal about their degree of acculturation. Recent immigrants behave just like Asians in Asia; those who have been here longer or who grew up in the United States behave more like their white counterparts. I have not encountered any distinctly Asian American pattern of response. What makes the experience of Asian American handicapped people unique is the duality of responses we elicit.

Regardless of racial or cultural background, most handicapped people have to learn to find a balance between the desire to attain physical independence and the need to take care of ourselves by not overtaxing our bodies. In my case, I've had to learn to accept the fact that leading an active life has its price. Between the ages of eight and eighteen, I walked without using crutches or braces but the effort caused my right leg to become badly misaligned. Soon after I came to the United States, I had a series of operations to straighten out the bones of my right leg; afterwards though my leg looked straighter and presumably better, I could no longer walk on my own. Initially my doctors fitted me with a brace, but I found wearing one cumbersome and soon gave it up. I could move around much more easily—and more important, faster—by using one crutch. One orthopedist after another warned me that using a single crutch was a bad practice. They were right. Over the years my spine developed a double-S curve and for the last twenty years I have suffered from severe, chronic back pains, which neither conventional physical therapy nor a lighter work load can eliminate.

The only thing that helps my backaches is a good massage, but the soothing effect lasts no more than a day or two. Massages are expensive, especially when one needs them three times a week. So I found a job that pays better, but at which I have to work longer hours, consequently increasing the physical strain on my body—a sort of vicious circle. When I was in my thirties, my doctors told me that if I kept leading the strenuous life I did, I would be in a wheelchair by the time I was forty. They were right on target: I bought myself a wheelchair when I was forty-one. But being the incorrigible character that I am, I use it only when I am *not* in a hurry!

It is a good thing, however, that I am too busy to think much about my handicap or my backaches because pain can physically debilitate as well as cause depression. And there are days when my spirits get rather low. What has helped me is realizing that being handicapped is akin to growing old at an accelerated rate. The contradiction I experience is that often my mind races along as though I'm only twenty while my body feels about sixty. But fifteen or twenty years hence, unlike my peers who will have to cope with aging for the first time, I shall be full of cheer because I will have already fought, and I hope won, that battle long ago.

Beyond learning how to be physically independent and, for some of us, living with chronic pain or other kinds of discomfort, the most difficult thing a handicapped person has to deal with, especially during puberty and early adulthood, is relating to potential sexual partners. Because American culture places so much emphasis on physical attractiveness, a person with a shriveled limb, or a tilt to the head, or the inability to speak clearly, experiences great uncertainty—indeed trauma—when interacting with someone to whom he or she is attracted. My problem was that I was not only physically handicapped, small, and short, but worse, I also wore glasses and was smarter than all the boys I knew! Alas, an insurmountable combination. Yet somehow I have managed to have intimate relationships, all of them with extraordinary men. Not surprisingly, there have also been countless men who broke my heart—men who enjoyed my company "as a friend," but who never found the courage to date or make love with me, although I am sure my experience in this regard is no different from that of many able-bodied persons.

The day came when my backaches got in the way of having an active sex life. Surprisingly that development was liberating because I stopped worrying about being attractive to men. No matter how headstrong I had been, I, like most women of my generation, had had the desire to be alluring to men in-grained into me. And that longing had always worked like a brake on my behavior. When what men think of me ceased to be compelling, I gained greater freedom to be myself.

I've often wondered if I would have been a different person had I not been physically handicapped. I really don't know, though there is no question that being handicapped has marked me. But at the same time I usually do not *feel* handicapped—and consequently, I do not *act* handicapped. People are therefore less likely to treat me as a handicapped person. There is no doubt, however, that the lives of my parents, sister, husband, other family members, and some close friends have been affected by my physical condition. They have had to learn not to hide me away at home, not to feel embarrassed by how I look or react to people who say silly things to me, and not to resent me for the extra demands my condition makes on them. Perhaps the hardest thing for those who live with handicapped people is to know when and how to offer help. There are no guidelines applicable to all situations. My advice is, when in doubt, ask, but ask in a way that does not smack of pity or embarrassment. Most important, please don't talk to us as though we are children.

So, has being physically handicapped been a handicap? It all depends on one's attitude. Some years ago, I told a friend that I had once said to an affirmative action compliance officer (somewhat sardonically since I do not believe in the head count approach to affirmative action) that the institution which employs me is triply lucky because it can count me as nonwhite, female and handicapped. He responded, "Why don't you tell them to count you four times? . . . Remember, you're short, besides!"

IF MEN COULD MENSTRUATE

45

Gloria Steinem

A white minority of the world has spent centuries conning us into thinking that a white skin makes people superior—even though the only thing it really does is make them more subject to ultraviolet rays and to wrinkles. Male human beings have built whole cultures around the idea that penis-envy is "natural" to women—though having such an unprotected organ might be said to make men vulnerable, and the power to give birth makes womb-envy at least as logical.

In short, the characteristics of the powerful, whatever they may be, are thought to be better than the characteristics of the powerless—and logic has nothing to do with it.

What would happen, for instance, if suddenly, magically, men could menstruate and women could not?

The answer is clear—menstruation would become an enviable, boast-worthy, masculine event:

Men would brag about how long and how much.

Boys would mark the onset of menses, that longed-for proof of manhood, with religious ritual and stag parties.

Congress would fund a National Institute of Dysmenorrhea to help stamp out monthly discomforts.

Sanitary supplies would be federally funded and free. (Of course, some men would still pay for the prestige of commercial brands such as John Wayne Tampons, Muhammad Ali's Rope-a-dope Pads, Joe Namath Jock Shields— "For Those Light Bachelor Days," and Robert "Baretta" Blake Maxi-Pads.)

Military men, right-wing politicians, and religious fundamentalists would cite menstruation ("*men*-struation") as proof that only men could serve in the Army ("you have to give blood to take blood"), occupy political office ("can women be aggressive without that steadfast cycle governed by the planet Mars?"), be priests and ministers ("how could a woman give her blood for our sins?"), or rabbis ("without the monthly loss of impurities, women remain unclean").

Male radicals, left-wing politicians, and mystics, however, would insist that women are equal, just different; and that any woman could enter their

From: *Ms.* VII (October 1978): 110. © Gloria Steinem. Reprinted by permission.

ranks if only she were willing to self-inflict a major wound every month ("you *must* give blood for the revolution"), recognize the preeminence of menstrual issues, or subordinate her selfness to all men in the Cycle of Enlightenment.

Street guys would brag ("I'm a three-pad man") or answer praise from a buddy ("Man, you lookin' *good!*") by giving fives and saying, "Yeah, man, I'm on the rag!"

TV shows would treat the subject at length. ("Happy Days": Richie and Potsie try to convince Fonzie that he is still "The Fonz," though he has missed two periods in a row.) So would newspapers. (SHARK SCARE THREAT-ENS MENSTRUATING MEN. JUDGE CITES MONTHLY STRESS IN PARDONING RAPIST.) And movies. (Newman and Redford in "Blood Brothers"!)

Men would convince women that intercourse was *more* pleasurable at "that time of the month." Lesbians would be said to fear blood and therefore life itself—though probably only because they needed a good menstruating man.

Of course, male intellectuals would offer the most moral and logical arguments. How could a woman master any discipline that demanded a sense of time, space, mathematics, or measurement, for instance, without that in-built gift for measuring the cycles of the moon and planets—and thus for measuring anything at all? In the rarefied fields of philosophy and religion, could women compensate for missing the rhythm of the universe? Or for their lack of symbolic death-and-resurrection every month?

Liberal males in every field would try to be kind: the fact that "these people" have no gift of measuring life or connecting to the universe, the liberals would explain, should be punishment enough.

And how would women be trained to react? One can imagine traditional women agreeing to all these arguments with a staunch and smiling masochism. ("The ERA would force housewives to wound themselves every month": Phyllis Schafly. "Your husband's blood is as sacred as that of Jesus—and so sexy, too!": Marabel Morgan.) Reformers and Queen Bees would try to imitate men, and *pretend* to have a monthly cycle. All feminists would explain endlessly that men, too, needed to be liberated from the false idea of Martian aggressiveness, just as women needed to escape the bonds of menses-envy. Radical feminists would add that the oppression of the nonmenstrual was the pattern for all other oppressions. ("Vampires were our first freedom fighters!") Cultural feminists would develop a bloodless imagery in art and literature. Socialist feminists would insist that only under capitalism would men be able to monopolize menstrual blood. . . .

In fact, if men could menstruate, the power justifications could probably go on forever.

If we let them.

FROM "KIKE" TO "JAP": *How Misogyny, Anti-Semitism, and Racism Construct the "Jewish American Princess"*

46

Evelyn Torton Beck

The stereotyping of the Jewish American woman as the JAP, which stands for Jewish American Princess, is an insult, an injury, and violence that is done to Jewish women. The term is used widely by both men and women, by both Jews and non-Jews. When gentiles use it, it is a form of anti-Semitism. When Jews use it, it is a form of self-hating or internalized anti-Semitism. It is a way of thinking that allows some Jewish women to harm other Jewish women who are just like them except for the fact that one is okay—she's *not* a JAP. The other is not okay—she's too JAPie. The seriousness of this term becomes evident when we substitute the words "too Jewish" for "too JAPie," and feel ourselves becoming considerably less comfortable.

When I speak on college campuses, young women frequently tell me that when someone calls them a Jew they are insulted because they know it's being said with a kind of hostility, but if someone calls them a JAP they don't mind because they frequently use this term themselves. They think the "J" in JAP really doesn't mean anything—it's just there. While everyone seems to know what the characteristics of a "Jewish American Princess" are, no one ever seems to think about what they are saying when they use the term. How is it that you don't have to be Jewish to be a JAP? If this is so, why is the word "Jewish" in the acronym at all? Words are not meaningless unless we choose to close our ears and pretend not to hear.

This subject is frequently trivialized, but when it is not, when we take it seriously, it makes us extremely tense. Why is that? I think it's because it takes us into several "war zones." It brings us in touch with Jew-hating, or anti-Semitism. It brings us in touch with misogyny, or woman-hating. And it brings us in touch with class-hatred, old money vs. nouveau riche. (Jews have classically been seen as intruders in the United States and have been resented for "making it.") It also puts us strongly in touch with racism. It is no accident

From: Evelyn Torton Beck, "From 'Kike' to 'JAP,'" in *Sojourner: The Women's Forum*, September 1988, pp. 18–20. Reprinted by permission.

that the acronym JAP is also the word used for our worst enemies in World War II—who were known as "the Japs." During World War II, posters and slogans saying "Kill Japs" were everywhere. It was a period in which slang terms were readily used in a pejorative way to identify many different minorities: "Japs," "Kikes," "Spics," "Wops," "Chinks" were commonplace terms used unthinkingly. And women were—and, unfortunately, still are—easily named "bitches," "sluts," and "cunts."

In such a climate, negative stereotypes easily overlap and elide. For example, in the popular imagination, Jews, "Japs," women, and homosexuals have all been viewed as devious, unreliable, and power hungry. What has happened in the decades following World War II is that the "Japs," whom we dehumanized when we dropped our atom bomb on them, have subliminally merged in the popular imagination with "kikes" and other foreign undesirables. (The fact that in the 1980s Japan poses a serious economic threat to the United States should not be overlooked either.) While efforts to eradicate slurs against ethnic minorities have made it not okay to use explicitly ethnic epithets, women still provide an acceptable target, especially when the misogyny is disguised as supposedly "good-natured" humor. In this insidious and circuitous way, the Jewish American woman carries the stigmas of the "kikes" and "Japs" of a previous era. And that is very serious.

The woman, the Jewish woman as JAP has replaced the male Jew as the scapegoat, and the Jewish male has not only participated, but has, in fact, been instrumental in creating and perpetuating that image. I want to show how some of the images of Jewish women created in American culture by Jewish men provided the roots of the "Jewish American Princess."

. . . The Jewish American Princess phenomenon is not new; I (as well as other Jewish feminists) have been talking about it for at least ten years now, but only recently has it been given wide public attention. One reason for this is that it is beginning to be seen in the light of increased anti-Semitism and racism, particularly on college campuses. Dr. Gary Spencer, who is a *male* professor of sociology at Syracuse University (and it is unfortunate that his being male gives him credibility over women saying the same things) closely examined the library and bathroom graffiti of his school and interviewed hundreds of students on his campus and has concluded that "JAP"-baiting is widespread, virulent, and threatening to all Jews, not "just" Jewish women (which we gather might have been okay or certainly considerably less serious).

Spencer discovered that nasty comments about "JAPs" led to more generally anti-Semitic graffiti that said among other slogans, "Hitler was right!" "Give Hitler a second chance!" and "I hate Jews." He also discovered that there were certain places in which Jewish women—JAPs—were not welcome:

for example, certain cafes where Jewish women were hassled if they entered. He also found that certain areas of the University were considered "JAP-free zones" and other areas (particular dorms) that were called "Jew havens." At The American University in Washington, D.C., largely Jewish residence halls are called "Tokyo Towers," making the racial overtones of "JAP" explicit. But let the parallels to Nazi occupied Europe not be lost upon us. Under the Nazis, movements of Jews were sharply restricted: there were many areas which Jews could not enter, and others (like ghettos and concentration camps) that they could not leave.

What I want to do now is to show how characteristics that have historically been attributed to Jews, primarily Jewish men, have been reinterpreted in terms of women: how misogyny combined with Jew-hating creates the Jewish American Princess. And I want you to remember that Jewish men have not only participated in this trashing, but they have not protected Jewish women when other men and women have talked about "JAPs" in this way. And this fact, I think, has made this an arena into which anyone can step—an arena that becomes a minefield when Jews step into it.

Jews have been said to be materialistic, money-grabbing, greedy, and ostentatious. Women have been said to be vain, trivial, and shallow; they're only interested in clothing, in show. When you put these together you get the Jewish-woman type who's only interested in designer clothes and sees her children only as extensions of herself. The Jew has been seen as manipulative, crafty, untrustworthy, unreliable, calculating, controlling, and malevolent. The Jewish Princess is seen as manipulative, particularly of the men in her life, her husband, her boyfriend, her father. And what does she want? Their *money!* In addition, she's lazy—she doesn't work inside or outside the home. She is the female version of the Jew who, according to anti-Semitic lore, is a parasite on society; contradictorily, the Jew has been viewed both as dangerous "communist" as well as non-productive "capitalist." The cartoon vision of the Jewish American Princess is someone who sucks men dry: she is an "unnatural mother" who refuses to nurture her children (the very opposite of the "Jewish mother" whose willingness to martyr herself makes *her* ludicrous). And she doesn't "put out" except in return for goods; she isn't really interested in either sexuality or lovingness. We live in a world climate and culture in which materialism is rampant, and Jewish women are taking the rap for it. The irony is they are taking the rap from non-Jews and Jewish men alike—even from some Jewish women.

Another way in which Jewish women are carrying the anti-Semitism that was directed in previous eras at Jewish men is in the arena of sexuality. Jews have been said to be sexually strange, exotic. There are many stereotypes of

Jewish men as lechers. The Jewish American Princess is portrayed as both sexually frigid (withholding) and as a nymphomaniac. Here we again see the familiar anti-Semitic figure of the Jew as controlling and insatiably greedy, always wanting more, combining with the misogynist stereotype of the insatiable woman, the woman who is infinitely orgasmic, who will destroy men with her desire. Like the Jew of old, the Jewish woman will suck men dry. But she is worse than "the Jew"—she will also turn on her own kind.

There are physical stereotypes as well: the Jew with the big hook nose, thick lips, and frizzy hair. The Jewish American Princess has had a nose job and her hair has been straightened, but she too has large lips (an image we immediately also recognize as racist). Jews are supposed to be loud, pushy, and speak with unrefined accents. Jewish American Princesses are said to come from Long Island and speak with funny accents: "Oh my Gawd!" The accent has changed from the lampooned immigrant speech of previous generations, but assimilation into the middle class hasn't helped the Jewish American Princess get rid of her accent. It doesn't matter how she speaks, because if it's Eastern and recognizably Jewish, it's not okay.

I also want to give you some idea of how widespread and what a money-making industry the Jewish American Princess phenomenon has become. There are greeting cards about the "JAP Olympics," with the JAP doing things like "bank-vaulting" instead of pole-vaulting. Or cross-country *"kvetching"* instead of skiing. In this card the definition of the Yiddish term *"kvetch"* reads: "an irritable whine made by a three-year-old child or a JAP at any age." So in addition to the all-powerful monster, you also have the infantilization of the Jewish woman. And there are the Bunny Bagelman greeting cards: Bunny has frizzy hair, big lips, is wearing ostentatious jewelry—and is always marked as a Jew in some way. One of her cards reads, "May God Bless you and keep you . . . rich!" Or Bunny Bagelman is a professional, dressed in a suit carrying a briefcase, but this image is undermined by the little crown she incongruously wears on her head bearing the initials "JAP." There is also a Halloween card with a grotesque female figure; the card reads, "Is it a vicious vampire? No, it's Bunny Bagelman with PM syndrome!" In analyzing these kinds of cartoons, you begin to see how sexism is absolutely intertwined with anti-Semitism.

Such attacks devalue Jewish women and keep them in line. An incident reported by Professor Spencer at Syracuse University makes this quite evident. At a basketball game, when women who were presumed to be "JAPs" stood up and walked across the floor at half-time (and it happened to Jewish and non-Jewish women), 2,000 students stood up, accusingly pointed their fingers at them, and repeatedly yelled, "JAP, JAP, JAP, JAP, JAP" in a loud chorus. This was so humiliating and frightening that women no longer got out of their seats to go to the bathroom or to get a soda. This is a form of public harassment that is guaranteed to control behavior and parallels a phenomenon called "punching" at the University of Dar El Salaam, Tanzania. Here, when women were "uppity" or otherwise stepped out of line, huge posters with their pictures on them were put all over campus, and no one was to speak to them. If you spoke to these women, you were considered to be like them. This is a very effective way of controlling people.

The threat of physical violence against Jewish women (in the form of "Slap-a-JAP" T-shirts and contests at bars) is evident on many Eastern college campuses. A disc jockey at The American University went so far as to sponsor a "fattest JAP-on-campus" contest. That this kind of unchecked verbal violence can lead to murder is demonstrated by lawyer Shirley Frondorf in a recent book entitled *Death of a "Jewish American Princess": The True Story of a Victim on Trial* (Villard Books, 1988). Frondorf shows how the murder of a Jewish woman by her husband was exonerated and the victim placed on trial because she was someone who was described by her husband as "materialistic, who shopped and spent, nagged shrilly and bothered her husband at work"—in other words, she was a "JAP" and therefore deserved what she got. This account demonstrates the dangers inherent in stereotyping and the inevitable dehumanization that follows.

One of the most aggressively sexual forms of harassment of Jewish women, which amounted to verbal rape, were signs posted at a college fair booth at Cornell University that read, "Make her prove she's not a JAP, make her swallow." Part of the mythology is that the Jewish woman will suck, but she won't swallow. So you see that as the degradation of woman *as woman* escalates, the anti-Semitism also gets increasingly louder. In a recent Cornell University student newspaper, a cartoon offered advice on how to "exterminate" JAPs by setting up a truck offering bargains, collecting the JAPs as they scurried in, and dropping them over a cliff. While the word "Jew" was not specifically mentioned, the parallels to the historical "rounding up" of Jews and herding them into trucks to be exterminated in the camps during World War II can hardly be ignored. This cartoon was created by a Jewish man.

This leads me directly to the third thing I want to discuss, namely, how and why Jewish men have participated in constructing and perpetuating the image of the Jewish American Princess as monster. How is it that the Jewish Mother (a mildly derogatory stereotype that nonetheless contained some warmth) has become the grotesque [image] that is the Jewish American Princess, who, unlike the Jewish Mother, has absolutely no redeeming features? Exactly how the Jewish Mother (created entirely by second-generation American men who had begun to mock the very nurturance they had relied upon for their success) gave birth to the Jewish American Princess is a long and complex story. This story is intertwined with the overall economic success of Jews as a class in the United States, the jealousy others have felt over this success, and the discomfort this success creates in Jews who are fearful of living out the stereotype of the "rich Jew." It is also a likely conjecture that middle-class American Jewish men view the large numbers of Jewish women who have successfully entered the work force as professionals as a serious economic and ego threat.

We find the origins of the Jewish American Princess in the fiction of American Jewish males of the last three decades. In the '50s, Herman Wouk's *Marjory Morningstar* (nee Morgenstern) leaves behind her immigrant background, takes a new name (one that is less recognizably Jewish), manipulates men, has no talent, and is only interested in expensive clothing. The postwar Jewish male, who is rapidly assimilating into American middle-class culture and leaving behind traditional Jewish values, is creating the Jewish woman—the materialistic, empty, manipulative Jewish woman, the Americanized daughter who fulfills the American Dream for her parents but is, at the same time, punished for it. It looks as if the Jewish woman was created in the image of the postwar Jewish male but viewed by her creator as grotesque. All the characteristics he cannot stand in himself are displaced onto the Jewish woman.

In the '60s, Philip Roth created the spoiled and whiny Brenda Potemkin in *Goodbye, Columbus* at the same time that Shel Silverstein created his image of the perfect Jewish mother as martyr. Some of you may remember this popular story from your childhood. A synopsis goes something like this: "Once there was a tree and she loved the little boy. And he slept in her branches, and loved the tree and the tree was happy. And as the boy grew older, he needed things from her. He needed apples, so she gave him apples, and she was happy. Then she cut off her branches because the boy needed them to build a house. And she was happy. Then finally he needed her trunk because he wanted to build a big boat for himself. And she was happy. The tree gave and gave of herself, and finally the tree was alone and old when the boy returned one more time. By now, the tree had nothing to give. But the boy/man is himself old now, and he doesn't need much except a place on which to sit. And the tree said, 'An old stump is good for sitting and resting on. Come boy and sit and rest on me.' And the boy did, and the tree was happy." This "positive" entirely self-*less* mother, created as a positive wish fantasy by a Jewish man, very easily tips over into its opposite, the monstrous woman, the self-absorbed "JAP" who is negatively self-less. She has no center. She *is* only clothes, money, and show.

In concluding, I want to bring these strands together and raise some questions. Obviously Jews need to be as thoughtful about consumerism as others, but we need to ask why the Jewish woman is taking the rap for the consumerism which is rampant in our highly materialistic culture in general. We need to think about the image of the Jewish American Princess and the father she tries to manipulate. What has happened to the Jewish Mother? Why has she dropped out of the picture? If (as is likely true of all groups) some middle-class Jewish women (and men) are overly focused on material things, what is the other side of that? What about the middle-class fathers who measure their own success by what material goods they are able to provide to their wives and children and who don't know how to show love in any other way? Someone who doesn't know how to give except through material goods could easily create a child who comes to expect material goods as a proof of love and self-worth, especially if sexist gender expectations limit the options for women. We need to look more closely at the relationship between the "monster" daughter and the father who helped create her.

This brings up another uncomfortable subject—incest in Jewish families. We have to look carefully at the image of the "little princess" who sits on Daddy's lap and later becomes this monstrous figure. (My father thought it appropriate for me to sit on his lap until the day he died, well into his '80s, and I do not believe he was unusual in his expectations.) There are enough stories of incest in which we know that the father who sexually abuses his daughter when she is a child becomes quite distant when she reaches adolescence and may continue to abuse her in psychological ways. And the JAP image is a real form of psychological abuse. We need to look at these things to understand that this phenomenon is not trivial, and to understand how it undermines *all* Jewish women and particularly harms young women coming of age. It cannot do Jewish men much good either to think of their sisters, daughters, mothers, and potential girlfriends with such contempt.

Last, I want to say that we have many false images of Jewish families. There *is* violence in Jewish families, just as there is violence in families of all groups. It is time to put the whole question of the Jewish American Princess into the context of doing away with myths of all kinds. The Jewish family is no more nor less cohesive than other families, although there is great pressure on Jewish families to pretend they are. Not all Jewish families are non-alcoholic; not all Jewish families are heterosexual; not all Jews are upper or middle class; and not all are urban or Eastern. It's important that the truth of Jewish women's (and also Jewish men's) lives be spoken. Beginning to take apart this image of the Jewish American Princess can make us look more closely at what it is that we, in all of our diversity as Jews, are; what we are striving towards; and what we hope to become.

Sexuality

A NEW POLITICS OF SEXUALITY

47

June Jordan

As a young worried mother, I remember turning to Dr. Benjamin Spock's *Common Sense Book of Baby and Child Care* just about as often as I'd pick up the telephone. He was God. I was ignorant but striving to be good: a good Mother. And so it was there, in that best-seller pocketbook of do's and don't's, that I came upon this doozie of a guideline: Do not wear miniskirts or other provocative clothing because that will upset your child, especially if your child happens to be a boy. If you give your offspring "cause" to think of you as a sexual being, he will, at the least, become disturbed; you will derail the equilibrium of his notions about your possible identity and meaning in the world.

It had never occurred to me that anyone, especially my son, might look upon me as an asexual being. I had never supposed that "asexual" was some kind of positive designation I should, so to speak, lust after. I was pretty surprised by Dr. Spock. However, I was also, by habit, a creature of obedience. For a couple of weeks I actually experimented with lusterless colors and dowdy tops and bottoms, self-consciously hoping thereby to prove myself as a lusterless and dowdy and, therefore, excellent female parent.

This essay was adopted from the author's keynote address to the Bisexual, Gay, and Lesbian Student Association at Stanford University on April 29, 1991. It was published in *The Progressive*, July 1991.

Years would have to pass before I could recognize the familiar, by then, absurdity of a man setting himself up as the expert on a subject that presupposed women as the primary objects for his patriarchal discourse—on motherhood, no less! Years passed before I came to perceive the perversity of dominant power assumed by men, and the perversity of self-determining power ceded to men by women.

A lot of years went by before I understood the dynamics of what anyone could summarize as the Politics of Sexuality.

I believe the Politics of Sexuality is the most ancient and probably the most profound arena for human conflict. Increasingly, it seems clear to me that deeper and more pervasive than any other oppression, than any other bitterly contested human domain, is the oppression of sexuality, the exploitation of the human domain of sexuality for power.

When I say sexuality, I mean gender: I mean male subjugation of human beings because they are female. When I say sexuality, I mean heterosexual institutionalization of rights and privileges denied to homosexual men and women. When I say sexuality, I mean gay or lesbian contempt for bisexual modes of human relationship.

The Politics of Sexuality therefore subsumes all of the different ways in which some of us seek to dictate to others of us what we should do, what we should desire, what we should dream about, and how we should behave ourselves, generally. From China to Iran, from Nigeria to Czechoslovakia, from Chile to California, the politics of sexuality—enforced by traditions of state-sanctioned violence plus religion and the law—reduces to male domination of women, heterosexist tyranny, and, among those of us who are in any case deemed despicable or deviant by the powerful, we find intolerance for those who choose a different, a more complicated—for example, an interracial or bisexual—mode of rebellion and freedom.

We must move out from the shadows of our collective subjugation—as people of color/as women/as gay/as lesbian/as bisexual human beings.

I can voice my ideas without hesitation or fear because I am speaking, finally, about myself. I am Black and I am female and I am a mother and I am bisexual and I am a nationalist and I am an antinationalist. And I mean to be fully and freely all that I am!

Conversely, I do not accept that any white or Black or Chinese man—I do not accept that, for instance, Dr. Spock—should presume to tell me, or any other woman, how to mother a child. He has no right. He is not a mother. My child is not his child. And, likewise, I do not accept that anyone—any woman or any man who is not inextricably part of the subject he or she dares to address—should attempt to tell any of us, the objects of her or his pre-sumptuous discourse, what we should do or what we should not do.

Recently, I have come upon gratuitous and appalling pseudoliberal pro-nouncements on sexuality. Too often, these utterances fall out of the mouths of men and women who first disclaim any sentiment remotely related to homophobia, but who then proceed to issue outrageous opinions like the following:

- That it is blasphemous to compare the oppression of gay, lesbian, or bisexual people to the oppression, say, of black people, or of the Palestinians.
- That the bottom line about gay or lesbian or bisexual identity is that you can conceal it whenever necessary and, so, therefore, why don't you do just that? Why don't you keep your deviant sexuality in the closet and let the rest of us—we who suffer oppression for reasons of our ineradicable and always visible components of our personhood such as race or gen-der—get on with our more necessary, our more beleaguered struggle to survive?

Well, number one: I believe I have worked as hard as I could, and then harder than that, on behalf of equality and justice—for African-Americans, for the Palestinian people, and for people of color everywhere.

And no, I do not believe it is blasphemous to compare oppressions of sexuality to oppressions of race and ethnicity: freedom is indivisible or it is nothing at all besides sloganeering and temporary, short-sighted, and short-lived advancement for a few. Freedom is indivisible, and either we are working for freedom or you are working for the sake of your self-interests and I am working for mine.

If you can finally go to the bathroom wherever you find one, if you can finally order a cup of coffee and drink it wherever coffee is available, but you cannot follow your heart—you cannot respect the response of your own hon-est body in the world—then how much of what kind of freedom does any one of us possess?

Or, conversely, if your heart and your honest body can be controlled by the state, or controlled by community taboo, are you not then, and in that case, no more than a slave ruled by outside force?

What tyranny could exceed a tyranny that dictates to the human heart, and that attempts to dictate the public career of an honest human body?

Freedom is indivisible; the Politics of Sexuality is not some optional "special-interest" concern for serious, progressive folk.

And, on another level, let me assure you: if every single gay or lesbian or bisexual man or woman active on the Left of American politics decided to stay home, there would be *no* Left left.

One of the things I want to propose is that we act on that reality: that we insistently demand reciprocal respect and concern from those who cheerfully depend upon our brains and our energies for their, and our, effective impact on the political landscape.

Last spring, at Berkeley, some students asked me to speak at a rally against racism. And I did. There were four or five hundred people massed on Sproul Plaza, standing together against that evil. And, on the next day, on that same plaza, there was a rally for bisexual and gay and lesbian rights, and students asked me to speak at that rally. And I did. There were fewer than seventy-five people stranded, pitiful, on that public space. And I said then what I say today: That was disgraceful! There should have been just one rally. One rally: freedom is indivisible.

As for the second, nefarious pronouncement on sexuality that now enjoys mass-media currency: the idiot notion of keeping yourself in the closet—that is very much the same thing as the suggestion that black folks and Asian-Americans and Mexican-Americans should assimilate and become as "white" as possible—in our walk/talk/music/food/values—or else. Or else? Or else we should, deservedly, perish.

Sure enough, we have plenty of exposure to white everything so why would we opt to remain our African/Asian/Mexican selves? The answer is that suicide is absolute, and if you think you will survive by hiding who you really are, you are sadly misled: there is no such thing as partial or intermittent suicide. You can only survive if you—who you really are—do survive.

Likewise, we who are not men and we who are not heterosexist—we, sure enough, have plenty of exposure to male-dominated/heterosexist this and that.

But a struggle to survive cannot lead to suicide: suicide is the opposite of survival. And so we must not conceal/assimilate/integrate into the would-be dominant culture and political system that despises us. Our survival requires that we alter our environment so that we can live and so that we can hold each other's hands and so that we can kiss each other on the streets, and in the daylight of our existence, without terror and without violent and sometimes fatal reactions from the busybodies of America.

Finally, I need to speak on bisexuality. I do believe that the analogy is interracial or multiracial identity. I do believe that the analogy for bisexuality is a multicultural, multi-ethnic, multiracial world view. Bisexuality follows from such a perspective and leads to it, as well.

Just as there are many men and women in the United States whose parents have given them more than one racial, more than one ethnic identity and cultural heritage to honor; and just as these men and women must deny no given part of themselves except at the risk of self-deception and the insanities that must issue from that; and just as these men and women embody the principle of equality among races and ethnic communities; and just as these men and women falter and anguish and choose and then falter again and then anguish and then choose yet again how they will honor the irreducible complexity of their God-given human being—even so, there are many men and women, especially young men and women, who seek to embrace the complexity of their total, always-changing social and political circumstance.

They seek to embrace our increasing global complexity on the basis of the heart and on the basis of an honest human body. Not according to ideology. Not according to group pressure. Not according to anybody's concept of "correct."

This is a New Politics of Sexuality. And even as I despair of identity politics—because identity is given and principles of justice/equality/freedom cut across given gender and given racial definitions of being, and because I will call you my brother, I will call you my sister, on the basis of what you *do* for justice, what you *do* for equality, what you *do* for freedom and *not* on the basis of you who are, even so I look with admiration and respect upon the new, bisexual politics of sexuality.

This emerging movement politicizes the so-called middle ground: bisexuality invalidates either/or formulation, either/or analysis. Bisexuality means I am free and I am as likely to want and to love a woman as I am likely to want and to love a man, and what about that? Isn't that what freedom implies?

If you are free, you are not predictable and you are not controllable. To my mind, that is the keenly positive, politicizing significance of bisexual affirmation:

To insist upon complexity, to insist upon the validity of all of the components of social/sexual complexity, to insist upon the equal validity of all of the components of social/sexual complexity.

This seems to me a unifying, 1990s mandate for revolutionary Americans planning to make it into the twenty-first century on the basis of the heart, on the basis of an honest human body, consecrated to every struggle for justice, every struggle for equality, every struggle for freedom.

WHERE HAS GAY LIBERATION GONE?   **48**
An Interview with Barbara Smith

Amy Gluckman and Betsy Reed

The links between homophobia and sexism have been analyzed a lot, but the linkages between homophobia and other forms of oppression, especially class and race oppression, haven't been thought about or analyzed as much. What connections do you see between, first of all, gay oppression and our economic system, and issues of class?

I think that's really a very complicated issue. I often say that, unlike racial oppression, lesbian and gay oppression is not economically linked, and is not structurally and historically linked, to the founding of this country and of capitalism in the United States. This country was not founded on homophobia; it was founded on slavery and racism and, before that, prior to the importation of slaves, it was founded on the genocide of the indigenous people who lived here, which also had profound racial consequences and rationales. I think that sometimes, certain lesbians and gay men get very upset when I and others—I'm not the only person who would say it—put that kind of analysis out because they think it means that I'm saying that lesbian and gay oppression is not serious. That's not what I'm saying at all.

Do you think that extends into the future? In other words, would you say that capitalism could go along its merry way and assimilate gay people and provide gay rights, but otherwise remain the same?

. . . As with all groups, I think that our economic system has the most implications for lesbians and gay men when their class position makes them vulnerable to that economic system. So in other words, it's not that in general being lesbian or gay puts you into a critical relationship to capitalism, it's that a large proportion of lesbians and gay men are poor or working class, but of course they're completely invisible the way the movement's politics are defined now.

From: Amy Gluckman and Betsy Reed, eds., *Homo Economics, Capitalism, Community, and Lesbian and Gay Life* (New York: Routledge, 1997), pp. 195–207. Reprinted by permission.

So much of this society is about consumerism. As long as lesbians and gay men are characterized as people who have huge amounts of disposable income and who are kind of fun and trendy—nice entertainment type people, k. d. langs and Martinas, just a little on the edge but not really that threatening—as long as they're characterized in that way, probably capitalism can incorporate them. If they begin to think about how extending lesbian and gay rights fully might shake up the patriarchal nuclear family and the economic arrangements that are tied to that, then that might be the point at which capitalists would say, "Well no, I don't think we can include that." As long as it's about k. d. lang and Cindy Crawford on the cover of *Vanity Fair*, capitalism doesn't have any problem with that because that's nothing but an image. . . .

. . . Right-wing organizations have used inflated income information about gays and lesbians to try to drum up homophobia, which is almost a direct appeal to people's economic frustration. As an organizer, have you encountered that? To what extent do you think that the backlash against lesbian and gay rights has to do with the degree to which people in this country under capitalism are economically exploited and might seek a hateful outlet for that?

What they say is that these people do not qualify as a disenfranchised group because look at their income levels, as if the only way you could be disenfranchised is by income or lack of access to it. When they tell the untruth that all of us are economically privileged, of course that fans it. I can't say that I have personally seen that myth being picked up. When I come into contact with straight people, it's usually Black people. And it's usually around Black issues.

I don't know that the economic thing is such an issue for heterosexual Black communities; I think it's the moral thing. It's not so much that those white gays are rich, it's that those white gays are sinners, they're going against God. And they're also white, so they must be racist. And if you are a Black lesbian or gay man, then you must be a racial traitor. Those are the kinds of things that I hear, not so much that they're so rich. But I think there's an assumption on the part of most people of color that *most* white people are better off than we are anyway. It's an assumption that in some cases is accurate and in some cases is a myth that is perpetuated to keep people apart who should be in solidarity with each other because of class. It's a myth that keeps people away from each other, because there are some real commonalities between white and poor and being of color and poor. . . .

Can you say more about the links that you see between racism and gay oppression?

All of the different kinds of oppression are tied to each other, particularly when it comes to the kinds of repression and oppression that are practiced against different groups of people. When you look at a profile of how people who are oppressed experience their oppression, you see such similar components: demonization, scapegoating, police brutality, housing segregation, lack of access to certain jobs and employment, even the taking away of children—custody. They have done that to poor women of color and to poor women from time out of mind. This is not a new phenomenon that women who had children couldn't keep them because the state intervened. It's just that this group of people—lesbians and some gay men—are now experiencing the same thing. There are many similarities in what we experience. And also the same people who are hounding the usual scapegoat target groups, they're hounding the lesbian and gay community too. Our enemies are the same. That to me is the major thing that should be pulling us together.

The militant right wing in this country has targeted the lesbian and gay community, but they have targeted other groups, too. And they're really being quite successful during this time period. Does the Oklahoma City bombing have anything to do with the fate of the lesbian and gay community? Absolutely it does! The Oklahoma bombing epitomizes what dire straits the country is in as a whole, and if we had a responsible lesbian/gay/bisexual/transgendered people movement, it would be asking questions like, "Okay, what is our movement supposed to be doing now, post–Oklahoma City?" That really should have been a wake-up call.

We need to look at why those people are so antigovernment. The press never explains; all it gave us after the bombing was, "They're antigovernment and they're very upset over Waco." The reason those people are so antigovernment is that they think this government is a Zionist conspiracy that privileges Black savages. That's what they think. It's never explained that the reason they're so antigovernment is because they're white supremacists, they're anti-Semitic, they're homophobic, and they're definitely opposed to women's freedom too.

You asked about race and lesbian and gay oppression. One of the things that I wanted to say early on is that the clearest responses that I have to that are out of my own experience as a person who has those identities linked. I think that the clearest answers come from those of us who simultaneously experience these oppression; however, identity politics has been so maligned during this period that I almost hesitate to bring it up, because I don't want people to think I'm saying that the only reason it makes a difference is because it's bothering *me*. Of course, I am concerned about these issues because in my own life and experience I know what struggle is about, I know what oppression is about, and I have seen and experienced suffering myself.

That's not the only reason I am an activist, though. I care about everyone who is under siege. Some people think that because I'm so positively pro-Black and because I speak out against racism at every turn, that I don't really care if white people are suffering. Quite to the contrary. I care about all people who are not getting a fair shake, who are not getting an equal chance to fulfill their maximum potential and to live without fear and to have the basic things that every human needs: shelter, clothing, quality health care, meaningful and fairly compensated work, love and caring and freedom to express and to create, all those things that make life worth living. How limited would my politics be if I was only concerned about people like me! Given who I am as a Black woman and a lesbian and a person from a working-class home who is a socialist opposed to the exploitation of capitalism, well maybe that isn't so narrow. But what if I only cared about other Black lesbians? I would be sitting on the head of a pin.

The best heroes and the best heroines that we have throughout history have been those people who have gone beyond the narrow expectations that their demographic profile would lead you to expect. . . .

Those privileged, white gay men you mentioned that you view as setting the agenda currently—suppose homophobia could be eased in some ways. What stake do they have in other forms of oppression being ended?

The systems of oppression really do tie together. The plans and the strategies for oppressing and repressing our various groups are startlingly similar, and a society that is unjust, it's like cancer or a bleeding ulcer or something. You can't contain it. I guess it was Martin Luther King who said that when you have injustice anywhere, you really have it everywhere. It poisons the body politic of the society as a whole, and therefore you can't have singular solutions. Those white gay men who have disposable income and who think that all they need to do is get rid of the most blatant homophobia in corporate, government, and military settings, in the legal system and on TV, and everything else will fall into place, that they'll have a nice life in their little enclaves—they're dreaming. Let's say they got rid of homophobia in those places that I mentioned—which is not really getting rid of homophobia, it's getting rid of it in places that make their lives difficult—but suppose they were able to do that. If they were living in or near a city that has imploded, like Los Angeles, because of racial and economic exploitation and oppression, then how free are they going to be? I don't know why people can't understand how interconnected our fates are as creatures on the planet. . . .

One of the arguments people make for single-issue activism is that the gay community is too diverse to agree on an entire multi-issue agenda. They say, "We'll never get anywhere if we have to agree on everything else, so we're just going to work on one thing that we do agree about."

. . . If gay rights were put in place tomorrow, my behind would still be on fire. I would still be in ultimate danger here, because racism is still in the saddle. And so is class oppression, and so is sexual oppression. So getting a middle-of-the-road, mainstream gay rights agenda passed, how's that going to stop me from being raped? How's that going to help me not get breast cancer? How is it going to help the environment not get poisoned? How is it going to help the children of my community to have a chance for a decent life?

THE BEAUTY MYTH

49

Naomi Wolf

. . . Beauty pornography looks like this: The perfected woman lies prone, pressing down her pelvis. Her back arches, her mouth is open, her eyes shut, her nipples erect; there is a fine spray of moisture over her golden skin. The position is female superior; the stage of arousal, the plateau phase just preceding orgasm. On the next page, a version of her, mouth open, eyes shut, is about to tongue the pink tip of a lipstick cylinder. On the page after, another version kneels in the sand on all fours, her buttocks in the air, her face pressed into a towel, mouth open, eyes shut. The reader is looking through an ordinary women's magazine. In an ad for Reebok shoes, the woman sees a naked female torso, eyes averted. In an ad for Lily of France lingerie, she sees a naked female torso, eyes shut; for Opium perfume, a naked woman, back and buttocks bare, falls facedown from the edge of a bed; for Triton showers, a naked woman, back arched, flings her arms upward; for Jogbra sports bras, a naked

---

female torso is cut off at the neck. In these images, where the face is visible, it is expressionless in a rictus of ecstasy. The reader understands from them that she will have to look like that if she wants to feel like that.

Beauty sadomasochism is different: In an ad for Obsession perfume, a well-muscled man drapes the naked, lifeless body of a woman over his shoulder. In an ad for Hermès perfume, a blond woman trussed in black leather is hanging upside down, screaming, her wrists looped in chains, mouth bound. In an ad for Fuji cassettes, a female robot with a playmate's body, but made of steel, floats with her genitals exposed, her ankles bolted and her face a steel mask with slits for the eyes and mouth. In an ad for Erno Laszlo skin care products, a woman sits up and begs, her wrists clasped together with a leather leash that is also tied to her dog, who is sitting up in the same posture and begging. In an American ad for Newport cigarettes, two men tackle one woman and pull another by the hair; both women are screaming. In another Newport ad, a man forces a woman's head down to get her distended mouth around a length of spurting hose gripped in his fist; her eyes are terrified. In an ad for Saab automobiles, a shot up a fashion model's thighs is captioned, "Don't worry. It's ugly underneath." In a fashion layout in *The Observer* (London), five men in black menace a model, whose face is in shock, with scissors and hot iron rods. In *Tatler* and *Harper's and Queen*, "designer rape sequences (women beaten, bound and abducted, but immaculately turned out and artistically photographed)" appear. In Chris von Wangenheim's *Vogue* layout, Doberman pinschers attack a model. Geoffrey Beene's metallic sandals are displayed against a background of S and M accessories. The woman learns from these images that no matter how assertive she may be in the world, her private submission to control is what makes her desirable. . . .

Sexual "explicitness" is not the issue. We could use a lot more of that, if explicit meant honest and revealing; if there were a full spectrum of erotic images of uncoerced real women and real men in contexts of sexual trust, beauty pornography could theoretically hurt no one. Defenders of pornography base their position on the idea of freedom of speech, casting pornographic imagery as language. Using their own argument, something striking emerges about the representation of women's bodies: the representation is heavily censored. Because we see many versions of the naked Iron Maiden, we are asked to believe that our culture promotes the display of female sexuality. It actually shows almost none. It censors representations of women's bodies, so that only the official versions are visible. Rather than seeing images *of* female desire or that cater *to* female desire, we see mock-ups of living mannequins, made to contort and grimace, immobilized and uncomfortable under hot lights, professional set-pieces that reveal little about female sexuality. In the United States and Great Britain, which have no tradition of public nakedness, women rarely—and almost never outside a competitive context—see what other *women* look like naked; we see only identical humanoid products based loosely on women's bodies. . . .

. . . Leaving aside the issue of what violent sexual imagery does, it is still apparent that there is an officially enforced double standard for men's and women's nakedness in mainstream culture that bolsters power inequities.

The practice of displaying breasts, for example, in contexts in which the display of penises would be unthinkable, is portrayed as trivial because breasts are not "as naked" as penises or vaginas; and the idea of half exposing men in a similar way is moot because men don't have body parts comparable to breasts. But if we think about how women's genitals are physically concealed, unlike men's, and how women's breasts are physically exposed, unlike men's, it can be seen differently: women's breasts, then correspond to men's penises as the vulnerable "sexual flower" on the body, so that to display the former and conceal the latter makes women's bodies vulnerable while men's are protected. Cross-culturally, unequal nakedness almost always expresses power relations: in modern jails, male prisoners are stripped in front of clothed prison guards; in the antebellum South, young black male slaves were naked while serving the clothed white masters at table. To live in a culture in which women are routinely naked where men aren't is to learn inequality in little ways all day long. So even if we agree that sexual imagery is in fact a language, it is clearly one that is already heavily edited to protect men's sexual—and hence social—confidence while undermining that of women. . . .

Why this flood of images now? They do not arise simply as a market response to deep-seated, innate desires already in place. They arise also—and primarily—to set a sexual agenda and to *create* their versions of desire. The way to instill social values, writes historian Susan G. Cole, is to eroticize them. Images that turn women into objects or eroticize the degradation of women have arisen to counterbalance women's recent self-assertion. They are welcome and necessary because the sexes have come too close for the comfort of the powerful; they act to keep men and women apart, wherever the restraints of religion, law, and economics have grown too weak to continue their work of sustaining the sex war.

Heterosexual love, before the women's movement, was undermined by women's economic dependence on men. Love freely given between equals is the child of the women's movement, and a very recent historical possibility, and as such very fragile. It is also the enemy of some of the most powerful interests of this society.

If women and men in great numbers were to form bonds that were equal, nonviolent, and sexual, honoring the female principle no less or more than the male, the result would be more radical than the establishment's worst nightmares of homosexual "conversions." A mass heterosexual deviation into tenderness and mutual respect would mean real trouble for the status quo since heterosexuals are the most powerful sexual majority. The power structure would face a massive shift of allegiances: from each relationship might emerge a doubled commitment to transform society into one based publicly on what have traditionally been women's values, demonstrating all too well the appeal for both sexes of a world rescued from male dominance. The good news would get out on the street: free women have more fun; worse, so do free men.

Male-dominated institutions—particularly corporate interests—recognize the dangers posed to them by love's escape. Women who love themselves are threatening; but men who love real women, more so. Women who have broken out of gender roles have proved manageable: those few with power are being retrained as men. But with the apparition of numbers of men moving into passionate, sexual love of real women, serious money and authority could defect to join forces with the opposition. Such love would be a political upheaval more radical than the Russian Revolution and more destabilizing to the balance of world power than the end of the nuclear age. It would be the downfall of civilization as we know it—that is, of male dominance; and for heterosexual love, the beginning of the beginning.

Images that flatten sex into "beauty," and flatten the beauty into something inhuman, or subject her to eroticized torment, are politically and socioeconomically welcome, subverting female sexual pride and ensuring that men and women are unlikely to form common cause against the social order that feeds on their mutual antagonism, their separate versions of loneliness.

Barbara Ehrenreich, Elizabeth Hess, and Gloria Jacobs, in *Re-Making Love*, point out that the new market of sexual products demands quick-turnover sexual consumerism. That point applies beyond the sexual accessories market to the entire economy of consumption. The last thing the consumer index wants men and women to do is to figure out how to love one another: the $1.5-trillion retail-sales industry depends on sexual estrangement between men and women, and is fueled by sexual dissatisfaction. Ads do not sell sex—that would be counterproductive, if it meant that heterosexual women and men turned to one another and were gratified. What they sell is sexual discontent.

Though the survival of the planet depends on women's values balancing men's, consumer culture depends on maintaining a broken line of communication between the sexes and promoting matching sexual insecurities. Harley-Davidsons and Cuisinarts stand in for maleness and femaleness. But sexual satisfaction eases the stranglehold of materialism, since status symbols no longer look sexual, but irrelevant. Product lust weakens where emotional and sexual lust intensifies. The price we pay for artificially buoying up this market is our heart's desire. The beauty myth keeps a gap of fantasy between men and women. That gap is made with mirrors; no law of nature supports it. It keeps us spending vast sums of money and looking distractedly around us, but its smoke and reflection interfere with our freedom to be sexually ourselves.

Consumer culture is best supported by markets made up of sexual clones, men who want objects and women who want to be objects, and the object desired ever-changing, disposable, and dictated by the market. The beautiful object of consumer pornography has a built-in obsolescence, to ensure that as few men as possible will form a bond with one woman for years or for a lifetime, and to ensure that women's dissatisfaction with themselves will grow rather than diminish over time. Emotionally unstable relationships, high divorce rates, and a large population cast out into the sexual marketplace are good for business in a consumer economy. Beauty pornography is intent on making modern sex brutal and boring and only as deep as a mirror's mercury, antierotic for both men and women.

But even more powerful interests than the consumer index depend on heterosexual estrangement and are threatened by heterosexual accord. The military is supported by nearly one third of the United States government's budget; militarism depends on men choosing the bond with one another over the bond with women and children. Men who loved women would shift loyalties back to the family and community from which becoming a man is one long exile. Serious lovers and fathers would be unwilling to believe the standard propaganda of militarism: that their wives and children would benefit from their heroic death. Mothers don't fear mothers; if men's love for women and for their own children led them to define themselves first as fathers and lovers, the propaganda of war would fall on deaf ears: the enemy would be a father and partner too. This percentage of the economy is at risk from heterosexual love. Peace and trust between men and women who are lovers would be as bad for the consumer economy and the power structure as peace on earth for the military-industrial complex.

Heterosexual love threatens to lead to political change: an erotic life based on nonviolent mutuality rather than domination and pain teaches firsthand its appeal beyond the bedroom. A consequence of female self-love is that the woman grows convinced of social worth. Her love for her body will be unqualified, which is the basis of female identification. If a woman loves her own body, she doesn't grudge what other women do with theirs; if she loves femaleness, she champions its rights. It's true what they say about women: Women *are* insatiable. We *are* greedy. Our appetites do need to be controlled if things are to stay in place. If the world were ours too, if we believed we could get away with it, we *would* ask for more love, more sex, more money, more commitment to children, more food, more care. These sexual, emotional, and physical demands *would* begin to extend to social demands: payment for care of the elderly, parental leave, child care, etc. The force of female desire would be so great that society would truly have to reckon with what women want, in bed and in the world.

The economy also depends on a male work structure that denies the family. Men police one another's sexuality, forbidding each other to put sexual love and family at the center of their lives; women define themselves as successful according to their ability to sustain sexually loving relationships. If too many men and women formed common cause, that definition of success would make its appeal to men, liberating them from the echoing wind-tunnel of competitive masculinity. Beauty pornography is useful in preventing that eventuality: when aimed at men, its effect is to keep them from finding peace in sexual love. The fleeting chimera of the airbrushed centerfold, always receding before him, keeps the man destabilized in pursuit, unable to focus on the beauty of the woman—known, marked, lined, familiar—who hands him the paper every morning.

The myth freezes the sexual revolution to bring us full circle, evading sexual love with its expensive economic price tag. The nineteenth century constrained heterosexuality in arranged marriages; today's urban overachievers sign over their sexual fate to dating services, and their libido to work: one survey found that many yuppie couples share mutual impotence. The last century kept men and women apart in rigid gender stereotypes, as they are now estranged through rigid physical stereotypes. In the Victorian marriage market, men judged and chose; in the stakes of the beauty market, men judge and choose. It is hard to love a jailer, women knew when they had no legal rights. But it is not much easier to love a judge. Beauty pornography is a war-keeping force to stabilize the institutions of a society under threat from an outbreak of heterosexual love. . . .

When they discuss [their bodies], women lean forward, their voices lower. They tell their terrible secret. It's my breasts, they say. My hips. It's my thighs. I hate my stomach. This is not aesthetic distaste, but deep sexual shame. The parts of the body vary. But what each woman who describes it shares is the conviction that *that* is what the pornography of beauty most fetishizes. Breasts, thighs, buttocks, bellies; the most sexually central parts of women, whose "ugliness" therefore becomes an obsession. Those are the parts most often battered by abusive men. The parts that sex murderers most often mutilate. The parts most often defiled by violent pornography. The parts that beauty surgeons most often cut open. The parts that bear and nurse children and feel sexual. A misogynist culture has succeeded in making women hate what misogynists hate. . . .

MAIDEN VOYAGE: *Excursion into Sexuality and Identity Politics in Asian America*

50

Dana Y. Takagi

The topic of sexualities—in particular, lesbian, gay, and bisexual identities— is an important and timely issue in that place we imagine as Asian America. All of us in Asian American Studies ought to be thinking about sexuality and Asian American history for at least two compelling reasons.

One, while there has been a good deal of talk about the "diversity" of Asian American communities, we are relatively uninformed about Asian American subcultures organized specifically around sexuality. There are Asian American gay and lesbian social organizations, gay bars that are known for Asian clientele, conferences that have focused on Asian American lesbian and gay experiences, and . . . electronic bulletin boards catering primarily to gay Asians, their friends, and their lovers. I use the term "subcultures" here rather loosely and not in the classic sociological sense, mindful that the term is somewhat inaccurate since gay Asian organizations are not likely to view themselves as a gay subculture within Asian America any more than they are likely to think of themselves as an Asian American subculture within gay America. If anything, I expect that many of us view ourselves as on the margins of both communities. That state of marginalization in both communities is what prompts this essay and makes the issues raised in it all the more urgent for all of us—gay, straight, or somewhere-in-between. . . .

To be honest, it is not clear to me exactly how we ought to be thinking about these organizations, places, and activities. On the one hand, I would argue that an organization like the Association of Lesbians and Gay Asians (ALGA) ought to be catalogued in the annals of Asian American history. But on the other hand, having noted that ALGA is as Asian American as Sansei Live! or the National Coalition for Redress and Reparation, the very act of including lesbian and gay experiences in Asian American history, which seems important in a symbolic sense, produces in me a moment of hesitation. Not because I do not think that lesbian and gay sexualities are deserving of a place in Asian American history, but rather, because the inscription of non-straight

From: Russell Leong, ed., *Asian American Sexualities: Dimensions of the Gay and Lesbian Experience.* (New York: Routledge, 1996), pp. 21–35. Reprinted by permission.

sexualities in Asian American history immediately casts theoretical doubt about how to do it. As I will suggest, the recognition of different sexual practices and identities that also claim the label Asian American presents a useful opportunity for rethinking and reevaluating notions of identity that have been used, for the most part, unproblematically and uncritically in Asian American Studies.

The second reason, then, that we ought to be thinking about gay and lesbian sexuality and Asian American Studies is for the theoretical trouble we encounter in our attempts to situate and think about sexual identity and racial identity. Our attempts to locate gay Asian experiences in Asian American history render us "uninformed" in an ironic double sense. On the one hand, the field of Asian American Studies is mostly ignorant about the multiple ways that gay identities are often hidden or invisible within Asian American communities. But the irony is that the more we know, the less we know about the ways of knowing. On the other hand, just at the moment that we attempt to rectify our ignorance by adding say, the lesbian, to Asian American history, we arrive at a stumbling block, an ignorance of how to add her. Surely the quickest and simplest way to add her is to think of lesbianism as a kind of ad hoc subject-position, a minority within a minority. But efforts to think of sexuality in the same terms that we think of race, yet simultaneously different from race in certain ways, and therefore, the inevitable "revelation" that gays/lesbians/bisexuals are like minorities but also different too, is often inconclusive, frequently ending in "counting" practice. While many minority women speak of "triple jeopardy" oppression—as if class, race, and gender could be disentangled into discrete additive parts—some Asian American lesbians could rightfully claim quadruple jeopardy oppression—class, race, gender, and sexuality. Enough counting. Marginalization is not as much about the quantities of experiences as it is about qualities of experience. And, as many writers, most notably feminists, have argued, identities whether sourced from sexual desire, racial origins, languages of gender, or class roots, are simply not additive.[1]

NOT COUNTING

A discussion of sexualities is fraught with all sorts of definition conundrums. What exactly does it mean, sexualities? The plurality of the term may be unsettling to some who recognize three (or two, or one) forms of sexual identity: gay, straight, bisexual. But there are those who identify as straight, but regularly indulge in homoeroticism, and, of course, there are those who claim the identity gay/lesbian, but engage in heterosexual sex. In addition, some people identify themselves sexually but do not actually have sex, and, there are those who claim celibacy as a sexual practice. For those who profess a form of sexual identity that is, at some point, at odds with their sexual practice or sexual desire, the idea of a single, permanent, or even stable sexual identity is confining and inaccurate. Therefore, in an effort to capture the widest possible range of human sexual practices, I use the term sexualities to refer to the variety of practices and identities that range from homoerotic to heterosexual desire. In this essay, I am concerned mainly with homosexual desire and the question of what happens when we try to locate homosexual identities in Asian American history.

Writing, speaking, acting queer. Against a backdrop of lotus leaves, sliding shoji panels, and the mountains of Guilin. Amid the bustling enclaves of Little Saigon, Koreatown, Chinatown, and Little Tokyo. Sexual identity, like racial identity, is one of many types of recognized "difference." If marginalization is a qualitative state of being and not simply a quantitative one, then what is it about being "gay" that is different from "Asian American?"

The terms "lesbian" and "gay," like "Third World," "woman," and "Asian American," are political categories that serve as rallying calls and personal affirmations. In concatenating these identities we create and locate ourselves in phrases that seem a familiar fit: black gay man, Third World woman, working-class Chicana lesbian, Asian American bisexual, etc. But is it possible to write these identities—like Asian American gay—without writing oneself into the corners that are either gay and only gay, or, Asian American and only Asian American? Or, as Trinh T. Minh-ha put it, "How do you inscribe difference without bursting into a series of euphoric narcissistic accounts of yourself and your own kind?"[2]

It is vogue these days to celebrate difference. But underlying much contemporary talk about difference is the assumption that differences are comparable things. For example, many new social movements activists, including those in the gay and lesbian movement, think of themselves as patterned on the "ethnic model."[3] And for many ethnic minorities, the belief that "gays are oppressed too" is a reminder of a sameness, a common political project in moving margin to center, that unites race-based movements with gays, feminists, and greens. The notion that our differences are "separate but equal" can be used to call attention to the specificity of experiences or to rally the troops under a collective banner. Thus, the concept of difference espoused in identity politics may be articulated in moments of what Spivak refers to as "strategic essentialism" or in what Hall coins "positionalities." But in the heat of local political struggles and coalition building, it turns out that not all differences are created equally. For example, Ellsworth recounts how differences of race, nationality, and gender, unfolded in the context of a relatively safe environment, the university classroom:

> Women found it difficult to prioritize expressions of racial privilege and oppression when such prioritizing threatened to perpetuate their gender oppression. Among international students, both those who were of color and those who were White found it difficult to join their voices with those of U.S. students of color when it meant a subordination of their oppressions as people living under U.S. imperialist policies and as students for whom English was a second language. Asian American women found it difficult to join their voices with other students of color when it meant subordinating their specific oppressions as Asian Americans. I found it difficult to speak as a White woman about gender oppression when I occupied positions of institutional power relative to all students in the class, men and women, but positions of gender oppression relative to students who were White men, and in different terms, relative to students who were men of color.[4]

The above example demonstrates the tensions between sameness and difference that haunt identity politics.

There are numerous ways that being "gay" is not like being "Asian." Two broad distinctions are worth noting. The first . . . is the relative invisibility of sexual identity compared with racial identity. While both can be said to be socially constructed, the former are performed, acted out, and produced, often in individual routines, whereas the latter tends to be more obviously "written" on the body and negotiated by political groups.[5] Put another way, there is a quality of voluntarism in being gay/lesbian that is usually not possible as an Asian American. One has the option to present oneself as "gay" or "lesbian," or alternatively, to attempt to "pass," or, to stay in "the closet," that is, to hide one's sexual preference.[6] However, these same options are not available to most racial minorities in face-to-face interactions with others.

As Asian Americans, we do not think in advance about whether or not to present ourselves as "Asian American," rather, that is an identification that is worn by us, whether we like it or not, and which is easily read off of us by others.

A second major reason that the category "gay" ought to be distinguished from the category "Asian American" is for the very different histories of each group. Studying the politics of being "gay" entails on the one hand, an analysis of discursive fields, ideologies, and rhetoric about sexual identity, and on the other hand, knowledge of the history of gays/lesbians as subordinated minorities relative to heterosexuals. Similarly, studying "Asian America" requires analysis of semantic and rhetorical discourse in its variegated forms, racist, apologist, and paternalist, and requires in addition, an understanding of the specific histories of the peoples who recognize themselves as Asian or Asian American. But the specific discourses and histories in each case are quite different. Even though we make the same intellectual moves to approach each form of identity, that is, a two-tracked study of ideology on the one hand, and history on the other, the particular ideologies and histories of each are very different.[7]

In other words, many of us experience the worlds of Asian America and gay America as separate places—emotionally, physically, intellectually. We sustain the separation of these worlds with our folk knowledge about the family-centeredness and supra-homophobic beliefs of ethnic communities. Moreover, it is not just that these communities know so little of one another, but, we frequently take great care to keep those worlds distant from each other. What could be more different than the scene at gay bars like "The End Up" in San Francisco, or "Faces"in Hollywood, and, on the other hand, the annual Buddhist church bazaars in the Japanese American community or Filipino revivalist meetings?[8] These disparate worlds occasionally collide through individuals who manage to move, for the most part, stealthily, between these spaces. But it is the act of deliberately bringing these worlds closer together that seems unthinkable. Imagining your parents, clutching bento box lunches, thrust into the smoky haze of a South of Market leather bar in San Francisco is no less strange a vision than the idea of Lowie taking Ishi, the last of his tribe, for a cruise on Lucas' Star Tours at Disneyland. "Cultural strain," the anthropologists would say. Or, as Wynn Young, laughing at the prospect of mixing his family with his boyfriend, said, "Somehow I just can't picture this conversation at the dinner table, over my mother's homemade barbecued pork: 'Hey, Ma. I'm sleeping with a sixty-year-old white guy who's got three kids, and would you please pass the soy sauce?'"[9]

Thus, "not counting" is a warning about the ways to think about the relationship of lesbian/gay identities to Asian American history. While it may seem politically efficacious to toss the lesbian onto the diversity pile, adding one more form of subordination to the heap of inequalities, such a strategy glosses over the particular or distinctive ways sexuality is troped in Asian America. Before examining the possibilities for theorizing "gay" and "Asian American" as nonmutually exclusive identities, I turn first to a fuller description of the chasm of silence that separates them.

SILENCES

The concept of silence is a doggedly familiar one in Asian American history. For example, Hosokawa characterized the Nisei as "Quiet Americans" and popular media discussions of the "model minority" typically describe Asian American students as "quiet" along with "hard working" and "successful." In the popular dressing of Asian American identity, silence has functioned as a metaphor for the assimilative and positive imagery of the "good" minorities.

More recently, analysis of popular imagery of the "model minority" suggests that silence ought to be understood as an adaptive mechanism to a racially discriminatory society rather than as an intrinsic part of Asian American culture.[10]

If silence has been a powerful metaphor in Asian American history, it is also a crucial element of discussions of gay/lesbian identity, albeit in a somewhat different way. In both cases, silence may be viewed as the oppressive cost of a racially biased or heterosexist society. For gays and lesbians, the act of coming out takes on symbolic importance, not just as a personal affirmation of "this is who I am," but additionally as a critique of expected norms in society, "we are everywhere." While "breaking the silence" about Asian Americans refers to crashing popular stereotypes about them, and shares with the gay act of "coming out" the desire to define oneself rather than be defined by others, there remains an important difference between the two.

The relative invisibility of homosexuality compared with Asian American identity means that silence and its corollary space, the closet, are more ephemeral, appear less fixed as boundaries of social identities, less likely to be taken-for-granted than markers of race, and consequently, more likely to be problematized and theorized in discussions that have as yet barely begun on racial identity. Put another way, homosexuality is more clearly seen as constructed than racial identity.[11] Theoretically speaking, homosexual identity does not enjoy the same privileged stability as racial identity. The borders that separate gay from straight, and, "in" from "out," are so fluid that in the final moment we can only be sure that sexual identities are . . . "less a matter of final discovery than a matter of perpetual invention."[12]

Thus, while silence is a central piece of theoretical discussions of homosexuality, it is viewed primarily as a negative stereotype in the case of Asian Americans. What seems at first a simple question in gay identity of being "in" or "out" is actually laced in epistemological knots.

For example, a common question asked of gays and lesbians by one another, or by straights, is, "Are you out?" The answer to that question (yes and no) is typically followed by a list of who knows and who does not (e.g., my coworkers know, but my family doesn't . . .). But the question of who knows or how many people know about one's gayness raises yet another question, "how many, or which, people need to know one is gay before one qualifies as 'out'?" Or as Fuss says, "To be out, in common gay parlance, is precisely to be no longer out; to be out is to be finally outside of exteriority and all the exclusions and deprivations such outsider-hood imposes. Or, put another way, to be out is really to be in—inside the realm of the visible, the speakable, the culturally intelligible."[13]

Returning to the issue of silence and homosexuality in Asian America, it seems that topics of sex, sexuality, and gender are already diffused through discussions of Asian America.[14] For example, numerous writers have disclosed, and challenged, the panoply of contradictory sexually charged images of Asian American women as docile and subservient on the one hand, and as ruthless mata-hari, dragon-lady aggressors on the other. And of course, Frank Chin's tirades against the feminization of Asian American men has been one reaction to the particular way in which Asian Americans have been historically (de)sexualized as racial subjects. Moving from popular imagery of Asian Americans, the people, to Asia, the nation, Rey Chow uses Bertolucci's blockbuster film, *The Last Emperor*, to illustrate what she calls, "the metaphysics of feminizing the other (culture)" wherein China is predictably cast as a "feminized, eroticized, space."[15]

That the topic of homosexuality in Asian American studies is often treated in whispers, if mentioned at all, should be some indication of trouble. It is noteworthy, I think, that in the last major anthology on Asian American women, *Making Waves*, the author of the essay on Asian American lesbians was the only contributor who did not wish her last name to be published.[16] Of course, as we all know, a chorus of sympathetic bystanders is chanting about homophobia, saying, "she was worried about her job, her family, her community . . ." Therefore, perhaps a good starting point to consider lesbian and gay identities in Asian American studies is by problematizing the silences surrounding homosexuality in Asian America.

It would be easy enough for me to say that I often feel a part of me is "silenced" in Asian American Studies. But I can hardly place all of the blame on my colleagues. Sometimes I silence myself as much as I feel silenced by them. And my silencing act is a blaring welter of false starts, uncertainties, and anxieties. For example, on the one hand, an omnipresent little voice tells me that visibility is better than invisibility, and therefore, coming out is an affirming social act. On the other hand, I fear the awkward silences and struggle for conversation that sometimes follow the business of coming out. One has to think about when and where to time the act since virtually no one has ever asked me, "Are you a lesbian?" Another voice reminds me that the act of coming out, once accomplished, almost always leaves me wondering whether I did it for myself or them. Not only that, but at the moment that I have come out, relief that is born of honesty and integrity quickly turns to new uncertainty. This time, my worry is that someone will think that in my coming out, they will now have a ready-made label for me, lesbian. The prospect that someone may think that they know me because they comprehend the category lesbian fills me with stubborn resistance. The category lesbian calls up so many different images of women who love other women that I do not think that any one—gay or straight—could possibly know or find me through that category alone. No wonder that I mostly find it easier to completely avoid the whole issue of sexual identity in discussions with colleagues.

There are so many different and subtle ways to come out. I am not much of a queer nation type, an "in your face" queer—I catalogue my own brand of lesbian identity as a kind of Asian American "take" on gay identity. I do not wear pink triangles, have photos of girls kissing in my living room, or, make a point of bringing up my girlfriend in conversation. In effect, my sexual identity is often backgrounded or stored somewhere in between domains of public and private. I used to think that my style of being gay was dignified and polite—sophisticated, civilized, and genteel. Work was work and home was home. The separation of work and home has been an easy gulf to maintain, less simple to bridge. Recently, however, I have come to think otherwise.

But all this talk about me is getting away from my point, which is that while it would be easy enough for me to say many of us feel "silenced," which alone might argue for inclusion of gay sexualities in discourse about the Asian American experience, that is not enough. Technically speaking then, the terms "addition" and "inclusion" are misleading. I'm afraid that in using such terms, the reader will assume that by adding gay/lesbian experiences to the last week's topics in a course on Asian American contemporary issues, or, by including lesbians in a discussion of Asian women, the deed is done. Instead, I want to suggest that the task is better thought of as just begun, that the topic of sexualities ought to be envisioned as a means, not an end, to theorizing about the Asian American experience.

NOTES

My special thanks to Russell Leong for his encouragement and commentary on this essay.

1. See Teresa de Lauretis, "Feminist Studies/Critical Studies: Issues, Terms, and Contexts," in *Feminist Studies/Critical Studies*, ed. Teresa de Lauretis (Bloomington: Indiana University Press, 1986), 1–19; bell hooks, *Yearning: Race, Gender and Cultural Politics* (Boston: South End Press, 1990); Trinh T. Minh-ha, *Woman, Native, Other* (Bloomington: Indiana University Press, 1989); Chandra Talpade Mohanty, "Under Western Eyes: Feminist Scholarship and Colonialist Discourses," in *Third World Women and the Politics of Feminism*, eds. Chandra Talpade Mohanty, Ann Russo, and Lourdes Torres (Bloomington: Indiana University Press, 1991), 52–80; Linda Alcoff, "Cultural Feminism versus Post-Structuralism: The Identity Crisis in Feminist Theory," *Signs*, 13: 3 (1988): 405–37.

2. Trin T. Minh-ha, 28.

3. Epstein (1987). Jeffrey Escoffier, editor of *Outlook* magazine made this point in a speech at the American Educational Research Association meetings in San Francisco, April 24, 1992.

4. See Elizabeth Ellsworth, "Why Doesn't This Feel Empowering? Working through the Repressive Myths of Critical Pedagogy," 59: 3 (1989): 297–324.

5. Of course there are exceptions, for example, blacks that "pass" and perhaps this is where homosexuality and racial identity come closest to one another, amongst those minorities who "pass" and gays who can also "pass."

6. I do not mean to suggest that there is only one presentation of self as lesbian. For example, one development recently featured in the *Los Angeles Times* is the evolution of "lipstick lesbians" (Van Gelder, 1991). The fashion issue has also been discussed in gay/lesbian publications. For example, Stein (1988) writing for *Outlook* has commented on the lack of correspondence between fashion and sexual identity, "For many, you can dress as femme one day and a butch the next. . . ."

7. Compare for example the histories: Ronald Takaki's *Strangers from a Different Shore: A History of Asian Americans* (Boston: Little, Brown, 1989) with Jonathan Katz's *Gay American History* (New York: Meridian, 1992), Michel Foucault's *The History of Sexuality* (New York: Vintage, 1980), and David Greenberg, *The Construction of Homosexuality* (Chicago: University of Chicago Press, 1988).

8. See Steffi San Buenaventura, "The Master and the Federation : A Filipino-American Social Movement in California and Hawaii," *Social Process in Hawaii* 33 (1991): 169–93.

9. Wynn Young, "Poor Butterfly," *Amerasia Journal* 17 : 2 (1991): 118.

10. See Keith Osajima, "Asian Americans as the Model Minority: An Analysis of the Popular Press Image in the 1960s and 1980s," *Reflections on Shattered Windows: Promises and Prospects for Asian American Studies*, eds. Gary Y. Okihiro, Shirley Hune, Arthur A. Hansen, and John M. Liu (Pullman: Washington State University Press, 1988), 165–74.

11. See Judith Butler, *Gender Trouble* (New York: Routledge, 1990).

12. Diana Fuss, "Inside/Out," *inside/out*, ed. Diana Fuss (New York: Routledge, 1991), 1–10.

13. Ibid.

14. Consider for example debates in recent times over intermarriage patterns, the controversy over Asian Americans dating white men, the Asian Men's calendar, and the continuation of discussions started over a decade ago about gender, assimilation, and nativism in Asian American literature.

15. See Rey Chow, *Woman and Chinese Modernity* (Minneapolis: University of Minnesota Press, 1991).

16. See Asian Women United, *Making Waves* (Boston: Beacon Press, 1989).

GETTING OFF ON FEMINISM

51

Jason Schultz

Minutes after my best friend told me he was getting married, I casually offered to throw a bachelor party in his honor. Even though such parties are notorious for their degradation of women, I didn't think this party would be much of a problem. Both the bride and groom considered themselves feminists, and I figured that most of the men attending would agree that sexism had no place in the celebration of this union. In fact, I thought the bachelor party would be a great opportunity to get a group of men together for a social event that didn't degenerate into the typical antiwomen, homophobic male-bonding thing. Still, ending one of the most sexist traditions in history—even for one night—was a lot tougher than I envisioned.

I have to admit that I'm not a *complete* iconoclast: I wanted to make the party a success by including at least some of the usual elements, such as good food and drink, great music, and cool things to do. At the same time, I was determined not to fall prey to traditional sexist party gimmicks such as prostitutes, strippers jumping out of cakes, or straight porn. But after nixing all the traditional lore, even *I* thought it sounded boring. What were we going to do except sit around and think about women?

"What about a belly dancer?" one of the ushers suggested when I confided my concerns to him. "That's not as bad as a stripper." I sighed. This was supposed to be an occasion for the groom and his male friends to get together, celebrate the upcoming marriage, and affirm their friendship and connection with each other as men. "What . . . does hiring a female sex worker have to do with any of that?" I shouted into the phone. I quickly regained my calm, but his suggestion still stung. We had to find some other way.

I wanted my party to be as "sexy" as the rest of them, but I had no idea how to do that in the absence of female sex workers. There was no powerful alternative image in our culture from which I could draw. I thought about renting some gay porn, or making it a cross-dressing party, but many of the guests were conservative, and I didn't want to scare anyone off. Besides, what would it say about a bunch of straight men if all we could do to be sexy was act queer for a night?

From: Rebecca Walker, ed., *To Be Real: Telling the Truth and Changing the Face of Feminism* (New York: Anchor Books, 1995), pp. 107–26. Reprinted by permission.

Over coffee on a Sunday morning, I asked some of the other guys what they thought was so "sexy" about having a stripper at a bachelor party.

"Well," David said, "it's just a gag. It's something kinda funny and sexy at the same time."

"Yeah," A.J. agreed. "It's not all that serious, but it's something special to do that makes the party cool."

"But *why* is it sexy and funny?" I asked. "Why can't we, as a bunch of guys, be sexy and funny ourselves?"

"'Cause it's easier to be a guy with other guys when there's a chick around. It gives you all something in common to relate to."

"Hmm. I think I know what you mean," I said. "When I see a stripper, I get turned on, but not in the same way I would if I was with a lover. It's more like going to a show or watching a flick together. It's enjoyable, stimulating, but it's not overwhelming or intimate in the same way that sex is. Having the stripper provides a common emotional context for us to feel turned on. But we don't have to do anything about it like we would if we were with a girl-friend, right?"

"Well, my girlfriend would kill me if she saw me checking out this strip-per," Greg replied. "But because it's kind of a male-bonding thing, it's not as threatening to our relationship. It's not because it's the stripper over her, it's because it's just us guys hanging out. It doesn't go past that."

Others agreed. "Yeah. You get turned on, but not in a serious way. It makes you feel sexy and sexual, and you can enjoy feeling that way with your friends. Otherwise, a lot of times, just hanging out with the guys is pretty boring. Especially at a bachelor party. I mean, that's the whole point, isn't it—to celebrate the fact that we're bachelors, and he"—referring to Robert, the groom—"isn't!"

Through these conversations, I realized that having a female sex worker at the party would give the men permission to connect with one another with-out becoming vulnerable. When men discuss sex in terms of actions—who they "did," and how and where they did it—they can gain recognition and validation of their sexuality from other men without having to expose their *feelings* about sex.

"What other kinds of things make you feel sexy like the stripper does?" I asked several of the guys.

"Watching porn sometimes, or a sexy movie."

A.J. said, "Just getting a look from a girl at a club. I mean, she doesn't even have to talk to you, but you still feel sexy and you can still hang out with your friends."

Greg added, "Sometimes just knowing that my girlfriend thinks I'm sexy, and then talking about her with friends, makes me feel like I'm the man. Or I'll hear some other guy talk about his girlfriend in a way that reminds me of mine, and I'll still get that same feeling. But that doesn't happen very often, and usually only when talking with one other guy.

This gave me an idea. "I've noticed that same thing, both here and at school with my other close guy friends. Why doesn't it happen with a bunch of guys, say at a party?"

"I don't know. It's hard to share a lot of personal stuff with guys," said Adam, "especially about someone you're seeing, if you don't feel comfortable. Well, not comfortable, because I know most of the guys who'll be at the party, but it's more like I don't want them to hassle me, or I might say something that freaks them out."

"Or you're just used to guys talking shit about girls," someone else added. "Like at a party or hanging out together. They rag on them, or pick out who's the cutest or who wants to do who. That's not the same thing as really talking about what makes you feel sexy."

"Hmm," I said. "So it's kind of like if I were to say that I liked to be tied down to the bed, no one would take me seriously. You guys would probably crack up laughing, make a joke or two, but I'd never expect you to actually join in and talk about being tied up in a serious way. It certainly wouldn't feel 'sexy,' would it? At least not as much as the stripper."

"Exactly. You talking about being tied down here is fine, 'cause we're into the subject of sex on a serious kick and all. But at a party, people are bullshitting each other and gabbing, and horsing around. The last thing most of us want is to trip over someone's personal taste or start thinking someone's a little queer."

"You mean queer as in homosexual?" I asked.

"Well, not really, 'cause I think everyone here is straight. But more of queer in the sense of perverted or different. I mean, you grow up in high school thinking that all guys are basically the same. You all want the same thing from girls in the same way. And when someone like you says you like to be tied down, it's kinda weird—almost like a challenge. It makes me have to respond in a way that either shows me agreeing that I also like to be tied down or not. And if someone's a typical guy and he says that, it makes you think he's different—not the same guy you knew in high school. And if he's not the same guy, then it challenges you to relate to him on a different level."

"Yeah, I guess in some ways it's like relating to someone who's gay," Greg said. "He can be cool and all, and you can get along totally great. But there's this barrier that's hard to cross over. It kinda keeps you apart. And that's not what you want to feel toward your friends, especially at a party like this one, where you're all coming together to chill."

As the bachelor party approached, I found myself wondering whether my friends and I could "come together to chill"—and affirm our status as sexual straight men—without buying into homophobic or sexist expressions. At the same time, I was doing a lot of soul-searching on how we could challenge the dominant culture's vision of male heterosexuality, not only by deciding against having a stripper at our party, but also by examining and redefining our own relationships with women.

SEX AND THE SENSITIVE MAN

According to the prevailing cultural view, "desirable" hetero men are inherently dominant, aggressive, and, in many subtle and overt ways, abusive to women. To be sexy and powerful, straight men are expected to control and contrive a sexuality that reinforces their authority. Opposing these notions of power subjects a straight guy to being branded "sensitive," submissive, or passive—banished to the nether regions of excitement and pleasure, the unmasculine, asexual, "vanilla" purgatory of antieroticism. Just as hetero women are often forced to choose between the images of the virgin and the whore, modern straight men are caught in a cultural tug-of-war between the Marlboro Man and the Wimp.

So where does that leave straight men who want to reexamine what a man is and change it? Can a good man be sexy? Can a sexy man be good? What is good sex, egalitarian sex? More fundamentally, can feminist women and men coexist comfortably, even happily, within the same theoretical framework—or the same bedroom?

Relationships with men remain one of the most controversial topics among feminists today. Having sex, negotiating emotional dependency, and/or raising children force many hetero couples to balance their desire to be together with the oppressive dynamics of sexism. In few other movements are the oppressor group and the oppressed group so intimately linked.

But what about men who support feminism? Shouldn't it be okay for straight feminist women to have sex with them? Straight men aren't always oppressive in their sexuality, are they?

You may laugh at these questions, but they hold serious implications for straight feminist sex. I've seen many relationships between opposite-sex activists self-destruct because critical assumptions about power dynamics and desires were made in the mind, but not in the bed. I've even been told that straight male feminists can't get laid without (A) feeling guilty; (B) reinforcing patriarchy; or (C) maintaining complete passivity during sexual activity. Each of these three options represents common assumptions about the sexuality of straight men who support feminism. Choice A, "feeling guilty," reflects the belief that straight male desire inherently contradicts the goals of feminism and fails to contribute to the empowerment of women. It holds that any man who enjoys sex with a woman must be benefiting from sexist male privilege, not fighting against it. In other words, het sex becomes a zero-sum game where if men gain, feminism loses.

Choice B represents the assumption that hetero male sex is inherently patriarchal. Beyond merely being of no help, as in Choice A, straight male sexuality is seen as part of the problem. Within this theory, one often hears statements such as "all heterosexual sex is rape." Even though these statements are usually taken out of context, the ideas behind them are problematic. In essence, they say that you can never have a male/female interaction that isn't caught up in oppressive dynamics. Men and women can never be together, especially in such a vulnerable exchange as sexuality, without being subject to the misdistribution of power in society.

The third choice, "maintaining complete passivity," attempts a logical answer to the above predicament. In order to come even close to achieving equality in heterosexuality (and still get laid), men must "give up" all their power through inactivity. A truly feminist man should take no aggressive or dominant position. He should, in fact, not act at all; he should merely lie back and allow the woman to subvert male supremacy through her complete control of the situation. In other words, for a man and a woman to share sexuality on a "level playing field," the man must remove all symptoms of his power through passivity, even though the causes of that inequality (including his penis!) still exist. . . .

Does it have to be this way? Must male heterosexuality always pose a threat to feminism? What about the sensitive guy? Wasn't that the male cry (whimper) of the nineties? Sorry, but all the media hype about sensitivity never added up to significant changes in behavior. Straight male sexuality still remains one of the most underchallenged areas of masculinity in America. Some men *did* propose a different kind of sexuality for straight men in the 1970s, one that emphasizes feelings and sensitivity and emotional connection. But these efforts failed to affect our ideas in any kind of revolutionary way. Now, instead of a "sexy" sensitive guy, men's magazines are calling for the emergence of the "Post-Sensitive Man," while scientific studies tell us that women prefer Clint Eastwood over Michael Bolton.

Why did sensitivity fail? Were straight women, even feminists, lying to men about what they wanted? The answer is "yes" and "no." I don't think sensitivity was the culprit. I think the problem was men's passivity, or more specifically, men's lack of assertiveness and power.

In much of our understanding, power is equated with oppression: images of white supremacists dominating people of color, men dominating women, and the rich dominating the poor underline the histories of many cultures and societies. But power need not always oppress others. One can, I believe, be powerful in a nonoppressive way.

In order to find this sort of alternative, we need to examine men's experience with power and sexuality further. Fortunately, queer men and women have given us a leg up on the process by reenergizing the debate about what is good sex and what is fair sex. Gay male culture has a long history of exploring nontraditional aspects of male sexuality, such as cross-dressing, bondage and dominance, and role playing. These dynamics force gay men to break out of a singular experience of male sexual desire and to examine the diversity within male sexuality in the absence of gender oppression. Though gay men's culture still struggles with issues such as the fetishizing of men of color and body fascism, it does invite greater exploration of diversity than straight male culture. Gay culture has broader and more inclusive attitudes about what is sexy and a conception of desire that accommodates many types of sex for many types of gay men. For straight men in our culture, there is such a rigid definition of "sexy" that it leaves us few options besides being oppressive, overbearing, or violent.

Part of the success that gay male culture enjoys in breaking out of monolithic notions of male sexuality lies in the acceptance it receives from its partners and peers. Camp, butch, leather, drag-queer culture is constantly affirming the powerful presence of alternative sexualities. Straight male culture, on the other hand, experiences a lack—a void of acceptance—whenever it tries to assert some image other than the sexist hetero male. Both publicly and in many cases privately, alternative straight male sexualities fail to compete for attention and acceptance among hetero men and women. . . .

We need to assert a new feminist sexuality for men, one that competes with the traditional paradigm but offers a more inviting notion of how hetero men can be sexual while tearing apart the oppressive and problematic ways in which so many of us have experienced sexuality in the past. We need to find new, strong values and ideas of male heterosexuality instead of passive identifies that try to distance us from sexist men. We need to stop trying to avoid powerful straight sexuality and work to redefine what our power means and does. We need to find strength and desire outside of macho, antiwomen ways of being masculine. . . .

Gay men and lesbians have engaged in a cultural dialogue around sexuality over the last twenty-five years; straight women are becoming more and more vocal. But straight men have been almost completely silent. This silence, I think, stems in large part from fear: our cultures tell us that being a "real" man means not being feminine, not being gay, and not being weak. They warn us that anyone who dares to stand up to these ideas becomes a sitting target to have his manhood shot down in flames.

BREAKING THE SILENCE

Not becoming a sitting target to have *my* manhood shot down was high on my mind when the evening of my best friend's bachelor party finally arrived. But I was determined not to be silent about how I felt about the party and about new visions for straight men within our society.

We decided to throw the party two nights before the wedding. We all gathered at my house, each of us bringing a present to add to the night's activities. After all the men had arrived, we began cooking dinner, breaking open beer and champagne, and catching up on where we had left off since we last saw each other.

During the evening, we continued to talk off and on about why we didn't have a stripper or prostitute for the party. After several rounds of margaritas and a few hands of poker, tension started to build around the direction I was pushing the conversation.

"So what don't you like about strippers?" David asked me.

This was an interesting question. I was surprised not only by the guts of David to ask it, but also by my own mixed feelings in coming up with an answer. "It's not that I don't like being excited, or turned on, per se," I responded. "In fact, to be honest, watching a female stripper is an exciting and erotic experience for me. But at the same time, it's a very uncomfortable one. I get a sense when I watch her that I'm participating in a misuse of pleasure, if that makes sense."

I looked around at my friends. I couldn't tell whether the confused looks on their faces were due to the alcohol, the poker game, or my answer, so I continued. "Ideally, I would love to sit back and enjoy watching someone express herself sexually through dance, seduction, flirtation—all the positive elements I associate with stripping," I said. "But at the same time, because so many strippers are poor and forced to perform in order to survive economically, I feel like the turn-on I get is false. I feel like I get off easy, sitting back as the man, paying for the show. No one ever expects me to get up on stage.

"And in that way, it's selling myself short sexually. It's not only saying very little about the sexual worth of the woman on stage, but the sexual worth of me as the viewer as well. By *only* being a viewer—just getting off as a member of the audience—the striptease becomes a very limiting thing, an imbalanced dynamic. If the purpose is for me to feel sexy and excited, but not to act on those feelings, I'd rather find a more honest and direct way to do it. So personally, while I would enjoy watching a stripper on one level, the real issues of economics, the treatment of women, and the limitation of my own sexual personae push me to reject the whole stripper thing in favor of something else."

"But what else do you do to feel sexy?" A. J. asked.

"That's a tough question," I said. "Feeling sexy often depends on the way other people act toward you. For me, right now, you guys are a huge way for me to feel sexy. [Some of the men cringe.] I'm not saying that we have to challenge our sexual identities, although that's one way. But we can cut through a lot of this locker-room macho crap and start talking with each other about how we feel sexually, what we think, what we like, etc. Watching the stripper makes us feel sexy because we get turned on through the dynamic between her performance and our voyeurism. We can find that same erotic connection with each other by re-creating that context between us. In such a case, we're still heterosexual—we're no more having sex with each other than we are with the stripper. But we're not relying on the imbalanced dynamic of sex work to feel pleasure as straight men." . . .

The guys were silent for a few seconds, but soon afterwards, the ice seemed to break. . . . They agreed that, as heterosexual men, we should be able to share with each other what we find exciting and shouldn't *need* a female stripper to feel sexy. In some ways it may have been the desire to define their own sexuality that changed their minds; in others it may have been a traditionally masculine desire to reject any dependency on women. In any case, other men began to speak of their own experiences with pleasure and desire, and we continued to talk throughout the night, exploring the joys of hot sex, one-night stands, and even our preferences for certain brands of condoms. We discussed the ups and down of monogamy versus "open" dating and the pains of long-distance relationships.

Some men continued to talk openly about their desire for straight pornography or women who fit the traditional stereotype of feminity. But others contradicted this, expressing their wish to move beyond that image of women in their lives. The wedding, which started out as the circumstance for our gathering, soon fell into the background of our thoughts as we focused away from institutional ideas of breeder sexuality and began to find common ground through our real-life experiences and feelings as straight men. In the end, we all toasted the groom, sharing stories, jokes, and parts of our lives that many of us had never told. Most importantly, we were able to express ourselves sexually without hiding who we were from each other.

Thinking back on the party, I realized that the hard part was figuring out what we all wanted and how to construct a different way of finding that experience. The other men there wanted it just as much as I did. The problem was that we had no ideas of what a different kind of bachelor party might look like. Merely eliminating the old ways of relating (i.e., the female sex workers) left a gap, an empty space which in many ways *felt* worse than the sexist connection that existed there before; we felt passive and powerless. Yet we found a new way of interacting—one that embraced new ideas and shared the risk of experiencing them.

Was the party sexy? Did we challenge the dominance of oppressive male sexuality? Not completely, but it was a start. I doubt anyone found my party as "sexy" as traditional ones might be, but the dialogue has to start somewhere. It's going to take a while to generate the language and collective tension to balance the cultural image of heterosexual male sexuality with true sexual diversity. Still, one of my friends from high school—who's generally on the conservative end of most issues—told me as he was leaving that of all the bachelor parties he had been to, this was by far the best one. "I had a great time," he said. "Even without the stripper." . . .

When it comes to sex, feminist straight men must become participants in the discourse about our own sexuality. We have to fight the oppressive images of men as biological breeders and leering animals. We must find ways in which to understand our diverse backgrounds, articulate desires that are not oppressive, and acknowledge the power we hold. We must take center stage when it comes to articulating our views in a powerful voice. I'm not trying to prescribe any particular form of sexuality or specify what straight men should want. But until we begin to generate our own demands and desires in an honest and equitable way for feminist straight women to hear, I don't think we can expect to be both good *and* sexy any time soon.

Violence and Social Control

MORE POWER THAN WE WANT:

Masculine Sexuality and Violence

52

Bruce Kokopeli and George Lakey

Masculine sexuality involves the oppression of women, competition among men, and homophobia (fear of homosexuality). Patriarchy, the systematic domination of women by men through unequal opportunities, rewards, punishments, and the internalization of unequal expectations through sex role differentiation, is the institution which organizes these behaviors. Patriarchy is men having more power, both personally and politically, than women of the same rank. This imbalance of power is the core of patriarchy, but definitely not the extent of it.

Sex inequality cannot be routinely enforced through open violence or even blatant discriminatory agreements—patriarchy also needs its values accepted in the minds of people. If as many young *women* wanted to be physicians as men, and as many young men *wanted* to be nurses as women, the medical schools and the hospitals would be hard put to maintain the masculine domination of health care; open struggle and the naked exercise of power would be necessary. Little girls, therefore, are encouraged to think "nurse" and boys to think "doctor."

Patriarchy assigns a list of human characteristics according to gender: women should be nurturant, gentle, in touch with their feelings, etc.; men

From: *Off Their Backs . . . and on Our Own Two Feet* (Philadelphia: New Society Publishers, 1983), pp. 17–24. Reprinted by permission.

should be productive, competitive, super-rational, etc. Occupations are valued according to these gender-linked characteristics, so social work, teaching, housework, and nursing are of lower status than business executive, judge, or professional football player.

When men do enter "feminine" professions, they disproportionately rise to the top and become chefs, principals of schools, directors of ballet, and teachers of social work. A man is somewhat excused from his sex role deviation if he at least dominates within the deviation. Domination, after all, is what patriarchy is all about.

Access to powerful positions by women (i.e., those positions formerly limited to men) is contingent on the women adopting some masculine characteristics, such as competitiveness. They feel pressure to give up qualities assigned to females (such as gentleness) because those qualities are considered inherently weak by patriarchal culture. The existence, therefore, of a woman like Indira Gandhi in the position of a dictator in no way undermines the basic sexist structure which allocates power to those with masculine characteristics.

Patriarchy also shapes men's sexuality so it expresses the theme of domination. Notice the masculine preoccupation with size. The size of a man's body has a lot to say about his clout or his vulnerability, as any junior high boy can tell you. Many of these schoolyard fights are settled by who is bigger than whom, and we experience in our adult lives the echoes of intimidation and deference produced by our habitual "sizing up" of the situation.

Penis size is part of this masculine preoccupation, this time directed toward women. Men want to have large penises because size equals power, the ability to make a woman "really feel it." The imagery of violence is close to the surface here, since women find penis size irrelevant to sexual genital pleasure. "Fucking" is a highly ambiguous word, meaning both intercourse and exploitation/assault.

It is this confusion that we need to untangle and understand. Patriarchy tells men that their need for love and respect can only be met by being masculine, powerful, and ultimately violent. As men come to accept this, their sexuality begins to reflect it. Violence and sexuality combine to support masculinity as a character ideal. To love a woman is to have power over her and to treat her violently if need be. The Beatles' song "Happiness Is a Warm Gun" is but one example of how sexuality gets confused with violence and power. We know one man who was discussing another man who seemed to be highly fertile—he had made several women pregnant. "That guy," he said, "doesn't shoot any blanks."

Rape is the end logic of masculine sexuality. Rape is not so much a sexual act as an act of violence expressed in a sexual way. The rapist's mind-set—that violence and sexuality *can* go together—is actually a product of patriarchal conditioning, for most of us men understand the same, however abhorrent rape may be to us personally.

In war, rape is astonishingly prevalent even among men who "back home" would not do it. In the following description by a marine sergeant who witnessed a gang rape in Vietnam, notice that nearly all the nine-man squad participated:

> They were supposed to go after what they called a Viet Cong whore. They went into her village and instead of capturing her, they raped her—every man raped her. As a matter of fact, one man said to me later that it was the first time he had ever made love to a woman with his boots on. The man who led the platoon, or the squad, was actually a private. The squad leader was a sergeant but he was a useless person and he let the private take over his squad. Later he said he took no part in the raid. It was against his morals. So instead of telling his squad not to do it, because they wouldn't listen to him anyway, the sergeant went into another side of the village and just sat and stared bleakly at the ground, feeling sorry for himself. But at any rate, they raped the girl, and then, the last man to make love to her, shot her in the head. [Vietnam Veterans Against the War, statement by Michael McClusker in *The Winter Soldier Investigation: An Inquiry Into American War Crimes.*]

Psychologist James Prescott adds to this account:

> What is it in the American psyche that permits the use of the word "love" to describe rape? And where the act of love is completed with a bullet in the head! [*Bulletin of the Atomic Scientists,* November 1975, p. 17.]

MASCULINITY AGAINST MEN: THE MILITARIZATION OF EVERYDAY LIFE

Patriarchy benefits men by giving us a class of people (women) to dominate and exploit. Patriarchy also oppresses men, by setting us at odds with each other and shrinking our life space.

The pressure to win starts early and never stops. Working-class gangs fight over turf; rich people's sons are pushed to compete on the sports field. British military officers, it is said, learned to win on the playing fields of Eton.

Competition is conflict held within a framework of rules. When the stakes are really high, the rules may not be obeyed; fighting breaks out. We men mostly relate through competition, but we know what is waiting in the wings. John Wayne is not a cultural hero by accident.

Men compete with each other for status as masculine males. Because masculinity equals power, this means we are competing for power. The ultimate proof of power/masculinity is violence. A man may fail to "measure up" to the macho stereotype in important ways, but if he can fight successfully with the person who challenges him on his deviance, he is still all right. The television policeman Baretta is strange in some ways: he is gentle with women and he cried when a man he loved was killed. However, he has what are probably the largest biceps in television and he proves weekly that he can beat up the toughs who come his way.

The close relationship between violence and masculinity does not need much demonstration. War used to be justified partly because it promoted "manly virtue" in a nation. Those millions of people in the woods hunting deer, in the National Rifle Association, and cheering on the bloodiest hockey teams are overwhelmingly men.

The world situation is so much defined by patriarchy that what we see in the wars of today is competition between various patriarchal ruling classes and governments breaking into open conflict. Violence is the accepted masculine form of conflict resolution. Women at this time are not powerful enough in the world situation for us to see mass overt violence being waged on them. But the violence is in fact there; it is hidden through its legitimization by the state and by culture.

In everyday middle-class life, open violence between men is of course rare. The defining characteristics of masculinity, however, are only a few steps removed from violence. Wealth, productivity, or rank in the firm or institution translate into power—the capacity (whether or not exercised) to dominate. The holders of power in even polite institutions seem to know that violence is at their fingertips, judging from the reactions of college presidents to student protest in the 1960s. We know of one urban "pacifist" man, the head of a theological seminary, who was barely talked out of calling the police to deal with a nonviolent student sit-in at "his" seminary!

Patriarchy teaches us at very deep levels that we can never be safe with other men (or perhaps with anyone!), for the guard must be kept up lest our vulnerability be exposed and we be taken advantage of. At a recent Quaker conference in Philadelphia, a discussion group considered the value of personal sharing and openness in the Quaker Meeting. In almost every case the women advocated more sharing and the men opposed it. Dividing by gender on that issue was predictable; men are conditioned by our life experience of masculinity to distrust settings where personal exposure will happen, especially if men are present. Most men find emotional intimacy possible only with women; many with only one woman; some men cannot be emotionally intimate with anyone.

Patriarchy creates a character ideal—we call it masculinity—and measures everyone against it. Many men as well as women fail the test and even men who are passing the test today are carrying a heavy load of anxiety about tomorrow. Because masculinity is a form of domination, no one can really rest secure. The striving goes on forever unless you are actually willing to give up and find a more secure basis for identity.

MASCULINITY AGAINST GAY MEN: PATRIARCHY FIGHTS A REAR GUARD ACTION

Homophobia is the measure of masculinity. The degree to which a man is thought to have gay feelings is the degree of his unmanliness. Because patriarchy presents sexuality as men over women (part of the general dominance theme), men are conditioned to have only that in mind as a model of sexual expression. Sex with another man must mean being dominated, which is very scary. A nonpatriarchal model of sexual expression as the mutuality of equals doesn't seem possible; the transfer of the heterosexual model to same-sex relations can at best be "queer," at worst, "perverted."

In the recent book *Blue Collar Aristocrats*, by E. E. LeMasters, a working-class tavern is described in which the topic of homosexuality sometimes comes up. Gayness is never defended. In fact, the worst thing you can call a man is homosexual. A man so attacked must either fight or leave the bar.

Notice the importance of violence in defending yourself against the charge of being a "pansy." Referring to your income or academic degrees or size of your car is no defense against such a charge. Only fighting will reestablish your respect as a masculine male. Because "gay" appears to mean "powerless," one needs to go to the masculine source of power—violence—for adequate defense.

Last year, the Argentinian government decided to persecute gays on a systematic basis. The Ministry of Social Welfare offered the rationale for this policy in an article in its journal, which also attacked lesbians, concluding that they should be put in jail or killed:

> As children they played with dolls. As they grew up, violent sports horrified them. As was to be expected, with the passage of time and the custom of listening to foreign mulattos on the radio, they became conscientious objectors. [*El Caudillo*, February 1975, excerpted in *Peace News*, July 11, 1975, p. 5]

The Danish government, by contrast with Argentina, has liberal policies on gay people. There is no government persecution and all government jobs are open to gays—except in the military and the diplomatic service! Two places where the nation-state is most keen to assert power are places where gays are excluded as a matter of policy.

We need not go abroad to see the connections between violence and homophobia. In the documentary film *Men's Lives*, a high school boy is interviewed on what it is like to be a dancer. While the interview is conducted, we see him working out, with a very demanding set of acrobatic exercises. The boy mentions that other boys think he must be gay. "Why is that?" the interviewer asks. "Dancers are free and loose," he replies; they are not big like football players; and "you're not trying to kill anybody."

Different kinds of homosexual behavior bring out different amounts of hostility, curiously enough. That fact gives us further clues to violence and female oppression. In prisons, for example, men can be respected if they fuck other men, but not if they are themselves fucked. (We use the word "fucked" intentionally for its ambiguity.) Often prison rapes are done by men who identify as heterosexual; one hole substitutes for another in this scene, for sex is in either case an expression of domination for the masculine mystique.

But for a man to be entered sexually, or to use effeminate gestures and actions, is to invite attack in prison and hostility outside. Effeminate gay men are at the bottom of the totem pole because they are *most like women*, which is nothing less than treachery to the Masculine Cause. Even many gay men shudder at drag queens and vigilantly guard against certain mannerisms because they, too, have internalized the masculinist dread of effeminacy.

John Braxton's report of prison life as a draft resister is revealing on this score. The other inmates knew immediately that John was a conscientious objector because he did not act tough. They also assumed he was gay, for the same reason. (If you are not masculine, you must be a pacifist and gay, for masculinity is a package which includes both violence and heterosexuality.)

A ticket of admission to masculinity, then, is sex with women, and bisexuals can at least get that ticket even if they deviate through having gay feelings as well. This may be why bisexuality is not feared as much as exclusive gayness among men. Exclusively gay men let down the Masculine Cause in a very important way—those gays do not participate in the control of women through sexuality. Control through sexuality matters because it is flexible; it usually is mixed with love and dependency so that it becomes quite subtle. (Women often testify to years of confusion and only the faintest uneasiness at their submissive role in traditional heterosexual relationships and the role sex plays in that.)

Now we better understand why women are in general so much more supportive of gay men than nongay men are. Part of it of course is that heterosexual men are often paralyzed by fear. Never very trusting, such men find gayness one more reason to keep up the defenses. But heterosexual women are drawn to active support for the struggles of gay men because there is a common enemy—patriarchy and its definition of sexuality as domination. Both heterosexual women and gay men have experienced first hand the violence of sexism; we all have experienced its less open forms such as put-downs and discrimination, and we all fear its open forms such as rape and assault.

Patriarchy, which links characteristics (gentleness, aggressiveness, etc.) to gender, shapes sexuality as well, in such a way as to maintain male power. The Masculine Cause draws strength from homophobia and resorts habitually to violence in its battles on the field of sexual politics. It provides psychological support for the military state and is in turn stimulated by it.

THE POLICE AND THE BLACK MALE   53

Elijah Anderson

The police, in the Village-Northton [neighborhood] as elsewhere, represent society's formal, legitimate means of social control. Their role includes protecting law-abiding citizens from those who are not law-abiding by preventing crime and by apprehending likely criminals. Precisely how the police fulfill the public's expectations is strongly related to how they view the neighborhood and the people who live there. On the streets, color-coding often works to confuse race, age, class, gender, incivility, and criminality, and it expresses itself most concretely in the person of the anonymous black male. In doing their job, the police often become willing parties to this general color-coding of the public environment, and related distinctions, particularly those of skin color and gender, come to convey definite meanings. Although such coding may make the work of the police more manageable, it may also fit well with their own presuppositions regarding race and class relations, thus shaping of-

---

Ed. note: This article is based on the author's field research in two city neighborhoods he calls Village-Northton.

From: Elijah Anderson, *Streetwise* (Chicago: University of Chicago Press, 1990), pp. 190–206. Copyright © 1990 University of Chicago Press. Reprinted by permission of the publisher and author.

ficers' perceptions of crime "in the city." Moreover, the anonymous black male is usually an ambiguous figure who arouses the utmost caution and is generally considered dangerous until he proves he is not. . . .

To be white is to be seen by the police—at least superficially—as an ally, eligible for consideration and for much more deferential treatment than that accorded to blacks in general. This attitude may be grounded in the backgrounds of the police themselves. Many have grown up in . . . "ethnic" neighborhoods. They may serve what they perceive as their own class and neighborhood interests, which often translates as keeping blacks "in their place"—away from neighborhoods that are socially defined as "white." In trying to do their job, the police appear to engage in an informal policy of monitoring young black men as a means of controlling crime, and often they seem to go beyond the bounds of duty. . . .

On the streets late at night, the average young black man is suspicious of others he encounters, and he is particularly wary of the police. If he is dressed in the uniform of the "gangster," such as a black leather jacket, sneakers, and a "gangster cap," if he is carrying a radio or a suspicious bag (which may be confiscated), or if he is moving too fast or too slow, the police may stop him. As part of the routine, they search him and make him sit in the police car while they run a check to see whether there is a "detainer" on him. If there is nothing, he is allowed to go on his way. After this ordeal the youth is often left afraid, sometimes shaking, and uncertain about the area he had previously taken for granted. He is upset in part because he is painfully aware of how close he has come to being in "big trouble." He knows of other youths who have gotten into a "world of trouble" simply by being on the streets at the wrong time or when the police were pursuing a criminal. In these circumstances, particularly at night, it is relatively easy for one black man to be mistaken for another. Over the years, while walking through the neighborhood I have on occasion been stopped and questioned by police chasing a mugger, but after explaining myself I was released.

Many youths, however, have reason to fear such mistaken identity or harassment, since they might be jailed, if only for a short time, and would have to post bail money and pay legal fees to extricate themselves from the mess. . . . When law-abiding blacks are ensnared by the criminal justice system, the scenario may proceed as follows. A young man is arbitrarily stopped by the police and questioned. If he cannot effectively negotiate with the officer(s), he may be accused of a crime and arrested. To resolve this situation he needs financial resources, which for him are in short supply. If he does not have money for any attorney, which often happens, he is left to a public defender who may be more interested in going along with the court system than in fighting for a poor black person. Without legal support, he may well wind up "doing time" even if he is innocent of the charges brought against him. The next time he is stopped for questioning he will have a record, which will make detention all the more likely.

Because the young black man is aware of many cases when an "innocent" black person was wrongly accused and detained, he develops an "attitude" toward the police. The street word for police is "the man," signifying a certain machismo, power, and authority. He becomes concerned when he notices "the man" in the community or when the police focus on him because he is outside his own neighborhood. The youth knows, or soon finds out, that he exists in a legally precarious state. Hence he is motivated to avoid the police, and his public life becomes severely circumscribed. . . .

Such scrutiny and harassment by local police make black youths see them as a problem to get beyond, to deal with, and their attempts affect their overall behavior. To avoid encounters with the man, some streetwise young men camouflage themselves, giving up the urban uniform and emblems that identify them as "legitimate" objects of police attention. They may adopt a more conventional presentation of self, wearing chinos, sweat suits, and generally more conservative dress. Some youths have been known to "ditch" a favorite jacket if they see others wearing one like it, because wearing it increases their chances of being mistaken for someone else who may have committed a crime.

But such strategies do not always work over the long run and must be constantly modified. For instance, because so many young ghetto blacks have begun to wear Fila and Adidas sweat suits as status symbols, such dress has become incorporated into the public image generally associated with young black males. These athletic suits, particularly the more expensive and colorful ones, along with high-priced sneakers, have become the leisure dress of successful drug dealers, and other youths will often mimic their wardrobe to "go for bad" in the quest for local esteem. Hence what was once a "square" mark of distinction approximating the conventions of the wider culture has been adopted by a neighborhood group devalued by that same culture. As we saw earlier, the young black male enjoys a certain power over fashion: whatever the collective peer group embraces can become "hip" in a manner the wider society may not desire. . . . These same styles then attract the attention of the agents of social control.

THE IDENTIFICATION CARD

Law-abiding black people, particularly those of the middle class, set out to approximate middle-class whites in styles of self-presentation in public, including dress and bearing. Such middle-class emblems, often viewed as "square,"

are not usually embraced by young working-class blacks. Instead, their connections with and claims on the institutions of the wider society seem to be symbolized by the identification card. The common identification card associates its holder with a firm, a corporation, a school, a union, or some other institution of substance and influence. Such a card, particularly from a prominent establishment, puts the police and others on notice that the youth is "somebody," thus creating an important distinction between a black man who can claim a connection with the wider society and one who is summarily judged as "deviant." Although blacks who are established in the middle class might take such cards for granted, many lower-class blacks, who continue to find it necessary to campaign for civil rights denied them because of skin color, believe that carrying an identification card brings them better treatment than is meted out to their less fortunate brothers and sisters. For them this link to the wider society, though often tenuous, is psychically and socially important. . . .

"DOWNTOWN" POLICE AND LOCAL POLICE

In attempting to manage the police—and by implication to manage themselves—some black youths have developed a working connection of the police in certain public areas of the Village-Northton. Those who spend a good amount of their time on these corners, and thus observing the police, have come to distinguish between the "downtown" police and the "regular" local police.

The local police are the ones who spend time in the area; normally they drive around in patrol cars, often one officer to a car. These officers usually make a kind of working peace with the young men on the streets; for example, they know the names of some of them and may even befriend a young boy. Thus they offer an image of the police department different from that displayed by the "downtown" police. The downtown police are distant, impersonal, and often actively looking for "trouble." They are known to swoop down arbitrarily on gatherings of black youths standing on a street corner; they might punch them around, call them names, and administer other kinds of abuse, apparently for sport. A young Northton man gave the following narrative about his experiences with the police.

> And I happen to live in a violent part. There's a real difference between the violence level in the Village and the violence level in Northton. In the nighttime it's more dangerous over there.
>
> It's so bad now, they got downtown cops over there now. They doin' a good job bringin' the highway patrol over there. Regular cops don't like that. You can

tell that. They even try to emphasize to us the certain category. Highway patrol come up, he leave, they say somethin' about it. "We can do our job over here." We call [downtown police] Nazis. They about six feet eight, seven feet. We walkin', they jump out. "You run, and we'll blow your nigger brains out." I hate bein' called a nigger. I want to say somethin' but get myself in trouble.

When a cop do somethin', nothing happen to 'em. They come from downtown. From what I heard some of 'em don't even wear their real badge numbers. So you have to put up with that. Just keep your mouth shut when they stop you, that's all. Forget about questions, get against the wall, just obey 'em. "Put all that out right there"—might get rough with you now. They snatch you by the shirt, throw you against the wall, pat you hard, and grab you by the arms, and say, "Get outta here." They call you nigger this and little black this, and things like that. I take that. Some of the fellas get mad. It's a whole different world. . . .

You call a cop, they don't come. My boy got shot, we had to take him to the hospital ourselves. A cop said, "You know who did it?" We said no. He said, "Well, I hope he dies if y'all don't say nothin'." What he say that for? My boy said, "I hope your mother die," he told the cop right to his face. And I was grabbin' another cop, and he made a complaint about that. There were a lot of witnesses. Even the nurse behind the counter said the cop had no business saying nothin' like that. He said it loud, "I hope he dies." Nothin' like that should be comin' from a cop.

Such behavior by formal agents of social control may reduce the crime rate, but it raises questions about social justice and civil rights. Many of the old-time liberal white residents of the Village view the police with some ambivalence. They want their streets and homes defended, but many are convinced that the police manhandle "kids" and mete out an arbitrary form of "justice." These feelings make many of them reluctant to call the police when they are needed, and they may even be less than completely cooperative after a crime has been committed. They know that far too often the police simply "go out and pick up some poor black kid." Yet they do cooperate, if ambivalently, with these agents of social control. . . .

. . . Stories about police prejudice against blacks are often traded at Village get-togethers. Cynicism about the effectiveness of the police mixed with community suspicion of their behavior toward blacks keeps middle-class Villagers from embracing the notion that they must rely heavily on the formal means of social control to maintain even the minimum freedom of movement they enjoy on the streets.

Many residents of the Village, especially those who see themselves as the "old guard" or "old-timers," who were around during the good old days when antiwar and antiracist protest was a major concern, sigh and turn their heads when they see the criminal justice system operating in the ways described here. They express hope that "things will work out," that tensions will ease, that crime will decrease and police behavior will improve. Yet as incivility and crime become increasing problems in the neighborhood, whites become less tolerant of anonymous blacks and more inclined to embrace the police as their heroes. . . .

Gentrifiers and the local old-timers who join them, and some traditional residents continue to fear, care more for their own safety and well-being than for the rights of young blacks accused of wrong-doing. Yet reliance on the police, even by an increasing number of former liberals, may be traced to a general feeling of oppression at the hands of street criminals, whom many believe are most often black. As these feelings intensify and as more yuppies and students inhabit the area and press the local government for services, especially police protection, the police may be required to "ride herd" more stringently on the youthful black population. Thus young black males are often singled out as the "bad" element in an otherwise healthy diversity, and the tensions between the lower-class black ghetto and the middle- and upper-class white community increase rather than diminish.

KOREAN AMERICANS VS. AFRICAN AMERICANS:

Conflict and Construction

54

Sumi K. Cho

The violence and destruction that followed the Rodney King verdict again exploded the myth of a shared consensus around American justice and democracy. The blind injustice of the Simi Valley "not guilty" verdict produced a rainbow coalition of people—old, young, of all colors—who had few or no reservations about looting stores owned primarily by Koreans and Latinos. It

Author's Note: I would like to thank Gil Gott, Pedro Noguera, Susan Lee, Ronnie Stevenson, Ann Park, Evageline Ordaz, Christopher Burroughs, and Bob Gooding-Williams for their helpful comments on this essay.

From Sumi Cho, "Korean Americans vs. African Americans: Conflict and Construction," in *Reading Rodney King/Reading Urban Uprising,* ed. Robert Gooding-Williams (New York: Routledge, 1993), pp. 196–211. Reprinted by permission of the author.

soon became clear that the nation's professional, academic, political, and business elite were ill-equipped to deal with the complexity of issues before them, captured in the simple but straightforward plea by Rodney King: "Can we all get along?"

King's question hints at the real question confronting U.S. society: who is the "we" that must get along? For too long, the political and academic tradition has defined U.S. race relations in terms of a Black/white binary opposition. For example, a CNN–*Time Magazine* poll taken immediately after the verdict surveyed "Americans" on their opinions regarding the verdict and the violence that followed. Yet the poll only sought the views of African Americans and whites regarding the future of race relations. The Black/white framing of race issues must give way to a fuller, more differentiated understanding of a multiracial, multiethnic society divided along the lines of race, class, gender, and other axes in order to explicate effectively the Los Angeles explosion and to contribute to the long-term empowerment of those who, for a short time, exercised the power of their own agency.

Dominating the current debate within the Black/white racial paradigm are, on the one hand, the human-capital theorists[1] who assert that the degradation of "family values" caused the fires in L.A. Dan Quayle's now infamous Murphy Brown speech, for example, pointed the finger at a television sitcom for contributing to the social and moral decay which lie behind the problems in South Central L.A. by portraying the single motherhood of a white woman from the professional class. On the other hand, there are structuralists[2] proposing that the U.S. adopt a "Marshall Plan" for its cities to address the institutional lockout of people of color from economic development.

While the structural explanations provide more illuminating insights than human-capital theories, they cannot fully explain the myriad of events in Los Angeles, particularly those that affected the Korean-American community. The portrayal of Asian Americans as the paragons of socioeconomic success contributed to the targeting of Korean Americans as a scapegoat by those above and below Koreans on the socioeconomic ladder during the L.A. riots. The King verdict and the failure of the U.S. economy to provide jobs and a decent standard of living for all of its peoples, the ostensible root causes of the rioting, were not the fault of Korean (or Latino) shopowners. If, as some have suggested, the Willie Horton imagery[3] is the myth and Rodney King's beating the practice, then likewise, the casting of Asian Americans as a model minority is the myth, and the looting and burning of Koreatown and Korean-owned stores is the practice.

Manipulation of Korean Americans into a "model minority" contributed to their "triple scapegoating" following the King verdict. The first layer of attack came from those who targeted Korean-owned stores for looting and arson. The second layer consisted of those in positions of power who were responsible for the sacrifice of Koreatown, Pico Union, and South Central Los Angeles to ensure the safety of wealthier, whiter communities. The final scapegoating came at the hands of the media, eager to sensationalize the events by excluding Korean perspectives from coverage and stereotyping the immigrant community. These three forces combined to blame the Korean-American community for the nation's most daunting economic and sociopolitical problems.

Rodney King's spontaneous reaction upon seeing the varied groups of color pitted against each other in L.A. reflected a deeper understanding of the commonalities among those groups than many intellectuals and politicians have shown: "I love people of color," he declared, although almost all the mainstream media edited out this statement. Further, his plea can be read as a challenge to the academic community and intellectual activists of racial and ethnic politics: "We're all stuck here for awhile. . . . Let's try to work it out."

Race-relations theorists must accept this challenge and go beyond the standard structuralist critique of institutional racism to incorporate the most difficult issues presented by the aftermath of the not-guilty verdicts in the Rodney King case. Specifically, intellectual activists must devise a new approach to understanding interethnic conflict between subordinated groups and work toward a proactive theory of social change. Such a theory would examine the structural conditions that influence patterns of conflict or cooperation between groups of color, racial ideologies, and how they influence group prejudices, as well as the roles of human agency, community education, and political leadership and accountability. Such a theory should strive to explain and resolve interethnic conflicts to unite subordinated groups.

KOREAN AMERICANS IN L.A.: MODEL MINORITY MYTH AND PRAXIS

The View from Below: The Politics of Resentment

The deteriorated socioeconomic conditions of neglected inner cities have led scholars to compare ghettos like South Central Los Angeles (which includes the Southeast and South Central city-planning areas) to South African "bantustans" that serve solely as "holding space for blacks and browns no longer of use to the larger economy."[4] According to 1990 census data, African Americans constitute 24.5 percent of South Central's residents, and Latinos account for 46 percent (with a rapidly growing Latino population). Only 3.6 percent of the total land mass is zoned for industry.[5] South Central residents lack self-determination and political-economic power. In this context, Korean Americans who open stores in the neighborhood are resented by long-deprived residents and are seen as "outsiders" exerting unfair control and power in the community. The interaction between the two racial groups is structured strictly by market relationships: one is the consumer, the other is the owner. This market structuring of group relations has influenced the "Korean/Black conflict" and contributed to the course of events following the not-guilty verdicts.

Koreans first were scapegoated by rioters of all colors who looted stores and later set them afire. Reports of the early activity following the not-guilty verdict in the Rodney King beating centered around a crowd that had gathered in front of Tom's Liquor Store at Florence and Normandie in the late afternoon of 29 April 1992. One of the first targets of the group was the "swap meets" because they were Korean-owned. The swap meets were essentially indoor flea markets that operated every day during the week and offered discounted prices on consumer items such as electronics and clothing. Whether these stores provided a service to an underserved community by offering low-priced goods where megaretailers refused to tread or whether they shamelessly promoted consumerism to an economically disadvantaged population was not at issue. The most oft-stated reason for the targeting of these stores and Korean stores in general was a familiar refrain: Korean owners were rude to African-American and Latino customers. One Latino interviewed on television was asked why people were looting Koreans. "Because we hate 'em. Everybody hates them," he responded.

Much has been said about the rudeness of Korean owners. Some of the major media outlets that covered the tensions between African Americans and Korean Americans attempted to reduce the conflict to "cultural differences" such as not smiling enough, not looking into another person's eyes, not placing change in a person's hands. Although Koreans wanted very badly to believe in this reductionism, one making an honest assessment must conclude that far too many Korean shopowners had accepted widespread stereotypes about African Americans as lazy, complaining criminals.

The dominant U.S. racial hierarchy and its concomitant stereotypes are transferred worldwide to every country that the United States has occupied militarily. Korean women who married American GIs and returned to the United States after the Korean War quickly discovered the social significance of marrying a white versus an African-American GI. American racial hierarchies were telegraphed back home. When Koreans immigrate to the U.S., internationalized stereotypes are reinforced by negative depictions of African

Americans in U.S. films, television shows, and other popular forms of cultural production. This stereotype, combined with the high crime rate inherent in businesses such as liquor or convenience stores (regardless of who owns them), produced the prejudiced, paranoid, bunker mentality of Soon Ja Du who shot Latasha Harlins, a 15-year-old African-American girl, in the back of the neck during a dispute over a bottle of orange juice. Since 1 January 1990, at least 25 Korean merchants have been killed by non-Korean gunmen.[6]

On the other hand, many African Americans also internalize stereotypes of Korean Americans. Asian Americans walk a fine line between being seen as model minorities and callous unfair competitors. The result is a split image of success and greed that goes together with callousness and racial superiority. Distinctions between Asian ethnicities are often blurred as are distinctions between Asians and Asian Americans. "Sins of the neighbor" are passed on to Koreans in the United States. For example, when then prime minister Yasuhiro Nakasone of Japan made his blundering remarks in 1986 that "the level of intelligence in the United States is lowered by the large number of Blacks, Puerto Ricans, and Mexicans who live there,"[7] this comment was often applied generally to represent the views of all Asians, including Koreans, although Korea has a long history of colonization at the hands of the Japanese.

In rationalizing the violence that followed the verdict, African-American leaders often repeated the myth that Korean immigrants unfairly compete with aspiring entrepreneurs from the Black community because Korean Americans receive preferential treatment over African Americans for bank and government loans. "Until banks make loans available to Blacks as much as other people, there's going to be resentment," observed John Murray of Cal-Pac, a Black liquor-store-owners' association.[8] In reality, however, banks and government lenders uniformly reject loan applications for businesses located in poor, predominantly minority neighborhoods such as South Central Los Angeles, regardless of the applicant's color. Korean immigrants rarely receive traditional financing. Those who open liquor stores and small businesses often come over with some capital and/or borrow from family and friends. At times, groups of Koreans will act as their own financial institutions through informal rotating credit associations known as "kyes" (pronounced "keh").

Thus, the ability to open stores largely depends upon a class variable, as opposed to a racial one. In fact, Korean merchants at times purchased their liquor and grocery stores from African-American owners. Prior to the Watts riot in 1965, most of the liquor-store owners in South Central Los Angeles were Jewish. After the riots, many Jewish owners wanted out. The easing of government-backed loans and the low selling price of the stores in the mid-1960s opened the way for African-American entrepreneurs. The labor-intensive, high-risk nature of work combined with the deregulation of liquor pricing in 1978 and subsequent price wars that drove many small stores out of business led to a sell-off of liquor stores by African Americans in the late 1970s and early 1980s. The sell-off coincided with a large increase in Korean immigration to the U.S. during the same time period. African-American merchants often made a solid profit selling to Koreans.[9] Korean immigrants, many of whom held college degrees, could not find jobs commensurate with their training and education in the new country due to structural discrimination, occupational downgrading, and general antiforeign sentiment. A disproportionate number of Korean families turned to small businesses as a living. According to U.S. Census Bureau statistics, the rate of self-employment among Koreans is higher than that of any other ethnic group.[10] This self-employment, or "self-exploitation" as some have termed the practice, extends not only to the "head of the household," but often to the entire family, including children. Given the high educational background of this group, high self-employment rates in mom-and-pop grocery stores can hardly be viewed as "success."[11]

Nevertheless, the politics of resentment painted Koreans as callous and greedy invaders who got easy bank loans. As this depiction ran unchecked, it became increasingly easy to consider violence against such a contemptible group. The popular rap artist, Ice Cube, warned Korean shopowners in his song entitled "Black Korea" (on his *Death Certificate* album) to "pay respect to the black fist, or we'll burn your store right down to a crisp." The scene was set for disaster and required simply a spark to ignite a highly flammable situation. When the scorching injustice of the verdict was announced, Korean-owned businesses were scapegoated as the primary target for centuries of racial injustice against African Americans—injustice that predated the bulk of Korean immigration which occurred only after the lifting of discriminatory immigration barriers in 1965.

The View from Above: "Racist Love" and the Model Minority

The police nonresponse to the initial outbreak of violence represented a conscious sacrificing of South Central Los Angeles and Koreatown, largely inhabited by African Americans, Chicanos, Latinos, and Korean Americans, to ensure the safety of affluent white communities. The LAPD's intentional neglect in failing to respond to the initial outbreak of rioting set up a foreseeable conflict between African Americans, Chicanos, Latinos, and Korean Americans with destructive and deadly consequences. When it became evident that no police would come to protect Koreatown, Koreans took up arms in self-defense against other minority groups. Rather than focus on whether these actions by the Korean community were right or wrong, one should question the allocation of police-protection resources and the absolute desertion of Koreatown and South Central L.A. by the police. When Deputy Chief Matthew Hunt, who commands officers in South Los Angeles, asked Chief of Police Daryl Gates for greater preparation prior to the announcement of the verdicts, the chief refused, claiming smugly that such preparations were unnecessary.

It was a badly kept secret that angry mobs bent upon destruction and violence would be descending on Koreatown. When Korean Americans called the LAPD and local and state officials for assistance and protection, there was no response. It was very clear that Koreatown was on its own. In the absence of police protection, the community harnessed its resources through Radio Korea (KBLA) in Los Angeles to coordinate efforts to defend stores from attack. People from as far away as San Bernadino and Orange County came to help. Radio Korea reported on the movement of the crowds and instructed volunteers where to go. In short, the radio station and individual volunteers served as the police force for Koreatown. It is bitterly ironic that some Koreans defending stores were later arrested by police for weapon-permit violations.

In stark contrast, when looters began to work over major shopping malls such as the Fox Hills Mall in Culver City, they were quickly stopped by the police, with "merchants and residents praising [police] efforts." [12] Likewise, the police made sure that the downtown business interests were secured. The west-side edition of the *Los Angeles Times* even boasted that the police forces in the predominantly white communities of Santa Monica, Beverly Hills, and West Hollywood "emerged remarkably unscathed by the riots." [13] Other areas, especially communities of color such as South Central, Pico Union, and Koreatown, effectively became a "no-person's-land." Enter at your own risk. There was no protection for these areas. Although residents paid taxes for police protection, the LAPD made a conscious choice to stay out of the neighborhoods of color that were most at risk during the postverdict rioting.

Although Korean Americans donated substantial amounts of money to the election campaigns of state and local officials, no political assistance was forthcoming in their time of need. "We've made significant contributions to the city council, as well as county, state and federal politicians, and they failed us," said Bong Hwan Kim, executive director of the L.A. Korean Youth Center. [14] Attempts to contact Governor Wilson for help from the National Guard were in vain. The National Guard did not arrive on the scene until there was little left to do. Wilson's explanation was that the National Guard was late because they had run out of ammunition. [15] During the initial outbreak of violence, it is now known Chief Gates was attending an expensive fundraiser on the west side while these communities burned. In response, he later stated at his first press conference following the verdict that police were needed to protect the firefighters. Unfortunately for Gates, the fire department was unwilling to go along with this excuse. In a stinging report to the Fire Commission,

L.A. Fire Chief Donald Manning complained that the LAPD dismissed their requests for protection during the outbreak of violence as "not a top priority," thereby delaying fire-department responses to the fires.[16] Moreover, Gates was stumped by an *L.A. Times* reporter who pressed him about the hours prior to the arsons, when some of the most sensationalized violence occurred outside of Tom's Liquor Store.

Many Koreans mistakenly believed that they would be taken care of if they worked hard, did not complain, and contributed handsomely to powerful politicians such as Governor Wilson and local officials. One hard lesson to be learned from the aftermath of the King verdict and the Korean-American experience is that a model minority is expediently forgotten and dismissed if white dominance or security is threatened. This point was also demonstrated through the Asian-American admissions scandals at the nation's premier institutions of higher education when Asian-American students began to outperform white students on SATs and GPAs. The rules of the game of "meritocracy" had to be changed to put a ceiling on Asian-American admissions, thereby producing the largest unstated affirmative-action policy benefiting whites.

The easy dismissal of Korean-American pleas for help is rooted in the relationship between the dominant power brokers and the model-minority group. The model-minority "thesis," popularized during the late 1960s and the mid-1980s during the Reagan administration, represents a political backlash to civil rights struggles. It is critical to understand that the model-minority thesis is no compliment for Asian Americans. It has been historically constructed to discipline activists in the Black Power, Black Panther, American Indian, United Farm Workers, and other radical social movements demanding institutional change.[17] Because it has this genesis, the embrace of Asian Americans as a model minority is an embrace of "racist love."[18] The basis of that love has a racist origin: to provide a public rationale for the ongoing subordination of non-Asian people of color. Because the embrace or love is not genuine, one cannot reasonably expect the architects truly to care about the health or well-being of the model minority. . . .

NOTES

1. The human-capital model is a variant of neoclassical economic theory applied to labor-market economics. The model attempts to explain a dependent variable such as income inequality as derivative of independent variables such as educational attainment, work experience, total number of hours worked per year, marital status, and English proficiency, among others. It explains Asian Americans' advancement by fore-

grounding their investments in human capital such as educational attainment. It would also explain the lack of mobility of certain groups such as Southeast Asians or African Americans who had not yet invested appropriately in human capital.

2. Structuralists emphasize the basic characteristics or "structures" of a social system. For example, internal colonial theory places the U.S. historical and contemporary actualities of race in the context of international colonialism that connects the Third World abroad with the Third World within the U.S. This framework emphasizes similarities in the patterns of racial domination and exploitation experienced by people of color and distinguishes their incorporation into the U.S. from the processes of immigration and assimilation by which European Americans are incorporated into a nation. See Robert Blauner, "Colonized and Immigrant Minorities," *Racial Oppression in America* (New York: Harper and Row, 1972), 51–82.

3. President Bush's 1988 television campaign centered around an ad attacking Michael Dukakis for being "soft on crime." To support this claim, the TV ads played upon white fears by focusing on close-up "mug" shots of Willie Horton, an African-American convict who committed subsequent crimes while on an early-release program implemented by Dukakis, then governor of Massachusetts.

4. Cynthia Hamilton, *Apartheid in an American City: The Case of the Black Community in Los Angeles* (Van Nuys, CA: Labor Community Strategy Center, 1991), 1.

5. Ibid. Internal colonial theory is useful in viewing South Central as a Third World colony, within a First World nation, that lacks political and economic self-determination, similar to the relationship between Third World nations abroad and First World ones.

6. Eui-Young Yu, "We Saw Our Dreams Burn for No Reason," *San Francisco Examiner*, 24 May 1992, editorial page.

7. Robert C. Toh, "Blacks Pressing Japanese to Halt Slurs, Prejudice," *Los Angeles Times*, 13 December 1990.

8. Susan Moffat, "Shopkeepers Fight Back," *Los Angeles Times*, 15 May 1992.

9. Ibid.

10. At 17 percent, self-employment rates of foreign-born Korean men top those of white men (10 percent) and other people of color. U.S. Bureau of the Census, 1980 Census of Population, vol. 2, Subject Reports, *Asian and Pacific Islander Population in the United States: 1980*, table 45A. A 1990 survey conducted by California State University sociologist Eui-Young Yu found that nearly 40 percent of Korean families own a business.

11. The long hours worked by more members of the family under dangerous conditions, combined with the high educational levels of those owning businesses, calls into question the definition of success applied to Korean Americans. Some have suggested that a blue-collar union job offers better working conditions and a more reliable income than engaging in small businesses.

12. Nancy Hill-Holtzman and Mathis Chazanov, "Police Credited for Heading Off Spread of Riots," *Los Angeles Times*, 7 May 1992.

13. Ibid.

14. Steven Chin, "Innocence Lost: L.A.'s Koreans Fight to be Heard," *San Francisco Examiner*, 9 May 1992.

15. California National Guard Brigadier General Daniel Brennan stated that bullets and grenades were not loaded for transport to L.A. because there were no lights on the parade ground where the ammunition is stored. Daniel Weintraub, "National Guard Official Cites Series of Delays," *Los Angeles Times*, 7 May 1992.

16. Rich Connell and Richard Simon, "Top LAPD Officer, Fire Chief Cite Flaws in Police Response," *Los Angeles Times*, 8 May 1992.

17. *At a time when it is being proposed that hundreds of billions be spent to uplift Negroes and other minorities,* the nation's 300,000 Chinese are moving ahead on their own . . . with no help from anyone else. Still being taught in Chinatown is the old idea that people should depend on their own efforts . . . not a welfare check . . . in order to reach America's 'promised land' (emphasis added). "Success Story of One Minority Group in U.S.," *U.S. News and World Report*, 26 December 1966, 73. Similarly, scholar Thomas Sowell argues in *Race and Economics* (New York: McKay and Co., 1975) that historic exclusion from U.S. political institutions paradoxically benefits Jews and Asian Americans. "[T]hose American ethnic groups that have succeeded best politically have not usually been the same as those who succeeded best economically. . . . [T]hose minorities that have pinned their greatest hopes on political action—the Irish and the Negroes, for example—have made some of the slower economic advances."

18. Frank Chin and Jeffrey Paul Chan, "Racist Love," in *Seen through Shuck* ed. Richard Kostelanetz (New York: Ballantine Books, 1972), 65–79.

FRATERNITIES AND RAPE ON CAMPUS   55

Patricia Yancey Martin and Robert A. Hummer

Many rapes, far more than come to the public's attention, occur in fraternity houses on college and university campuses, yet little research has analyzed fraternities at American colleges and universities as rape-prone contexts (cf. Ehrhart and Sandler 1985). . . .

Author's note: We gratefully thank Meena Harris and Diane Mennella for assisting with data collection. The senior author thanks the graduate students in her fall 1988 graduate research methods seminar for help with developing the initial conceptual framework. Judith Lorber and two anonymous *Gender & Society* referees made numerous suggestions for improving our article and we thank them also.

From: *Gender & Society* 3 (December 1980): 457–73. Reprinted by permission of Sage Publications, Inc.

Gary Tash, writing as an alumnus and trial attorney in his fraternity's magazine, claims that over 90 percent of all gang rapes on college campuses involve fraternity men (1988, p. 2). Tash provides no evidence to substantiate this claim, but students of violence against women have been concerned with fraternity men's frequently reported involvement in rape episodes (Adams and Abarbanel 1988). Ehrhart and Sandler (1985) identify over 50 cases of gang rapes on campus perpetrated by fraternity men, and their analysis points to many of the conditions that we discuss here. Their analysis is unique in focusing on conditions in fraternities that make gang rapes of women by fraternity men both feasible and probable. They identify excessive alcohol use, isolation from external monitoring, treatment of women as prey, use of pornography, approval of violence, and excessive concern with competition as precipitating conditions to gang rape (also see Merton 1985; Roark 1987).

The study reported here confirmed and complemented these findings by focusing on both conditions and processes. We examined dynamics associated with the social construction of fraternity life, with a focus on processes that foster the use of coercion, including rape, in fraternity men's relations with women. Our examination of men's social fraternities on college and university campuses as groups and organizations led us to conclude that fraternities are a physical and sociocultural context that encourages the sexual coercion of women. We make no claims that all fraternities are "bad" or that all fraternity men are rapists. Our observations indicated, however, that rape is especially probable in fraternities because of the kinds of organizations they are, the kinds of members they have, the practices their members engage in, and a virtual absence of university or community oversight. Analyses that lay blame for rapes by fraternity men on "peer pressure" are, we feel, overly simplistic (cf. Burkhart 1989; Walsh 1989). We suggest, rather, that fraternities create a sociocultural context in which the use of coercion in sexual relations with women is normative and in which the mechanisms to keep this pattern of behavior in check are minimal at best and absent at worst. We conclude that unless fraternities change in fundamental ways, little improvement can be expected.

METHODOLOGY

Our goal was to analyze the group and organizational practices and conditions that create in fraternities an abusive social context for women. We developed a conceptual framework from an initial case study of an alleged gang rape at Florida State University that involved four fraternity men and an 18-year-old coed. The group rape took place on the third floor of a fraternity house and ended with the "dumping" of the woman in the hallway of a neighboring fraternity house. According to newspaper accounts, the victim's blood-alcohol concentration, when she was discovered, was .349 percent, more than three times the legal limit for automobile driving and an almost lethal amount. One law enforcement officer reported that sexual intercourse occurred during the time the victim was unconscious: "She was in a life-threatening situation" (*Tallahassee Democrat* 1988b). When the victim was found, she was comatose and had suffered multiple scratches and abrasions. Crude words and a fraternity symbol had been written on her thighs (*Tampa Tribune* 1988). When law enforcement officials tried to investigate the case, fraternity members refused to cooperate. This led, eventually, to a five-year ban of the fraternity from campus by the university and by the fraternity's national organization.

In trying to understand how such an event could have occurred, and how a group of over 150 members (exact figures are unknown because the fraternity refused to provide a membership roster) could hold rank, deny knowledge of the event, and allegedly lie to a grand jury, we analyzed newspaper articles about the case and conducted open-ended interviews with a variety of respondents about the case and about fraternities, rapes, alcohol use, gender relations, and sexual activities on campus. Our data included over 100 newspaper articles on the initial gang rape case; open-ended interviews with Greek (social fraternity and sorority) and non-Greek (independent) students (N = 20); university administrators (N = 8, five men, three women); and alumni advisers to Greek organizations (N = 6). Open-ended interviews were held also with judges, public and private defense attorneys, victim advocates, and state prosecutors regarding the processing of sexual assault cases. Data were analyzed using the grounded theory method (Glaser 1978; Martin and Turner 1986). In the following analysis, concepts generated from the data analysis are integrated with the literature on men's social fraternities, sexual coercion, and related issues.

FRATERNITIES AND THE SOCIAL CONSTRUCTION OF MEN AND MASCULINITY

Our research indicated that fraternities are vitally concerned—more than with anything else—with masculinity (cf. Kanin 1967). They work hard to create a macho image and context and try to avoid any suggestion of "wimpishness," effeminacy, and homosexuality. Valued members display, or are willing to go along with, a narrow conception of masculinity that stresses competition, athleticism, dominance, winning, conflict, wealth, material possessions, willingness to drink alcohol, and sexual prowess vis-à-vis women.

Valued Qualities of Members

When fraternity members talked about the kind of pledges they prefer, a litany of stereotypical and narrowly masculine attributes and behaviors was recited and feminine or woman-associated qualities and behaviors were expressly denounced (cf. Merton 1985). Fraternities seek men who are "athletic," "big guys," good in intramural competition, "who can talk college sports." Males "who are willing to drink alcohol," "who drink socially," or "who can hold their liquor" are sought. Alcohol and activities associated with the recreational use of alcohol are cornerstones of fraternity social life. Non-drinkers are viewed with skepticism and rarely selected for membership.[1]

Fraternities try to avoid "geeks," nerds, and men said to give the fraternity a "wimpy" or "gay" reputation. Art, music, and humanities majors, majors in traditional women's fields (nursing, home economics, social work, education), men with long hair, and those whose appearance or dress violate current norms are rejected. Clean-cut, handsome men who dress well (are clean, neat, conforming, fashionable) are preferred. One sorority woman commented that "the top ranking fraternities have the best looking guys."

One fraternity man, a senior, said his fraternity recruited "some big guys, very athletic" over a two-year period to help overcome its image of wimpiness. His fraternity had won the interfraternity competition for highest grade-point average several years running but was looked down on as "wimpy, dancy, even gay." With their bigger, more athletic recruits, "our reputation improved; we're a much more recognized fraternity now." Thus a fraternity's reputation and status depends on members' possession of stereotypically masculine qualities. Good grades, campus leadership, and community service are "nice" but masculinity dominance—for example, in athletic events, physical size of members, athleticism of members—counts most.

Certain social skills are valued. Men are sought who "have good personalities," are friendly, and "have the ability to relate to girls" (cf. Longino and Kart 1973). One fraternity man, a junior, said: "We watch a guy [a potential pledge] talk to women . . . we want guys who can relate to girls." Assessing a pledge's ability to talk to women is, in part, a preoccupation with homosexuality and a conscious avoidance of men who seem to have effeminate manners or qualities. If a member is suspected of being gay, he is ostracized and informally drummed out of the fraternity. A fraternity with a reputation as wimpy or tolerant of gays is ridiculed and shunned by other fraternities. Militant heterosexuality is frequently used by men as a strategy to keep each other in line (Kimmel 1987).

Financial affluence or wealth, a male-associated value in American culture, is highly valued by fraternities. In accounting for why the fraternity involved in the gang rape that precipitated our research project had been recognized recently as "the best fraternity chapter in the United States," a university official said: "They were good-looking, a big fraternity, had lots of BMWs [expensive, German-made automobiles]." After the rape, newspaper stories described the fraternity members' affluence, noting the high number of members who owned expensive cars (*St. Petersburg Times* 1988).

The Status and Norms of Pledgeship

A pledge (sometimes called an associate member) is a new recruit who occupies a trial membership status for a specific period of time. The pledge period (typically ranging from 10 to 15 weeks) gives fraternity brothers an opportunity to assess and socialize new recruits. Pledges evaluate the fraternity also and decide if they want to become brothers. The socialization experience is structured partly through assignment of a Big Brother to each pledge. Big Brothers are expected to teach pledges how to become a brother and to support them as they progress through the trial membership period. Some pledges are repelled by the pledging experience, which can entail physical abuse; harsh discipline; and demands to be subordinate, follow orders, and engage in demeaning routines and activities, similar to those used by the military to "make men out of boys" during boot camp.

Characteristics of the pledge experience are rationalized by fraternity members as necessary to help pledges unite into a group, rely on each other, and join together against outsiders. The process is highly masculinist in execution as well as conception. A willingness to submit to authority, follow orders, and do as one is told is viewed as a sign of loyalty, togetherness, and unity. Fraternity pledges who find the pledge process offensive often drop out. Some do this by openly quitting, which can subject them to ridicule by brothers and other pledges, or they may deliberately fail to make the grades necessary for initiation or transfer schools and decline to reaffiliate with the fraternity on the new campus. One fraternity pledge who quit the fraternity he had pledged described an experience during pledgeship as follows:

> This one guy was always picking on me. No matter what I did, I was wrong. One night after dinner, he and two other guys called me and two other pledges into the chapter room. He said, "Here, X, hold this 25 pound bag of ice at arms' length 'til I tell you to stop." I did it even though my arms and hands were killing me. When I asked if I could stop, he grabbed me around the throat and lifted me off the floor. I thought he would choke me to death. He cussed me and called me all kinds of names. He took one of my fingers and twisted it until it nearly broke. . . . I stayed in the fraternity for a few more days, but then I decided to quit. I hated it. Those guys are sick. They like seeing you suffer.

Fraternities' emphasis on toughness, withstanding pain and humiliation, obedience to superiors, and using physical force to obtain compliance contributes to an interpersonal style that de-emphasizes caring and sensitivity but fosters intragroup trust and loyalty. If the least macho or most critical pledges drop out, those who remain may be more receptive to, and influenced by, masculinist values and practices that encourage the use of force in sexual relations with women and the covering up of such behavior (cf. Kanin 1967).

Norms and Dynamics of Brotherhood

Brother is the status occupied by fraternity men to indicate their relations to each other and their membership in a particular fraternity organization or group. Brother is a male-specific status; only males can become brothers, although women can become "Little Sisters," a form of pseudomembership. "Becoming a brother" is a rite of passage that follows the consistent and often lengthy display by pledges of appropriately masculine qualities and behaviors. Brothers have a quasi-familial relationship with each other, are normatively said to share bonds of closeness and support, and are sharply set off from non-members. Brotherhood is a loosely defined term used to represent the bonds that develop among fraternity members and the obligations and expectations incumbent upon them (cf. Marlowe and Auvenshine [1982] on fraternities' failure to encourage "moral development" in freshman pledges).

Some of our respondents talked about brotherhood in almost reverential terms, viewing it as the most valuable benefit of fraternity membership. One senior, a business-school major who had been affiliated with a fairly high-status fraternity throughout four years on campus, said:

> Brotherhood spurs friendship for life, which I consider its best aspect, although I didn't see it that way when I joined. Brotherhood bonds and unites. It instills values of caring about one another, caring about community, caring about ourselves. The values and bonds [of brotherhood] continually develop over the four years [in college] while normal friendships come and go.

Despite this idealization, most aspects of fraternity practice and conception are more mundane. Brotherhood often plays itself out as an overriding concern with masculinity and, by extension, femininity. As a consequence, fraternities comprise collectivities of highly masculinized men with attitudinal qualities and behavioral norms that predispose them to sexual coercion of women (cf. Kanin 1967; Merton 1985; Rapaport and Burkhart 1984). The norms of masculinity are complemented by conceptions of women and femininity that are equally distorted and stereotyped and that may enhance the probability of women's exploitation (cf. Ehrhart and Sandler 1985; Sanday 1981, 1986).

Practices of Brotherhood

Practices associated with fraternity brotherhood that contribute to the sexual coercion of women include a preoccupation with loyalty, group protection and secrecy, use of alcohol as a weapon, involvement in violence and physical force, and an emphasis on competition and superiority.

Loyalty, Group Protection, and Secrecy.   Loyalty is a fraternity preoccupation. Members are reminded constantly to be loyal to the fraternity and to their brothers. Among other ways, loyalty is played out in the practices of group protection and secrecy. The fraternity must be shielded from criticism. Members are admonished to avoid getting the fraternity in trouble and to bring all problems "to the chapter" (local branch of a national social fraternity) rather than to outsiders. Fraternities try to protect themselves from close scrutiny and criticism by the Interfraternity Council (a quasi-governing body composed of representatives from all social fraternities on campus), their fraternity's national office, university officials, law enforcement, the media, and the public. Protection of the fraternity often takes precedence over what is procedurally, ethically, or legally correct. Numerous examples were related to us of fraternity brothers' lying to outsiders to "protect the fraternity."

Group protection was observed in the alleged gang rape case with which we began our study. Except for one brother, a rapist who turned state's evidence, the entire remaining fraternity membership was accused by university and criminal justice officials of lying to protect the fraternity. Members consistently failed to cooperate even though the alleged crimes were felonies, involved only four men (two of whom were not even members of the local chapter), and the victim of the crime nearly died. According to a grand jury's findings, fraternity officers repeatedly broke appointments with law enforcement officials, refused to provide police with a list of members, and refused to cooperate with police and prosecutors investigating the case (*Florida Flambeau*, 1988).

Secrecy is a priority value and practice in fraternities, partly because full-fledged membership is premised on it (for confirmation, see Ehrhart and Sandler 1985; Longino and Kart 1973; Roark 1987). Secrecy is also a boundary-maintaining mechanism, demarcating in-group from out-group, us from them. Secret rituals, handshakes, and mottoes are revealed to pledge brothers as they are initiated into full brotherhood. Since only brothers are supposed to know a fraternity's secrets, such knowledge affirms membership in the fraternity and separates a brother from others. Extending secrecy tactics from protection of private knowledge to protection of the fraternity from criticism is a predictable development. Our interviews indicated that individual members knew the difference between right and wrong, but fraternity norms that emphasize loyalty, group protection, and secrecy often overrode standards of ethical correctness.

Alcohol as Weapon.   Alcohol use by fraternity men is normative. They use it on weekdays to relax after class and on weekends to "get drunk," "get crazy," and "get laid." The use of alcohol to obtain sex from women is pervasive—in other words, it is used as a weapon against sexual reluctance. According to several fraternity men whom we interviewed, alcohol is the major tool used to gain sexual mastery over women (cf. Adams and Abarbanel 1988; Ehrhart and Sandler 1985). One fraternity man, a 21-year-old senior, described alcohol use to gain sex as follows: "There are girls that you know will fuck, then some you have to put some effort into it. . . . You have to buy them drinks or find out if she's drunk enough. . . ."

A similar strategy is used collectively. A fraternity man said that at parties with Little Sisters: "We provide them with 'hunch punch' and things get wild. We get them drunk and most of the guys end up with one." "'Hunch punch,'" he said, "is a girls' drink made up of overproof alcohol and powdered Kool-Aid, no water or anything, just ice. It's very strong. Two cups will do a number on a female." He had plans in the next academic term to surreptitiously give hunch punch to women in a "prim and proper" sorority because "having sex with prim and proper sorority girls is definitely a goal." These women are a challenge because they "won't openly consume alcohol and won't get openly drunk as hell." Their sororities have "standards committees" that forbid heavy drinking and easy sex.

In the gang rape case, our sources said that many fraternity men on campus believed the victim had a drinking problem and was thus an "easy make." According to newspaper accounts, she had been drinking alcohol on the evening she was raped; the lead assailant is alleged to have given her a bottle of wine after she arrived at his fraternity house. Portions of the rape occurred in a shower, and the victim was reportedly so drunk that her assailants had difficulty holding her in a standing position (*Tallahassee Democrat*, 1988a). While raping her, her assailants repeatedly told her they were members of another fraternity under the apparent belief that she was too drunk to know the difference. Of course, if she was too drunk to know who they were, she was too drunk to consent to sex (cf. Allgeier 1986; Tash 1988).

One respondent told us that gang rapes are wrong and can get one expelled, but he seemed to see nothing wrong in sexual coercion one-on-one.

He seemed unaware that the use of alcohol to obtain sex from a woman is grounds for a claim that a rape occurred (cf. Tash 1988). Few women on campus (who also may not know these grounds) report date rapes, however; so the odds of detection and punishment are slim for fraternity men who use alcohol for "seduction" purposes (cf. Byington and Keeter 1988; Merton 1985).

Violence and Physical Force.   Fraternity men have a history of violence (Ehrhart and Sandler 1985; Roark 1987). Their record of hazing, fighting, property destruction, and rape has caused them problems with insurance companies (Bradford 1986; Pressley 1987). Two university officials told us that fraternities "are the third riskiest property to insure behind toxic waste dumps and amusement parks." Fraternities are increasingly defendants in legal actions brought by pledges subjected to hazing (Meyer 1986; Pressley 1987) and by women who were raped by one or more members. In a recent alleged gang rape incident at another Florida university, prosecutors failed to file charges, but the victim filed a civil suit against the fraternity nevertheless (*Tallahassee Democrat*, 1989).

Competition and Superiority.   Interfraternity rivalry fosters in-group identification and out-group hostility. Fraternities stress pride of membership and superiority over other fraternities as major goals. Interfraternity rivalries take many forms, including competition for desirable pledges, size of pledge class, size of membership, size and appearance of fraternity house, superiority in intramural sports, highest grade-point averages, giving the best parties, gaining the best or most campus leadership roles, and, of great importance, attracting and displaying "good looking women." Rivalry is particularly intense over members, intramural sports, and women (cf. Messner 1989).

FRATERNITIES' COMMODIFICATION OF WOMEN

In claiming that women are treated by fraternities as commodities, we mean that fraternities knowingly, and intentionally, *use* women for their benefit. Fraternities use women as bait for new members, as servers of brother's needs, and as sexual prey.

Women as Bait.   Fashionably attractive women help a fraternity attract new members. As one fraternity man, a junior, said, "They are good bait." Beautiful, sociable women are believed to impress the right kind of pledges and give the impression that the fraternity can deliver this type of woman to its members. Photographs of shapely, attractive coeds are printed in fraternity brochures and videotapes that are distributed and shown to potential pledges. The women pictured are often dressed in bikinis, at the beach, and are pictured hugging the brothers of the fraternity. One university official says such recruitment materials give the message: "Hey, they're here for you, you can have whatever you want," and, "we have the best looking women. Join us and you can have them too." Another commented: "Something's wrong when males join an all-male organization as the best place to meet women. It's so illogical."

Fraternities compete in promising access to beautiful women. One fraternity man, a senior, commented that "the attraction of girls [i.e., a fraternity's success in attracting women] is a big status symbol for fraternities." One university official commented that the use of women as a recruiting tool is so well entrenched that fraternities that might be willing to forgo it say they cannot afford to unless other fraternities do so as well. One fraternity man said, "Look, if we don't have Little Sisters, the fraternities that do will get all the good pledges." Another said, "We won't have as good a rush [the period during which new members are assessed and selected] if we don't have these women around."

In displaying good-looking, attractive, skimpily dressed, nubile women to potential members, fraternities implicitly, and sometimes explicitly, promise sexual access to women. One fraternity man commented that "part of what being in a fraternity is all about is the sex." . . .

Women as Servers.   The use of women as servers is exemplified in the Little Sister program. Little Sisters are undergraduate women who are rushed and selected in a manner parallel to the recruitment of fraternity men. They are affiliated with the fraternity in a formal but unofficial way and are able, indeed required, to wear the fraternity's Greek letters. Little Sisters are not full-fledged fraternity members, however, and fraternity national offices and most universities do not register or regulate them. Each fraternity has an officer called Little Sister Chairman who oversees their organization and activities. The Little Sisters elect officers among themselves, pay monthly dues to the fraternity, and have well-defined roles. Their dues are used to pay for the fraternity's social events, and Little Sisters are expected to attend and hostess fraternity parties and hang around the house to make it a "nice place to be." One fraternity man, a senior, described Little Sisters this way: "They are very social girls, willing to join in, be affiliated with the group, devoted to the fraternity." Another member, a sophomore, said: "Their sole purpose is social— attend parties, attract new members, and 'take care' of the guys."

Our observations and interviews suggested that women selected by fraternities as Little Sisters are physically attractive, possess good social skills, and are willing to devote time and energy to the fraternity and its members.

. . . The title of Little Sister reflects women's subordinate status; fraternity men in a parallel role are called Big Brothers. Big Brothers assist a sorority primarily with the physical work of sorority rushes, which, compared to fraternity rushes, are more formal, structured, and intensive. Sorority rushes take place in the daytime and fraternity rushes at night so fraternity men are free to help. According to one fraternity member, Little Sister status is a benefit to women because it gives them a social outlet and "the protection of the brothers." The gender-stereotypic conceptions and obligations of these Little Sister and Big Brother statuses indicate that fraternities and sororities promote a gender hierarchy on campus that fosters subordination and dependence in women, thus encouraging sexual exploitation and the belief that it is acceptable.

Women as Sexual Prey.   Little Sisters are a sexual utility. Many Little Sisters do not belong to sororities and lack peer support for refraining from unwanted sexual relations. One fraternity man (whose fraternity has 65 members and 85 Little Sisters) told us they had recruited "wholesale" in the prior year to "get lots of new women." The structural access to women that the Little Sister program provides and the absence of normative supports for refusing fraternity members' sexual advances may make women in this program particularly susceptible to coerced sexual encounters with fraternity men.

Access to women for sexual gratification is a presumed benefit of fraternity membership, promised in recruitment materials and strategies and through brothers' conversations with new recruits. One fraternity man said: "We always tell the guys that you get sex all the time, there's always new girls. . . . After I became a Greek, I found out I could be with females at will." A university official told us that, based on his observations, "no one [i.e., fraternity men] on this campus wants to have 'relationships.' They just want to have fun [i.e., sex]." Fraternity men plan and execute strategies aimed at obtaining sexual gratification, and this occurs at both individual and collective levels.

Individual strategies include getting a woman drunk and spending a great deal of money on her. As for collective strategies, most of our undergraduate interviewees agreed that fraternity parties often culminate in sex and that this outcome is planned. One fraternity man said fraternity parties often involve sex and nudity and can "turn into orgies." . . .

When asked about the women who come to such a party, he said: "Some

Little Sisters just won't go. . . . The girls who do are looking for a good time, girls who don't know what it is, things like that."

Other respondents denied that fraternity parties are orgies but said that sex is always talked about among the brothers and they all know "who each other is doing it with." One member said that most of the time, guys have sex with their girlfriends "but with socials, girlfriends aren't allowed to come and it's their [members'] big chance [to have sex with other women]." The use of alcohol to help them get women into bed is a routine strategy at fraternity parties.

CONCLUSIONS

In general, our research indicated that the organization and membership of fraternities contribute heavily to coercive and often violent sex. Fraternity houses are occupied by same-sex (all men) and same-age (late teens, early twenties) peers whose maturity and judgment is often less than ideal. Yet fraternity houses are private dwellings that are mostly off-limits to, and away from scrutiny of, university and community representatives, with the result that fraternity house events seldom come to the attention of outsiders. Practices associated with the social construction of fraternity brotherhood emphasize a macho conception of men and masculinity, a narrow, stereotyped conception of women and femininity, and the treatment of women as commodities. Other practices contributing to coercive sexual relations and the cover-up of rapes include excessive alcohol use, competitiveness, and normative support for deviance and secrecy (cf. Bogal-Allbritten and Allbritten 1985; Kanin 1967).

Some fraternity practices exacerbate others. Brotherhood norms require "sticking together" regardless of right or wrong; thus rape episodes are unlikely to be stopped or reported to outsiders, even when witnesses disapprove. The ability to use alcohol without scrutiny by authorities and alcohol's frequent association with violence, including sexual coercion, facilitates rape in fraternity houses. Fraternity norms that emphasize the value of maleness and masculinity over femaleness and femininity and that elevate the status of men and lower the status of women in members' eyes undermine perceptions and treatment of women as persons who deserve consideration and care (cf. Ehrhart and Sandler 1985; Merton 1985).

Androgynous men and men with a broad range of interests and attributes are lost to fraternities through their recruitment practices. Masculinity of a narrow and stereotypical type helps create attitudes, norms, and practices that predispose fraternity men to coerce women sexually, both individually and collectively (Allgeier 1986; Hood 1989; Sanday 1981, 1986). Male athletes on campus may be similarly disposed for the same reasons (Kirshenbaum 1989; Telander and Sullivan 1989).

Research into the social contexts in which rape crimes occur and the social constructions associated with these contexts illuminate rape dynamics on campus. Blanchard (1959) found that group rapes almost always have a leader who pushes others into the crime. He also found that the leader's latent homosexuality, desire to show off to his peers, or fear of failing to prove himself a man are frequently an impetus. Fraternity norms and practices contribute to the approval and use of sexual coercion as an accepted tactic in relations with women. Alcohol-induced compliance is normative, whereas, presumably, use of a knife, gun, or threat of bodily harm would not be because the woman who "drinks too much" is viewed as "causing her own rape" (cf. Ehrhart and Sandler 1985).

Our research led us to conclude that fraternity norms and practices influence members to view the sexual coercion of women, which is a felony crime, as sport, a contest, or a game (cf. Sato 1988). This sport is played not between men and women but between men and men. Women are the pawns or prey in the interfraternity rivalry game; they prove that a fraternity is successful or prestigious. The use of women in this way encourages fraternity men to see women as objects and sexual coercion as sport. Today's societal norms support young women's right to engage in sex at their discretion, and coercion is unnecessary in a mutually desired encounter. However, nubile young women say they prefer to be "in a relationship" to have sex while young men say they prefer to "get laid" without a commitment (Muehlenhard and Linton 1987). These differences may reflect, in part, American puritanism and men's fears of sexual intimacy or perhaps intimacy of any kind. In a fraternity context, getting sex without giving emotionally demonstrates "cool" masculinity. More important, it poses no threat to the bonding and loyalty of the fraternity brotherhood (cf. Farr 1988). Drinking large quantities of alcohol before having sex suggests that "scoring" rather than intrinsic sexual pleasure is a primary concern of fraternity men.

Unless fraternities' composition, goals, structures, and practices change in fundamental ways, women on campus will continue to be sexual prey for fraternity men. As all-male enclaves dedicated to opposing faculty and administration and to cementing in-group ties, fraternity members eschew any hint of homosexuality. Their version of masculinity transforms women, and men with womanly characteristics, into the out-group. "Womanly men" are ostracized; feminine women are used to demonstrate members' masculinity. Encouraging renewed emphasis on their founding values (Longino and Kart 1973), service orientation and activities (Lemire 1979), or members' moral development (Marlowe and Auvenshine 1982) will have little effect on fraternities' treatment of women. A case for or against fraternities cannot be made by studying individual members. The fraternity qua group and organization is at issue. Located on campus along with many vulnerable women, embedded in a sexist society, and caught up in masculinist goals, practices, and values, fraternities' violation of women—including forcible rape—should come as no surprise.

NOTE

1. Recent bans by some universities on open-keg parties at fraternity houses have resulted in heavy drinking before coming to a party and an increase in drunkenness among those who attend. This may aggravate, rather than improve, the treatment of women by fraternity men at parties.

References

Adams, Aileen, and Gail Abarbanel. 1988. *Sexual Assault on Campus: What Colleges Can Do.* Santa Monica, CA: Rape Treatment Center.

Allgeier, Elizabeth. 1986. "Coercive Versus Consensual Sexual Interactions." G. Stanley Hall Lecture to American Psychological Association Annual Meeting, Washington, D.C., August.

Blanchard, W. H. 1959. "The Group Process in Gang Rape." *Journal of Social Psychology* 49:259–66.

Bogal-Allbritten, Rosemarie B., and William L. Allbritten. 1985. "The Hidden Victims: Courtship Violence Among College Students." *Journal of College Student Personnel* 43:201–4.

Bradford, Michael. 1986. "Tight Market Dries Up Nightlife at University." *Business Insurance* (March 2): 2, 6.

Burkhart, Barry. 1989. Comments in Seminar on Acquaintance/Date Rape Prevention: A National Video Teleconference, February 2.

Burkhart, Barry R., and Annette L. Stanton. 1985. "Sexual Aggression in Acquaintance Relationships." Pp. 43–65 in *Violence in Intimate Relationships*, edited by G. Russell. Englewood Cliffs, NJ: Spectrum.

Byington, Diane B., and Karen W. Keeter. 1988. "Assessing Needs of Sexual Assault Victims on a University Campus." Pp. 23–31 in *Student Services: Responding to Issues and Challenges.* Chapel Hill: University of North Carolina Press.

Ehrhart, Julie K., and Bernice R. Sandler. 1985. *Campus Gang Rape: Party Games?* Washington, D.C.: Association of American Colleges.

Farr, K. A. 1988. "Dominance Bonding Through the Good Old Boys Sociability Network." *Sex Roles* 18:259–77.

Florida Flambeau. 1988. "Pike Members Indicted in Rape." (May 19):1, 5.

Glaser, Barney G. 1978. *Theoretical Sensitivity: Advances in the Methodology of Grounded Theory.* Mill Valley, CA: Sociology Press.

Hood, Jane. 1989. "Why Our Society Is Rape-Prone." *New York Times,* May 16.

Hughes, Michael J., and Roger B. Winston, Jr. 1987. "Effects of Fraternity Membership on Interpersonal Values." *Journal of College Student Personnel* 45:405–11.

Kanin, Eugene J. 1967. "Reference Groups and Sex Conduct Norm Violations." *The Sociological Quarterly* 8:495–504.

Kimmel, Michael, ed. 1987. *Changing Men: New Directions in Research on Men and Masculinity.* Newbury Park, CA: Sage.

Kirshenbaum, Jerry. 1989. "Special Report, an American Disgrace: A Violent and Unprecedented Lawlessness Has Arisen Among College Athletes in All Parts of the Country." *Sports Illustrated* (February 27): 16–19.

Lemire, David. 1979. "One Investigation of the Stereotypes Associated with Fraternities and Sororities." *Journal of College Student Personnel* 37:54–57.

Longino, Charles F., Jr., and Cary S. Kart. 1973. "The College Fraternity: An Assessment of Theory and Research." *Journal of College Student Personnel* 31:118–25.

Marlowe, Anne F., and Dwight C. Auvenshine. 1982. "Greek Membership: Its Impact on the Moral Development of College Freshmen." *Journal of College Student Personnel* 40:53–57.

Martin, Patricia Yancey, and Barry A. Turner. 1986. "Grounded Theory and Organizational Research." *Journal of Applied Behavioral Science* 22:141–57.

Merton, Andrew. 1985. "On Competition and Class: Return to Brotherhood." *Ms.* (September): 60–65, 121–22.

Messner, Michael. 1989. "Masculinities and Athletic Careers." *Gender & Society* 3:71–88.

Pressley, Sue Anne. 1987. "Fraternity Hell Night Still Endures." *Washington Post* (August 11): B1.

Rapaport, Karen, and Barry R. Burkhart. 1984. "Personality and Attitudinal Characteristics of Sexually Coercive College Males." *Journal of Abnormal Psychology* 93: 216–21.

Roark, Mary L. 1987. "Preventing Violence on College Campuses." *Journal of Counseling and Development* 65:367–70.

Sanday, Peggy Reeves. 1981. "The Socio-Cultural Context of Rape: A Cross-Cultural Study." *Journal of Social Issues* 37:5–27.

———. 1986. "Rape and the Silencing of the Feminine." Pp. 84–101 in *Rape,* edited by S. Tomaselli and R. Porter. Oxford: Basil Blackwell.

St. Petersburg Times. 1988. "A Greek Tragedy." (May 29):1F, 6F.

Sato, Ikuya. 1988. "Play Theory of Delinquency: Toward a General Theory of 'Action.'" *Symbolic Interaction* 11:191–212.

Tallahassee Democrat. 1988a. "FSU Fraternity Brothers Charged" (April 27):1A, 12A.

———. 1988b. "FSU Interviewing Students About Alleged Rape" (April 24):1D.

———. 1989. "Woman Sues Stetson in Alleged Rape" (March 19):3B.

Tampa Tribune. 1988. "Fraternity Brothers Charged in Sexual Assault of FSU Coed." (April 27): 6B.

Tash, Gary B. 1988. "Date Rape." *The Emerald of Sigma Pi Fraternity* 75(4):1–2.

Telander, Rick, and Robert Sullivan. 1989. "Special Report, You Reap What You Sow." *Sports Illustrated* (February 27): 20–34.

The Tomahawk. 1988. "A Look Back at Rush, A Mixture of Hard Work and Fun" (April/May): 3D.

Walsh, Claire. 1989. Comments in Seminar on Acquaintance/Date Rape Prevention: A National Video Teleconference, February 2.

V

Making a Difference

What in your life do you care about so much that it would spur you to work for social change? Is it your family, your children, your neighborhood, a concern for a social issue, or an ethical framework that requires not just talk but action? Most people think that people who work for social change must be somehow extraordinary like Martin Luther King or other heroic leaders. But most people who engage in social activism are ordinary, everyday people who decide to take action about something that touches their lives.

Why don't we know more about these people? Dominant social institutions and the ideologies that defend them obscure the individual and collective political activism of everyday individuals. By making the political activism of everyday people from historically marginalized groups invisible, social institutions suppress the strength of these groups and render them more easily exploited. Investigating forms of power used by historically marginalized groups offers one way of rethinking social change and reconceptualizing how people work to make a difference. As Black feminist theorist bell hooks suggests, "Sexism has never rendered women powerless. It has either suppressed their strength or exploited it. Recognition of that strength, that power, is a step women together can take toward liberation."[1] African Americans, Native Americans, women, Latinos, gays and lesbians, and members of the poor and working class have never been powerless. The question is how to identify and use forms of power that often go unrecognized.

The political activism of individuals from historically privileged groups is hidden as well. Many people who currently benefit within existing race, class, and gender arrangements have little interest in studying—much less show-casing—the ideas and actions of those who not only see the unfairness of race, class, and gender relations in the United States, but who try to make a differ-ence. The expected political protest from those harmed by existing race, class, and gender relations is routinely discredited. Stereotyping people of color, women, working-class and poor people, and gays, lesbians, and bisexuals as angry minorities, strident feminists, disgruntled trade unionists, or hyper-sensitive homosexuals effectively recasts political protest in negative terms and works to strip their protest of meaning. These tactics remain less effective against those who benefit from existing structures of race, class, gender, and sexuality but who decide to use their privilege in new ways. In some ways, individuals who are White, middle class, male and/or straight and who reject current social arrangements that privilege them may be more threatening than other groups because they often have the power to make change. Con-sider how different things would be if White American children learned ways to be "White" different from those they are routinely taught or if men learned models of masculinity that do not demean women. Relegating political protest of this sort to the margins remains vitally important to dominant interests.

Despite their invisibility on both sides of privilege, most of the people who try to make a difference are people just like you. Often some sort of cata-lyst spurs them to action. The needless death of a relative, the placement of a toxic dump site in their neighborhood, losing a job when an employer closes shop, a police beating, or hearing a sermon about Christian brotherhood one Sunday morning in the familiarity of a racially segregated church have all been catalysts for activism. Because people engaging in social activism come from all walks of life, their paths into social activism can be quite unexpected, and the strategies they select for making a difference can be richly diverse. The articles in "Making a Difference" explore the various ways that people across race, class, and gender groups participate in social activism. They also provide new visions about what is possible. Collectively, these authors offer inclusive thinking, reconstructed knowledge, and a distinctive perspective on the kinds of questions we might consider in working to make a difference.

WHO ARE THE ACTIVISTS?

When we ask students in our classes to describe an "activist," they often think of a "radical"-looking young person, typically with strange hair and unusual clothing, who gives fiery speeches and engages in losing causes. People who work for social change rarely fit this stereotype and actually reflect a much broader spectrum of folks. Often they remain invisible because we do not label their activities as activism.

There is no "typical" activist. Almost anyone can make a difference in the context of her or his everyday life. Celene Krauss's article "Women of Color on the Front Line" describes how the majority of participants in the environmental justice movement are mothers who became involved out of concern for their children's health. Her article highlights how women often become politically involved—through concerns over family, children, and quality of life in their neighborhoods. Still, women's activism often is overlooked. Although academic credentials, positions of authority, and economic resources can do much to help individuals challenge hierarchies of race, class, and gender, Krauss suggests that the individual and collective actions of ordinary people form the bedrock for many social movements.

Charon Asetoyer's article "From the Ground Up," points to how much women and men can accomplish through persistence. When Asetoyer found herself moving to the Yankton Sioux reservation in South Dakota, she did not explicitly search out activist activities. Instead, they found her. Through conversations with her few neighbors, she learned of problems that women and children were having in her new community. Health issues came up first, and these were the issues that she and others initially addressed. Over time, the informal discussions about women's health issues on the reservation grew to the point where her group was able to purchase a building and organize the Native Women's Reproductive Rights Coalition, involving women from eight states. Asetoyer's article provides an historical overview of how one individual decided to respond in the context of her everyday life, instead of throwing up her hands in defeat.

WORKING FOR CHANGE IN MANY PLACES

People decide to work for change from diverse social locations. Workplaces, families, schools, and the government—the social institutions we examined in Part III—all represent potential locations for social activism. But there are many others as well. The media, health care organizations, the penal system, and recreational organizations also constitute potential sites of social activism.

People often think that, without a group of some sort, they cannot make a difference. Again, this narrow view of activists and activism obscures the tremendous impact that even one individual can have when she or he tries. Enid Lee's efforts as one teacher developing an antiracist, multicultural curriculum within her own classroom reflect this choice to work from a position of authority inside an existing social institution. In her interview with Barbara

Miner, Lee offers some solid advice to teachers who want to make a difference in their own classrooms. While acknowledging the difficulties of working in conditions with inadequate funding and uncooperative colleagues, Lee firmly believes that attitude and commitment still matter. As she points out, "If you don't take multicultural education or antiracist education seriously, you are actually promoting a monocultural or racist education. There is no neutral ground on this issue."

Whether by choice or by necessity, other individuals and groups work outside workplaces, schools, and other formal social institutions. Members of "The Boys and Girls of (Union) Summer" spent their time confronting what they deemed unjust labor conditions in workplaces where they themselves were not employed. The familiar boycotts, picketing, leafleting, and other direct-action strategies long associated with social movements of all types typically constitute actions taken outside some institution. Although activities such as these can be trivialized in the media, it is important to remember that direct action from outsider locations represents one important way to make a difference.

Sometimes institutional policies and practices become the focus of change. Because existing institutions are structured around race, class, and gender oppression, there are limits to the kinds of change that can be achieved from inside them. Furthermore, even when African Americans, Native Americans, women, Asian Americans, the poor, gays and lesbians, Latinos, and other historically marginalized groups gain power within them, change does not automatically follow. Replacing one type of dominant group with another of a different color or gender may improve the lives of some people without addressing the fundamental inequalities that pervade existing social institutions.

At other times, the goal is to enlist support from these same social institutions for a social agenda. This acknowledges that some social problems cannot be solved by ordinary citizens: institutional involvement is often needed. Despite the cynicism that many people express regarding the proper role of government, many social issues cannot be addressed without government intervention of some sort. Thus, actions of the federal government were required to desegregate the American South in the 1950s and 1960s. While government cannot solve social issues created through current race, class, and gender arrangements, social issues such as homelessness, environmental concerns, violence against children, poverty, racial discrimination, and gender discrimination require government partnership of some sort to support the actions of everyday citizens who care about these problems.

STRATEGIES: WHAT WORKS?

We are often asked, what strategies make a difference? The range of strategies that individuals and groups use in working for social change is immense. Some engage in individualistic strategies. One well-placed individual with a clear sense of what is possible and why it is important to try can accomplish amazing things. Others engage in collective actions where a small group of like-minded individuals organize around some concern that is important to them. By engaging in strategic resistance, a few people can make a big difference if they use their resources wisely. Charon Asetoyer's article on building Native American health services from the ground up illustrates this process. Still other activists participate in national and global mass movements.

Regardless of the strategies selected, one important issue that surfaces repeatedly involves the importance of building coalitions and working across differences. Bringing ordinary people together who have long been separated by race, class, and gender (among other barriers) raises the necessity of building coalitions across differences—and the difficulty of doing so. Using the image of the barred room in "Coalition Politics: Turning the Century," Bernice Johnson Reagon explores the creation of nurturing spaces and communities where individuals try and recover from the damaging effects of race, class, and gender oppression. "Nationalism is crucial to a people if you are going to ever impact as a group in your interest," suggests Reagon. On some college campuses, African American organizations, for example, offer a space where Black students can retreat from the difficulties of desegregating what many experience as hostile White campuses, learn ways of supporting one another, and identify the positive qualities of difference based on group identity. These spaces are valuable, but Reagon also points to the limits of their value. Nationalism and the celebration of difference that may accompany it become limiting if a group does not look beyond its own "room" and learn to appreciate differences of other groups. Within the category African American student, tremendous heterogeneity exists. Flattening differences within groups to create a unidimensional unity can rob Black student organizations of their creativity. Moreover, over time, if it fails to build coalitions with other campus and community groups, even the most dedicated Black student group may find itself increasingly isolated and ineffective.

Coalition building operates best when individuals and groups share a concern for a social issue while recognizing the diversity of ways that people experience, think about, and want to deal with that issue. People from across race, class, and gender categories engage in social issues regarding a range of social issues. Part IV of this volume explored American identity, sexuality, and violence as three important social issues that benefited from a race-, class-, and gender-inclusive analysis. There are many more such issues, and working for change involves bringing a inclusive framework to ideas and actions that address these social issues.

As Reagon also points out, coalition building requires hard work that creates tensions and discomforts. This tension can be further aggravated when groups of different power within hierarchies of race, class, and gender try to build coalitions. As the concrete suggestions in John Anner's article "Having the Tools at Hand: Building Successful Multicultural Social Justice Organizations" remind us, people can learn to work together across all sorts of differences. Through case studies of several grass-roots groups who realized that their political effectiveness would be increasingly compromised if they didn't learn to deal with difference, Anner develops a list of models and mechanisms that can be used to build successful multicultural social justice organizations. Despite the difficulties, "change means growth, and growth can be painful," observes Audre Lorde in her essay included in Part II of this volume. "But we sharpen self-definition by exposing the self in work and struggle together with those whom we define as different from ourselves, although sharing the same goals."

TOWARD A POLITICS OF EMPOWERMENT

Power is typically equated with domination and control over people or things. Social institutions depend on this version of power to reproduce hierarchies of race, class, and gender. Exploration of the experiences of African Americans, Latinos, women, Native Americans, Asian Americans, and the poor reveals much about how dominant groups exercise power, but centering on the experiences of historically marginalized groups also reveals much about resistance, because members of these groups engage in individual acts of resistance and organized political activism to challenge race, class, and gender oppression.

Re-envisioning and exercising power to bring about social change requires a sense of purpose and a vision that encourages us to look beyond what already exists. We must learn to imagine what is possible. For example, what type of environmental policies would result if working-class White women and women of color were central in the environmental movement's decision-making processes? If teachers tried to implement antiracist education from kindergarten through post-graduate education, how might American national identity be different? How might health care be organized if Native American women's ideas were central to planning? If men and women truly learned to work together, how might economic security be better provided for all?

Trying to make a difference takes time. One can take the long view or focus on the here and now. These are not either/or choices but two ways of looking at the same thing. Many people become cynical because they expect change to be immediate and dramatic rather than long-term and continual. How people deal with this disappointment is revealing. The White working-class women who became concerned about the effects of environmental dumping on their children initially trusted the government to provide solutions. When it did not, they could easily have become cynical and thrown in the towel. Instead, they tried to find another way because too much was at stake. Their initial idealism became tempered by pragmatism. Taking a long view, seeing the connectedness among all sorts of people, and involving people across race, class, and gender—and beyond—will be necessary to bringing lasting change. Clearly this will not happen tomorrow. But if people do not try to make a difference, "it doesn't just happen" at all.

NOTE

1. bell hooks, *From Margin to Center* (Boston: South End Press, 1984), p. 93.

Political Activism

COALITION POLITICS: *Turning the Century**

56

Bernice Johnson Reagon

I've never been this high before. I'm talking about the altitude. There is a lesson in bringing people together where they can't get enough oxygen, then having them try to figure out what they're going to do when they can't think properly. I'm serious about that. There probably are some people here who can breathe, because you were born in high altitudes and you have big lung cavities. But when you bring people in who have not had the environmental conditioning, you got one group of people who are in a strain—and the group of people who are feeling fine are trying to figure out why you're staggering around, and that's what this workshop is about this morning.

I wish there had been another way to graphically make me feel it because I belong to the group of people who are having a very difficult time being here. I feel as if I'm gonna keel over any minute and die. That is often what it feels like if you're *really* doing coalition work. Most of the time you feel threatened to the core and if you don't, you're not really doing no coalescing.

I'm Bernice Reagon. I was born in Georgia, and I'd like to talk about the fact that in about twenty years we'll turn up another century. I believe that we are positioned to have the opportunity to have something to do with what makes it into the next century. And the principles of coalition are directly related to that. You don't go into coalition because you just *like* it. The only reason you would consider trying to team up with somebody who could possibly kill you, is because that's the only way you can figure you can stay alive.

---

* Based on a presentation at the West Coast Women's Music Festival 1981, Yosemite National Forest, California.

From: Barbara Smith (ed.), *Home Girls: A Black Feminist Anthology* (New York: Kitchen Table Press, 1983), pp. 356–68. Reprinted by permission.

A hundred years ago in this country we were just beginning to heat up for the century we're in. And the name of the game in terms of the dominant energy was technology. We have lived through a period where there have been things like railroads and telephones, and radios, TVs and airplanes, and cars, and transistors, and computers. And what this has done to the concept of human society and human life is, to a large extent, what we in the latter part of this century have been trying to grapple with. With the coming of all that technology, there was finally the possibility of making sure no human being in the world would be unreached. You couldn't find a place where you could hide if somebody who had access to that technology wanted to get to you. Before the dawning of that age you had all these little cute villages and the wonderful homogeneous societies where everybody looked the same, did things the same, and believed the same things, and if they didn't, you could just kill them and nobody would even ask you about it.

We've pretty much come to the end of a time when you can have a space that is "yours only"—just for the people you want to be there. Even when we have our "women-only" festivals, there is no such thing. The fault is not necessarily with the organizers of the gathering. To a large extent it's because we have just finished with that kind of isolating. There is no hiding place. There is nowhere you can go and only be with people who are like you. It's over. Give it up.

Now every once in awhile there is a need for people to try to clean out corners and bar the doors and check everybody who comes in the door, and check what they carry in and say, "Humph, inside this place the only thing we are going to deal with is X or Y or Z." And so only the X's or Y's or Z's get to come in. That place can then become a nurturing place or a very destructive place. Most of the time when people do that, they do it because of the heat of trying to live in this society where being an X or Y or Z is very difficult, to say the least. The people running the society call the shots as if they're still living in one of those little villages, where they kill the ones they don't like or put them in the forest to die. (There are some societies where babies are born and if they are not wanted for some reason they are put over in a corner. They do that here too, you know, put them in garbage cans.) When somebody else is running a society like that, and you are the one who would be put out to die, it gets too hard to stay out in that society all the time. And that's when you find a place, and you try to bar the door and check all the people who come in. You come together to see what you can do about shouldering up all of your energies so that you and your kind can survive.

There is no chance that you can survive by staying *inside* the barred room. (Applause) That will not be tolerated. The door of the room will just be painted red and then when those who call the shots get ready to clean house, they have easy access to you.

But that space while it lasts should be a nurturing space where you sift out what people are saying about you and decide who you really are. And you take the time to try to construct within yourself and within your community who you would be if you were running society. In fact, in that little barred room where you check everybody at the door, you act out community. You pretend that your room is a world. It's almost like a play, and in some cases you actually grow food, you learn to have clean water, and all of that stuff, you just try to do it all. It's like, "If *I* was really running it, this is the way it would be."

Of course the problem with the experiment is that there ain't nobody in there but folk like you, which by implication means you wouldn't know what to do if you were running it with all of the other people who are out there in the world. Now that's nationalism. I mean it's nurturing, but it is also nationalism. At a certain stage nationalism is crucial to a people if you are going to ever impact as a group in your own interest. Nationalism at another point becomes reactionary because it is totally inadequate for surviving in the world with many peoples. (Applause)

Sometimes you get comfortable in your little barred room, and you decide you in fact are going to live there and carry out all of your stuff in there. And you gonna take care of everything that needs to be taken care of in the barred room. If you're white and in the barred room and if everybody's white, one of the first things you try to take care of is making sure that people don't think that the barred room is a racist barred room. So you begin to talk about racism and the first thing you do is say, "Well, maybe we better open the door and let some Black folks in the barred room." Then you think, "Well, how we gonna figure out whether they're X's or not?" Because there's nothing in the room but X's. (Laughter) You go down the checklist. You been working a while to sort out who you are, right? So you go down the checklist and say, "If we can find Black folk like that we'll let them in the room." You don't really want Black folks, you are just looking for yourself with a little color to it.

And there are those of us Black folk who are like that. So if you're lucky you can open the door and get one or two. Right? And everything's wonderful. But no matter what, there will be one or two of us who have not bothered to be like you and you know it. We come knocking on your door and say, "Well, you let them in, you let me in too." And we will break your door down trying to get in. (Laughter) As far as we can see we are also X's. Cause you didn't say, "THIS BARRED ROOM IS FOR WHITE X'S ONLY." You just said it was for X's. So everybody who thinks they're an X comes running to get into the room. And because you trying to take care of everything in this room, and you know you're not racist, you get pressed to let us all in.

The first thing that happens is that the room don't feel like the room anymore. (Laughter) And it ain't home no more. It is not a womb no more. And you can't feel comfortable no more. And what happens at that point has to do with trying to do too much in it. You don't do no coalition building in a womb. It's just like trying to get a baby used to taking a drink when they're in your womb. It just don't work too well. Inside the womb you generally are very soft and unshelled. You have no covering. And you have no ability to handle what happens if you start to let folks in who are not like you.

Coalition work is not work done in your home. Coalition work has to be done in the streets. And it is some of the most dangerous work you can do. And you shouldn't look for comfort. Some people will come to a coalition and they rate the success of the coalition on whether or not they feel good when they get there. They're not looking for a coalition; they're looking for a home! They're looking for a bottle with some milk in it and a nipple, which does not happen in a coalition. You don't get a lot of food in a coalition. You don't get fed a lot in a coalition. In a coalition you have to give, and it is different from your home. You can't stay there all the time. You go to the coalition for a few hours and then you go back and take your bottle wherever it is, and then you go back and coalesce some more.

It is very important not to confuse them—home and coalition. Now when it comes to women—the organized women's movement—this recent thrust—we all have had the opportunity to have some kind of relationship with it. The women's movement has perpetuated a myth that there is some common experience that comes just cause you're women. And they're throwing all these festivals and this music and these concerts happen. If you're the same kind of women like the folk in that little barred room, it works. But as soon as some other folk check the definition of "women" that's in the dictionary (which you didn't write, right?) they decide that they can come because they are women, but when they do, they don't see or hear nothing that is like them. Then they charge, "This ain't no women's thing!" (Applause) Then if you try to address that and bring them in, they start to play music that ain't even women's music! (Laughter and hoots) And you try to figure out what happened to your wonderful barred room. It comes from taking a word like "women" and using it as a code. There is an in-house definition so that when you say "women only" most of the time that means you had better be able— if you come to this place—to handle lesbianism and a lot of folks running around with no clothes on. And I'm being too harsh this morning as I talk to you, but I don't want you to miss what I'm trying to say. Now if you come and you can't handle that, there's another term that's called "woman-identified." They say you might be a woman but you're not woman-identified, and we only want women who are "woman-identified." That's a good way to leave a lot of women out of your room.

So here you are and you grew up and you speak English and you know about this word "woman" and you know you one, and you walk into this "woman-only" space and you ain't there. (Laughter) Because "woman" in that space does not mean "woman" from your world. It's a code word and it traps, and the people that use the word are not prepared to deal with the fact that if you put it out, everybody that thinks they're a woman may one day want to seek refuge. And it ain't no refuge place! And it's not safe! It should be a coalition! It may have been that in its first year the Michigan National "Women-Only" festival was a refuge place. By the fourth year it was a place of coalition, and it's not safe anymore. (Applause) It ain't safe for nobody who comes. When you walk in there you in trouble—and everybody who comes is trying to get to their home there. At this festival [Yosemite] they said: whatever you drink, bring it with you—tea, honey, you know, whatever it is—and we will provide hot water. Now I understand that you got here and there was no hot water. Can't get nothing! That is the nature of coalition. (Laughter) You have to give it all. It is not to feed you; you have to feed it. And it's a monster. It never gets enough. It always wants more. So you better be sure you got your home someplace for you to go to so that you will not become a martyr to the coalition. Coalition *can* kill people; however, it is not by nature fatal. You do not have to die because you are committed to coalition. I'm not so old, and I don't know nothing else. But you do have to know how to pull back, and you do have to have an old-age perspective. You have to be beyond the womb stage.

None of this matters at all very much if you die tomorrow—that won't even be cute. It only matters if you make a commitment to be around for another fifty more years. There are some grey haired women I see running around occasionally, and we have to talk to those folks about how come they didn't commit suicide forty years ago. Don't take everything they say because some of the stuff they gave up to stay around ain't worth considering. But be sure you get on your agenda some old people and try to figure out what it will be like if you are a raging radical fifty years from today.

Think about yourself that way. What would you be like if you had white hair and had not given up your principles? It might be wise as you deal with coalition efforts to think about the possibilities of going for fifty years. It calls for some care. I'm not gonna be suicidal, if I can help it. Sometimes you don't even know you just took a step that could take your head off cause you can't know everything when you start to coalesce with these people who sorta look like you in just one aspect but really they belong to another group. That is really the nature of women. It does not matter at all that biologically we have being women in common. We have been organized to have our primary cultural signals come from some other factors than that we are women. We are not from our base acculturated to be women people, capable of crossing our first people boundaries—Black, White, Indian, etc.

Now if we are the same women from the same people in this barred room, we never notice it. That stuff stays wherever it is. It does not show up until somebody walks into the room who happens to be a woman but really is also somebody else. And then out comes who we really are. And at that point you are not a woman. You are Black or you are Chicana or you are Disabled or you are Racist or you are White. The fact that you are a woman is not important at all and it is not the governing factor to your existence at that moment. I am now talking about bigotry and everybody's got it. I am talking about turning the century with some principles intact. Today wherever women gather together it is not necessarily nurturing. It is coalition building. And if you feel the strain, you may be doing some good work. (Applause) So don't come to no women's festival looking for comfort unless you brought it in your little tent. (Laughter) And then if you bring it in your tent don't be inviting everybody in because everybody ain't your company, and then you won't be able to stand the festival. Am I confusing you? Yes, I am. If coalition is so bad, and so terrible, and so uncomfortable, why is it necessary? That's what you're asking. Because the barred rooms will not be allowed to exist. They will all be wiped out. That is the plan that we now have in front of us.

Now these little rooms were created by some of the most powerful movements we have seen in this country. I'm going to start with the Civil Rights movement because of course I think that that was the first one in the era we're in. Black folks started it, Black folks did it, so everything you've done politically rests on the efforts of my people—that's my arrogance! Yes, and it's the truth; it's my truth. You can take it or leave it, but that's the way I see it. So once we did what we did, then you've got women, you've got Chicanos, you've got the Native Americans, and you've got homosexuals, and you got all of these people who also got sick of somebody being on their neck. And maybe if they come together, they can do something about it. And I claim all of you as coming from something that made me who I am. You can't tell me that you ain't in the Civil Rights movement. You are in the Civil Rights movement that we created that just rolled up to your door. But it could not stay the same, because if it was gonna stay the same it wouldn't have done you no good. Some of you would not have caught yourself dead near no Black folks walking around talking about freeing themselves from racism and lynching. So by the time our movement got to you it had to sound like something you knew about. Like if I find out you're gay, you gonna lose your job.

There were people who came South to work in the movement who were not Black. Most of them were white when they came. Before it was over, that category broke up—you know, some of them were Jewish, not simply white, and some others even changed their names. Say if it was Mary when they came South, by the time they were finished it was Maria, right? It's called finding yourself. At some point, you cannot be fighting oppression and be oppressed yourself and not feel it. Within the Black movement there was also all of the evils of the society, so that anything that was happening to you in New York or the West Coast probably also happened to you in another way, within the movement. And as you became aware of that you tried to talk to these movement people about how you felt. And they say, "Well let's take that up next week. Because the most important thing now is that Black people are being oppressed and we must work with that." Watch these mono-issue people. They ain't gonna do you no good. I don't care who they are. And there are people who prioritize the cutting line of the struggle. And they say the cutting line is this issue, and more than anything we must move on this issue and that's automatically saying that whatever's bothering you will be put down if you bring it up. You have to watch these folks. Watch these groups that can only deal with one thing at a time. On the other hand, learn about space within coalition. You can't have everybody sitting up there talking about everything that concerns you at the same time or you won't get no place. . . .

It must become necessary for all of us to feel that this is our world. And that we are here to stay and that anything that is here is ours to take and to use in our image. And watch that "our"—make it as big as you can—it ain't got nothing to do with that barred room. The "our" must include everybody you have to include in order for you to survive. You must be sure you understand that you ain't gonna be able to have an "our" that don't include Bernice Johnson Reagon, cause I don't plan to go nowhere! That's why we have to have coalitions. Cause I ain't gonna let you live unless you let me live. Now there's danger in that, but there's also the possibility that we can both live—if you can stand it. . . .

THE BOYS AND GIRLS
OF (UNION) SUMMER

57

Marc Cooper

Twenty-one-year-old Nicola Grunthal is no less than a living 4-foot, 10-inch trophy for the newly invigorated A.F.L.-C.I.O. It was only two summers ago that she was on the fast track to a lifetime of privilege, studying to be a diplomat and spending her vacation interning at the White House. But now she's changed her plans. She still wants to finish Harvard Law School, but "no way I'm going into corporate America," she says. "I want to work with labor designing international campaigns against multinational corporations. I'm juiced on that idea. Really juiced."

Grunthal had been slowly drifting toward a career in social activism since her self-described "disillusionment" with mainstream politics, but her new direction jelled only as she was completing her three-week stint in the A.F.L.-C.I.O.'s first-ever "Union Summer." That program—often compared to the civil rights movement's Freedom Summer of 1964—has thrust more than 1,000 mostly young people onto the front lines of the U.S. labor movement. In forty-one separate three-week "waves" in twenty-two different cities coast to coast, the new recruits are given a place to sleep, a light varnishing of labor history, a stipend of $210 a week and are then thrown raw into local organizing drives.

"Put simply, we want to inject a massive dose of class consciousness into youth politics," says Andy Levin, the 36-year-old head of Union Summer, himself a grad of Harvard Law. "Yes, we want to recruit new blood. But more important, we want to transform the politics of the next generation of activists." Since the days of Vietnam, says Levin, "most young progressives have been either antilabor or have ignored labor," falling into single-issue or strict identity politics. "Our message is that labor is where it's at in the fight for social justice in the nineties."

A bitter, protracted campaign by the United Steelworkers against Japanese-owned Bridgestone/Firestone, based here in Nashville, is where Grunthal and nine other Union Summer (U/S) enlistees were sent to test their mettle. In July 1994, more than 4,000 members of the United Rubber Work-

From: *The Nation* 19 (August 12/19, 1996): 18–20. Reprinted with permission.

ers went on strike after the company demanded rotating twelve-hour shifts, cuts in benefits and lower pay for new hires. A half-year later the company permanently replaced about half the strikers—a move recently deemed illegal by the National Labor Relations Board (but still being contested by Bridgestone). Ten months after the strike began, the union saw no option but to return to work without a contract, though not everyone was taken back.

Shortly afterward, the cash-flush Steelworkers merged with the defeated rubber union and things began to turn around. A tenacious worldwide corporate campaign and consumer boycott were launched to embarrass and pressure Bridgestone/Firestone. Slowly, some striking workers were hired back. But as the second anniversary of the walkout loomed, some 500 workers were still out of a job, the company continued to refuse union terms for a contract and the Steelworkers were escalating their fight.

When I met Grunthal and the rest of the "Nashville Ten"—as these Union Summer kids call themselves—they were in their final week of service, feverishly working for a successful conclusion to the Steelworkers' July 12–13 "International Days of Rage." Of the Union Summer group, eight are between 18 and 21, the youngest having just graduated from high school. One is 22; the oldest, a schoolteacher, is 27. And there they were, front and center in what Steelworkers corporate campaign organizer Doug Niehouse calls "the biggest coordinated international hit ever on a corporation." During the Days of Rage, Argentine and Brazilian Bridgestone workers were to stage stoppages, while other unions in France, Italy, Japan and Turkey would be protesting company policy. Throughout the United States, union workers planned to set up picket lines around 950 Bridgestone-owned MasterCare garages. In Nashville alone, there would be protests at sixteen separate MasterCares, a march on company headquarters and a rally the following day to be keynoted by the new A.F.L.-C.I.O. president, John Sweeney.

"That's why we're working fifteen hours a day," says 20-year-old Amanda Johnson, a junior at the University of New Mexico. "It's exciting and immediate. You know you are doing something right there."

Far from being simple grunts, the U/S kids have been given tangible organizing responsibilities. One group has been charged with pulling together a prayer breakfast to recruit local clergy into the campaign. Another has organized two neighborhood "blitzes," where the entire group will be going door to door in areas where Bridgestone/Firestone execs live, asking their neighbors to call the company and urge a settlement. They're out every morning at 7:30 leafleting customers at the MasterCare centers. At night they listen to "educational modules"—talks and videos on labor history and organizing techniques. In between it all, at 9:30 sharp each morning, they sit in as equals in Local 1055's "war room," where campaign strategy and logistics are hammered out. Bunked in collectively at a Vanderbilt University chapel, they take turns spending the night in sleeping bags in the union-run tent city outside the doors of the company headquarters.

It's a lot of responsibility, but the consensus is that the Nashville Ten are holding up admirably. "I'd trade these kids in one minute for most of our own union organizers," says Dick Blin, on loan from the Paperworkers to handle media. "There's so much deadwood around us, so-called organizers who make a hundred Gs a year and can't remember the last time they talked to a worker. These kids, on the other hand, are great, fearless and full of energy."

Perhaps this Nashville site is extraordinary compared with others, its success owing to the militancy and professionalism of the union Local. The achievement of this group is no doubt also facilitated by its immediate supervisor, a powerhouse named Joe Geevarghese, himself only 23 and now a full-time Steelworkers staff organizer (after graduating from Vanderbilt last year as student body president). Andy Levin admits that at least one site fell apart in internecine squabbling and that not all have had such meaty work to do as here in Nashville (even among this group one college junior decided to go home five days early, upset over the "confrontational" tactic of the neighborhood blitz). But after I called around to other sites and looked over the first-person report faxed around the U/S network, it seems safe to say that in general Union Summer has been successful in bringing young people into meaningful organizing work.

In Los Angeles, the youth activists were so adept at organizing county home-care workers that when the program was over, four U/S participants were hired to keep working for the Service Employees International Union. In Watsonville, California, U/S detachments are involved in the country's largest single organizing effort, the drive to unionize 20,000 strawberry workers. In Akron, Ohio, an all-female contingent worked with the Service Employees to help foster a community-based poor people's organization. In posh Hilton Head, South Carolina, U/S activists worked in a campaign to unionize low-wage hotel workers. Some of those same students are already pulling together other U/S alumni from several Southern states to head a regional campaign to fight privatization of government service jobs. Union Summer volunteers are signed up for a train ride through the South aimed at organizing textile workers. In Washington, D.C., a couple dozen Union Summer recruits wound up working on political campaigns for the left-of-center New Party.

For all the media comparisons with Freedom Summer, Union Summer, in principle, has much more in common with the community organizing program started in 1963, at Tom Hayden's urging, by the incipient Students for a Democratic Society. After receiving an initial grant of $5,000 from the United Auto Workers, the early student radicals came up with the Economic

Research and Action Project (ERAP). The concept was to deploy in a dozen urban centers during summer vacation groups of ten to twenty students who would organize impoverished residents around a host of social justice issues. For two years the ERAPs came and went with little tangible success, until they disappeared completely when S.D.S. shifted its focus to the horrors of Vietnam. Even so, the experience deeply marked the core group of 500 or so ERAP activists. "Hardly anyone on the 'outside' can imagine the completeness of transformation or the depth of [the students'] commitment," wrote Andrew Kopkind after visiting the Newark site in 1965. In fact, in the turbulent handful of years that immediately followed, much of the cream of the radical movement grew directly out of the ERAP projects.

These are, of course, different times and a different generation of students. "The problem is, there's no transference between generations," says Geevarghese, the Steelworkers organizer. That analysis was borne out by hours of conversation I had with the Nashville U/S group during the week I spent with them. There was an unquestionable sense of disjuncture with history, even their own personal histories. Four of the ten had come from union families, but all four said that until this experience they had never known what their union parent really did. Nor was there much evidence of the ideological fervor that drove the original S.D.S.ers.

By contrast with the mood of their sixties forerunners, this group's is one of quiet skepticism and sincere confusion and curiosity about the present and the future rather than anger and indignation. Then again, thirty years ago the disillusionment provoked by Bull Connor's fire hoses and Lyndon Johnson's B-52s was shockingly fresh and created a sense of betrayal that was immediately radicalizing. With the exception of the schoolteacher, the Nashville participants were all born after Nixon's resignation and were preschoolers or in kindergarten when the Great Communicator was sworn in.

A majority of Union Summer participants are women, 52 percent are of color and just under half come from union families. But the lure of labor work seems strong, even on top-drawer college campuses. Almost two dozen A.F.L.-C.I.O. recruiters combed more than a hundred campuses this past spring, and union officials predict that as many as 1,000 students will eventually apply this year to become full-time organizers—more than twice the number of 1995. At Swarthmore alone, thirty applications were filed.

The Union Summer program and the stepped-up recruitment of young organizers is central to the strategic plan of the new A.F.L.-C.I.O. leadership under Sweeney. Indeed, Union Summer was part of Sweeney's election platform and is integral to his vow to create a new "culture of organizing" inside America's flagging labor movement. The decline in union membership to its current level of only 10.4 percent of the private-sector work force can be traced not only to aggressive corporate strategy and a stacked legislative deck but also to a sort of narcolepsy that has beset the official union culture in recent decades.

Now, Sweeney has brought the seven-year-old Organizing Institute into the heart of the labor federation and has named its respected leader, Richard Bensinger, to head the newly created A.F.L.-C.I.O. Organizing Department. Armed with a $20 million budget, Bensinger has a mandate to jump-start massive campaigns that will bring tens of thousands of new, mostly young workers into unions.

"But don't think for a moment that it's all clear sailing within the A.F.L.," says one veteran organizer. "There's still plenty of tension within the A.F.L. between Sweeney and the old guard, who still have plenty of sinecures and power. And there's plenty of tension within the Sweeney camp as well about how far they are really going to go. Don't forget, Sweeney himself is hardly a radical. No doubt, however, a successful Union Summer is going to give a lot of ammo to those who want to take organizing seriously."

In any case, the mathematics are daunting. U/S head Andy Levin estimates that to just stand still at 10 percent of the labor force, unions will have to recruit 300,000 new members a year. To get back to unionization levels of a few decades ago, a million or more new union members a year for several years would be needed. "That's why Union Summer is so crucial," he says. "The people we are bringing in—youth, women, minorities—look a lot more like the work force we need to organize than the older generation does. They're our future."

Is the A.F.L.-C.I.O., however, the best future for a new generation of radical youth? While Sweeney's administration is a world removed from the dinosaur regime of predecessor Lane Kirkland, there are real limits to how far a "reformed" labor movement might be willing to go. "In most union activity it's still taboo to focus on social class," says Joe Geevarghese. But it's also understood that Sweeney is a transitional figure whose program will either fail and let the A.F.L. drift back to sleep, or will in the relatively near future open the door to a more militant leadership under a younger group of supporters he has brought in. The more space that younger allies like Levin are given, and the better they perform, the better chance there is for the more hopeful course of the federation. But no matter where the A.F.L.-C.I.O. eventually goes, Union Summer has to be viewed as positive. The contemporary left is too quick to politicize the way goods are consumed (choosing politically correct coffee, cosmetics and culture) rather than the way they are produced. Union Summer begins to reverse the focus.

It's the final day of the Steelworkers' Days of Rage campaign, and everything has gone like clockwork. Faxed reports confirm the international work stoppages and protests against Bridgestone/Firestone. The previous after-noon, the Nashville Union Summer activists helped lead a raucous march of union members and supporters onto the corporate campus. This morning, each U/S activist led a protest at a local MasterCare garage. And now with 2,500 or so union members rallying under a stifling sun at a lakeside campsite, the Union Summer crew has been invited into an air-conditioned cabin for a private meeting with John Sweeney. The air inside the room is electric. Certainly not because they are meeting with the suspendered and balding 62-year-old leader of the A.F.L.-C.I.O., who thanks them for their hard work. But rather because the kids themselves are charged up and downright giddy.

It's the charge you get from directly confronting power, no matter on how small or symbolic a scale. And for most, if not all, of the activists in the room, the morning's MasterCare demonstrations had been their baptism in direct action. As Sweeney went around the room asking the organizers to relate their personal experience, they bubbled. Acting on their own authority, at least four of the nine activists "went in"—led their groups past police lines into brief, nonviolent occupations of the targeted MasterCare service centers. It didn't change the world. The coolheaded Nashville police force didn't even arrest anyone, but a psychic barricade has been crossed. "What a rush," said Grun-thal. "What an unbelievable rush."

FROM THE GROUND UP

58

Charon Asetoyer

My husband is Dakota Sioux from the Yankton Sioux reservation, located in the South Central part of South Dakota. His father passed away in 1983, and in 1985 we moved to his reservation in order to fulfill our traditional commit-ment. It is a tradition to hold a memorial on the anniversary of a loved one's death for the following four years. It was to be a temporary stay, until our commitment was over. We had planned to move to Oklahoma so I could work for my tribe.

From: *Women's Review of Books* 11 (July 1994): 22. Reprinted by permission.

After we had moved to South Dakota, I applied for a job with the Yankton Sioux Tribe. The tribe is small and jobs were held by members of the tribe; it was not likely that I would be hired. I was turned down. But word got around that Clarence Rockboy had married a Comanche woman and had returned home with her, and people were curious to meet me.

Women in particular started to drop by, and I took that opportunity to invite others to come and talk. Before long a group of women were meeting regularly in my home to talk about "women's things." We moved the meetings to an empty bedroom in our basement. Out conversations centered around problems that women and children were having in the community and what we might do to address them.

By 1986 we were incorporated as a nonprofit organization. My small basement bedroom turned into an office. Women would stop in to have coffee and talk about their problems at home. Early one morning a woman came by to ask me what I knew about Fetal Alcohol Syndrome (FAS). She told me that she had gone on drinking during her pregnancy and that her child, now a teenager, was having many learning problems. She wanted to know where she could get the child tested for FAS. She was frustrated and had nowhere to turn for support or help. Before long we were meeting on a regular basis; she began bringing other mothers to join in. Soon we had a community task force on FAS, the beginning of our first project, "Women and Children in Alcohol."

It was natural that health issues should come up first: in reservation communities across the U.S., health problems are devastating, and some of the worst are on the Yankton Sioux Reservation. The infant mortality rate has been as high as 23.8 per thousand. Diabetes affects 70 percent of the people over the age of 40, and the rates of secondary health conditions caused by diabetes, such as kidney failure, heart attacks, stroke, amputations and blindness, are alarmingly high. Alcoholism is out of control and contributes to many of these health problems, including domestic abuse and fetal alcohol births—one out of four children has FAS. Many health problems are related to poor nutrition, caused by years and even generations of eating government commodity foods that are high in salt, sugars and fat and low in fiber.

Native activists have brought the Indian Health Service under fire for years for their neglect and mistreatment of Native women. In the 1960s sterilization abuses surfaced, as they did again in the eighties when it was revealed that the Indian Health Service had used Depo-Provera on Native women who were developmentally disabled before it was FDA approved. Today the Indian Health Service is still abusing Native women's reproductive rights by limiting our contraception choices. They are promoting contraceptives like Depo Provera and Norplant, as well as sterilization. When the Hyde Amendment took away federal funding for abortion, Native women lost their access to abortion altogether: the Indian Health Service, our primary health care provider, is a division of the U.S. Public Service—that is, the Federal government.

The direction our programs have taken has been determined by the women who have come through our doors. I was sitting at my desk one afternoon working on an article for our newsletter when the phone rang. "Is it true that a tubal ligation can be temporary and that I can have it reversed when I want to have a child?" the caller asked. When I asked her why she wanted the information, she said an Indian Health Service doctor had recommended that she have her tubes tied. The procedure, he hold her, could be reversed later if she wanted. Incidents like this spurred us to create a reproductive health program.

Other issues evolved after we started doing workshops in the community. One Saturday morning a young woman came to my front door, banging for someone to let her in. I opened the door and she ran in. I recognized the look of fright and helplessness on her face, and locked the door behind her. During my first marriage I too had been a victim of domestic violence. There I was, standing in the hallway of my home faced with something that I had wanted to leave behind. Only this time I was not the one in need of someone to take me in.

In the course of time we created a program to assist women like this one and their children. We started out with a safe house program, sheltering women in local motels. In 1991, after a long court battle over zoning, the Native American Community Board opened a shelter for women and children fleeing from domestic violence and sexual assault. It now houses eighteen women and children and has provided services for over 900 women and children over the past two and a half years.

As more concerns were brought forward, we needed additional programming and additional staff. The office had grown too small for the organization. Several months earlier I had visited the National Black Women's Health Project in Atlanta, Georgia. The Project was located in a large house which seemed ideal: there was room to grow in, space for program work, a kitchen, a yard, and a strong sense of pride among the staff in their unique working environment. It all seemed so right: why not for us too? We wanted to do nutrition work to address diabetics, so we needed a place to prepare food; we wanted to work with children, so we needed space for them as well as for office work. And if we could buy a house we would no longer have to pay rent to a landlord.

Owning our own place seemed like a good idea but no more than a dream at that point, so I put it in the back of my mind. Who would fund a group of women to buy a house anyway, especially a group of Native American women living on a reservation in South Dakota? I had no idea what the cost of real estate might be in Lake Andes, South Dakota. But I kept thinking about the idea. A friend mentioned a small house she knew of, a fixer-upper, around $5,000 to $7,000. I went to look at it: it did seem to suit our needs.

Shortly after, I was invited to a conference sponsored by the National Women's Health Network, a health advocacy organization based in Washington, D.C. At a networking reception the first evening, I met another woman of color, Luz Alveolus Martinets, the director of the National Latina Health Organization. We talked about the work we were doing, and I eventually told her about our dream of buying a house for our project. When I told her what it would cost, she thought I meant the down payment. But $6,000 was the price for the entire house.

Luz told me that most of the women attending the conference were working women and could afford to make a contribution toward the house. All I had to do was to get up during lunch the next day, when we all had to introduce ourselves, and make an appeal for donations. At lunch I tried to avoid Luz: it seemed too hard to stand up in a crowded room of women I really did not know and ask for money. But Luz found me. Still, when it was my turn to introduce myself, I started to sit down without making my request for donations. Luz kept encouraging me to go ahead and make my request. Before long, all the women in the room wanted to know what we were arguing about. So I went ahead and asked for their help. Then I saw Byllye Avery of the National Black Women's Health Project, across the room, reach into her pocketbook and pull out her checkbook. Byllye held up the first check and announced to the room, "I donate $100." The pocketbooks opened and the checks piled up. It was like a dream: the house was soon to become a reality. I left the conference with about half of the money we needed.

Soon after we purchased the house, people from the community came to help clean, paint and panel the walls. A few weeks of hard work transformed the little brick house across from my home into a warm environment where women and children could come together to share in health education activities and to organize around issues confronting our community. In February 1988, the doors opened on the first Native American Women's Health Education Resource Center based on a reservation, located in Lake Andes, South Dakota on the Yankton Sioux Reservation.

Not long after ther Resource Center opened a young Yankton Sioux woman walked in, introduced herself and asked if she could do her college internship with us. She was interested in working with children. It was time, we decided, for us to examine how we could provide services for children with special needs. The Child Development Program was our answer.

The Program brought children with special needs together to work, play and learn in an after-school program. Fetal alcohol-affected children, latchkey children, gifted children, and children from highly stressed families participate in the program. Today the program has evolved into a Dakota Cultural program in which elders of the Yankton Sioux tribe work with children, teaching them Dakota language, values and spirituality.

An Adult Education Program was not something we had even thought about developing—until one day a call came in from one of the local high school counselors, asking if a sixteen-year-old pregnant student could continue her schooling at the Resource Center with tutoring from our staff. "She's getting too big to sit in class all day long," he said. We said "Sure," and the next week we had a student. Soon after that a mother of one of the children in our Child Development Program asked if we could show her how to use a computer. She said that she already knew how to type, and she could get a job if she knew how to use a computer. This was the beginning of a program that offered job readiness skills and GED completion.

As more women heard about the work and the services that the Resource Center was providing, a feeling of trust was established. Women turned to us for help concerning sterilization abuse, and for information about sexually transmitted diseases, AIDS, family planning and abortion. A high school student came into the Resource Center one afternoon wanting information about pregnancy, AIDS and condoms. She asked if we could get her a home pregnancy test kit, since she thought that she might be pregnant, but if she went down to the local drugstore for it, the clerk would recognize her and call her mother.

It was not long before we were going into the schools with AIDS and contraception information, and conducting AIDS workshops in the community. We hired four high school students, two girls and two boys, and trained them to be AIDS peer counselors, so they could go back into the community and schools to give out accurate information to other young people.

It became obvious to us that no one else in our community, and no other Native women's groups on other reservations, were doing this kind of work. As women turned to us for information about abortion and issues of reproductive health, we knew that we needed to organize our reproductive health work more formally. Our next move was to bring Native women together in a project that would address our unmet health care needs and the reproductive abuses committed by the Indian Health Service. In 1990, 36 Native women met in Pierre, South Dakota, to form the first Native Women's Reproductive Rights Coalition. The Coalition has developed a reproductive rights agenda that defines the issues of reproductive rights and health as Native women ourselves understand them.

Over the past three years the Native Women's Reproductive Rights Coalition has grown to involve over 150 women from eight states, and it continues to grow each year. It organizes Native women to address such concerns as

Norplant and Depo-Provera abuse, environmental issues, abortion, and the oppressive effects of organized religion in reservation communities.

Another central part of the Coalition's work has been to include the teachings of our elders. The passing down of traditional methods of birthing, child spacing, abortion, parenting and "becoming a woman" ceremonies is an important issue for Native women and has become a major mission of the Coalition. Through its work, combined with the direct services of our Resource Center, we can bring into the national and international arena practices and policies that we ourselves have shaped and that truly represent the issues confronting us.

TAKING MULTICULTURAL, ANTIRACIST EDUCATION SERIOUSLY: *An Interview with Enid Lee*

59

Barbara Miner

What do you mean by a multicultural education?

The term *multicultural education* has a lot of different meanings. The term I use most often is *antiracist education*.

Multicultural or antiracist education is fundamentally a perspective. It's a point of view that cuts across all subject areas, and addresses the histories and experiences of people who have been left out of the curriculum. Its purpose is to help us deal equitably with all the cultural and racial differences that you find in the human family. It's also a perspective that allows us to get at explanations for why things are the way they are in terms of power relationships, in terms of equality issues.

So when I say multicultural or antiracist education, I am talking about equipping students, parents, and teachers with the tools needed to combat racism and ethnic discrimination, and to find ways to build a society that includes all people on an equal footing.

From: David Levine, Robert Lowe, Bob Peterson, and Rita Tenorio, eds. *Rethinking Schools: An Agenda for Change* (New York: New Press, 1995), pp. 9–16. Reprinted by permission.

It also has to do with how the school is run in terms of who gets to be involved with decisions. It has to do with parents and how their voices are heard or not heard. It has to do with who gets hired in the school.

If you don't take multicultural education or antiracist education seriously, you are actually promoting a monocultural or racist education. There is not neutral ground on this issue.

Why do you use the term antiracist education instead of multicultural education?

Partly because in Canada, multicultural education often has come to mean something that is quite superficial: the dances, the dress, the dialect, the dinners. And it does so without focusing on what those expressions of culture mean: the values, the power relationships that shape the culture.

I also use the term *antiracist education* because a lot of multicultural education hasn't looked at discrimination. It has the view, "People are different and isn't that nice," as opposed to looking at how some people's differences are looked upon as deficits and disadvantages. In antiracist education, we attempt to look at—and change—those things in school and society that prevent some differences from being valued.

Oftentimes, whatever is white is treated as normal. So when teachers choose literature that they say will deal with a universal theme or story, like childhood, all the people in the stories are of European origin; it's basically white culture and civilization. That culture is different from others, but it doesn't get named as different. It gets named as normal.

Antiracist education helps us move that European perspective over to the side to make room for other cultural perspectives that must be included.

What are some ways your perspective might manifest itself in a kindergarten classroom, for example?

It might manifest itself in something as basic as the kinds of toys and games that you select. If all the toys and games reflect the dominant culture and race and language, then that's what I call a monocultural classroom even if you have kids of different backgrounds in the class.

I have met some teachers who think that just because they have kids from different races and backgrounds, they have a multicultural classroom. Bodies of kids are not enough.

It also gets into issues such as what kind of pictures are up on the wall? What kinds of festivals are celebrated? What are the rules and expectations in the classroom in terms of what kinds of languages are acceptable? What kinds of interactions are encouraged? How are the kids grouped? These are just some of the concrete ways in which a multicultural perspective affects a classroom.

How does one implement a multicultural or antiracist education?

It usually happens in stages. Because there's a lot of resistance to change in schools, I don't think it's reasonable to expect to move straight from a monocultural school to a multiracial school.

First there is this surface stage in which people change a few expressions of culture in the school. They make welcome signs in several languages and have a variety of foods and festivals. My problem is not that they start there. My concern is that they often stop there. Instead, what they have to do is move very quickly and steadily to transform the entire curriculum. For example, when we say classical music, whose classical music are we talking about? European? Japanese? And what items are on the tests? Whose culture do they reflect? Who is getting equal access to knowledge in the school? Whose perspective is heard? Whose is ignored?

The second stage is transitional and involves creating units of study. Teachers might develop a unit on Native Americans, or Native Canadians, or people of African background. And they have a whole unit that they study from one period to the next. But it's a separate unit and what remains intact is the main curriculum, the main menu. One of the ways to assess multicultural education in your school is to look at the school organization. Look at how much time you spend on which subjects. When you are in the second stage you usually have a two- or three-week unit on a group of people or an area that's been omitted in the main curriculum.

You're moving into the next stage of structural change when you have elements of that unit integrated into existing units. Ultimately, what is at the center of the curriculum gets changed in its prominence. For example, civilizations. Instead of just talking about Western civilization, you begin to draw on what we need to know about India, Africa, China. We also begin to ask different questions about why and what we are doing. Whose interest is it in that we study what we study? Why is it that certain kinds of knowledge get hidden? In mathematics, instead of studying statistics with sports and weather numbers, why not look at employment in light of ethnicity?

Then there is the social change stage, when the curriculum helps lead to changes outside of the school. We actually go out and change the nature of the community we live in. For example, kids might become involved in how the media portray people and start a letter-writing campaign about news that is negatively biased. Kids begin to see this as a responsibility that they have to change the world.

I think about a group of elementary school kids who wrote to the manager of the store about the kinds of games and dolls that they had. That's a long way from having some dinner and dances that represent an "exotic" form of life.

In essence, in antiracist education we use knowledge to empower people and to change their lives.

Teachers have limited money to buy new materials. How can they begin to incorporate a multicultural education even if they don't have a lot of money?

We do need money and it is a pattern to underfund antiracist intiatives so that they fail. We must push for funding for new resources because some of the information we have is downright inaccurate. But if you have a perspective, which is really a set of questions that you ask about your life, and you have the kids ask, then you can begin to fill in the gaps.

Columbus is a good example. It turns the whole story on its head when you have the children try to find out what the people who were on this continent might have been thinking and doing and feeling when they were being "discovered," tricked, robbed, and murdered. You might not have that information on hand, because that kind of knowledge is deliberately suppressed. But if nothing else happens, at least you shift your teaching, to recognize the native peoples as human beings, to look at things from their view.

There are other things you can do without new resources. You can include, in a sensitive way, children's backgrounds and life experiences. One way is through interviews with parents and with community people, in which they can recount their own stories, especially their interactions with institutions like schools, hospitals, and employment agencies. These are things that often don't get heard.

I've seen schools inviting grandparents who can tell stories about their own lives, and these stories get to be part of the curriculum later in the year. It allows excluded people, it allows humanity, back into the schools. One of the ways that discrimination works is that it treats some people's experiences, lives, and points of view as though they don't count, as though they are less valuable than other people's.

I know we need to look at materials. But we can also take some of the existing curriculum and ask kids questions about what is missing, and whose interest is being served when things are written in the way they are. Both teachers and students must alter that material.

How can a teacher who knows little about multiculturalism be expected to teach multiculturally?

I think the teachers need to have the time and encouragement to do some reading and to see the necessity to do so. A lot has been written about multiculturalism. It's not like there's no information.

You also have to look around at what people of color are saying about their lives and draw from those sources. You can't truly teach this until you reeducate yourself from a multicultural perspective. But you can begin. It's an ongoing process.

Most of all, you have to get in touch with the fact that your current education has a cultural bias, that it is an exclusionary, racist bias and that it needs to be purged. A lot of times people say, "I just need to learn more about those other groups." And I say, "No, you need to look at how the dominant culture and biases affect your view of nondominant groups in society." You don't have to fill your head with little details about what other cultural groups eat and dance. You need to take a look at your culture, what your idea of normal is, and realize it is quite limited and is in fact just reflecting a particular experience. You have to realize that what you recognize as universal is, quite often, exclusionary. To be really universal, you must begin to learn what Africans, Asians, Latin Americans, the aboriginal peoples, and all silenced groups of Americans have had to say about the topic.

How can one teach multiculturally without making white children feel guilty or threatened?

Perhaps a sense of being threatened or feeling guilty will occur. But I think it is possible to have kids move beyond that.

First of all, recognize that there have always been white people who have fought against racism and social injustice. White children can proudly identify with these people and join in that tradition of fighting for social justice.

Second, it is in their interest to be opening their minds and finding out how things really are. Otherwise, they will constantly have an incomplete picture of the human family.

The other thing is, if we don't make clear that some people benefit from racism, then we are being dishonest. What we have to do is talk about how young people can use that from which they benefit to change the order of things so that more people will benefit.

If we say we are all equally discriminated against on the basis of racism, or sexism, that's not accurate. We don't need to be caught up in the guilt of our benefit, but should use our privilege to help change things.

I remember a teacher telling me last summer that after she listened to me on the issue of racism, she felt ashamed of who she was. And I remember wondering if her sense of self was founded on a sense of superiority. Because if that's true, then she is going to feel shaken. but if her sense of self is founded on working with people of different colors to change things, then there is no need to feel guilt or shame.

Where does an antisexist perspective fit into a multicultural perspective?

In my experience, when you include racism as just another of the "isms," it tends to get sidetracked or omitted. That's because people are sometimes uncomfortable with racism, although they may be comfortable with class and gender issues. I like to put racism in the foreground, and then include the others by example and analysis.

I certainly believe that sexism—and ageism and heterosexism and class issues—have to be taken up. But in my way of thinking I don't list them under multicultural education.

For me, the emphasis is race. I've seen instances where teachers have replaced a really sexist set of materials with nonsexist materials—but the new resources included only white people. In my judgment, more has been done in the curriculum in terms of sexism that in terms of racism.

Of course, we must continue to address sexism in all its forms, no question about that. But we cannot give up on the fight against racism either. As a black woman, both hurt my heart. . . .

What can school districts do to further multicultural education?

Many teachers will not change the curriculum if they have no administrative support. Sometimes, making these changes can be scary. You can have parents on your back, kids who can be resentful. You can be told you are making the curriculum too political.

What we are talking about here is pretty radical; multicultural education is about challenging the status quo and the basis of power. You need administrative support to do that.

In the final analysis, multicultural or antiracist education is about allowing educators to do the things they have wanted to do in the name of their profession: to broaden the horizons of the young people they teach, to give them skills to change a world in which the color of their skin defines their opportunities, where some human beings are treated as if they are just junior children.

Maybe teachers don't have this big vision all the time. But I think those are the things that a democratic society is supposed to be about.

When you look at the state of things in the United States and Canada, it's almost as if many parts of the society have given up on decency, doing the right thing, and democracy in any serious way. I think that antiracist education gives us an opportunity to try again.

Unfortunately, I feel that this educational movement is going to face a serious challenge. The 1980s were marked by very conservative attitudes, and some of the gains of the social change movements in the 1960s and 1970s were rolled back.

A major struggle is taking place in the 1990s to regain those victories of the 1960s and 1970s. I think that antiracist education can help us do that. But the conservative forces are certainly not going to allow this to happen without a battle. So we'd better get ready to fight.

WOMEN OF COLOR
ON THE FRONT LINE

60

Celene Krauss

Toxic waste disposal is a central focus of women's grass-roots environmental activism. Toxic waste facilities are predominantly sited in working-class and low-income communities and communities of color, reflecting the dispro-portionate burden placed on these communities by a political economy of growth that distributes the costs of economic growth unequally. Spurred by the threat that toxic wastes pose to family health and community survival, female grass-roots activists have assumed the leadership of community envi-ronmental struggles. As part of a larger movement for environmental justice, they constitute a diverse constituency, including working-class housewives and secretaries, rural African American farmers, urban residents, Mexican American farm workers, and native Americans.

These activists attempt to differentiate themselves from what they see as the white, male, middle-class leadership of many national environmental or-ganizations. Unlike the more abstract, issue-oriented focus of national groups, women's focus is on environmental issues that grow out of their concrete, im-mediate experiences. Female blue-collar activists often share a loosely defined ideology of environmental justice and a critique of dominant social institu-tions and mainstream environmental organizations, which they believe do not address the broader issues of inequality underlying environmental hazards. At the same time, these activists exhibit significant diversity in their conceptual-ization of toxic waste issues, reflecting different experiences of class, race, and ethnicity.

This [essay] looks at the ways in which different working-class women formulate ideologies of resistance around toxic waste issues and the process by which they arrive at a concept of environmental justice. Through an analysis of interviews, newsletters, and conference presentations, I show the voices of white, African American, and Native American female activists and the re-sources that inform and support their protests. What emerges is an environ-mental discourse that is mediated by subjective experiences and interpreta-

From: Robert D. Bullard, ed., *Unequal Protection: Environmental Justice and Communities of Color* (San Francisco: Sierra Club Books, 1994), pp. 256–71. Reprinted by permission.

tions and rooted in the political truths women construct out of their identities as housewives, mothers, and members of communities and racial and ethnic groups.

THE SUBJECTIVE DIMENSION
OF GRASS-ROOTS ACTIVISM

Grass-roots protest activities have often been trivialized, ignored, and viewed as self-interested actions that are particularistic and parochial, failing to go beyond a single-issue focus. This view of community grass-roots protests is held by most policymakers as well as by many analysts of movements for progressive social change.

In contrast, the voices of blue-collar women engaged in protests regarding toxic waste issues tell us that single-issue protests are about more than the single issue. They reveal a larger world of power and resistance, which in some measure ends up challenging the social relations of power. This challenge becomes visible when we shift the analysis of environmental activism to the experiences of working-class women and the subjective meanings they create around toxic waste issues.

In traditional sociological analysis, this subjective dimension of protest has often been ignored or viewed as private and individualistic. Feminist theory, however, helps us to see its importance. For feminists, the critical reflection on the everyday world of experience is an important subjective dimension of social change. Feminists show us that experience is not merely a personal, individualistic concept. It is social. People's experiences reflect where they fit in the social hierarchy. Thus, blue-collar women of differing backgrounds interpret their experiences of toxic waste problems within the context of their particular cultural histories, starting from different assumptions and arriving at concepts of environmental justice that reflect broader experiences of class and race.

Feminist theorists also challenge a dominant ideology that separates the "public" world of policy and power from the "private" and personal world of everyday experience. By definition, this ideology relegates the lives and concerns of women relating to home and family to the private, nonpolitical arena, leading to invisibility of their grass-roots protests about issues such as toxic wastes. As Ann Bookman has noted in her important study of working-class women's community struggles, women's political activism in general, and working-class political life at the community level in particular, remain "peripheral to the historical record . . . where there is a tendency to privilege male political activity and labor activism."[1] The women's movement took as its central task the reconceptualization of the political itself, critiquing this dominant ideology and constructing a new definition of the political located in the everyday world of ordinary women rather than in the world of public policy. Feminists provide a perspective for making visible the importance of particular, single-issue protests regarding toxic wastes by showing how ordinary women subjectively link the particulars of their private lives with a broader analysis of power in the public sphere.

Social historians such as George Rudé have pointed out that it is often difficult to understand the experience and ideologies of resistance because ordinary working people appropriate and reshape traditional beliefs embedded within working-class culture, such as family and community. This point is also relevant for understanding the environmental protests of working-class women. Their protests are framed in terms of the traditions of motherhood and family; as a result, they often appear parochial or even conservative. As we shall see, however, for working-class women, these traditions become the levers that set in motion a political process, shaping the language and oppositional meanings that emerge and providing resources for social change.

Shifting the analysis of toxic waste issues to the subjective experience of ordinary women makes visible a complex relationship between everyday life and the larger structures of public power. It reveals the potential for human agency that is hidden in a more traditional sociological approach and provides us with a means of seeing "the sources of power which subordinated groups have created."[2]

The analysis presented in this [essay] is based on the oral and written voices of women involved in toxic waste protests. Interviews were conducted at environmental conferences such as the First National People of Color Environmental Leadership Summit, Washington, D.C., 1991, and the World Women's Congress for a Healthy Planet, Miami, Florida, 1991, and by telephone. Additional sources include conference presentations, pamphlets, books, and other written materials that have emerged from this movement. This research is part of an ongoing comparative study that will examine the ways in which experiences of race, class, and ethnicity mediate women's environmental activism. Future research includes an analysis of the environmental activism of Mexican American women in addition to that of the women discussed here.

TOXIC WASTE PROTESTS AND THE RESOURCE OF MOTHERHOOD

Blue-collar women do not use the language of the bureaucrat to talk about environmental issues. They do not spout data or marshal statistics in support of their positions. In fact, interviews with these women rarely generate a lot of discussion about the environmental problem per se. But in telling their stories about their protest against a landfill or incinerator, they ultimately tell larger stories about their discovery or analysis of oppression. Theirs is a political, not a technical, analysis.

Working-class women of diverse racial and ethnic backgrounds identify the toxic waste movement as a women's movement, composed primarily of mothers. Says one woman who fought against an incinerator in Arizona and subsequently worked on other anti-incinerator campaigns throughout the state, "Women are the backbone of the grass-roots groups; they are the ones who stick with it, the ones who won't back off." By and large, it is women, in their traditional role as mothers, who make the link between toxic wastes and their children's ill health. They discover the hazards of toxic contamination: multiple miscarriages, birth defects, cancer deaths, and so on. This is not surprising, as the gender-based division of labor in a capitalist society gives working-class women the responsibility for the health of their children.

These women define their environmental protests as part of the work that mothers do. Cora Tucker, an African American activist who fought against uranium mining in Virginia and who now organizes nationally, says:

> It's not that I don't think that women are smarter, [she laughs] but I think that we are with the kids all day long. . . . If Johnny gets a cough and Mary gets a cough, we try to discover the problem.

Another activist from California sums up this view: "If we don't oppose an incinerator, we're not doing our work as mothers."

For these women, family serves as a spur to action, contradicting popular notions of family as conservative and parochial. Family has a very different meaning for these women than it does for the middle-class nuclear family. Theirs is a less privatized, extended family that is open, permeable, and attached to community. This more extended family creates the networks and resources that enable working-class communities to survive materially given few economic resources. The destruction of working-class neighborhoods by economic growth deprives blue-collar communities of the basic resources of survival; hence the resistance engendered by toxic waste issues. Working-class women's struggles over toxic waste issues are, at root, issues about survival. Ideologies of motherhood, traditionally relegated to the private sphere, become political resources that working-class women use to initiate and justify their resistance. In the process of protest, working-class women come to reject the dominant ideology, which separates the public and private arenas.

Working-class women's extended network of family and community serves as the vehicle for spreading information and concern about toxic waste issues. Extended networks of kinship and friendship become political resources of opposition. For example, in one community in Detroit, women discovered patterns of health problems while attending Tupperware parties. Frequently, a mother may read about a hazard in a newspaper, make a tentative connection between her own child's ill health and the pollutant, and start telephoning friends and family, developing an informal health survey. Such a discovery process is rooted in what Sarah Ruddick has called the everyday practice of mothering.[3] Through their informal networks, they compare notes and experiences and develop an oppositional knowledge used to resist the dominant knowledge of experts and the decisions of government and corporate officials.

These women separate themselves from "mainstream" environmental organizations, which are seen as dominated by white, middle-class men and concerned with remote issues. Says one woman from Rahway, New Jersey: "The mainstream groups deal with safe issues. They want to stop incinerators to save the eagle, or they protect trees for the owl. But we say, what about the people?"

Another activist implicitly criticizes the mainstream environmental groups when she says of the grass-roots Citizens' Clearinghouse for Hazardous Wastes:

> Rather than oceans and lakes, they're concerned about kids dying. Once you've had someone in your family who has been attacked by the environment—I mean who has had cancer or some other disease—you get a keen sense of what's going on.

It is the traditional, "private" women's concerns about home, children, and family that provide the initial impetus for blue-collar women's involvement in issues of toxic waste. The political analyses they develop break down the public-private distinction of dominant ideology and frame a particular toxic waste issue within broader contexts of power relationships.

THE ROLE OF RACE, ETHNICITY, AND CLASS

Interviews with white, African American, and Native American women show that the starting places for and subsequent development of their analyses of toxic waste protests are mediated by issues of class, race, and ethnicity.

White working-class women come from a culture in which traditional women's roles center on the private arena of family. They often marry young; although they may work out of financial necessity, the primary roles from which they derive meaning and satisfaction are those of mothering and taking care of family. They are revered and supported for fulfilling the ideology of a patriarchal family. And these families often reflect a strong belief in the existing political system. The narratives of white working-class women involved in toxic waste issues are filled with the process by which they discover the injustice of their government, their own insecurity about entering the public sphere of politics, and the constraints of the patriarchal family, which, ironically prevent them from becoming fully active in the defense of their family, especially in their protest. Their narratives are marked by a strong initial faith in "their" government, as well as a remarkable transformation as they become disillusioned with the system. They discover "that they never knew what they were capable of doing in defense of their children."

For white working-class women, whose views on public issues are generally expressed only within family or among friends, entering a more public arena to confront toxic waste issues is often extremely stressful. "Even when I went to the PTA," says one activist, "I rarely spoke. I was so nervous." Says another: "My views have always been strong, but I expressed them only in the family. They were not for the public." A strong belief in the existing political system is characteristic of these women's initial response to toxic waste issues. Lois Gibbs, whose involvement in toxic waste issues started at Love Canal, tells us, "I believed if I had a problem I just had to go to the right person in government and he would take care of it."

Initially, white working-class women believe that all they have to do is give the government the facts and their problem will be taken care of. They become progressively disenchanted with what they view as the violation of their rights and the injustice of a system that allows their children and family to die. In the process, they develop a perspective of environmental justice rooted in issues of class, the attempt to make democracy real, and a critique of the corporate state. Says one activist who fought the siting of an incinerator in Sumter County, Alabama: "We need to stop letting economic development be the true God and religion of this country. We have to prevent big money from influencing our government."

A recurring theme in the narratives of these women is the transformation of their beliefs about government and power. Their politicization is rooted in the deep sense of violation, betrayal, and hurt they feel when they find that their government will not protect their families. Lois Gibbs sums up this feeling well:

> I grew up in a blue-collar community. We were very into democracy. There is something about discovering that democracy isn't democracy as we know it. When you lose faith in your government, it's like finding out your mother was fooling around on your father. I was very upset. It almost broke my heart because I really believed in the system. I still believe in the system, only now I

believe that democracy is of the people and by the people, that people have
to move it, it ain't gonna move by itself.

Echoes of this disillusionment are heard from white blue-collar women
throughout the country. One activist relates:

We decided to tell our elected officials about the problems of incineration because we didn't think they knew. Surely if they knew that there was a toxic waste
dump in our county they would stop it. I was politically naive. I was real surprised because I live in an area that's like the Bible Belt of the South. Now I think
the God of the United States is really economic development, and that has got
to change.

Ultimately, these women become aware of the inequities of power as it is
shaped by issues of class and gender. Highly traditional values of democracy
and motherhood remain central to their lives. But in the process of politicization through their work on toxic waste issues, these values become transformed into resources of opposition that enable women to enter the public
arena and challenge its legitimacy. They justify their resistance as a way to
make democracy real and to protect their children.

White blue-collar women's stories are stories of transformations: transformations into more self-confident and assertive women; into political activists who challenge the existing system and feel powerful in that challenge; into
wives and mothers who establish new relationships with their spouses (or get
divorced) and new, empowering relationships with their children as they provide role models of women capable of fighting for what they believe in.

African American working-class women begin their involvement in toxic
waste protests from a different place. They bring to their protests a political
awareness that is grounded in race and that shares none of the white blue-
collar women's initial trust in democratic institutions. These women view government with mistrust, having been victims of racist policies throughout their
lives. Individual toxic waste issues are immediately framed within a broader
political context and viewed as environmental racism. Says an African American activist from Rahway, New Jersey:

When they sited the incinerator for Rahway, I wasn't surprised. All you have to
do is look around my community to know that we are a dumping ground for all
kinds of urban industrial projects that no one else wants. I knew this was about
environmental racism the moment that they proposed the incinerator.

An African American woman who fought the siting of a landfill on the South
Side of Chicago reiterates this view: "My community is an all-black commu-

nity isolated from everyone. They don't care what happens to us." She describes her community as a "toxic doughnut":

> We have seven landfills. We have a sewer treatment plant. We have the Ford Motor Company. We have a paint factory. We have numerous chemical companies and steel mills. The river is just a few blocks away from us and is carrying water so highly contaminated that they say it would take seventy-five years or more before they can clean it up.

This activist sees her involvement in toxic waste issues as a challenge to traditional stereotypes of African American women. She says, "I'm here to tell the story that all people in the projects are not lazy and dumb!"

Some of these women share experiences of personal empowerment through their involvement in toxic waste issues. Says one African American activist:

> Twenty years ago I couldn't do this because I was so shy. . . . I had to really know you to talk with you. Now I talk. Sometimes I think I talk too much. I waited until my fifties to go to jail. But it was well worth it. I never went to no university or college, but I'm going in there and making speeches.

However, this is not a major theme in the narratives of female African American activists, as it is in those of white blue-collar women. African American women's private work as mothers has traditionally extended to a more public role in the local community as protectors of the race. As a decade of African American feminist history has shown, African American women have historically played a central role in community activism and in dealing with issues of race and economic injustice. They receive tremendous status and recognition from their community. Many women participating in toxic waste protests have come out of a history of civil rights activism, and their environmental protests, especially in the South, develop through community organizations born during the civil rights movement. And while the visible leaders are often male, the base of the organizing has been led by African American women, who, as Cheryl Townsend Gilkes has written, have often been called "race women," responsible for the "racial uplift" of their communities.[4]

African American women perceive that traditional environmental groups only peripherally relate to their concerns. As Cora Tucker relates:

> This white woman from an environmental group asked me to come down to save a park. She said that they had been trying to get black folks involved and that they won't come. I said, "Honey, it's not that they aren't concerned, but when their babies are dying in their arms they don't give a damn about a park."

> I said, "They want to save their babies. If you can help them save their babies, then in turn they can help you save your park." And she said, "But this is a real immediate problem." And I said, "Well, these people's kids dying is immediate."

Tucker says that white environmental groups often call her or the head of the NAACP at the last minute to participate in an environmental rally because they want to "include" African Americans. But they exclude African Americans from the process of defining the issues in the first place. What African American communities are doing is changing the agenda.

Because the concrete experience of African Americans' lives is the experience and analysis of racism, social issues are interpreted and struggled with within this context. Cora Tucker's story of attending a town board meeting shows that the issue she deals with is not merely the environment but also the disempowerment she experiences as an African American woman. At the meeting, white women were addressed as Mrs. So-and-So by the all-white, male board. When Mrs. Tucker stood up, however, she was addressed as "Cora":

> One morning I got up and I got pissed off and I said, "What did you call me?" He said, "Cora," and I said, "The name is Mrs. Tucker." And I had the floor until he said "Mrs. Tucker." He waited five minutes before he said "Mrs. Tucker." And I held the floor. I said, "I'm Mrs. Tucker," I said, "Mr. Chairman, I don't call you by your first name and I don't want you to call me by mine. My name is Mrs. Tucker. And when you want me, you call me Mrs. Tucker." It's not that— I mean it's not like you gotta call me Mrs. Tucker, but it was the respect.

In discussing this small act of resistance as an African American woman, Cora Tucker is showing how environmental issues may be about corporate and state power, but they are also about race. For female African American activists, environmental issues are seen as reflecting environmental racism and linked to other social justice issues, such as jobs, housing, and crime. They are viewed as part of a broader picture of social inequity based on race. Hence, the solution articulated in a vision of environmental justice is a civil rights vision— rooted in the everyday experience of racism. Environmental justice comes to mean the need to resolve the broad social inequities of race.

The narratives of Native American women are also filled with the theme of environmental racism. However, their analysis is laced with different images. It is a genocidal analysis rooted in the Native American cultural identification, the experience of colonialism, and the imminent endangerment of their culture. A Native American woman from North Dakota, who opposed a landfill, says:

> Ever since the white man came here, they keep pushing us back, taking our lands, pushing us onto reservations. We are down to 3 percent now, and I see this as just another way for them to take our lands, to completely annihilate our races. We see that as racism.

Like that of the African American women, these women's involvement in toxic waste protests is grounded from the start in race and shares none of the white blue-collar women's initial belief in the state. A Native American woman from southern California who opposed a landfill on the Rosebud Reservation in South Dakota tells us:

> Government did pretty much what we expected them to do. They supported the dump. People here fear the government. They control so many aspects of our life. When I became involved in opposing the garbage landfill, my people told me to be careful. They said they annihilate people like me.

Another woman involved in the protest in South Dakota describes a government official's derision of the tribe's resistance to the siting of a landfill:

> If we wanted to live the life of Mother Earth, we should get a tepee and live on the Great Plains and hunt buffalo.

Native American women come from a culture in which women have had more empowered and public roles than is the case in white working-class culture. Within the Native American community, women are revered as nurturers. From childhood, boys and girls learn that men depend on women for their survival. Women also play a central role in the decision-making process within the tribe. Tribal council membership is often equally divided between men and women; many women are tribal leaders and medicine women. Native American religions embody a respect for women as well as an ecological ethic based on values such as reciprocity and sustainable development: Native Americans pray to Mother Earth, as opposed to the dominant culture's belief in a white, male, Anglicized representation of divinity.

In describing the ways in which their culture integrates notions of environmentalism and womanhood, one woman from New Mexico says:

> We deal with the whole of life and community; we're not separated, we're born into it—you are it. Our connection as women is to the Mother Earth, from the time of our consciousness. We're not environmentalists. We're born into the struggle of protecting and preserving our communities. We don't separate ourselves. Our lifeblood automatically makes us responsible; we are born with it.

> Our teaching comes from a spiritual base. This is foreign to our culture. There
> isn't even a word for dioxin in Navajo.

In recent years, Native American lands have become common sites for commercial garbage dumping. Garbage and waste companies have exploited the poverty and lack of jobs in Native American communities and the fact that Native American lands, as sovereign nation territories, are often exempt from local environmental regulations. In discussing their opposition to dumping, Native American women ground their narratives in values about land that are inherent in the Native American community. They see these projects as violating tribal sovereignty and the deep meaning of land, the last resource they have. The issue, says a Native American woman from California, is

> protection of the land for future generations, not really as a mother, but for the
> health of the people, for survival. Our tribe bases its sovereignty on our land
> base, and if we lose our land base, then we will be a lost people. We can't afford
> to take this trash and jeopardize our tribe.
>
> If you don't take care of the land, then the land isn't going to take care of
> you. Because everything we have around us involves Mother Earth. If we don't
> take care of the land, what's going to happen to us?

In the process of protest, these women tell us, they are forced to articulate more clearly their cultural values, which become resources of resistance in helping the tribe organize against a landfill. While many tribal members may not articulate an "environmental" critique, they well understand the meaning of land and their religion of Mother Earth, on which their society is built.

CONCLUSION

The narratives of white, African American, and Native American women involved in toxic waste protests reveal the ways in which their subjective, particular experiences lead them to analyses of toxic waste issues that extend beyond the particularistic issue to wider worlds of power. Traditional beliefs about home, family, and community provide the impetus for women's involvement in these issues and become a rich source of empowerment as women reshape traditional language and meanings into an ideology of resistance. These stories challenge traditional views of toxic waste protests as parochial, self-interested, and failing to go beyond a single-issue focus. They show that single-issue protests are ultimately about far more and reveal the experiences of daily life and resources that different groups use to resist. Through environmental protests, these women challenge, in some measure, the social relations of race, class, and gender.

These women's protests have different beginning places, and their analyses of environmental justice are mediated by issues of class and race. For white blue-collar women, the critique of the corporate state and the realization of a more genuine democracy are central to a vision of environmental justice. The definition of environmental justice that they develop becomes rooted in the issue of class. For women of color, it is the link between race and environment, rather than between class and environment, that characterizes definitions of environmental justice. African American women's narratives strongly link environment justice to other social justice concerns, such as jobs, housing, and crime. Environmental justice comes to mean the need to resolve the broad social inequities of race. For Native American women, environmental justice is bound up with the sovereignty of the indigenous peoples.

In these women's stories, their responses to particular toxic waste issues are inextricably tied to the injustice they feel as mothers, as working-class women, as African Americans, and as Native Americans. They do not talk about their protests in terms of single issues. Thus, their political activism has implications far beyond the visible, particularistic concern of a toxic waste dump site or the siting of a hazardous waste incinerator.

NOTES

1. Sandra Morgen, "'It's the Whole Power of the City Against Us!': The Development of Political Consciousness in a Women's Health Care Coalition," in *Women and the Politics of Empowerment*, eds. Ann Bookman and Sandra Morgen (Philadelphia: Temple University Press, 1988), p. 97.

2, Sheila Rowbotham, *Women's Consciousness, Man's World* (New York: Penguin, 1973).

3. See Sara Ruddick, *Maternal Thinking: Towards a Politics of Peace* (New York: Ballantine, 1989).

4. Cheryl Townsend Gilkes, "Building in Many Places: Multiple Commitments and Ideologies in Black Women's Community Work," in *Women and the Politics of Empowerment*, op. cit.

HAVING THE TOOLS AT HAND:

61

Building Successful Multicultural Social Justice Organizations

John Anner

The long-awaited age of the true international city is fully upon us, although not quite in the cheerful Disney World format expected by the advocates of multiculturalism. Urban America has become a staggeringly diverse melange of cultures and languages; it's no longer hyperbole to talk about the "global village," not with millions of the global villagers themselves living just down the street.

The changing demographic composition of many areas of the country is coupled with other shifts in the global political economy that are having severe negative effects on both urban and rural America. In brief, the unrestrained ability of international corporations to shift production and jobs just about anywhere in the world has meant disrupted communities, declining wages, and lower standards of living for most Americans—and increasing social decay and conflict.

Social justice organizations and movements have to contend with new constituents and new conditions. They have to figure out how to mobilize and organize people who may share geographic proximity and hard times but little else, in a political climate that has shifted the blame for worsening economic conditions onto the immigrants, the poor, and the powerless. And they have to figure out how to develop effective strategies for defending their constituents in an era in which progressive government-financed social legislation is close to politically impossible.

On the whole, the traditional defenders of low-income and working-class Americans have done a very poor job of building a sophisticated multicultural response to the changes outlined above, which is one reason things are so terrible. Unions, long the main bulwark against attacks on people's livelihoods, are barely awakening from a long period of political sleep. Progressive academics, professionals, leaders of national civil rights organizations,

From: John Anner, ed., *Beyond Identity Politics: Emerging Social Justice Movements in Communities of Color* (Boston: South End Press, 1996), pp. 153–66. Reprinted by permission.

advocates, lobbyists, and politicians appear to be helpless, timid, and out of ideas, endlessly fighting symbolic battles. And for whatever reason, there is no credible leftist political party in America.

It's more or less an article of faith among progressive political thinkers that we need a new mass-based struggle for social justice featuring leadership from people of color and a diverse membership. I believe . . . that there is an emerging "second generation" of grass-roots community and labor organizations that have developed over the past five to ten years. . . . The energy and mutual respect characteristic of the politics of identity have been used to revitalize and strengthen class-based organizing and to build strong relations between communities normally divided by race, language, and culture.

NO MYSTERY TO MULTICULTURALISM

There is no mystery to building effective multicultural social justice organizations, say many activists and organizers. A variety of organizations provide training and consultations, and, as this [essay] indicates, there are specific methods available to activists and organizers. "What matters most of all," says diversity trainer Guadalupe Guajardo, of Technical Assistance for Community Services (TACS) in Portland, Oregon, "is being able to listen and learn, and having the tools at hand to make changes."

This [essay] focuses on the specific tactics and strategies that social justice organizers use to build multicultural organizations. . . . Organizers can begin to confront the larger economic and political forces devastating low- and middle-income communities by building social justice organizations that can adapt to rapid changes in the demographic and cultural environment. While these organizations may not lead a national social justice movement, if and when it develops, they will train its political leaders, provide models of effective organizing methods, educate and develop memberships, keep people involved on a regular basis in community work, and test the local power structure for weak points.

A movement can draw on this base and use it to inspire large numbers of people to take action. This phenomenon occurred in California's anti-Proposition 187 movement. Established community, student, labor, civil rights, professional, and other groups sounded the alarm and laid the groundwork, but in the end a movement exploded outside the boundaries these organizations created. Indeed, one of the biggest protest marches, which took place in Los Angeles, occurred despite attempts by some "Stop 187" organizations to prevent it. A movement, when one develops, pulls in political, community, and labor organizations, but also inspires large numbers of previously uninvolved people.

At the same time, the efficacy and longevity of a movement rests on the foundation created through the daily back-breaking work of local grass-roots organizing. This work must include constant attention to developing multicultural memberships and alliances.

The models and mechanisms being used to build successful multicultural social justice organizations include:

- Building personal relationships between members from different backgrounds.
- Actively engaging in solidarity campaigns, actions, and activities with social justice organizations in other communities.
- Challenging bigoted statements and attitudes when they arise.
- Holding regular discussions, forums, "educationals," and workshops to enhance people's understandings of other communities and individuals.
- Working to change the culture of the organization so that members see themselves as "members of the community" first instead of members of a particular part of the community.
- Developing issues, tactics, and campaigns that are relevant to different communities and that reveal fundamental areas of common interest.
- Conducting antiracism training to get people to confront and deal with their biases.
- Examining and changing the organization's practices in order to hire, promote, and develop people of color.
- Confronting white privilege and nationalism.
- Hiring, recruiting, and training more people of color for leadership positions.

IT DOESN'T JUST HAPPEN

Elsa Barboza is an organizer with South Los Angeles-based Action for Grass-roots Empowerment and Neighborhood Development Alternatives (AGENDA). AGENDA has generally focused on the African-American community, and most of the staff and organizers are African-American. In 1994, says Barboza, "we decided to move towards organizing in the Latino community for the simple reason that we have a lot of new immigrants from Central America in the neighborhoods. We wanted to [make AGENDA] an authentic multicultural organization, but we learned an important lesson: it doesn't just happen."

AGENDA organizers quickly found that bringing monolingual Spanish-speaking members to the general membership and committee meetings was not effective in involving the new Latino members. AGENDA staff decided instead to form a separate organization known as the Latino Organizing Committee. The plan is to build leadership in the Committee, take on a few organizing campaigns around issues of particular concern to those members, and "agitate and educate" the South Los Angeles Latino community. "In the beginning we used the same strategies and tactics for everyone," says Barboza, but it simply didn't work in terms of attracting more Latinos to the group.

One solution is to define issues and campaigns so as to make them relevant to all members of the organization. For example, People United for a Better Oakland (PUEBLO) conducted a campaign in 1994 to force Highland Hospital to hire more professional translators. At first glance, this would not seem to be an issue of major concern to the native English speakers in the group. However, PUEBLO organizers and leaders successfully argued that nobody can get adequate medical care if some doctors and nurses are being called away from their duties in order to translate for a patient who doesn't speak English.

AGENDA has taken a different tack, working instead to research and develop campaigns that are of particular concern to the Spanish-speaking members. At some point in the future, when the Latino Organizing Committee has had enough experience organizing, the two organizations will be merged. In the meantime, AGENDA is conducting what they call "educationals" with all their African-American members to "demystify" the changing demographics of Los Angeles, and show how low-income communities of color face similar challenges and problems.

In terms of specific mechanisms to overcome prejudice-based resistance on the part of current members, "we don't have any 'diversity training' going on just yet," says Barboza, "but we'll probably need it in the future." She says that people get past any initial resistance pretty quickly, however, and see through to the larger self-interest that unites people of color. All the members come together for general membership meetings and selected planning meetings. Translation is conducted by an interpreter who sits with the Spanish-speaking members.

. . . [A]nother community group that started out all African-American and gradually changed as the demographics of the neighborhood changed is Direct Action for Rights and Equality (DARE) in Providence, Rhode Island. DARE is a powerful ten-year-old organization well known at City Hall; part of that power, say staff members, comes from the group's representative membership. In the beginning, however, the few Latino and Asian members seldom participated in DARE's activities. Looking back, says former organizer Libero Della Piana, three things stand out as "holding back" other members: language, culture, and the issues DARE worked on.

In contrast to AGENDA, DARE set out from the beginning to avoid creating two different organizations. "We did have a Comité Latino where meetings are conducted only in Spanish," says Della Piana, "but we didn't want it to become a separate group because we didn't think we'd be able to later turn it into a cohesive organization."

Della Piana says that organizers who want to build multicultural organizations have to ask themselves if they are making everyone feel welcome and wanted. Translating all written materials is critical, he says, because it not only makes all members feel part of the organization but is a way of visibly showing that the group is serious about being multiracial. At this point, Providence is changing again, experiencing a big influx of immigrants from South Asia, and DARE "desperately needs" an organizer who speaks one or more of the languages of the Indian subcontinent.

IN THE HEAT OF THE STRUGGLE

Perhaps the most common way that multicultural groups deal with diversity in their memberships is through a political ideology that emphasizes how the struggle at hand transcends differences of race, age, gender, or sexuality. The Committee Against Anti-Asian Violence (CAAAV) is a New York-based community group that has been active and visible on issues ranging from hate crimes to police brutality and economic exploitation. The group's 2,500 members are diverse, but all Asian. "We are a pan-Asian organization," says staff organizer Saleem Osman, "so in our group there is no hostility, only solidarity, because we are all working together for the same things." Although it is certainly true that there is a good deal of mutual dislike and often active hostility among the large variety of immigrant communities that have found themselves living side by side in urban America, it does seem to be the case that these differences can be set aside in the heat of the struggle.

An additional factor might be that the members of multicultural groups tend to be self-selecting. If someone is really not happy being part of a diverse membership organization, they tend to simply not join in the first place, or leave as the group changes. Finally, differences that seemed of intense importance in the region of origin (Chinese versus Japanese, Salvadoran versus Honduran, North Indian versus South Indian) start to lose their meaning once the reality of life in the United States sinks in and the different groups find that more unites them than separates them.

On the other hand, says Osman, working in coalition with other groups requires a lot of internal education. "Our members, who are taxi drivers, garment workers, and vendors mostly, need to be challenged to look at [their prejudices]." Prejudice against Latinos and African Americans among CAAAV members was clearly revealed during city council hearings on a bill that would have granted taxi drivers the right to refuse to pick up any individual based on their appearance. CAAAV organizers knew that drivers were supportive of the bill because it gave them the right to refuse rides to African-American men.

"We had to do a lot of work with the members on that one," says Osman. "That was a law aimed directly at African-Americans. So we said 'no, we won't support it because it's racist.' But first we put together a video and used it to educate drivers" in day-long training sessions.

CAAAV then took the issue one step further, appearing at public hearings to argue against the law, sometimes surprising African-American civil rights groups. "The best way to overcome prejudices between [communities of color]," says Osman, "is to work together in solidarity with each other to build unity." For this reason, says Osman, CAAAV has actively sought to build a working relationship with Puerto Rican and African-American groups active in the fight against police brutality.

Organizations such as CAAAV, AGENDA, and PUEBLO frequently are called upon by other local groups to attend public hearings or demonstrations or turn out their members for community events. Wendall Chin, an organizer for a multiracial group called San Francisco Anglers for Environmental Rights (SAFER) says that an important way to overcome racial prejudices is to get the members involved in soliciting secondary support from other groups and in joining those groups' actions. SAFER is an environmental justice organization sponsored by Communities for a Better Environment that organizes low-income anglers who fish in the San Francisco Bay for food as opposed to sport. SAFER has joined forces with Texas-based Fuerza Unida in their campaign against the Levi Strauss Company, which is headquartered in San Francisco. Similarly, they have been actively involved with PUEBLO and Asian Immigrant Women Advocates (AIWA), both of which are located in Oakland, not too far from SAFER's main area of operations.

The relationship between external and internal politics is not always as easy and obvious as staff and leaders would like; this is especially true when the senior staff is white. "One of the reasons we had such trouble with the NTC [National Toxics Coalition]," says Sonia Peña, "is that the people in charge figured that because they were part of the struggle and did good political work, they therefore could not be considered racist or sexist or authoritarian or any of those things and didn't have to deal with the issues when they came up in the organization." Peña was on the NTC board at the time of its demise due to intractable internal problems in 1993; she is the lead organizer for the multiracial community group Denver Action for a Better Community.

The anti-AIDS activist group ACT UP and the gay visibility network Queer Nation also started to come apart at the seams in the mid-1990s, in part because of their inability to cope with demands by members of color that the particular needs of their communities receive greater attention. Similarly, at the 1995 National People's Action (NPA) conference in Washington, D.C., a group of Latino delegates stormed the stage and took over the microphone from executive director Gale Cinotta. Led by Juan Mireles, they demanded that the NPA start translating meetings and conference plenaries into Spanish and that a minimum of 25 percent of the delegates to the next year's convention be Latino. Mireles told freelance writer Daniel Cordes that part of the problem is that the NPA—as a predominantly white and African-American organization—doesn't see the need to organize around issues of particular concern to Latinos or to figure out mechanisms to bring in a more diverse membership.

For both NPA and NTC, shared concerns among the membership about declining neighborhoods, redlining, and economic justice were not enough to paper over conflicts among staff and/or members of different races. Some formal mechanism is needed for "surfacing" these conflicts, letting them come out into the open, and resolving them as a group.

This can be as simple as not letting bigoted comments pass without comment. When prejudices are aired openly in a meeting or other event, say organizers, it can poison relations unless it is dealt with openly. "I remember one time at a meeting where we had whites, Blacks, and three or four Mexican members," says SAFER organizer Wendell Chin," and an outspoken white leader commented that she was glad that [the anti-immigrant California ballot initiative] Proposition 187 had passed because of the problems too many immigrants were causing. I looked over at the Mexican [members] and they weren't saying anything. So I had to step in and intervene."

"It's always better if the challenge comes from inside the membership of the group," says PUEBLO organizer Danny HoSang, "because then the members who are being challenged don't feel like they are being singled out by the staff. But is nobody speaks up, the staff organizer has to do it. It's not only about consciousness-raising. All the members of the group need to feel like they are welcome and valued."

Sometimes a more involved process is required. When tensions surfaced between Asian and Black participants in a year-long program called A New

Collaboration for Hands-On Relationships (ANCHOR), program director Rinku Sen decided to skip the actual program for that week and hold a series of discussions about the differences between the two communities. "Just because we call ourselves people of color doesn't mean we have the same backgrounds," says Sen. "Someone who was born in Cambodia and moved into a Black neighborhood in Oakland might experience racism as a daily fact of life. But that experience itself might be different than it would be for a Black person, and of course it is filtered through their history and current expectations."

SPEAKING MY LANGUAGE

Multicultural organizations are usually multilingual. PUEBLO, DARE, and other community groups have invested in a number of simultaneous translating machines. "They are a costly but highly effective tool," says PUEBLO lead organizer Rosi Reyes. The machines allow members to be seated anywhere in the room, instead of being segregated in one area, while the translator speaks quietly into a transmitter worn over the head. Receivers are smaller than a pack of cigarettes, with tiny earphones.

Some groups break meetings into segments, with some parts translated and others held in one language. Monica Russo, an organizer with a Florida local of the UNITE, uses this technique with a membership that speaks English, Creole, and Spanish. Day-long trainings, for example, feature monolingual sessions combined with collective meals and informal periods.

Most organizers warn, however, that translation demands more than simply literally transcribing what is being said. In order to transmit the real meaning and invite participation, says Peter Cervantes-Gautschi, director of the Portland, Oregon Workers Organizing Committee, "the critical thing is that the [translator] has to be into the movement. Because if [that person] doesn't really understand what we're trying to accomplish, then they are not going to get it right."

FROM WHITE TO RAINBOW

"There are two main obstacles to building multiracial social justice movements," says Libero Della Piana, who edits *Race File* at the Applied Research Center, "nationalism and white racism." Although many barriers divide people of color from each other, diversifying white organizations presents a different and more difficult set of challenges. "People of color generally understand racism as institutional," says Guadalupe Guajardo, "while to white people white privilege is virtually invisible, and they see prejudice as being something personal.

"Sometimes the hardest thing is to get people to face the reality that racism does exist even though polite people don't make racist comments anymore. But the only way to reach people and get them to examine how they benefit [from white privilege] is to start from the assumption that people are basically good. If you call folks racist, you will just make them defensive and won't get anywhere."

When working with an all-white group that wants to become multicultural, TACS trainers lay out five concrete areas to examine and consider changing: (1) the recruitment process, including where it is done and what the qualification requirements are; (2) planning, i.e., who is at the table when plans are being made; (3) decisionmaking, i.e., who is involved when decisions are made, both in formal and informal settings; (4) resource allocation, including money, access, and power; (5) promotion and leadership. Promoting people of color to leadership positions is vital, but only if these individuals have a base of support, are going to be given power and resources, and are held accountable.

"People need solutions," says Guajardo. "We help them find a new model based on an alliance-building or partnership idea, instead of seeing diversity as a win/lost kind of thing."

A number of other organizations around the country conduct similar trainings, ranging from "racial awareness training" to "diversity management," "prejudice reduction," and "coalition building."

CHANGING THE MIX

A criticism frequently directed at national community organizing networks is that the staff organizers and directors are mostly white, while the people they organize are predominantly people of color. This is certainly true for NPA, the Association of Community Organizations for Reform Now (ACORN), and to a lesser extent the Industrial Areas Foundation (IAF). The same criticism is leveled at labor organizations and the progressive press. There are more people of color working at the average metropolitan daily newspaper than at all the left-wing magazines in the country combined.

This situation—at least on the community organizing side of things—is responsible for at least three significant trends in grass-roots social justice organizing that appeared in the 1980s and developed throughout the 1990s. First is the rapid growth of independent organizations in communities of color not connected to any of the community organizing networks or unions. . . .

Second is the formation of organizer training programs specifically designed to train organizers of color to work in community and labor organizations comprised of people of color. Of these, by far the most prominent is the Center for Third World Organizing (CTWO), which has trained several hundred organizers of color through ten years of programs such as the Minority Activist Apprenticeship Program (MAAP) and the Community Partnership Program. Along with the organizer training programs, CTWO has also developed a model of community organizing that relies on organizers of color.

Finally, many emerging social justice organizations have made an explicit commitment to leadership diversity. The New Party constitution, for example, states that 50 percent or more of the leaders of local chapters must be people of color and 50 percent must be women. Greater diversity in the organizing staff can bring immediate rewards for community and labor organizations; it is now pretty much accepted by labor leaders that white men are the least effective organizers.

Teamsters Local 175 in Seattle, Washington, wanted to organize Asian women working in private postal facilities. The notion that these women could not be organized, says organizer Michael Laslett, was partly due to the fact that nobody had ever tried, and partly due to the expected barriers of culture, race, and language. He brought on two Asian-American interns from the AFL-CIO Organizing Institute to go into one particular shop, which resulted in a victorious union drive. "Having Asian organizers is what made organizing this company possible," he said.

In 1994, the United Electrical Workers (UE) were able to organize Mexican workers at a SteelTech factory in Milwaukee by importing an organizer from a sister union in Mexico. Labor organizers of color can be pretty scarce. A perhaps apocryphal story has it that the Oil, Chemical, and Atomic Workers Union at one time employed the only Vietnamese labor organizer in the country, and used to lend him out to other unions that needed a Vietnamese-speaking organizer.

FLUID IDENTITIES

The strategies for diversifying that work with this generation may not work with the next one, however. According to Libero Della Piana, when push comes to shove, the color of the organizer matters less than the person's commitment, especially when working with youth. "Of course it helps to have a diverse organizing staff," he comments, "but in the end that's not what it's about. Members respect the staff that's willing to do the work. It's a delicate balance."

In fact, say some youth organizers, most of what has been outlined above is based on strict definitions that don't fit with the mixed identities and intensely multicultural lives of urban youth. These young people don't need to be taught how to get along with other cultures, nor do they necessarily identify with the racial categories that guide the previous generation.

"It's pretty wild," says Next Generation co-director Mike Perez, who used to work at the Oakland-based youth program Encampment for Citizenship. "Race and class intersect in different ways for young people than they do for their parents." Many young people believe they can choose their racial identity based on how they feel. "I didn't have enough attitude to be a Black girl," one white high school student told *YO!* editor Nell Bernstein, explaining why she dressed like a *cholita*, or Latina gansta girl.

"You have white kids coming in saying they are just as much a 'nigga' as any Black kid since they come from the same 'hood," says Perez. "You have Asian kids dressing hip-hop and talking [African-American dialect]. Other Black kids talk and act 'white,' in the eyes of their friends. Identities are very fluid, but it's what being young is about now."

Della Piana agrees: "During the campaign to stop Prop 187, the main organizations active in Oakland tried to define it as a Latino issue. The high school and junior high students who organized and marched in the streets refused to see it that way. Their notion of what the fight was about was based on who was in and who was out.

"Kids base their identity on the basis of a complicated formula of territory, race, class, and aspirations," he continues. "There's a connection between kids that nobody understands; they can relate to each other [across racial boundaries] in some way adults have a hard time with."

It is probably true, as some youth organizers argue, that the intense problems that race, gender, and sexuality caused for older political groups and movements will not be as much in evidence as younger people start to move into leadership positions in social justice organizations. Perhaps "the fire next time" will burn brightly in rainbow patterns. But if we want social justice organizing to move beyond the limitations of identity struggles into an enlightened next phase, with justice, community, democracy, and true solidarity on the top of the agenda, we would do well to remember what Elsa Barboza said: It doesn't just happen.